Abiotic Stress and Plant Physiology

Abiotic Stress and Plant Physiology

Volume 1: Metabolic Activities

A. Bhattacharya

Former Principal Scientist (Plant Physiology)
Indian Institute of Pulses Research
Kanpur, India

NEW INDIA PUBLISHING AGENCY

New Delhi – 110 034

NEW INDIA PUBLISHING AGENCY
101, Vikas Surya Plaza, CU Block, LSC Market
Pitam Pura, New Delhi 110 034, India
Phone: + 91 (11) 27 34 17 17 Fax: + 91 (11) 27 34 16 16
Email: info@nipabooks.com
Web: www.nipabooks.com

Feedback at feedbacks@nipabooks.com

ISBN: 978-93-85516-90-0

Composed, Designed and Printed in India

Preface

Agriculture *per say* continue to be dependent of prevailing environmental condition and various factors of abiotic stresses are serious threats to agriculture. The situation assumes its importance in India as this country is heavily dependent on arrival, period and duration of monsoon for its agricultural production. According to FAO statistics, more than 800 million hectares of land throughout the world are currently salt-affected, including both saline and sodic soils equating to more than 6% of the world's total land area. Continuing salinization of arable land is expected to have overwhelming global impact, resulting in a 30% loss of agricultural land over the next 25 years and up to 50% loss by 2050. Overall, it has been estimated that the world is losing at least 3 ha of arable land every minute due to soil salinity. Abiotic stress leads to a series of morphological, physiological, biochemical and molecular changes in plants that adversely affect growth and productivity. A frequent result is protein dysfunction. Understanding the mechanisms of protein folding stability and how this knowledge can be utilized is one of the most challenging strategies for aiding organisms undergoing stress conditions. As a more comprehensive view of these processes evolves, applications to reducing plant stress are emerging.

As we move forward, emerging information and novel approaches must continuously be applied in a timely and effective manner by both the research and applied agricultural communities. One promising approach to improving the ability of plants to cope with Abiotic stress is to combine utilization of the vast biodiversity of crop plants and their wild relatives with the rapidly emerging genetic and molecular techniques. Global programs, such as the Global Partnership Initiative for Plant Breeding Capacity Building (GIPB), aim to select and distribute seed crops and cultivars with tolerance to abiotic stresses in order to facilitate sustainable use of plant genetic resources for food and agriculture. While much has been achieved in recent years in developing plants genetically engineered for resistance to herbicides, pests and diseases, production of plants engineered for tolerance to abiotic stress has not progressed as rapidly and applications in canola, rice and maize, for example, have only recently begun to be commercialized. This is due largely to the more complex genetic mechanisms involved in tolerance to abiotic stresses.

Abiotic stress factors frequently constrain the growth and productivity of major crop species such as cereals. The single greatest abiotic stress factor that limits crop growth worldwide is water availability. While genetic increases in yield potential are best expressed in optimum environments, they are also associated with enhanced yields under drought and nitrogen deficiency. These gains are especially relevant given that further large increases in the area under irrigation are not expected, and land deterioration associated with intensive agriculture threatens those areas already irrigated.

This book is intended to cover all known factors of abiotic stresses and their respective effects on some of the important aspects of physiological processes in plants. Thus, evaluation of the effects of pesticides on the physiological parameters is important.

Suggestions and advice are most welcome for the improvements in structural and for additions of current technologies as well as methodologies. A feeling of successfulness would prevail in my efforts in publishing this book if it is known that how this text has become meaningful to our readers. Suggestions from researchers, teachers and students, who use the book, would be highly appreciated. Hope this book shall serve the need of students, faculty members and researchers. The suggestions and advice are most welcome for the improvements in structural and for additions of current techniques. I sincerely hope the review chapters of this part of the book shall meet the requirement of post graduate students as well as faculty members of plant physiology and will be glad to accept constructive criticism and suggestions from the faculty.

A. Bhattacharya

Former Principal Scientist (Plant Physiology)
Indian Institute of Pulses Research
Kanpur, India

Contents

1

Abiotic Stress and Metabolic Responses in Plants

A. Bhattacharya

Environmental stress can disrupt cellular structures and impair key physiological functions. Drought, salinity, and low temperature stress impose an osmotic stress that can lead to turgor loss. Membranes may become disorganized, proteins may undergo loss of activity or be denatured, and often excess levels of reactive oxygen species are produced leading to oxidative damage. As a consequence, inhibition of photosynthesis, metabolic dysfunction, and damage of cellular structures contribute to growth perturbances, reduced fertility, and premature senescence. Responses to environmental stresses occur at all levels of organization. Cellular responses to stress include adjustments of the membrane system, modifications of the cell wall architecture, and changes in cell cycle and cell division. In addition, plants alter metabolism in various ways, including production of compatible solutes that are able to stabilize proteins and cellular structures and/or to maintain cell turgor by osmotic adjustment, and redox metabolism to remove excess levels of ROS and re-establish the cellular redox balance. At the molecular level, gene expression is modified upon stress and epigenetic regulation plays an important role in the regulation of gene expression in response to environmental stress.

Accumulation of amino acids has been observed in many studies (on plants exposed to abiotic stress). This increase might stem from amino acid production and/or from enhanced stress-induced protein breakdown. While the overall accumulation of amino acids upon stress might indicate cell damage in some species, increased levels of specific amino acids have a beneficial effect during stress acclimation. For many years, the capacity to accumulate proline has been correlated with stress tolerance. Proline is considered to act as an osmolyte, a ROS scavenger, and a molecular chaperone stabilizing the structure of proteins, thereby protecting cells from damage caused by stress. Proline accumulates in many plant species in response to different environmental stresses including drought, high salinity, and heavy metals. Interestingly, heat stress did not lead to proline accumulation in tobacco and Arabidopsis plants and induced proline accumulation rendered plants more sensitive to heat. Proline levels are determined by the balance between biosynthesis and catabolism. Proline is produced in the cytosol or chloroplasts from glutamate, which is reduced to glutamate-semialdehyde by Δ-1-pyrroline-5-carboxylate synthetase. Pyrroline-5-carboxylate can spontaneously converted then further reduced to proline. Proline is degraded in mitochondria by proline dehydrogenase and dehydrogenase to glutamate. Stress conditions stimulate proline synthesis while proline catabolism is enhanced during recovery from stress.

In their native habitats, plant frequently grow under challenging conditions such as drought (deficient precipitation, drying winds), salinity, frost, high temperature, floods or high luminosity. These conditions are collectively known as abiotic stress and each of them may delay growth and development, reduce plant productivity and, in more extreme circumstances, reduce plant death (Jiang and Zhang, 2002; Osturk *et al.* 2002; Xiong *et al.* 2002; Rabbani, *et al.* 2003; Gracia *et al.* 2007). Under abiotic stress, changes in gene expression such as induction and/or repression of gene occur, and these changes may directly regulated by stress conditions or may result from secondary stresses and/or as a response to injuries to metabolic and cellular functions (Bray, 2002). In addition, genotypes that differ in tolerance to abiotic stress(s) should present quantitative and qualitative differences in gene expression. Thus, a specific physiological response to abiotic stress results from a combination of previous molecular events, activated by the perception of the stress signaling molecules. It is necessary to understand how these molecular events are activated / deactivated and how they interact. Crop plants can be genetically modified with the introduction of genes which are able to confer tolerance, thus maintaining plant productivity under adverse conditions (Bray 2002, Rabbani *et al.* 2003, Rodrigues *et al.* 2009, Zingaretti *et al.* 2011).

Stress in physical terms is defined as mechanical force per unit area applied to an object. In response to the applied stress, an object undergoes a change in the dimension, which is also known as strain. As plants are sessile, it is tough to measure the exact force exerted by stresses and therefore in biological terms it is difficult to define stress. A biological condition, which may be stress for one plant may be optimum for another plant. The most practical definition of a biological stress is an adverse force or a condition, which inhibits the normal functioning and well being of a biological system such as plants (Jones and Jones, 1989). Abiotic stress is defined as the negative impact of non-living factors on the living organisms in a specific environment. The non-living variable must influence the environment beyond its normal range of variation to adversely affect the population performance or individual physiology of the organism in a significant way (Rolf *et al.* 2004). Abiotic stress comes in many forms. The most common of the stressors are the easiest for people to identify, but there are many other, less recognizable abiotic stress factors which affect environments constantly (Palta *et al.* 2006). The most basic stressors include: high winds, extreme temperatures, drought, flood, other natural disasters, such as tornadoes, and wild fires (Palta *et al.* 2006). Lesser-known stressors generally occur on a smaller scale. They include: poor edaphic conditions like rock content and pH levels, high radiation, compaction, contamination, and other, highly specific conditions like rapid rehydration during seed germination.

Plants are continuously affected by a variety of environmental factors. Environmental factors, such as extreme temperatures (heat, cold, freezing), drought (deficient precipitation, drying winds), and contamination of soils by high salt concentration, are major abiotic environmental stressors that limit plant growth and development, and thus agronomical yield, and play a major role in determining the geographic distribution of plant species. These different adverse but not necessarily lethal conditions are generally known as stress. Whereas biotic environmental factors are other organisms such as symbionts, parasites, pathogens, herbivores, and competitors, abiotic factors include parameters and resources which determine plant growth like temperature, relative humidity, light, availability of water, mineral nutrients, and CO_2, as well as wind, ionizing radiation, or pollutants (Schulze *et al.* 2002). The effect each abiotic factor has on the plant depends on its quantity or intensity. For optimal growth, the plant requires a certain quantity of each abiotic environmental factor. Any deviation from such optimal external conditions, that is, an excess or deficit in the chemical or physical environment, is regarded as abiotic stress and adversely affects plant growth, development, and/or productivity (Bray *et al.* 2000). Abiotic stress factors include, for example, extreme temperatures (heat, cold, and freezing), too high or too low irradiation, water logging, drought, inadequate mineral nutrients in the soil, and excessive soil salinity. As especially drought and salt stress are becoming more and more serious threats to agriculture and the natural status of the environment they are recurring features of nearly all the world's climatic regions since various critical environmental threats with global implications have linkages to water crises (Gleick, 2000). These threats are collaterally catalyzed by global climate change and population growth.

The latest scientific data confirm that the earth's climate is rapidly changing. Due to rising concentrations of CO_2 and other atmospheric trace gases, global temperatures have increased by about 1°C over the course of the last century, and will likely rise even more rapidly in coming decades (IPCC 2007). Scientists predict that temperatures could rise by another 3–9°C by the end of the century with far-reaching effects. Increased drought and salinization of arable land are expected to have devastating global effects (Wang *et al.* 2003). Abiotic stress is already the primary reason of crop loss worldwide, reducing average yields for most major crop plants by more than 50% (Bray *et al.* 2000, Wang *et al.* 2003). It will soon become even more severe as desertification will further increase and the current amount of annual loss of arable area may double by the end of the century because of global warming (Vinocur and Altman 2005). Simultaneously, rapid population growth increasingly generates pressure on existing cultivated land and other resources (Ericson *et al.* 1999). Population migration to those arid and semiarid areas increases the problems of water shortage and worsens the situation of land degradation in the destination, and in turn causes severe problems of poverty, social instability, and population health threats (Moench, 2002). Water scarcity and desertification could critically undermine efforts for sustainable development, introducing new threats to human health, ecosystems, and national economies of various countries. Therefore, solutions to these problems are desperately needed, such as the improvement of salt and drought tolerance of crops, which in turn requires a detailed knowledge about salt and drought tolerance mechanisms in plants. The viability of plants in both dry and saline habitats depends on their ability to cope with (I) water deficit due to a low water potential of the soil and (II) restriction of CO_2 uptake. Plants growing on saline soils are additionally confronted with (III) ion toxicity and nutrient imbalance.

Water deficit causes detrimental changes in cellular components because the biologically active conformation and thus the correct functioning of proteins and bio-membranes depends on an intact hydration shell. As a consequence, severe osmotic stress can lead to an impairment of amino acid synthesis, protein metabolism, and the dark reaction of photosynthesis or respiration and can cause the breakdown of the osmotic system of the cell (Larcher, 2001, Schulze *et al.* 2002). Water deficit can be counteracted by compatible solutes, organic compounds which are highly soluble and do not interfere with cellular metabolism. They serve as a means for osmotic adjustment and also function as chaperons by attaching to proteins and membranes, thus preventing their denaturation. This protective function of compatible solutes can also alleviate ion specific effects of salt stress caused by ion toxicity and ion imbalance such as the precipitation of proteins due to changes in charge or the destruction of membranes caused by alterations of the membrane potential.

Regarding the restriction of CO_2 uptake (II), the negative effects of osmotic stress force plants to minimize water loss, growth depends on the ability to find the best tradeoff between a low transpiration and a high net photosynthetic rate (Koyro, 2006). However, various plant species show a clearly reduced assimilation rate under osmotic stress conditions due to stomatal closure

(Huchzermeyer and Koyro 2005). A consequence can be an excessive production of reactive oxygen species (ROS) which are highly destructive to lipids, nucleic acids, and proteins (Türkan and Demiral 2009, Geissler *et al.* 2010). However, generated ROS can be scavenged by the antioxidative system which includes nonenzymatic antioxidants and antioxidative enzymes (Blokhina *et al.* 2003). Ion toxicity (III) on saline habitats is caused by ion specific effects on membranes and proteins. On one hand, changes of the ionic milieu lead to alterations of the membrane potential and thus to a destruction of biomembranes (Schulze *et al.* 2002). On the other hand, the hydration and charge of proteins are negatively influenced, so that their precipitation is promoted, but their activity is reduced (Kreeb, 1996). These effects of salt stress can be alleviated by the protective chaperone function of compatible solutes. When looking at drought and salt tolerance of plants in the face of global climate change, another important aspect should be considered. Compared to salinity and drought, elevated atmospheric CO_2 concentrations have contrary effects on plants. They often improve photosynthesis while reducing stomatal resistance in C_3 plants, thus increasing water use efficiency, but decreasing photorespiration and oxidative stress (Urban, 2003, Kirschbaum, 2004, Rogers *et al.* 2004). Furthermore, more energy can be provided for energy-dependent tolerance mechanisms such as the synthesis of compatible solutes and antioxidants. Therefore, the salt and drought tolerance and the productivity of these plants can be enhanced under elevated CO_2 (Wullschleger *et al.* 2002, Urban 2003), increasing their future suitability as crops. Against the background described earlier, this review uncovers how compatible solutes and antioxidants alleviate environmental stress, especially drought and salt stress, and the role elevated CO_2 concentrations can play in this context.

Environmental stress can disrupt cellular structures and impair key physiological functions (Larcher, 2003). Drought, salinity, and low temperature stress impose an osmotic stress that can lead to turgor loss. Membranes may become disorganized, proteins may undergo loss of activity or be denatured, and often excess levels of reactive oxygen species (ROS) are produced leading to oxidative damage. As a consequence, inhibition of photosynthesis, metabolic dysfunction, and damage of cellular structures contribute to growth perturbances, reduced fertility, and premature senescence.

Different plant species are highly variable with respect to their optimum environments, and a harsh environmental condition, which is harmful for one plant species, might not be stressful for another (Munns and Tester, 2008). This is also reflected in the multitude of different stress-response mechanisms. Two major strategies can be distinguished: stress avoidance and stress tolerance (Levitt, 1980). Stress avoidance includes a variety of protective mechanisms that delay or prevent the negative impact of a stress factor on a plant. For example, cacti have constitutively adapted their morphology, physiology, and metabolism to hot and arid climates. Adaptation is stable and inherited. On the other hand, stress tolerance is the potential of a plant to acclimate to a stressful condition. For example, in summer, trees and herbaceous plants in northern latitudes cannot withstand freezing. Exposure to chilling temperatures, however, induces hardening and acclimated plants survive winter temperatures far below freezing. Plants can increase their resistance to various stresses including heat, saline, and drought conditions in response to a period of gradual exposure to these constraints. Acclimation is plastic and reversible. The physiological modifications induced during acclimation are diverse and are usually lost when the adverse environmental condition does not persist (Krasensky and Jonak, 2012).

Responses to environmental stresses occur at all levels of organization. Cellular responses to stress include adjustments of the membrane system, modifications of the cell wall architecture, and changes in cell cycle and cell division. In addition, plants alter metabolism in various ways, including production of compatible solutes (*e.g.* proline, raffinose, and glycine betaine) that are able to stabilize proteins and cellular structures and/or to maintain cell turgor by osmotic adjustment, and redox metabolism to remove excess levels of ROS and re-establish the cellular redox balance (Valliyodan and Nguyen, 2006, Munns and Tester, 2008, Janska *et al.* 2010). At the molecular level, gene expression is modified upon stress (Chinnusamy *et al.* 2007, Shinozaki and Yamaguchi-Shinozaki, 2007) and epigenetic regulation plays an important role in the regulation of gene expression in response to environmental stress (Hauser *et al.* 2011, Khraiwesh *et al.* 2011). Stress-inducible genes comprise genes involved in direct protection from stress, including the synthesis of osmoprotectants, detoxifying enzymes, and transporters, as well as genes that encode regulatory proteins such as transcription factors, protein kinases, and phosphatases.

1.1. Complexity of the Metabolic Response

Metabolic adjustments in response to unfavorable conditions are dynamic and multifaceted and not only depend on the type and strength of the stress, but also on the cultivar and the plant species. Traditionally, metabolic studies focused on single metabolites or groups of metabolites. Recent advances in metabolic profiling allow increasingly comprehensive metabolite analyses, illustrating the complexity of metabolic adjustments to stress. The metabolic profiles of different plant species, including rice, poplar, *Vitis vinifera*, *Thellungiella halophila*, and *Arabidopsis thaliana*, have been analysed after exposure to drought (Rizhsky *et al.* 2004; Cramer *et al.* 2007; Urano *et al.* 2009), salinity (Janz *et al.* 2010; Lugan *et al.* 2010), and temperature stress (Usadel *et al.* 2008; Espinoza *et al.* 2010; Caldana *et al.* 2011).

Plants respond to stress by a progressive adjustment of their metabolism with sustained, transient, early- and late-responsive metabolic alterations. For example, raffinose and proline accumulate to high levels over the course of several days of salt exposure, drought, or cold, whereas central carbohydrate metabolism changes rapidly in a complex, time-dependent manner. Some metabolic changes are common to salt, drought, and temperature stress, whereas others are specific. For example, levels of amino acids, sugars, and sugar alcohols typically increase in response to different stress conditions. Notably, proline accumulates upon drought, salt, and low temperature but not upon high temperature stress (Cramer *et al.* 2007; Gagneul *et al.* 2007; Kempa *et al.* 2008; Sanchez *et al.* 2008; Usadel *et al.* 2008; Urano *et al.* 2009; Lugan *et al.* 2010). In most studies, organic acids and TCA-cycle intermediates decreased in glycophytes after salt stress (Gagneul *et al.* 2007; Zuther *et al.* 2007; Sanchez *et al.* 2008), but increased in response to temperature or drought stress (Kaplan *et al.* 2004; Usadel *et al.* 2008; Urano *et al.* 2009). A direct comparison of the metabolic profile of *Arabidopsis thaliana* acclimated to either low or high temperatures showed a significant overlap, but also major differences in metabolic adjustments (Kaplan and Guy, 2004). A greater number of metabolite levels changed specifically in response to cold than to heat, pointing towards a strong impact of cold on plant metabolism. Similarly, metabolic rearrangements after drought were more profound than after heat stress and the metabolic profile of *Arabidopsis* plants exposed to a combination of drought and heat was more similar to that of drought-stressed than heat-stressed plants. Exposure of drought-stressed plants to heat also induced unique metabolic responses highlighting the complexity of metabolic adjustments in natural environments (Rizhsky *et al.* 2004).

Comparison of the metabolome of stress-tolerant accessions, cultivars, or species, with that of stress-sensitive ones, further show that, the role of metabolism in natural stress tolerance is multifaceted and diverse (Gong *et al.* 2005, Hannah *et al.* 2006, Zuther *et al.* 2007, Janz *et al.* 2010, Korn *et al.* 2010, Lugan *et al.* 2010). Generally, stress tolerant plants have higher levels of stress-related metabolites under normal growth conditions and/or accumulate larger amounts of protective metabolites, such as proline and soluble sugars, under unfavourable conditions, indicating that their metabolism is prepared for adverse growth conditions. For example, analyses of the metabolic profile of *Arabidopsis* accessions with different freezing tolerances point to a crucial role of compatible solutes, including proline and raffinose, in freezing tolerance (Hannah *et al.* 2006, Korn *et al.* 2010). Similarly, comparison of the metabolome of cultivars or species with different salt tolerance indicates a beneficial role of compatible solutes under salinity conditions (Gong *et al.* 2005, Zuther *et al.* 2007, Janz *et al.* 2010, Lugan *et al.* 2010). Interestingly, metabolism of salt-tolerant plants appeared to be pre-adapted to saline environments by constitutively high levels of some protective metabolites such as proline and raffinose. Consistent with classical studies, these recent metabolomic analyses illustrate that plants have developed a whole range of strategies to adapt their metabolism to unfavourable growth conditions and that enhanced stress resistance is not restricted to a single compound or mechanism. Different plant species accumulate different metabolites (*e.g.* trehalose, proline, glycine betaine) and there is no absolute requirement for the accumulation of a specific metabolite for acclimation to stress. In some cases, the flux through a metabolic pathway, rather than the accumulation of a specific metabolite *per se*, might contribute to stress tolerance (Krasensky and Jonak, 2012).

Several metabolites/metabolic pathways that contribute to stress acclimation also play a role in development. Appropriate proline levels have been shown to be important for embryo and flower development (Mattioli *et al.* 2008, 2009, Szekely *et al.* 2008). γ-Amino butyric acid (GABA) regulates pollen tube growth and guidance (Palanivelu *et al.* 2003). Polyamines regulate several developmental processes including embryogenesis, meristem, flower, and gametophyte development (Gupta and Kaur, 2005, Deeb *et al.* 2010, Zhang *et al.* 2011a, b). Trehalose metabolism is essential for proper embryo maturation as well as for vegetative and inflorescence development (Eastmond *et al.* 2002, van Dijken *et al.* 2004, Satoh-Nagasawa *et al.* 2006).

Most environmental stresses are affecting on the production of active oxygen species in plants, causing oxidative stress (Bartosz, 1997). Also, there is growing evidence that in plants subjected to environmental stress. The balance between the production of activated oxygen species and the quenching activity of antioxidant is upset, which often results in oxidative damage (Karpinski*et al.* 1997). Environmental stress causes significant crop losses. The stresses are numerous and often crop or location-specific. They include increased UV-B radiation, water, high salinity, temperature extremes, hypoxia (restricted oxygen supply in waterlogged and compacted soil), mineral nutrient deficiency, metal toxicity, herbicides, fungicides, air pollutants, light, temperature and topography. It is apparent that many environmental stresses exert at least part of their effect by causing oxidative damage (Smirnoff, 1995). Consequently, the antioxidant defense system of plants has been attracting considerable interest (Alscher *et al.* 1997). Characterization of mutants and transgenic plants with altered expression of antioxidant is a potentially powerful approach to understanding the functioning of the antioxidant system and its role in protecting plants against stress, and significant progress is now being made in this area (Ali and Alqurainy, 2006).

Atmospheric oxygen has been recognized for more than 100 years as the agent responsible for the deterioration of organic materials exposed to air. The parallel role of oxygen, a molecule essential form many forms of life, as a destructive (toxic) agent for living tissues has been discovered much more recently. Even under optimal conditions many metabolic processes produce active oxygen species. Among the four major active oxygen species [superoxide radical O^-_2, hydrogen peroxide H_2O_2, hydroxyl radical OH and singlet oxygen $1O_2^-$] H_2O_2 and the hydroxyl radical are most active, toxic and destructive (Smirnoff, 1993). In plants the most important of these are driven by or associated with light dependent events.

Photosynthetic cells are prone to oxidative stress because they contain an array of photosensitizing pigments and they both produce and consume oxygen. The photosynthetic electron transport system is the major source of active oxygen species in plant tissues (Asada, 1997), have the potential to generate singlet oxygen $1O_2$ and superoxide O^{-2}. Olga *et al.* (2003) concluded that generation of reactive oxygen species (ROS) is characteristic for hypoxia and especially for reoxygenation. Of the ROS, hydrogen peroxide (H_2O_2) and superoxide (O_2^-) are both produced in a number of cellular reactions, including the iron-catalysed Fenton reaction, and by various enzymes such as lipoxygenases, peroxidases, NADPH oxidase and xanthine oxidase. The main cellular components susceptible to damage by free radicals are lipids (peroxidation of unsaturated fatty acids in membranes), proteins (denaturation), carbohydrates and nucleic acids. Consequences of hypoxia-induced oxidative stress depend on tissue and/or species (*i.e.* their tolerance to anoxia), on membrane properties, on endogenous antioxidant content and on the ability to induce the response in the antioxidant system. Effective utilization of energy resources (starch, sugars) and the switch to anaerobic metabolism and the preservation of the redox status of the cell are vital for survival (Gout *et al.* 2001). The formation of ROS is prevented by an antioxidant system: low molecular mass antioxidants (ascorbic acid, glutathione, and tocopherols), enzymes regenerating the reduced forms of antioxidants, and ROS-interacting enzymes such as superoxide dismutase (SOD), peroxidases and catalases (Gout *et al.* 2001). In plant tissues many phenolic compounds (in addition to tocopherols) are potential antioxidants: flavonoids, tannins and lignin precursors may work as ROS-scavenging compounds (Olga *et al.* 2003). Antioxidants act as a cooperative network, employing a series of redox reactions. Interactions between ascorbic acid and glutathione, and ascorbic acid and phenolic compounds are well known. Under oxygen deprivation stress some contradictory results on the antioxidant status have been obtained.

Experiments on overexpression of antioxidant production do not always result in the enhancement of the antioxidative defense, and hence increased antioxidative capacity does not always correlate positively with the degree of protection (Gout *et al.* 2001). Factors which possibly affect the effectiveness of antioxidant protection under oxygen deprivation as well as under other environmental stresses and such aspects as compartmentalization of ROS formation and antioxidant localization, synthesis and transport of antioxidants, the ability to induce the antioxidant defense and cooperation (and/or compensation) between different antioxidant systems are the determinants of the competence of the antioxidant system (Olga *et al.* 2003).

The vast metabolic diversity observed in plants is the direct result of continuous evolutionary processes. There are more than 200,000 known plant secondary metabolites, representing a vast reservoir of diverse functions. When the environment is adverse and plant growth is affected, metabolism is profoundly involved in signaling, physiological regulation, and defense responses. At the same time, in feedback, abiotic stresses affect the biosynthesis, concentration, transport, and storage of primary and secondary metabolites. Metabolic adjustments in response to abiotic stressors involve fine adjustments in amino acid, carbohydrate, and amine metabolic pathways. Proper activation of early metabolic responses helps cells restore chemical and energetic imbalances imposed by the stress and is crucial to acclimation and survival. Time-series experiments have revealed that metabolic activities respond to stress more quickly than transcriptional activities do. In order to study and map all the simultaneous metabolic responses and, more importantly, to link these responses to a specific abiotic stress, integrative and comprehensive analyses are required. Metabolomics is the systematic approach through which qualitative and quantitative analysis of a large number of metabolites is increasing our knowledge of how complex metabolic networks interact and how they are dynamically modified under stress adaptation and tolerance processes. A vast amount of research has been done using metabolomic approaches to (*i*) characterize metabolic responses to abiotic stress, (*ii*) to discover novel genes and annotate gene function, and, (*iii*) more recently, to identify metabolic quantitative trait loci. The integration of the collected metabolic data concerning abiotic stress responses is helping in the identification of tolerance traits that may be transferable to cultivated crop species.

Owing to their sessile lifestyle, plants have evolved numerous adaptations to help them cope with unavoidable abiotic and biotic stresses that are imposed upon them in their natural environment. For example, plants frequently alter their growth and developmental patterns so as to alleviate some of the unfavorable environmental changes that they are exposed to. Thus, increased allocation of biomass to roots typically occurs in dry or mineral nutrient-deficient conditions. There is little doubt that this type of developmental plasticity is a key aspect to the survival of plants in extreme environments. However, it is equally apparent that certain biochemical and metabolic adaptations of plants also facilitate their growth and/or survival under sub-optimal environmental conditions. For instance, in contrast to animals, plants can often accomplish the same step in a metabolic pathway in several different ways. This so-called *'metabolic flexibility'* is perhaps best exemplified by a wide variety of genetic engineering experiments that have partially or fully eliminated individual enzymes traditionally considered to be essential, and yet the resulting transgenic plants were able to grow and develop more or less normally (Plaxton and Podestá, 2006).

1.2. Abiotic Stress and Metabolites

Amino acids

Accumulation of amino acids has been observed in many studies on plants exposed to abiotic stress (Kempa *et al.* 2008, Sanchez *et al.* 2008, Usadel *et al.* 2008, Lugan *et al.* 2010). This increase might stem from amino acid production and/or from enhanced stress-induced protein breakdown. While the overall accumulation of amino acids upon stress might indicate cell damage in some species (Widodo *et al.* 2009), increased levels of specific amino acids have a beneficial effect during stress acclimation.

The accumulation of certain amino acids especially proline, prevents the development of severe osmotic stress. Amino acids also have several other roles in plants, for example they regulate ion transport and stomatal opening and affect the synthesis and activity of enzymes, gene expression, and redox homeostasis, helping the plants to cope with the harmful effects of osmotic stress (Rai, 2002). Osmotic stress, which can be induced by limited water supply or osmotic agents, alters both the amino acid pattern and the concentrations of individual amino acids. Osmotic stress induced by polyethylene glycol resulted in an increase in the concentrations of aspartic acid, glutamine, asparagines, threonine, serine, alanine and proline during the first day of treatment in the roots of maize (Ogawa and Yamauchi, 2006). After dehydration the amino acid contents changed to differing extents in the desiccation-tolerant young leaves and sensitive old leaves of the resurrection plant *Sporobolus stapfianus* (Martinelli *et al.* 2007), indicating that these changes depended on the level of stress tolerance.

Proline plays a very important role as an osmoprotectant in the adaptation to osmotic stress (Verbruggen and Hermans, 2008). Accumulation induced by osmotic stress is based on increased synthesis and decreased degradation

(Yoshiba *et al.* 1997). Osmotic stress resulted in a greater increase in proline content in a drought-tolerant rice genotype than in a sensitive one, due to the greater expression of the gene encoding pyrroline-5-carboxylate synthase and activity of this enzyme involved in the proline synthesis (Cloudhary *et al.* 2005). Further evidence for the protective role of proline was found in transgenic plants, where its overproduction increased tolerance to osmotic stress (Verdoy *et al.* 2006). Besides proline, the involvement of other amino acids in the response to osmotic stress was also demonstrated. Arginine was found to function as a compatible solute, improving stress tolerance in yeast (Xu *et al.* 2011). Under hyperosmotic conditions the transcription of genes encoding the enzymes involved in arginine biosynthesis increased, while those involved in degradation decreased. Both exogenous arginine and the overproduction of two enzymes required for its synthesis increased tolerance. In addition, osmotic stress resulted in the increased expression of asparagine synthase genes in sunflower (Herrera-Rodríguez *et al.* 2007) and in wheat (Wang *et al.* 2005). Overexpression of glutamine synthase increased tolerance to osmotic stress in rice (Cai *et al.* 2009). These observations show that the changes in amino acid concentrations induced by osmotic stress may be due to the altered expression of genes encoding the enzymes involved in their metabolism (Kovacs *et al.* 2012).

Proline

For many years, the capacity to accumulate proline has been correlated with stress tolerance. Proline is considered to act as an osmolyte, a ROS scavenger, and a molecular chaperone stabilizing the structure of proteins, thereby protecting cells from damage caused by stress (Verbruggen and Hermans, 2008; Szabados and Savoure, 2010). Proline accumulates in many plant species in response to different environmental stresses including drought, high salinity, and heavy metals. Interestingly, heat stress did not lead to proline accumulation in tobacco and *Arabidopsis* plants and induced proline accumulation rendered plants more sensitive to heat (Dobra *et al.* 2010; Lv *et al.* 2011). Proline levels are determined by the balance between biosynthesis and catabolism (Szabados and Savoure, 2010). Proline is produced in the cytosol or chloroplasts from glutamate, which is reduced to glutamate-semialdehyde (GSA) by Δ-1-pyrroline-5-carboxylate synthetase (P5CS). GSA can spontaneously convert to pyrroline-5-carboxylate (P5C), which is then further reduced by P5C reductase (P5CR) to proline. Proline is degraded in mitochondria by proline dehydrogenase (ProDH) and P5C dehydrogenase (P5CDH) to glutamate. Stress conditions stimulate proline synthesis while proline catabolism is enhanced during recovery from stress. Over-expression of P5CS in tobacco and petunia led to increased proline accumulation and enhanced salt and drought tolerance (Hong *et al.* 2000; Yamada *et al.* 2005), whereas *Arabidopsis* P5CS1 knock-out plants were impaired in stress-induced proline synthesis and were hypersensitive to salinity (Szekely *et al.* 2008). Consistently, ProDH antisense *Arabidopsis* accumulated more proline and showed enhanced tolerance to freezing and high salinity (Nanjo *et al.* 1999b). It is discussed that, in an alternative pathway, mitochondrial P5C can be produced by δ-ornithine aminotransferase (δ-OAT) from ornithine (Miller *et al.* 2009). Over-expression of *Arabidopsis* δ-OAT has been shown to enhance proline levels and to increase the stress tolerance of rice and tobacco (Qu *et al.* 2005) even though *Arabidopsis* plants deficient in δ-OAT accumulated proline in response to stress and showed a salt stress tolerance similar to the wild type (Funck *et al.* 2008).

Proline is an amino acid known to occur widely in higher plants and in response to environmental stresses (especially salt/osmotic stress), and normally accumulates in large quantities. Under salt/osmotic conditions, it contributes to the stabilization of proteins, membranes and subcellular structures in cytosol, and protecting cellular functions by scavenging reactive oxygen species. Furthermore, it is known to induce expression of salt stress responsive genes, which possess proline-responsive elements (Kishor *et al.* 2005, Ashraf and Foolad, 2007). Accumulation of proline could be due to *de novo* synthesis or decreased degradation, or both, and it is synthesized from glutamate and ornithine. In plants, the main pathway is from glutamate, which is converted to proline by two successive reductions catalyzed by pyrroline-5-carboxylate synthases (P5CS) and pyrroline-5-carboxylate reductases (P5CR), respectively. Ornithine is the alternative precursor for proline, which can be transaminated to P5C by Orn-d-aminotransferase (OAT), a mitochondrial located enzyme (Kishor *et al.* 2005, Verbruggen and Hermans, 2008).

Several attempts were made to increase the level of proline accumulation in plants by transferring the genes associated with the biosynthetic pathway. Proline-level increase by overexpressing P5C enzymes were observed in several studies and in different plant species. In tobacco, the increase in expression of GK74 and GPR promoted the overproduction of proline (Stein *et al.* 2011*)*. The activities of P5CS and P5CR were significantly enhanced in the leaves of *Morus albas* with decreasing leaf water potentials (Chaitanya *et al.* 2009). In cactus pear, salt stress increased the expression of P5CS and induces proline accumulation (Silva-Ortega *et al,* 2008). On the other hand, proline levels decreased when P5C enzymes are silenced. A P5CS in *Arabidopsis* was knocked-out, reducing the proline synthesis (Székely *et al.* 2008). The down-regulation of proline dehydrogenase (ProDH) also increases proline levels. The transcription of antisensing ProDH improved the production of proline in tobacco

(Stein *et al.* 2011). In the mulberry, the activities of proline dehydrogenase were reduced with progressive increase in water stress (Chaitanya *et al.* 2009). Also, there are studies about the proline alternative pathway. The activities of ornithine transaminase were increased in *Morus alba* with low leaf water potentials (Chaitanya *et al.* 2009). In cashew plants, it was reported that under salt-induced stress the levels of ornithine were increased (Rocha *et al.* 2012). In *Hibiscus tiliaceus* a contradictory result was reported. Formate dehydrogenase, negatively associated with the accumulation of proline in response to osmotic stress, was found highly up-regulated during salt stress (Yang *et al.* 2011).

γ-Aminobutyric acid (GABA)

The non-protein amino acid γ-aminobutyric acid (GABA) rapidly accumulates to high levels under different adverse environmental conditions (Kempa *et al.* 2008, Renault *et al.* 2010). GABA is mainly synthesized from glutamate in the cytosol by glutamate decarboxylase (GAD) and then transported to the mitochondria. GABA transaminase (GABA-T) and succinic semialdehyde dehydrogenase (SSADH) convert GABA into succinate that feeds into the TCA-cycle (Fait *et al.* 2008). GABA metabolism has been associated with carbon–nitrogen balance and ROS scavenging (Song *et al.* 2010; Liu *et al.* 2011). A functional GABA shunt is important for stress tolerance. Salt stress enhances the activity of enzymes involved in GABA metabolism (Renault *et al.* 2010). *Arabidopsis* mutants defective in GABA-T were hypersensitive to ionic stress and showed increased levels of amino acid (including GABA), while carbohydrate levels were decreased (Renault *et al.* 2010). Disruption of the SSADH gene led to the accumulation of ROS associated with dwarfism and hypersensitivity to UV-B and heat stress (Bouche *et al.* 2003).

The pathway that converts glutamate to succinate through GABA is called "GABA shunt". The first step is a direct and irreversible decarboxylation of glutamate by glutamate decarboxylase (GAD, EC 4.1.1.15), a cytosolic enzyme. Typically, GABA levels in plant tissues is low, but it increases many times in response to various stimuli, including thermal shock, mechanical stimulation, drought, hypoxia and phytohormones (Shelp *et al.* 1999), showing that GABA is intensely and rapidly produced as a result of abiotic and biotic stresses. The GABA shunt has been associated with many physiological responses such as cytosolic pH regulation, the carbon influx into the Krebbs cycle, nitrogen metabolism, and protection against oxidative stress, osmoregulation, and cell signaling (Bouche and Fromm, 2004). It has been suggested that GABA synthesis induced by stress is the result of cytosolic acidity and the consequent GAD stimulation. However, it is unlikely that the numerous environmental factors that stimulate GABA accumulation are all mediated by a decrease in the cytosolic pH. It is known that stress factors such as mechanical or cold shock that stimulate GABA levels, increase cytosolic calcium levels which induce Ca^{2+}/calmodulin-dependent glutamate decarboxylase (GAD) activity and, therefore, induce GABA synthesis. According to Shelp *et al.* (1999), studies carried out in different petunia organs showed the presence of differential expression patterns of mRNA and GAD protein, suggesting that GAD activity is regulated on a transcriptional as well as translational level (Shelp *et al.*1999). The Arabidopsis (AtProT2) and tomato (LeProT1) transporters carry GABA as well as other components related to stress such as proline and glycine-betaine. The AtProT2 transporter can be induced by water and salinity stress. The findings have indicated that GABA may have a role as compatible osmolyte, considering that it is also highly water-soluble and has no toxic effect. In high concentrations, GABA presents cryoprotectant properties, stabilizing and protecting isolated thylakoids against frost in the presence of salt, as well as possessing hydroxyl radical elimination activities (Shelp *et al.*1999, Bouché and Fromm 2004). According to Shelp *et al.* (1999), whether GABA has a specific role (as osmolyte or osmoprotectant) under water stress or it is metabolized (to produce proline), remains unknown.

1.3. Oxidative Stress and Reactive Oxygen Species (ROS)

Oxygen is essential to aerobic life existence, but toxic by-products called reactive oxygen species (ROS) such as singlet oxygen (O_2), superoxide radicals (O_2^-), hydrogen peroxide (H_2O_2) and hydroxyl radicals (OH) are generated in all aerobic cells during normal cellular metabolism (respiration, photosynthesis, photorespiration and beta oxidation of fatty acids) in mitochondria, chloroplasts and peroxisomes (Xiong and Zhu 2002; Apel and Hirt, 2004; Gill and Tuteja, 2010). However, ROS levels increase as a consequence of various environment injuries to which plants are exposed, such as temperature, oxygen deprivation (Blokhina *et al.* 2003), high light intensity (Apel and Hirt 2004), water stress (Jiang and Zhang, 2002), salinity stress (Hernández et al. 2000), mechanical stress or even pollution.

ROS can react with a variety of biomolecules, alternating or blocking their biological functions, causing damage to cell components such as membrane lipids, deactivating enzymes (denaturation), carbohydrates, nucleic acids and to the photosystem II complex (Arora *et al.* 2002, Blokhina *et al.* 2003, Apel and Hirt, 2004). The injuries caused by ROS are known as oxidative stress and constitute one of the main damage factors in plants exposed to different environmental stresses (Kwon *et al.* 2002). According to Chen and Polle (2010), ABA, Ca^{2+} and ROS are involved in abiotic stress sensing, like soil salinity, with higher or faster activation of defenses in tolerant than susceptible poplar species. In order to the reduce toxic effects of ROS, plants use highly regulated enzymatic and non-enzymatic mechanisms to maintain a balance between the production and destruction of ROS to maintain cell redox homeostasis (Blokhina *et al.* 2003, Sairam and Tyagi, 2004). The term antioxidant can be considered to describe any compound able to extinguish ROS without undergoing conversion into a harmful radical. Antioxidant enzymes are those that catalyze such reactions or involved in ROS metabolism. Consequently, antioxidants and antioxidant enzymes interrupt cascades of uncontrolled oxidation (Noctor and Foyer 1998).

Therefore, plants possess the ability to fight against oxidative stress using ROS-eliminating systems such as superoxide dismutase (SOD, EC1.15.1.1), catalase (CAT, EC1.11.1.6), ascorbate peroxidase (APX, EC1.11.1.1), as well as antioxidant compounds of low molecular weight including ascorbate (ASC), glutathione, phenolic compounds and polyamines. In addition to the ROS-eliminating enzymes, SOD and APX, enzymes such as monodehydroascorbate reductase (MDHAR, EC1.6.5.4), dehydroascorbate reductase (DHAR) and glutathione reductase (GR, EC1.8.1.7), which are necessary for the regeneration of ascorbate and glutathione are also involved.

1.4. Abiotic Stress and Enzyme Activity

Superoxide dismutase (SOD)

Superoxide dismutase (SOD) is the first enzyme reported as being able to decompose a free radical. This enzyme catalyzes the superoxide radical as shown in the flowing reaction, $O^{2-} + O^{2-} + 2H + O^{2+} = H_2O_2$ reaction (Netto, 2001). Netto (2001) analyzed sugarcane ESTs (SUCEST databank) and found 5 isoforms of superoxide dismutase (SOD). The high number of isoforms may indicate that the superoxide radical can be very toxic to plants, though this free radical is not very reactive. Moreover, the author suggested that superoxide radical toxicity occurs due to its ability to react with nitric oxide to form peroxynitrite, which is a strong oxidant. Furthermore, superoxide (O^{2-}) and hydrogen peroxide (H_2O_2), in spite of not being very toxic, in the presence of trace amounts of Fe^{2+} and Fe^{3+} through the Haber-Weiss reaction, forms the hydroxyl radical (OH). This radical can cause damage to chlorophyll, proteins, DNA, lipids and other important macromolecules, thus can cause fatally affect plant growth and development, and finally and hence ultimately affect plant productivity (Sairam and Tyagi, 2004).

Catalase (CAT)

The catalase enzyme protects cells from hydrogen peroxide that may be generated from the reaction catalyzed by superoxide dismutase, through the beta-oxidation of fatty acids in peroxisomes or through other processes like the Mehler reaction in chloroplasts, in which H_2O_2 is generated under normal metabolism, by electron transport in mitochondria and photorespiration in peroxisomes (Neill *et al.* 2002). This hemeprotein catalyzes the reaction $2H_2O_2 + O_2 + 2H_2O$ (Gupta *et al.* 1998, Netto, 2001). Plant catalases derive from a common ancestor gene and can be divided into three distinct groups (CAT1, CAT2, and CAT3). Moreover, similar clusters of catalases were identified in the sugarcane expressed sequence tags data base (SUCEST) databank (Netto, 2001).

Ascorbate peroxidase (APX)

Ascorbate peroxidase (APX) is another key enzyme for the control of hydrogen peroxide concentrations due to its ability to catalyze the decomposition of hydrogen peroxide at the expense of ascorbate (ASC). The sequence of this protein containing heme distinguishes itself from other peroxidases and different APX forms occur in chloroplasts, cytosol, mitochondria, peroxisomes and glyoxysomes. Catalases (CAT) convert hydrogen peroxide (H_2O_2) into water and molecular oxygen. These enzymes have extremely high catalytic levels, but low affinity with the substrate, since the reaction requires two H_2O_2 molecules to activate its site. An alternative way of breaking down H_2O_2 is through peroxidases, which are found in every cell and have high affinity with H_2O_2 compared to CAT. Peroxidases, however, require a reducer since they reduce H_2O_2 to water.

In plant cells, the most important reduced substrate for H_2O_2 detoxification is ascorbate. APX uses two ascorbate molecules to reduce H_2O_2 to water, with the simultaneous generation of two monodehydroascorbate molecules (MDHA) (Noctor and Foyer, 1998). Two enzymes are involved in reduced ASC regeneration, called monodehydroascorbate reductase (MDHAR), which uses NADPH directly in recycling ASC and dehydroascorbate reductase (DHAR).

Although the interrelation of water stress, ABA, ROS and the antioxidant defense system had been studied in several plant species remains unclear which signals stimulate the increase of antioxidant enzymes, that are essential to the defense against oxidative stress. According to Apel and Hirt (2004), the generation of ROS into cell compartments, such as the mitochondria or chloroplasts results in changes to the nuclear transcriptomes, indicating which information should be transmitted from these organelles to the nucleus, but the identity of the transmitted signal remains unknown (Gill and Tuteja, 2010). There are three possible ways to indicate how ROS could affect gene expressions. ROS sensors could be activated, inducing signal cascades that ultimately culminate gene expression. Another alternative would be that signaling pathways could directly oxidized by ROS. And finally, ROS could change gene expression, affecting and modifying the activity of transcription factors.

According to Netto (2001), plants have developed different systems to confront the toxic effects of reactive oxygen (ROS) and nitrogen (RNS) species. The first antioxidant defense line involves the prevention of ROS formation. The second is formed of antioxidant enzymes and compounds of low molecular weight. Moreover, if the first antioxidant line of defense fails to prevent the formation of reactive species, antioxidant compounds break down reactive species, thus avoiding the generation of oxidative injuries in the biomolecules (Netto, 2001). According to Hoekstra (2002), drought and desiccation tolerant cells undergo less oxidative damages than Abiotic stress and cells that are sensitive to these stresses. This could be the result of an effective decrease in ROS production or of the activation of effective antioxidant systems, or both.

1.5. Abiotic Stress and Polyamines

Polyamines (PA) are small aliphatic molecules positively charged at cellular pH. Various stresses, such as drought, salinity and cold, modulate PA levels, and high PA levels have been positively correlated with stress tolerance (Cuevas *et al.* 2008; Groppa and Benavides, 2008; Usadel *et al.* 2008; Quinet *et al.* 2010; Alcazar *et al.* 2011). Putrescine, spermidine, and spermine are the most common PAs in higher plants. Putrescine is produced from either ornithine or arginine by ornithine decarboxylase (ODC) and arginine decarboxylase (ADC), respectively. Putrescine is converted to spermidine by spermidine synthase (SPDS) and then to spermine by spermine synthase (SPMS). Spermidine and spermine are substrates of polyamine-oxidases (PAOs), which catalyse the back-conversion to putrescine.

PAs have been implicated in protecting membranes and alleviating oxidative stress (Groppa and Benavides, 2008; Alcazar *et al.* 2010b; Hussain *et al.* 2011) but their specific function in stress tolerance is not well understood. Analyses of transgenic plants and of mutants involved in PA metabolism clearly showed a positive role of PAs in stress tolerance. Plants deficient in ADC1 or ADC2 had reduced putrescine levels and were hypersensitive to stress (Urano *et al.* 2004; Cuevas *et al.* 2008), whereas constitutive or stress-induced over-expression of ADC led to higher putrescine levels and enhanced drought and freezing tolerance (Capell *et al.* 2004; Alcazar *et al.* 2010b). *Arabidopsis* plants over-expressing SPDS produced higher amounts spermidine and were more resistant to drought, salt, and cold stress (Kasukabe *et al.* 2004). Plants deficient in SPMS were unable to synthesize spermine and are hypersensitive to salinity (Yamaguchi *et al.* 2006). Furthermore, the introduction of ornithine decarboxylase (ODC) from mouse enhanced polyamine levels and the tolerance of tobacco to salt stress (Kumriaa and Rajam, 2002).

In early days, Richards and Coleman (1952) observed the presence of a predominant unknown ninhydrin positive spot that accumulated in barley plants exposed to potassium starvation. After isolation and crystallization, this compound was identified as Put. Later on it was shown that K^+-deficient shoots fed with L-^{14}C-arginine produced labelled Put in a more rapid way compared to feeding with labelled ornithine. These results suggested that decarboxylation of arginine was the main way of accumulation of Put under K^+-deficiency (Smith and Richards, 1964). The relevance of the ADC pathway in plant responses to abiotic stress was later on established by Flores and Galston (1982). Further work in different plant species has shown that polyamine accumulation occurs in response to several adverse environmental conditions, including salinity, drought, chilling, heat, hypoxia, ozone, UV-B and UV-C, heavy metal toxicity, mechanical wounding and herbicide treatment. However, the physiological significance of these responses remained unclear, ant it had to be evaluated whether elevated polyamine levels were a result of stress-induced injury or a protective response to abiotic stress.

Classical approaches, using exogenous polyamine application and/or inhibitors of enzymes involved in polyamine biosynthesis, pointed to a possible role of these compounds in plant adaptation/defence to several environmental stresses. More recent studies using either transgenic overexpression or loss-of-function mutants support this protective role of PAs in plant response to abiotic stress (Alcázar *et al.* 2006b; Kusano *et al.* 2008; Gill and Tuteja, 2010). Indeed, heterologous overexpression of *ODC*, *ADC*, *SAMDC* and *SPDS* from different animal and plant sources in rice, tobacco and tomato has shown tolerance traits against a broad spectrum of stress conditions. Enhanced tolerance always correlated with elevated levels of Put and/or Spd and Spm. Antisense silencing of ethylene biosynthesis genes ACC synthase and ACC oxidase in tobacco has also shown to favour the Xux of SAM to polyamines (Wi and Park 2002), which leads to an increased tolerance to salt, drought and a broad spectrum of other abiotic stresses (Kasukabe *et al.* 2004; Wi *et al.* 2006). Similar results have been obtained by homologous overexpression of polyamine biosynthetic genes in *Arabidopsis*. Thus, overexpression of *SAMDC1* in Arabidopsis leads to elevated Spm levels and enhanced tolerance to various abiotic stress conditions. The *SAMDC1* overexpressing plants show elevated levels of ABA due to the induction of *NCED3*, a key enzyme involved in ABA biosynthesis. High Put levels induced by homologous overexpression of *ADC1*enhances freezing tolerance in *Arabidopsis* (Altabella *et al.* 2009). Similarly, elevated levels of Put by overexpressing *ADC2* produces drought tolerance in *Arabidopsis*, which may be related to reduction of water loss by the induction of stomata closure (Alcázar *et al.* 2010a).

The results obtained from loss-of-function mutations in polyamine biosynthetic genes further support the protective role of polyamines in plant response to abiotic stress. For example, EMS mutants of *Arabidopsis thaliana spe1-1* and *spe2-1* (which map to *ADC2*) displaying reduced ADC activity are deficient in polyamine accumulation after acclimation to high NaCl concentrations and exhibit more sensitivity to salt stress (Kasinathan and Wingler, 2004). Insertion mutants affecting the *Arabidopsis ADC2* gene present also altered responses to abiotic stress. Thus, *ADC2* induction by osmotic stress is impaired in the loss of function mutant *en9*, obtained by PCR screening of an En-1 mutagenized *Arabidopsis* population for insertions at the *ADC2* locus (Soyka and Heyer 1999). In the same way, a Ds insertion mutant (*adc2-1*), the Put content of which is diminished up to 75% of the control, is more sensitive to salt stress, whereas salt-induced injury was partly reverted by the addition of exogenous Put (Urano *et al.* 2004). Other *ADC1* (*adc1-2*, *adc1-3*) and *ADC2* (*adc2-3, adc2-4*) mutant alleles are more sensitive to freezing, and this phenotype is partially rescued by adding exogenous Put (Cuevas *et al.* 2008). Moreover, *acl5/spms Arabidopsis* double mutants that do not produce Spm are hypersensitive to salt and drought stresses, and the phenotype is mitigated by application of exogenous Spm (Kusano *et al.* 2007). All these examples illustrate that genetic modification of the polyamine biosynthetic pathway has been useful to discern the function of polyamines in plant responses to abiotic stress in both crops and model plants. Collectively, the available results indicate that elevated polyamine levels represent a stress-induced protective response.

1.6. Polyamines and ABA in Drought, Salt and Cold Stresses

The use of *Arabidopsis* has opened new perspectives in functional dissection of the polyamine metabolic pathway and its role in the control of abiotic stress responses (Alcázar *et al.* 2006b; Kusano *et al.* 2008; Takahashi and Kakehi, 2009; Gill and Tuteja, 2010). Annotation of the *Arabidopsis* genome has allowed a full compilation of the polyamine biosynthetic pathway. Transcript profiling by using Q-RT-PCR has revealed that water stress induces the expression of *ADC2*, *SPDS1* and *SPMS* genes (Alcázar *et al.* 2006a). The expression of some of these genes is also induced by ABA treatment (Perez-Amador *et al.* 2002; Urano *et al.* 2003). To get a further insight into ABA regulation of polyamine pathway, the expression of *ADC2*, *SPDS1* and *SPMS* was analysed in the ABA-deWcient (*aba2-3*) and ABA-insensitive (*abi1-1*) mutants subjected to water stress (Alcázar *et al.* 2006a). These three genes display reduced transcriptional induction in the stressed *aba2-3* and *abi1-1* mutants compared to the wild type, indicating that ABA modulates polyamine metabolism at the transcription level by up-regulating the expression of *ADC2*, *SPDS1* and *SPMS* genes under water stress conditions (Alcázar *et al.* 2006a). In addition, Put accumulation in response to drought is also impaired in the *aba2-3* and *abi1-1* mutants compared to wild-type plants. This result is further supported by metabolomic studies showing that polyamine responses to dehydration are also impaired in *nced3* mutants (Urano *et al.* 2009). All these observations support the conclusion that up-regulation of PA-biosynthetic genes and accumulation of Put under water stress are mainly ABA-dependent responses. Under salt stress conditions, there is a rapid increase in the expression of *ADC2* and *SPMS*, which is maintained during the 24-h treatment and results in increased Put and Spm levels (Urano *et al.* 2003). Spm-deficient mutants are sensitive to salt, whilst the addition of Spm suppresses the salt sensitivity, suggesting a protective role of this polyamine to high salinity (Yamaguchi *et al.* 2006). It is likely that polyamine responses to salt stress are also ABA-dependent, since both *ADC2* and *SPMS* are induced by ABA. In fact, stress-responsive, drought-responsive (DRE), low temperature-responsive (LTR) and ABA-responsive elements (ABRE and/or ABRE-related motifs) are present in the promoters of the polyamine biosynthetic genes (Alcázar *et al.* 2006b). This reinforces the view that in response to drought and salt treatments, the expression of some of the genes involved in polyamine biosynthesis are regulated by ABA. Transcript profiling has also revealed that cold enhances the expression of *ADC1*, *ADC2* and *SAMDC2* genes (Urano *et al.* 2003; Cuevas *et al.* 2008, 2009).

Free Put levels are increased on cold treatment and this correlates with the induction of *ADC* genes. Surprisingly, the levels of free Spd and Spm remain constant or even decrease in response to cold treatment. The absence of correlation between enhanced *SAMDC2* expression and the decrease of Spm levels may be a result of increased Spm catabolism (Cuevas *et al.* 2008). Since double mutants completely devoid of ADC activity are not viable in Arabidopsis (Urano *et al.* 2005), two independent mutant alleles for both *ADC1* and *ADC2* were used to study their response to freezing. The *adc1* and *adc2* mutations caused higher sensitivity to freezing conditions, in both acclimated and non-acclimated plants, whilst addition of Put complemented this stress sensitivity (Cuevas *et al.* 2008, 2009). Reduced expression of *NCED3* and several ABA-regulated genes was detected in the *adc1* mutants at low temperature.

Complementation analyses of *adc1* mutants with ABA and reciprocal complementation of *aba2-3* mutant with Put supported the conclusion that this diamine controls the levels of ABA in response to cold by modulating ABA biosynthesis at the transcriptional level (Cuevas *et al.* 2008, 2009). All these results suggest that Put and ABA are integrated in a positive feedback loop, in which ABA and Put reciprocally promote each other's biosynthesis in response to abiotic stress (Fig. 1). This highlights a novel mode of action of polyamines as regulators of ABA biosynthesis.

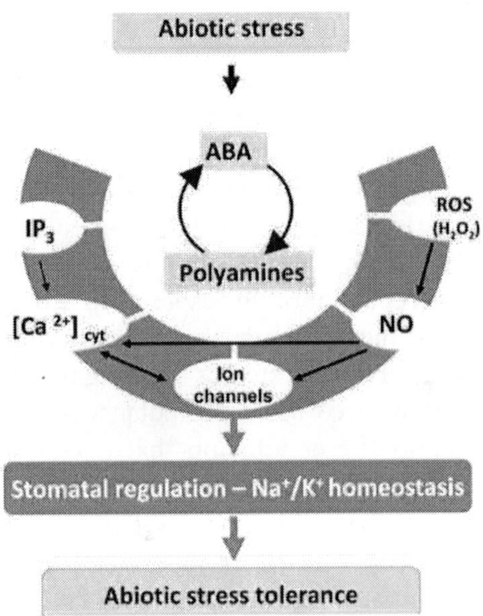

Fig. 1: Simplified model for the integration of polyamines with ABA, ROS (H_2O_2), NO, Ca^{2+} homeostasis and ion channel signalling in the abiotic stress response

1.7. Interplay between ABA, Polyamines, ROS (H_2O_2) and NO in Stomata Regulation

Abscisic acid (ABA) is an endogenous anti-transpirant that reduces water loss through stomatal pores on the leaf surface. Enhanced biosynthesis of ABA occurs in response to water deWcit, resulting in the redistribution and accumulation of ABA in guard cells. This results in the release of water, eZux and inXux of ions, and loss of turgor of guard cells, causing a closure of stomata (Bray, 1997). The ABA signaling pathway in stomata regulation involves many diVerent components such as ABA receptors, G-proteins, protein kinases and phosphatases, transcript-tion factors and secondary messengers, including Ca^{2+}, reactive oxygen species (ROS) and NO (Kuppusamy *et al.* 2009). It has been reported that Put, Spd and Spm also regulate stomatal responses by reducing their aperture and inducing closure (Liu *et al.* 2000; An *et al.* 2008). It is therefore likely that polyamines participate in ABA-mediated stress responses involved in stomatal closure. In this regard, evidences point to an interplay between polyamines with ROS generation and NO signalling in ABA-mediated stress responses (Yamasaki and Cohen, 2006) (Fig.1). The generation of ROS is tightly linked to polyamine catabolic processes, since amino oxidases generate H_2O_2, which is a ROS associated with plant defence and abiotic stress responses (Cona *et al.* 2006).

Furthermore, polyamines are reported to promote the production of NO in *Arabidopsis* (Tun *et al.* 2006). Both H_2O_2 and NO are involved in the regulation of stomatal movements in response to ABA, in such a way that NO generation depends on H_2O_2 production (Neill *et al.* 2008). In *Arabidopsis* guard cells, the production of H_2O_2 induced by ABA arises from superoxide generated by isoforms of NAD(P)H oxidases encoded by the *AtrbohD* and *AtrbohF* genes that are involved in ROS-dependent activation of Ca^{2+} channels and cytosolic Ca^{2+} increase (Desikan *et al.* 2004; She *et al.* 2004). Besides NADPH oxidases, apoplastic amino oxidases are also sources of ROS production (Cona *et al.* 2006). Indeed, ABA has been reported to activate Put catabolism and H_2O_2 production through DAO activities during the induction of stomatal closure in *Vicia faba* guard cells (An *et al.* 2008). ABA and Put promote an enhancement of Ca^{2+} concentration in guard cells, and this increase is impaired by DAO inhibitors. This suggests that the eVect of H_2O_2 from DAO-catalysed Put oxidation in guard cells is mediated by Ca^{2+} ions. In contrast to the eVects of Put, Spd and Spm did not contribute to ABA-promoted H_2O_2 generation in *V. faba* guard cells (An *et al.* 2008) despite the fact that the three polyamines induce stomatal closure (Liu *et al.* 2000). It was previously hypothesized that NO production in plants was mediated through the action of either nitric oxide synthase (NOS)-like or nitrate reductase (NR) activities. However, recent data argue against the involvement of NOS-like activity in H_2O_2-induced NO synthesis in guard cells (Bright *et al.* 2006; Neill *et al.* 2008). Tun *et al.* (2006) demonstrated that Spd and Spm induce rapid biosynthesis of NO, but Put application had little or no effect. The promotion by Spd and Spm of the 14-3-3-dependent inhibition of phosho-NR (Athwal and Huber, 2002), which down-regulates nitrate assimilation and NO production from nitrite, suggests the involvement of other sources for Spd and Spm-induced NO production (Yamasaki and Cohen 2006). The occurrence of a still unknown enzyme responsible for direct conversion of polyamines to NO thus cannot be ruled out. In any case, polyamines appear to regulate stomatal closure by activating the biosynthesis of signalling molecules (H_2O_2 and NO) through diVerent routes (Yamasaki and Cohen 2006). Altogether, the available data indicate those polyamines, ROS (H_2O_2) and NO act synergistically in promoting ABA responses in guard cells (Alcazar *et al.* 2010b).

1.8. Abiotic Stress and Glycine Betaine

The accumulation of low molecular weight water-soluble compounds known as "compatible solutes" or "osmolytes" is the common strategy adopted by many organisms to combat the environmental stresses. The most common compatible solutes are betaines, sugars (mannitol, sorbitol, and trehalose), polyols, polyamines, and amino acid (proline). Their accumulation is favored under water-deficit or salt stress as they provide stress tolerance to cell without interfering cellular machinery (Chen and Murata, 2002). The tolerant or sensitive species show differential stress tolerance depending on the levels of accumulation of these compounds during abiotic stresses. Genes participating in the biosynthesis of different kinds of compatible solutes have been identified from varied sources. Genetic engineering with these endogenous or ectopic genes has therefore, been used successfully to synthesize compatible solutes in target organisms and improvement of stress tolerance (Umezawa *et al.* 2006; Chen and Murata, 2008). A correlation between accumulation of sugars like raffinose family oligosaccharides (RFO), trehalose, fructan, galactinol and sugar alcohols like mannitol, D-ononitol and abiotic stress tolerance has also been reported (Taji *et al.* 2002; Bartels and Sunkar, 2005). Their biosynthetic genes have also been proven useful to improve abiotic stress tolerance in transgenic plants (Taji *et al.* 2002; Abebe *et al.* 2003; Kawakami *et al.* 2008). Similar to sugars, accumulation of proline is a common physiological response in many plants under biotic and abiotic stresses (Siripornadulsil *et al.* 2002). The main proline biosynthetic gene pyrroline-5-carboxylate synthase (P5CS) has been identified from different species and extensively used to enhance the levels of proline 0for abiotic stress tolerance in transgenic plants (Verbruggen and Hermans, 2008). Further, its stress-inducible expression has been shown as a better strategy than constitutive expression in order to minimize the undesirable effects in transgenic plants (Su and Wu, 2004; Molinari *et al.* 2007).

Among the nitrogenous compounds, polyamines accumulate in a variety of plants in response to abiotic stresses like salt and drought (Groppa and Benavides, 2008). Three polyamines namely, putrescine (diamine), spermine (triamine) and spermidine (tetramine) are found to be involved in variety of physiological functions in all organisms. Genes involved in polyamines metabolism have been cloned and often used to alter polyamines levels in transgenic plants for conferring abiotic stress tolerance (Alcázar *et al.* 2010a). However, it is not the higher accumulation of putrescine which increases plant stress tolerance, rather it is the conversion of putrescine into spermidine and spermine, polyamines responsible for stress tolerance (Capell *et al.* 2004). Another nitrogenous compound, glycinebetaine (GB) is a quaternary amine with zwitterionic nature and its natural accumulation is associated with abiotic stress tolerance in varied organisms. Like other compatible solutes, GB biosynthetic genes have also been widely used to improve abiotic stress tolerance in transgenic plants (Chen and Murata, 2011).

Although compatible solutes fall in different bio-chemical groups, similar roles have been assigned to them in plant protection against stresses. However, a precise role of compatible solutes, including GB, in abiotic stress tolerance is largely unknown and two basic functions attributed to these solutes are osmotic adjustment and cellular compatibility. Osmotic adjustment occurs through concentration dependent effects on osmotic pressure to absorb more water from surroundings. In cellular compatibility mechanism, these compounds replace water in biochemical reactions thereby, maintaining normal metabolism during stress (Bohnert and Jensen, 1996). One major issue with compatible solutes is their lower accumulation in transgenic plants as compared with their natural accumulators. At such low levels, compatible solutes might not contribute significantly to osmotic adjustment. Therefore, these compounds are also suggested to be involved in ROS scavenging, macromolecules (nucleic acids, proteins, lipids) protection, and act as reservoir of carbon and nitrogen source (Bohnert and Jensen, 1996, Umezawa *et al.* 2006). In addition, new aspects of their functionality, especially GB, are emerging fast. Present review highlights the new emerging roles of GB in protecting plants against environmental stresses.

Glycine betaine (GB) is a quaternary ammonium compound that occurs in a wide variety of plants, but its distribution among plants is sporadic (Cromwell and Rennie, 1953). *Arabidopsis* and many crop species do not accumulate GB. In plants that produce GB naturally, abiotic stress, such as cold, drought, and salt stress, enhances GB accumulation (Chen and Murata, 2011). Glycine betain can be synthesized from choline and glycine (Chen and Murata, 2011). Introduction of the GB biosynthesis pathway genes into non-accumulators improved their ability to tolerate abiotic stress conditions (Park *et al.* 2004; Waditee *et al.* 2005; Bansal *et al.* 2011), pointing to the beneficial role of GB in stress tolerance. Targeting GB production to the chloroplasts led to a higher tolerance of plants against stress than in the cytosol (Park *et al.* 2007). In salt-tolerant plant species, GB can also accumulate to osmotically significant levels (Rhodes and Hanson, 1993). It has been suggested that GB protects photosystem II, stabilizes membranes, and mitigates oxidative damage (Chen and Murata, 2011).

1.9. Abiotic Stress and Carbohydrates

Fructans

Plants accumulate carbohydrates such as starch and fructans as storage substances that can be mobilized during periods of limited energy supply or enhanced energetic demands. While most plant species use starch as their main storage carbohydrate, several angiosperms, mainly from regions with seasonal cold and dry periods, accumulate fructans (Hendry, 1993). Accumulation of fructans might be advantageous, due to their high water solubility, their resistance to crystallization at freezing temperatures, and the fact that fructan synthesis functions normally under low temperatures (Livingston *et al.* 2009).

Furthermore, fructans can stabilize membranes (Valluru and Van den Ende, 2008) and might indirectly contribute to osmotic adjustment upon freezing and dehydration by the release of hexose sugars (Spollen and Nelson, 1994; Olien and Clark, 1995).

Fructans are branched fructose polymers that are synthesized in the vacuole by fructosyltransferases, including 1-SST and 6-SFT, which transfer fructose from sucrose to the growing fructan chain (Livingston *et al.* 2009). The introduction of fructosyltransferases to fructan non-accumulating tobacco and rice plants stimulated fructan production associated with enhanced tolerance to low-temperature stress and drought (Li *et al.* 2007; Kawakami *et al.* 2008).

Starch, mono- and disaccharides

Starch is composed of glucose polymers arranged into osmotically inert granules. It serves as the main carbohydrate store in most plants and can be rapidly mobilized to provide soluble sugars. Its metabolism is very sensitive to changes in the environment. In addition to diurnal fluctuations in starch levels, salt and drought stress generally leads to a depletion of starch content and to the accumulation of soluble sugars in leaves (Basu *et al.* 2007; Kempa *et al.* 2008). Sugars that accumulate in response to stress can function as osmolytes to maintain cell turgor and have the ability to protect membranes and proteins from stress damage (Kaplan and Guy, 2004).

Starch is produced in plastids from excess sugars during photosynthesis and involves ADP-glucose pyrophosphorylase (AGPase), starch synthases, branching enzymes, and debranching enzymes. Phosphorylation of starch granules by glucan-water dikinase (GWD) and phosphoglucan-water dikinase (PWD) stimulates starch degradation. β-amylases produce maltose from glucans. In the cytosol, maltose is converted to glucose and, subsequently, fructose and sucrose are formed (Tetlow *et al.* 2004; Kotting *et al.* 2010).

Hydrolysis of starch by the β-amolytic pathway represents the predominant pathway of starch degradation in leaves under normal growth conditions and may also be involved in stress-induced starch hydrolysis. *Arabidopsis sex1* (*starch excess 1*) mutants, that are impaired in GWD activity, were compromised in cold-induced malto-oligosaccharide, glucose, and fructose accumulation during the early stages of cold acclimation and *sex1* plants that have been briefly pre-exposed to cold showed a reduced freezing tolerance (Yano *et al.* 2005). Osmotic stress was shown to enhance total â-amylase activity and to reduce light-stimulated starch accumulation in wild-type *Arabidopsis* but not in *bam1* (*bmy7*) mutants, which appeared to be hypersensitive to osmotic stress (Valerio *et al.* 2011). Similarly, BMY8 (BAM3) antisense plants accumulated high starch levels, were impaired in cold-induced maltose, glucose, fructose, and sucrose accumulation, and showed a reduced tolerance of photosystem II to low temperature stress (Kaplan and Guy, 2004).

Trehalose

The non-reducing disaccharide trehalose accumulates to high amounts in some desiccation-tolerant plants, for example, *Myrothamnus flabellifolius* (Bianchi *et al.* 1993). At sufficient levels, trehalose can function as an osmolyte and stabilize proteins and membranes (Paul *et al.* 2008). In most angiosperms however, trehalose is present in trace amounts and abiotic stress increases the levels of trehalose only moderately (Panikulangara *et al.* 2004; Rizhsky *et al.* 2004; Guy *et al.* 2008).

In plants, trehalose is synthesized in a two-step process (Paul *et al.* 2008). Trehalose-6-phosphate synthase (TPS) generates trehalose-6-phosphate (T6P) from UDP-glucose and glucose-6-phosphate followed by dephosphorylation to trehalose by trehalose-6-phosphate phosphatase (TPP). Transgenic expression of trehalose biosynthesis genes showed that enhanced trehalose metabolism can positively regulate tolerance to abiotic stress, even though only a limited increase in trehalose content could be observed, excluding a direct protective role of trehalose in these plants. Heterologous expression of genes involved in the trehalose pathway from *E. coli* or *Saccharomyces cerevisiae* enhanced tolerance to drought, salt, and low temperature stress in several plant species (Iordachescu and Imai, 2008). Over-expression of different isoforms of TPS from rice conferred enhanced resistance to salinity, cold, and/or drought (Li *et al.* 2011). *Arabidopsis* plants constitutively over-expressing AtTPS1 showed a small increase in trehalose and T6P levels and were more tolerant to drought (Avonce *et al.* 2004). Consistently, loss of TPS5 function, a TPS with a TPP domain, reduced the basal thermotolerance of *Arabidopsis* (Suzuki *et al.* 2008). TPP activity and trehalose levels transiently increased in response to cold stress in rice (Panikulangara *et al.* 2004) and over-expression of OsTPP1 rendered rice more tolerant to salinity and cold even though no increase in the trehalose content could be observed (Ge *et al.* 2008). These genetically engineered plants provide evidence that enhanced trehalose metabolism can positively regulate stress tolerance, but the precise role of T6P and trehalose during abiotic stress remains to be further elucidated.

Raffinose family oligosaccharides

Raffinose family oligosaccharides (RFO) such as raffinose, stachyose, and verbascose accumulate in various plant species during seed desiccation (Peterbauer and Richter, 2001) and in leaves of plants experiencing environmental stress like cold, heat, drought or high salinity (Kempa *et al.* 2008; Usadel *et al.* 2008). RFOs have been implicated in membrane protection and radical scavenging (Nishizawa *et al.* 2008). Biosynthesis of RFO is initiated by the formation of galactinol from *myo*-inositol and UDP-galactose by galactinol synthase (GolS). Sequential addition of galactose units, provided by galactinol, to sucrose leads to the formation of raffinose and higher order RFO (Peterbauer and Richter, 2001).

The role of the RFO pathway in acquiring stress tolerance is not yet fully understood. *Arabidopsis* plants over-expressing *Arabidopsis* GolS1 or GolS2 accumulated high levels of galactinol and raffinose and were more tolerant to drought and salinity stress (Nishizawa *et al.* 2008). Similarly, constitutive and stress-inducible expression of GolS1 from the resurrection plant *Boea hygrometrica* in tobacco conferred enhanced drought tolerance (Wang *et al.* 2009a). By contrast, increased raffinose content in *Arabidopsis* plants constitutively expressing cucumber GolS did not enhance freezing tolerance (Zuther *et al.* 2004) and neither *gols*1 knock-out plants nor knock-out mutants in the raffinose synthase gene, that were both impaired in raffinose accumulation, showed any obvious increase in sensitivity to heat or freezing stress, respectively (Panikulangara *et al.* 2004; Zuther *et al.* 2004).

Polyols

Polyols are implicated in stabilizing macromolecules and in scavenging hydroxyl radicals, thereby preventing oxidative damage of membranes and enzymes (Shen *et al.* 1997a). Accumulation of the straight-chain polyols, mannitol and sorbitol, has been correlated with stress tolerance in several plants species (Stoop *et al.* 1996). Expression of mannitol-1-phosphate dehydrogenase (mtlD) from *E. coli*, which catalyses the reversible conversion of fructose-6-phosphate to mannitol-1-phosphate in *Arabidopsis*, tobacco, poplar, and wheat induced mannitol accumulation and enhanced tolerance to salinity and/or water deficit (Abebe *et al.* 2003; Chen *et al.* 2005). Similarly, photosystem II was less affected by salinity in persimmon trees that accumulated sorbitol by over-expression of sorbitol-6-phosphate dehydrogenase (S6PDH) from apple (Gao *et al.* 2001).

The cyclic polyols *myo*-inositol, and its methylated derivatives D-ononitol and D-pinitol, accumulate upon salt stress in several halotolerant plant species (Sengupta *et al.* 2008). L-*myo*-inositol-1-phosphate synthase (MIPS) forms *myo*-inositol-1-P from glucose-6-P, which is dephosphorylated by *myo*-inositol 1-phosphate phosphatase (IMP) to form *myo*-inositol. Inositol *O*-methyltransferase (IMT) methylates inositol to form D-ononitol, which is epimerized to D-pinitol. In line with a stress-protective role, over-expression of

MIPS and IMT from halotolerant plants increased cyclic polyols levels and salt stress tolerance of tobacco (Majee *et al.* 2004; Patra *et al.* 2010).

1.10. Impact of Environmental Stresses

Ultraviolet stress

Sunlight contains energetic short wavelength ultraviolet (UV) photons which are potentially detrimental because of their destructive interactions with many cellular molecules, such as the amino acids of essential proteins, nucleic acids bases or membrane lipids (Smirnoff, 1993). Intense light has long been known to disrupt metabolic processes in plants, including photosynthesis, respiration glucose assimilation, and phosphorylation (Egneus *et al.* 1975). Approximately 4% of the total energy contained in sunlight occurs in the ultraviolet region (wavelengths shorter than 400 nm). The intensity of UV irradiance at the earth's surface varies greatly with season, time of day, latitude) ozone layer thickness, altitude, and cloud cover. Distinctions are sometimes made between the UV-A (400-320nm) and the UV-B (320-290nm) regions. In both cases, however, the fundamental mechanisms of photochemical damage are similar although different receptor molecules (chromophores) may be involved. UV-B damages DNA by causing oxidative cross linking between adjacent pyrimidine bases forming cyclobutane pyrimidine dimers and pyrimidine pyrimidone dimmers (Chamnongpol *et al.* 1997). Unless repaired, these block transcription and replication. Repair is achieved by light-activated photolyases which reduce the dimers in a light-dependent manner. Mutants deficient in photolyase activity have been isolated in rice (Hidema *et al.* 1997) and *Arabidopsis* (uvr2) (Landry *et al.* 1997). Both are UV-B sensitive and unable to repair cyclobutane pyrimidine dimers. The photolyase (PHR1) from Arabidopsis has been cloned by PCR using primers based on animal type II photolyases. Furthermore, PHR1 and uvr2 were shown to be the same by PCR, the mutant having a single base pair deletion (Ahmad *et al.* 1997). Identification of the photolyase gene opens the way for investigating the consequences of its over expression on UV-B resistance. There have been many reports on deleterious physiological effects on plants exposed to high levels of UV-B which may increase if stratospheric ozone concentrations decrease. The destructive action of UV irradiation results from both direct and indirect mechanisms involving endogenous sensitizers and the generation of active oxygen species. Physiological and biochemical effects of UV-B radiation include effects on enzymes, stomatal, resistance concentrations of chlorophyll, protein and lipid, reduction in leaf area, and tissue damage (Doke *et al.* 1994).

Some plants, however, appears to be quite resistant to increased UV irradiation. The differential susceptibility of plants to UV stress is clearly an important factor in their competitive relationships in terrestrial ecosystems (Strid, 1993), experiments with agriculturally important species pairs grown in pots have indicated that significant effects on biomass production took place when UV-B was present either at ambient or artificial increased levels. In the UV region of the electromagnetic spectrum, the energy of such photons is sufficient to break covalent bonds, although it is unusual for their energy converts the target molecules in its ground state to an electronically excited state whose excess energy manifests itself in a different and often quite unstable electron configuration. The initial excited state, a short- lived singlet having, fully paired electrons, may be deactivated by fluorescence (emission of a photon having a longer wavelength than the exciting radiation) and return to the ground state, it may react with neighboring molecules (although this is not common with singlet since their lifetime are normally too short for them to diffuse over very many molecular diameters) or it may undergo internal rearrangement to a longer lived excited state. The triplet state is much more likely to react chemically with surrounding molecules (Strid, 1993; Doke *et al.* 1994).

Water stress

Water stress is perhaps the most prevalent cause of crop yield loss but also the most difficult to tackle because of the strong link between transpiration and photosynthesis. Gene expression and signal transduction in water stressed plants has been recently reviewed (Shinozaki and Yamaguchi-Shinozaki, 1997). There is evidence for a nitrogen-activated protein kinase type system in plants analogous to that involved in yeast osmoregulation. In support of such a system, a protein kinase is rapidly activated in maize roots exposed to low water potential (Conley *et al.* 1997). The role of dehydrins, late embryogenesis proteins and related proteins, which accumulate in seeds and water-stressed vegetative tissues, has been reviewed (Close, 1997). Transgenic rice expressing HVA1, a gene encoding a late embryogenesis abundant protein from barley, has increased tolerance to drought and NaC1 as shown by simple growth analysis (Xu *et al.* 1996). Aquaporins (water channel proteins) are clearly involved in controlling water movement between cells (Maurel, 1997) and may be a target for manipulating water flow through the plant with potential for improving water relations and water use efficiency.

On exposure to osmotic stress as a result of drought, high salinity and low temperature plants accumulate a range of metabolically benign solutes, collectively known as compatible solutes or osmolytes. Their primary function is turgor maintenance but they may have other protective effects on macromolecules in dehydrating cells. The solutes accumulated vary between species and include proline, betaines, dimethylsulfonioproprionate (DMSP), polyols (mannitol, sorbitol, and pinitol), trehalose, and fructans. Over the past five years, a number of transgenic plants have been produced in which overaccumulation occurs (*e.g.* proline) or in which the ability to accumulate osmolytes not previously present has been introduced. The results suggest that they can improve plant growth during osmotic stress even at osmotically-insignificant levels. Glycine betaine (GB) is accumulated by a taxonomically restricted range of species. In higher plants, it is synthesised from choline via betaine aldehyde using choline monooxygenase (CMO) and betaine aldehyde dehydrogenase (BALDH) in chloroplasts (Rathinasabapathi *et al.* 1997).

Some bacteria, however, convert choline to betaine in a one step reaction catalysed by choline oxidase. CodA, which encodes choline oxidase from Arthrobacter globiformis, has been expressed in Arabidopsis, a non-betaine accumulator. The gene had a chloroplast targeting sequence. The average leaf concentration was low but, if chloroplast localised, would be 50 mM, which approaches an osmotically-significant concentration. The transgenic plants, judged by photographs and measurement of plant length, were more tolerant to NaCl and continuous light at 5°C (Hayashi *et al.* 1997). Cytoplasmic targeting had less effect on tolerance. The same group also showed that the cyanobacterium Synechococcus, transformed with the same gene, was also more tolerant to low temperature-induced photo-inhibition and provided evidence that GB affected membrane phase transitions and accelerated recovery of photo-inhibition (Deshnium *et al.* 1997).

There is a report that expression of *Atriplex hortensis* BALDH in rice increases GB content and salinity tolerance (Guo *et al.* 1997). Application of GB in foliar sprays is reported to improve the growth of water-stressed tobacco in laboratory experiments and maize, sorghum and soybean crops in the field (Agboma *et al.* 1997a, b, c). These interesting observations need more investigation, perhaps, to rule out the possibility that improved growth is caused by improved nitrogen supply. Contrary to this, exogenous GB is apparently toxic to *Brassica napus*, a non accumulator (Gibon *et al.* 1997). Higher plant BALDH has been cloned and this has now been followed by cloning of CMO. Like BALDH, CMO expression is upregulated by NaCI. It has a Rieske-type [2Fe-2S] cluster and it is ferredoxin-dependent and, therefore, it represents a novel type of plant oxidase (Rathinasabapathi *et al.* 1997). It is suggested that metabolic engineering of GB synthesis with plant BALDH and CMO would be preferable to using choline oxidase because the plant-derived genes may also have promotors, which could drive increased expression during water stress (Rathinasabapathi *et al.* 1997). Furthermore, CMO uses chloroplast ferredoxin as reductant, thus linking timing of high stress in the light to betaine synthesis (Rathinasabapathi *et al.* 1997). BALDH is equally efficient at catalysing oxidation of 3-dimethylsulfoniopro-prionaldehyde to DMSE a compatible solute of even narrower distribution than GB (Vojtechova *et al.* 1997) and ability to synthesise DMSP could have evolved by co-opting this enzyme. Transgenic tobacco expressing beet BALDH also had enhanced dehydrogenase activity towards 3-dimethylsulfonioproprionaldehyde and two other aldehydes confirming the view that BALDH is a multisubstrate enzyme (Trossat *et al.* 1997). BALDI-I expression may be more widespread than GB accumulation. BALDH is expressed in rice (non-GB accumulator) and the rice gene has high homology to barley BALDH. Unlike BALDH in betaine accumulating plants the rice enzyme is located in peroxisomes. Feeding betaine aldehyde dehydroganse to rice plants increases their GB content, and on the basis of photographic evidence, improves the growth of the plants at high salinity and low water content (Nakamura *et al.* 1997). Evidence suggests that drought causes oxidative damage through generation of oxygen radicals or inhibition of antioxidant systems in plant (Jagtap and Bhargava, 1995; Zhang and Kirkham, 1994; 1996).

Salinity stress

High salt concentration normally impair the cellular electron transport within the different sub-cellular compartments and lead to the generation of reactive oxygen species (ROS) such as singlet oxygen, superoxide, hydrogen peroxide and hydroxyl radicals (Hernandez *et al.* 1993; 1995). Excess of ROS triggers phytotoxic reactions such as lipid peroxidation, protein degradation and DNA mutation (Fridovich, 1986; Davies, 1987). Mannitol is accumulated by a wide range of species in response to salinity (Stoop *et al.* 1996). Mannitol synthesizing ability was introduced into transgenic tobacco by the *E. coli* mtlD gene encoding mannitol dehydrogenase. These plants accumulated modest amounts of mannitol but were said to be more salt-tolerant (Tarczynski *et al.* 1993). Shen *et al.* (1997a) have produced transgenic tobacco in which mtlD expression is targeted to the chloroplast. A concentration of 100mM was estimated. It has been suggested that compatible solutes, including mannitol, could be antioxidants by scavenging hydroxyl radicals (OH) (Orthen *et al.* 1994). This might be significant for plants exposed to drought and high salinity as there is strong evidence that oxidative generation of active oxygen species increases under such conditions (Biehler and Fock, 1996). The chloroplast-targetted mannitol accumulating tobacco has been used to test this hypothesis. Mannitol accumulation did not affect photosynthesis. The transgenic plants were more tolerant to OH^- generated in chloroplasts by methyl viologen treatment. The OH^- content of transgenic plants was also lower (Shen *et al.* 1997a) suggesting that the protection could result from OH^- scavenging by mannitol. In a further paper (Shen *et al.* 1997b), the same group have provided convincing evidence that a key target for OH^- produced by illuminated thylakoids is the Calvin cycle enzyme phosphoribulokinasc (regulated by thiol-disulfide interconversion). This inactivation, in mixtures containing thylakoids and phosphoribulokinase, was prevented by mannitol and could explain the in vivo protective effect of mannitol. Further support for the efficiency of mannitol acting as an OH^- scavenger in vivo has been provided by transformation of *Saccaromyces cerevisiae* with the same mtlD gene. A mutant unable to grow at high osmolarity because of inability to synthesize glycerol, the normal osmoregulatory solute of yeast, had this ability restored by introduction of mannitol synthesis capacity. Furthermore, the transgenic yeast was also more tolerant to chemically-generated OH^- in the growth medium (Chaturvedi *et al.* 1997). These studies suggest osmolytes could indeed have multiple functions and could explain the protective effects observed at osmotically insignificant concentrations (Garcia *et al.* 1997).

Metal stress

Accumulation of phytotoxic metal results from industrial and agricultural practices. Zinc, copper and cadmium are widespread pollutants resulting in stunted growth, chlorosis and necrosis (Oncel *et al.* 2000). Copper Cu^{2+} ions cause light mediated lipid peroxidation and pigment bleaching (Rousos *et al.* 1986; Sandmann and Gonzalea, 1989). Prolonged exposure to CuSO4 resulted in chlorophyll bleaching in rye and the endogenous CAT level declined (Streb *et al.* 1993). Thus enhanced susceptibility to photooxidative damage was related to the rapid loss of CAT activity, Cu^{2+} ions are redox active and catalyze fentonperoxides also originate from the induction of lipoxygenase in the presence of Cu^{2+}. This enzyme is known to initiate lipid peroxidation. Cadmium treatment, decrease the chlorophyll and heme levels of germinating mung bean seedlings by induction of lipoxgenase with the simultaneous inhibition of the antioxidative enzyme SOD and CAT (Somashekaraiah *et al.* 1992). Such inhibition results from binding of the metal to the important sulfhydryl groups of enzymes, which exacerbates the phytotoxic action of metals (van Assche and Clijsters, 1990). Pang *et al.* (2002) studied improving the plant ability to resist lead stress, effect of 0.05 mg/L La $(NO_3)_3$ on the activities of catalase (CAT), superoxide dismutase (SOD), the level of malondialdehyde (MDA) in wheat seedlings under lead stress were studied. The effect of La^{+3} on plant growth, chlorophyll content in wheat seedlings after adding 0, 50, 100 mg/l $Pb(NO_3)_3$ to the nutrient solution for 12 days was observed. The plants were grown in nutrient solution in a strictly controlled climate growth room. Effects of La^{+3} (with La treatment) compared with check groups was evidently observed. The activities of SOD and CAT in root were enhanced 0.45–1.69 times and 33.20–77.77% respectively and MDA content was reduced 11.05–27.49% in root after treatments from the second day till the end of the experiment. The activities of SOD and CAT was found to be increased slightly ($P < 0{:}05$) and MDA content decreased in shoot and root of wheat seedlings by La^{+3} under lead stress within five days after treatments compared with Pb^1 and Pb^2 groups. It was assumed that antioxidant enzymes was found to be increased by $La(NO_3)_3$, the antioxidant potential of the wheat seedlings to resist lead stress enhanced. It is suggested that La^{+3} could be used to resist lead stress at the beginning under stress while the stress was not so serious (Ali and Alqurainy, 2006).

Herbicide stress

Several herbicides have been found to generate active oxygen species, either by direct involvement in radical production or by inhibition of biosynthetic pathways. The generation of the hydrocarbon gas ethane, the production of malonaldehyde and changes in electrolytic conductivity has frequently been used as sensitive markers for herbicidal action in plants. The bipyridinium and diphenyl eather herbicide have been the most insensitively investigated in terms of their oxidative action in plants. The bipyridinium herbicides generate oxygen radicals directly in the light. Compounds such as paraquat (also known as methyl viologen) induce light dependent oxidative damage in plants. Members of this group are cused as sensitive markers for herbicide action in plants (Peleg *et al.* 1992). The bipyridinium and diphenyl ether healled total kill herbicids (Dodge, 1971). The dicationic nature of these compounds facilitates their reduction to radical cation. The PSI mediated reduction of the paraquat di-cation results in the formation of a mono-cation radical which then reacts with molecular oxygen to produce O^-_2 with the subsequent production of other toxic species, such as H_2O_2 and OH (Elstner *et al.* 1988). The diphenyl ethers, cylic imides and lutidine derivative, act by inhibition biosynthetic pathways with the subsequent accumulation of reactive radical-forming intermediates. These compounds cause severe toxicological problems and results in peroxidation of membrane lipids and general cellular oxidation. The mode of action of these herbicides is based on the ability to induce the abnormal accumulation of photosensitizing tetrapyrroles specifically protoporphyrin IX (Sandmann and Gonzalea 1989). This pigment is able to cause light dependent generation. These herbicides can also catalyze the oxidation of protoporphyrinogen to protoporphyrin IX. The penultimate step of both heme and chlorophyll biosynthesis is recorded (Witkowski and Hailing 1989). It is somewhat anomalous that the reaction product protoporphyrin IX accumulates in condition where the enzyme which catalyses its formation is expected to be inhibited. Other compounds such as diuron, that block photosynthetic electron transport and inhibitors of cartenoids biosynthesis, such as norflurazon, initiate photooxidative processes most probably via the generation of $1O_2$ (Halliwell, 1987). Herbicides which block photosynthesis cause increased excitation energy transfer from triplet chlorophyll to oxygen while those inhibit carotenoid biosynthesis eliminate important quenchers of the triplet chlorophyll and $1O_2$.

Fungicides stress

A number of agricultural chemicals such as fungicides (Mackay *et al.* 1987) and herbicides (Dixon *et al.* 1996) have been shown to possess antiozonant properties. However, most of these investigations have concentrated on the effectiveness of agrochemicals in preventing plants from ozone injury, and relatively little work has focused on the physiological and biochemical modes of action. The strobilurins are an important new class of fungicides with a unique mode of action which targets mitochondrial respiration in fungi. Few studies on the physiologic effects of strobilurins on and in plants showed that strobilurins increased grain yields, dry matter, chlorophyll and protein contents and delayed senescence (Mercer and Ruddock, 1998).

Air pollutant stress

Atmospheric pollutants such as ozone (O_3) and sulfur dioxide (SO_2) have been implicated in free radical formation (Asada, 1980; Mehlhorn, 1990) and are considered to be one of the major factors influencing modem forest decline. Ozone, which originates from a natural photochemical degradation of nitrous oxides (NO_2), seems to be a greater threat to plants than pollution (Heagle, 1989). Mehlhorn *et al.* (1990) suggested that the phytotoxicity of O_2 is due to its oxidizing potential and the consequent formation of radicals that induce free radical chain reactions. The O_3 concentration in the intercellular air spaces of leaves is close to ozone (Grimes *et al.* 1983). Ozone is, thus unlikely to reach the chloroplast but it nevertheless, causes pigment bleaching and lipid peroxidation (Heath, 1987). Stimulation of both synthesis and degradation of the PSII-DI protein occurs in spruce trees following O_3 treatment (Godde and Buchhold, 1992; Lutz *et al.* 1992) and a decrease in the activity and quantity of rubisco has been found in poplar following exposure to O_3 (Landry and Pell, 1993). Exposure to SO_2 results in tissue damage and release of stress ethylene from both photosynthetic and non-photosynthetic tissues (Peiser and Yang, 1979, 1985). Fumigation with SO_2 causes a shift in cytoplasmic pH. The prodon concentration of the cytoplasm is one of the most important factors regulating cellulase activity When cells are exposed to SO_2 an appreciable acidification of the cytoplasm occur because this gas reacts with water to form sulfurous acid which may then be converted into sulphuric acid (Laisk *et al.* 1988a, b; Veljovic-Jovanovic *et al.* 1993). These results, in loss of photosynthetic function caused by inhibition of the activity of SH-containing light-activated enzymes of the inhibition of the activity of SH- containing light-activated enzymes of the chloroplast (Shimazaki and Sugahara, 1980; Convello *et al.* 1989; Tanaka *et al.* 1982). The oxidation of sulfite to sulfate in the chloroplast also gives rise to the formation of O^-_2 (Asada, 1980). The oxidation of sulfite is initiated by light and is mediated by photosynthetic electron transport. Navari-Izzo *et al.* (1992) reported that the degradation of membrane lipid components possibly by de-esterification rather than peroxidation with SO_2. They found no evidence to support the view that free radical attack on polyunsaturated fatty acids occurred at low pollutant concentrations.

Light stress

A wealth of evidence shows that antioxidants are responsive to photooxidative stress (Alscher *et al.* 1997; Foyer *et al.* 1997). In bacteria, signal transduction systems involved in responses to oxidative stress have been identified (Demple, 1991), however, very little is known in plants. An important step forward has been made in identifying a possible mechanism of detecting oxidative stress in chloroplasts of leaves exposed to high levels of light. Transfer of *Arabidopsis* leaves from low light to high light causes rapid induction of mRNA for two nuclear-encoded cytosolic ascorbate peroxidase genes (APX1 and APX2). Also, within this 15 minute period the ratio of reduced to oxidized glutathione decreases, indicating that the leaves are under oxidative stress as a result of exposure to excess excitation energy. The induction of the APX genes was prevented by treatment with 3-(3,4-dichlorophenyl)-l,l-dimethylurea, which blocks photosynthetic electron transport before plastoquinone (PQ). Conversely, 2,5-dibromo-3-methyl-6- isopropyl-pbenzoquinone, which blocks electron transport at the cytochrome b/f complex, causes higher APX expression in low light. The results can be interpreted as suggesting that the redox state of PQ has a role in acting as a sensor of excess light. Interestingly, glutathione feeding also blocked the response and it was suggested that extra glutathione swamps a signal generated by the redox state of the endogenous pool (Karpinski *et al.* 1997). These observations require further investigation because it is clear that there is a rapid signal transduction process leading to induction of APX when leaves are exposed to excessive light.

Evidence of similar redox signaling in controlling light-mediated phosphorylation of light harvesting complexes and expression of photosynthesis enzymes exists (Escoubas *et al.* 1995; Vener *et al.* 1997). Signal transduction via increased cytosolic Ca^{2+} has been suggested (Price *et al.* 1994) and this observation has been extended by demonstration of elevated cytosolic Ca^{2+} in stomatal guard cells by methyl viologen and hydrogen peroxide treatment, which then causes closure (McAinsh *et al.* 1996).

Temperature stress

Various tolerance mechanisms have been suggested on the basis of the biochemical and physiological changes related to chilling injury (Oncel *et al.* 2000). Levitt (1980) has suggested that a major target of chilling injury is cell membranes. As temperature is reduced, a specific temperature determined by the ratio of saturated to unsaturated fatty acids accelerates the conversion of lipids of a liquid-crystalline condition into that of a solid condition in plant cell membranes (Quinn, 1988). The conversion of fatty acid may give rise to chilling resistance at lower temperatures in the plant cells. However some plants, which show a similar fatty acid ratio under chilling conditions, are very sensitive to chilling injury compared to others, thus other mechanisms may also be necessary for chilling injury. In previous studies it has been suggested that oxidative stress induced by chilling stress may play a pivotal role for chilling injury in plant cells (O'Kane *et al.* 1996).

Dong and Chin (2000) have investigated the antioxidant defense system and chilling stress-induced changes of antioxidant enzymes in the leaves of cucumber (*Cucumis sativus* L.). Chilling stress preferentially enhanced the activities of the superoxide dismutase (SOD), ascorbate peroxidase (APX), glutathione reductase (GR) and peroxidase specific to guaiacol, whereas it induced the decrease of catalase activity. In order to analyze the changes of antioxidant enzyme isoforms against chilling stress, foliar extracts were subjected to native PAGE. Leaves of cucumber had four isoforms of Mn-SOD and two isoforms of Cu: Zn-SOD. Fe-SOD isoform was not observed in this plant. Expression of Cu: Zn-SOD and Mn-SOD was preferentially enhanced by chilling stress. Expression of Mn SOD-2 and -4 was enhanced after 48 h of the poststress period. Five APX isoforms were presented in the leaves of cucumber. The intensities of APX-4 and -5 were enhanced by chilling stress, whereas that of APX-3 was significantly increased in the poststress periods after chilling stress. Gel stained for GR activity revealed six isoforms in the plant. Activation levels for most of GR isoforms were higher in the stressed-plants than the control and poststressed-plants, but that of GR-1 isoform was significantly higher in the post stressed plants than chilling stressed-plants. Dong and Chin (2000) results collectively suggest that chilling stress activates the enzymes of an SOD: ascorbate-glutathione cycle under catalase deactivation in the leaves of cucumber, but the response timing of enzyme isoforms against various environmental stresses is not the same for all isoforms of antioxidant enzymes.

Heat stress is a major abiotic factor limiting growth of temperate plant species in many areas during summer months and may become a threat as global warming occurs (Fry and Huang, 2004). Heat stress induces changes in various metabolites such as organic acids, amino acids, and carbohydrates, which have important functions involved in photosynthesis and respiration (Merewitz *et al.* 2012). These compounds are involved in vital metabolic functions within the plant such as regulating plant water relations, signaling, protein synthesis as well as stress defense (Sairam *et al.* 2000; Zobayed *et al.* 2005). Carbon di-oxide concentrations in the atmosphere are more than 100 $\mu mol \cdot mol^{-1}$ higher since the beginning of the industrialization era and the concentration is predicted to rise at a rate of ≈ 2 $\mu mol \cdot mol^{-1}$ per year (IPCC, 2007). Increasing evidence has shown that elevated CO_2 may promote plant growth and mitigate heat stress damagc in various plant species, particularly C_3 species (Hamerlynck *et al.* 2000; Kirkham, 2011; Qaderi *et al.* 2006) including perennial turfgrass species such as tall fescue (Yu *et al.* 2012). Various studies report elevated CO_2 promotes shoot growth, root growth, and photosynthesis but inhibits dark respiration and photorespiration rates (Leakey *et al.* 2006; Reddy *et al.* 2010). However, the metabolites associated with the improvement in plant growth and changes in photosynthesis and respiration metabolic processes resulting from elevated CO_2, particularly under heat stress, are not well documented.

Metabolic profiling is an effective and quantitative method to elucidate mechanisms of abiotic stress tolerance, including heat stress (Kaplan *et al.* 2004; Mayer *et al.* 1990). Mayer *et al.* (1990) reported an increase in the abundance of γ-aminobutyric acid (GABA), β-alanine, alanine, and proline in cowpea (*Vigna unguiculata*) as a result of heat shock under ambient CO_2 conditions. Du *et al.* (2011) reported that heat stress at ambient CO_2 induced significant accumulation of multiple metabolites, including threonic acid, galacturonic acid, gluconic acid, succinic acid, proline, aspartic acid, serine, valine, methylmaleic acid, fructose, galactose, xylose, glucose, mannose, and sucrose in kentucky bluegrass (*Poa pratensis*). Most of the previous studies examined metabolic changes in response to heat stress without consideration of the interaction with CO_2. Very few studies investigated changes in metabolite accumulation in response to heat stress under elevated CO_2 conditions despite some studies that reported the positive effects of elevated CO_2 on plant growth under detrimental environments such as water dehydration and fertilizer deficiency (Kirkham, 2011).

Although, numerous reports have been published that transgenic plants overproducing proline exhibited increased tolerance to various environmental stresses, such as freezing, high salinity, and drought (Nanjo *et al.* 1999a; Hong *et al.* 2000; Kumar *et al.* 2010). On the contrary, transgenic Arabidopsis (*Arabidopsis thaliana*) plants with an antisense AtP5CS cDNA displayed lower proline content and decreased tolerance to osmotic stress (Nanjo *et al.* 1999b). Similarly, lower salt tolerance was observed in the P5CS1 insertional mutant of Arabidopsis with impaired proline accumulation (Székely *et al.* 2008).

Compared with other stresses, however, only a few reports demonstrated that proline accumulated during heat stress. In the first leaf of barley (*Hordeum vulgare*) and radish (*Raphanus sativus*), proline content showed a slight increase under 41°C heat stress (Chu *et al.* 1974). In chickpea (*Cicer arietinum*), heat treatment resulted in a marginal increase in proline content compared with the control, and the increase was more significant after salicylic acid pretreatment (Chakraborty and Tongden, 2005). In suspensions cells of cowpea (*Vigna unguiculata*), the proline synthesis rate was increased 2.7-fold when cells had been transferred from 26°C to 42°C (Mayer *et al.* 1990). In a salt-tolerant strain of tobacco (*Nicotiana tabacum*) suspension cells, NrEs-1, a transient initial rise in proline content was reported at high temperature or at high temperature plus salinity, but high proline content could last a longer time in response to salinity solely (Kuznetsov and Shevyakova, 1997). In *Arabidopsis*, proline accumulation was not observed during heat stress (Hua *et al.* 2001; Rizhsky *et al.* 2004). This has posed a question whether proline accumulation is beneficial in all kinds of stresses, or is it only so in a specific set of stresses where proline accumulates naturally? It has been reported that high temperature exaggerated the toxic effect of exogenous Pro (Rizhsky *et al.* 2004), raising the possibility that the presence of high concentrations of proline during high temperature is inhibitory. In supporting this hypothesis, Proline dehydro-genase insertional mutants, with higher Pro content, exhibited lower thermotolerance compared with the wild type (Larkindale and Vierling, 2008), the p5cdh mutant, unable to completely oxidize proline was also demonstrated to bemore sensitive to heat shock (Miller *et al.* 2009). However, direct evidence for the effect of proline accumulation on thermotolerance is still lacking (Lv *et al.* 2011).

Topography stress

High mountain plants must have a very effective carbon assimilation mechanism due to a very short growing period. The extreme climatic conditions of high mountain zone, high irradiance, low temperature, rapid temperature change and a reduced CO_2 partial pressure creates unfavorable conditions for photosynthesis (Oncel *et al.* 2004; Germino and Smith, 2000). Previous studies revealed that alpine plants are highly efficient in photosynthesis at low temperatures and are also adapted to high irradiance (Germino and Smith, 2000).

The biggest damage caused by the high light intensity in plants is the inactivation of D1 protein (located on PSII) and the catalase (CAT) enzyme. Low temperature stress has a similar effect upon PSII and CAT (Streb *et al.* 1998). The alpine plants have a much more effective protection mechanism against oxidative damage compared with the plants growing in lower altitude regions (Streb *et al.* 1998; Polle *et al.* 1999). The steppe plants are also affected by the combination of high light intensity, high temperature and drought stresses. The steppe regions have very poor vegetation and a very short vegetative period since these conditions limit plant growth (Streb *et al.* 1997; Lamont and Lamont, 2000). The high light intensity, high temperature and the temperature difference between night and day increases the generation of reactive oxygen species and thus the risk of oxidative damage. The plants may have developed two strategies to adapt to these severe conditions: antioxidant protection and avoidance from oxidative stress (Streb *et al.* 1997). The damage from the electron transfer system results in the formation of free radicals (singlet oxygen, superoxide radical and hydroxyl radicals). It is necessary to activate the biochemical protection mechanism of the plant in order to eliminate these extremely hazardous radicals (Larson, 1988).

The antioxidant protection requires high amounts of carotenoids, ascorbic acid, atocopherol, glutathione, phenolics and flavinoids (Schoner and Krause, 1990) and the increased activities of CAT, superoxide dismutase (SOD) ascorbate peroxidase, dehydroascorbate reductase and glutathione reductase (GR) enzymes (Bowler *et al.* 1992). The antioxidant defense mechanism protects the unsaturated membrane lipids, nucleic acids, enzymes and other cellular structures from the harmful effects of free radicals (Foyer *et al.* 1994; Caasi-Lit *et al.* 1997). The tolerance of *Homogyne alpina*, an alpine plant, to light stress is explained by the presence of a stable CAT enzyme. *Ranunculus glacialis* growing at the same altitude has a weak antioxidant system (Streb *et al.* 1998; Wildi and Lütz, 1996). The activity of antioxidant enzymes and amount of carotenoids in *Retama raetam*, a desert plant, has been found to be higher than in non-desert plants (Streb *et al.* 1997).

Lack of oxygen stress

Lack of oxygen or anoxia is a common environmental challenge which plants have to face throughout their life. Winter ice encasement, seed imbibitions, spring floods and excess of rainfall are examples of natural conditions leading to root hypoxia or anoxia. Low oxygen concentration can also be a normal attribute of a plants' natural environment. Wetland species and aquatic plants have developed adaptative structural and metabolic features to combat oxygen deficiency. A decrease in adenylate energy charge, cytoplasmic acidification, anaerobic fermentation, elevation in cytosolic Ca^{2+} concentration, changes in the redox state and a decrease in the membrane barrier function, are the main features caused by lack of oxygen (Drew, 1997; Vartapetian and Jackson, 1997; Tadege *et al.* 1999). Regulation of anoxic metabolism is complex and not all the features are well established. In the recent paper by Gout *et al.* (2001) a cytoplasmic acidification process has been temporally resolved in sycamore *(Acer pseudoplatanus)* cell culture by NMR (nuclear magnetic resonance). The immediate response of cytoplasmic pH was solely dependent on proton-releasing metabolization of the nucleoside triphosphate pool, the long term regulation (after 20 min of anoxia) involves lactate synthesis, succinate, malate, amino acid metabolism and ethanolic fermentation (Gout *et al.* 2001).

Under natural conditions anoxic stress includes several transition states (hypoxia, anoxia and reoxygenation) characterized by different O_2 concentrations (Table 1). Excessive generation of reactive oxygen species (ROS), *i.e.* under oxidative stress, is an integral part of many stress situations, including hypoxia. Hydrogen peroxide accumulation under hypoxic conditions has been shown in the roots and leaves of *Hordeum vulgare* (Kalashnikov *et al.* 1994) and in wheat roots (Biemelt *et al.* 2000). The presence of H_2O_2 in the apoplast and in association with the plasma membrane has been visualized by transmission electron microscopy under hypoxic conditions in four plant species (Blokhina *et al.* 2001).

Table 1: ROS scavenging and detoxifying enzymes

Enzymes	EC Number	Reaction catalysed
Superoxide dismutase	1.15.1.1.	$O_2^- + O_2^+ \leftrightarrow 2H_2O + O_2$
Catalase	1.11.1.6.	$2H_2O_2 \leftrightarrow O_2 + 2H_2O$
Glutathion peroxidase	1.11.1.12.	2GSH + PUFA-OOH ↔ GSSG + PUFA + $2H_2O$
Glutathion-S-transferase	2.5.1.18.	RX +GSH ↔ HX + R-S-GSH*
Phospholipid-hydroxide glutathione peroxidase	1.11.1.9.	2GSH + PUFA-OOH (H_2O_2) ↔ GSSG + $2H_2O$**
Ascorbate peroxidase	1.11.1.11.	AA + H_2O_2 ↔ DHA + $2H_2O$
Guiacol type peroxidase	1.11.1.7.	Donor + H_2O_2 ↔ Oxidised donor + $2H_2O$***
Monodehydroascorbate reductase	1.6.5.4.	NADH + 2MDHA ↔ NAD^+ + 2AA
Dehydroascorbate reductase	1.8.5.1.	2GSH + DHA ↔ GSSG + AA
Glutathion reductase	1.6.4.2.	NADPH + GSSG ↔ NADP + 2GSH

*R may be an aliphatic, aromatic or heterocyclic group, X may be a sulfate, nitrite or halide group,**Reaction with H_2O_2 is slow, ***AA acts as electron donor

In these experiments H_2O_2 was probably of enzymatic origin considering the low oxygen concentration in the system and the positive effects of the various inhibitors of H_2O producing enzymes. Indirect evidence of ROS formation (*i.e.* lipid peroxidation products) under low oxygen has been detected (Blokhina *et al.* 1999; Yan *et al.* 1996; Chirkova *et al.* 1998). The phenomenon of cross-tolerance to various environmental stresses suggests the existence of a common factor, which provides crosstalk between different signaling pathways. ROS have recently been considered as possible signaling molecules in the detection of the surrounding oxygen concentration (Semenza, 1999). It has been suggested also that ROS and oxygen concentration (including hypoxia) can be sensed via the same mechanism. Several models employ direct sensing of oxygen (via haemoglobin or protein SH oxidation) or ROS sensing. There are two models which suggest either a decrease in ROS under oxygen deprivation (low NADPH oxidase activity) or an increase in ROS due to the inhibition of the mitochondrial electron transport chain.

Molecular oxygen is relatively unreactive (Elstner, 1987) due to its electron configuration. Activation of oxygen (*i.e.* the first univalent reduction step) is energy dependent and requires an electron donation. The subsequent one-electron reduction steps are not energy dependent and can occur spontaneously or require appropriate e^-/H^+ donors. In biological systems transition metal ions (Fe^{2+}, Cu^+) and semiquinones can act as e-donors. Four-electron reduction of oxygen in the respiratory electron transport chain (ETC) is always accompanied with a partial one- to three-electron reduction, yielding the formation of ROS. This term includes not only free radicals (superoxide radical, O_2^-, and hydroxyl radical, OH), but also molecules such as hydrogen peroxide (H_2O_2), singlet oxygen ($1O_2$) and ozone (O_3), Both O_2^- and the hydroperoxyl radical $HO_2^{-\cdot}$ undergo spontaneous dismutation to produce H_2O_2. Although H_2O_2 is less reactive than O_2^-, in the presence of reduced transition metals such as Fe^{2+} in a chelated form (which is the case in biological systems), the formation of OH can occur in the Fenton reaction. A potential route for the formation of a damaging species from a photochemical activated triplet state is the transfer of triplet energy to molecular oxygen. The product of the energy transfer reaction is singlet oxygen $1O_2$. The chlorophyll pigments associated with the electron transport system are the primary source of $1O_2$. The $1O_2$ may also arise as a by product of lipoxygenase activity, like the hydroxyl radical, $1O_2$ is highly destructive reacting with most biological molecules at near diffusion controlled rate (Chirkova, 1988). Several classes of biological molecules are susceptible to attack by $1O_2$ including several protein amino acids (cycteine, methionine, tryptophan and histidine) which react with it at quit rapid rate (Foyer *et al.*

1990). Polyunsaturated fatty acids also react at much slower rate, increasing with the number of double bonds in the molecule, to form lipid hydoperoxides and O_2^- radicals (Lynch and Thompson, 1984) this take place by enzyme lipoxygenase. Primary radical (R') formed as a result of UV irradiation lead to formation of lipid radicals (L'). Lipid radicals react with O_2 (a reaction limited only by the rate of diffusion) to produce lipid peroxy radicals (LOO'). This peroxide is likely contributors to damage and dysfunction of cell and organelle membranes.

Non-photochemical routes for oxidative damage in plants usually involve the interaction of molecular oxygen with free radicals to produce new, potentially harmful free radical species containing oxygen. This type of reaction may occur directly, or it may be promoted by enzyme catalysts normally present in the plant cell such as the enzyme lipoxygenase (Galliard and Chan, 1980). Atmospheric oxygen is unusual in that its ground state has two unpaired electrons, it is a triplet state with considerable diradical character. These penults are to enter into energetically favorable chain reactions with many organic free radicals. The formation of organic (usually carbon centered) free radicals R' from non radical precursors is called the "initiation phase" of the autooxidation. This process, which is often quite slow, results in the characteristic lag period of a radical chain reaction. In the propagation phase of the reaction there is a build up of peroxy radicals, ROO^-, and the subsequent reaction of peroxy radicals with compounds (R'H) having extractable hydrogen atoms. The new radicals are then available for further reaction with molecular oxygen. Finally when all the oxygen or active hydrogen species are used up the 'termination phase' begins. In this phase, the radicals recombine with each other to produce inactive, nonradical products. Synthetic organic chemists have created many effective inhibitors of oxidative damage for rubber, hydrocarbon fuels, plastics foodstuffs and many other materials.

1.11. Abiotic Stresses and its Impact on Agriculture

Today, in a world of 7 billion people, agriculture is facing great challenges to ensure a sufficient food supply while maintaining high productivity and quality standards. In addition to an ever increasing demographic demand, alterations in weather patterns due to changes in climate are impacting crop productivity globally. Unfavorable climate (resulting in abiotic stresses) not only causes changes in agro-ecological conditions, but indirectly affects growth and distribution of incomes, and thus increasing the demand for agricultural production (Schmidhuber and Tubiello, 2007). Adverse climatic factors, such as water scarcity (drought), extreme temperatures (heat, freezing), photon irradiance, and contamination of soils by high ion concentration (salt, metals), are the major growth stressors that significantly limit productivity and quality of crop species worldwide. As has been pointed out, current achievements in crop production have been associated with management practices that have degraded the land and water systems FAO (2011). Soil and water salinity problems exist in crop lands in China, India, the United States, Argentina, Sudan, and many other countries in Western and Central Asia. Globally, an estimated 34 million irrigated hectares are salinized (FAO, 2012) and the global cost of irrigation-induced salinity is equivalent to an estimated US$11 billion per year (FAO, 2005).

A promising strategy to cope with adverse scenario is to take advantage of the flexibility that biodiversity (genes, species, ecosystems) offers and increase the ability of crop plants to adapt to abiotic stresses. The Food and Agricultural Organization (FAO) of the United Nations promotes the use of adapted plants and the selection and propagation of crop varieties adapted or resistant to adverse conditions FAO (2007). Global programs, such as the Global Partnership Initiative for Plant Breeding Capacity Building (GIPB), aim to select and distribute crops and cultivars with tolerance to abiotic stresses for sustainable use of plant genetic resources for food and agriculture GIPB (2012).

Plant responses to abiotic stress

Through the history of evolution, plants have developed a wide variety of highly sophisticated and efficient mechanisms to sense, respond, and adapt to a wide range of environmental changes. When in adverse or limiting growth conditions, plants respond by activating tolerance mechanisms at multiple levels of organization (molecular, tissue, anatomical, and morpho-logical), by adjusting the membrane system and the cell wall architecture, by altering the cell cycle and rate of cell division, and by metabolic tuning Atkinson and Urwin (2012). At a molecular level, many genes are induced or repressed by abiotic stress, involving a precise regulation of extensive stress-gene networks (Delano-Frier *et al.* 2011; Grativol *et al.* 2012). Products of those genes may function in stress response and tolerance at the cellular level. Proteins involved in biosynthesis of osmoprotectant compounds, detoxification enzyme systems, proteases, transporters, and chaperones are among the multiple protein functions triggered as a first line of direct protection from stress. In addition, activation of regulatory proteins (*e.g.*, transcription factors, protein phosphatases, and kinases) and signaling molecules are essential in the concomitant regulation of signal transduction and stress-responsive gene expression (Wang *et al.* 2009b; Krasensky and Jonak, 2012). Early plant response mechanisms prevent or alleviate cellular damage caused by the stress and re-establish homeostatic conditions and allow continuation of growth (Peleg *et al.* 2011). Equilibrium recovery of the energetic, osmotic, and redox imbalances imposed by the stressor are the first targets of plant immediate responses.

Observed tolerance responses towards abiotic stress in plants are generally composed of stress-specific response mechanisms and also more general adaptive responses that confer strategic advantages in adverse conditions. General response mechanisms related to central pathways are involved in energy maintenance and include calcium signal cascades (Reddy *et al.* 2011; Pan *et al.* 2012), reactive oxygen species scavenging/signaling elements (Ahmad *et al.* 2010; Loiacono and Tullio, 2012), and energy deprivation (energy sensor protein kinase, SnRK1) signaling (Baena-Gonzalez and Sheen, 2008). Induction of these central pathways is observed during plant acclimation towards different types of stress. For example, protein kinase SnRK1is a central metabolic regulator of the expression of genes related to energy-depleting conditions, but this kinase also becomes active when plants face different types of abiotic stress such as drought, salt, flooding, or nutrient depravation (Hay *et al.* 2007; Ghillebert *et al.* 2011; Cho *et al.* 2012). SnRK1 kinases modify the expression of over 1000 stress-responsive genes allowing the re-establishment of homeostasis by repressing energy consuming processes, thus promoting stress tolerance (Baena-Gonzalez, 2010; Cho *et al.* 2012). The optimization of cellular energy resources during stress is essential for plant acclimation, energetically expensive processes are partially arrested, such as reproductive activities, translation, and some biosynthetic pathways. For example, nitrogen and carbon assimilation are impaired in maize during salt stress and potassium-deficiency stress, the synthesis of free amino acids, chlorophyll, and protein are also affected (Holcik and Sonenberg, 2005; Qu *et al.* 2011). Once energy-expensive processes are curtailed, energy resources can be redirected to activate protective mechanisms. This is exemplified by the decrease in *de novo* protein synthesis in *Brassica napus* seedlings, *Glycine max, Lotus japonicas*, and *Medicago truncatula* during heat stress accompanied by an increased translation of heat shock proteins (Dhaubhadel *et al.* 2002; Soares-Cavalcanti *et al.* 2012).

Plant respiratory metabolism represents a central feature of plant metabolic flexibility

Plants utilize sucrose and starch as the principle substrates for respiration, and can fully oxidize these fuels to CO_2 and H_2O via the standard pathways of glycolysis, the citric acid cycle, and the mitochondrial electron transport chain (miETC). However, there are at least four remarkable attributes in the organization and associated bioenergetic features of plant respiratory metabolism that are not commonly seen in other organisms.

Remarkable Attribute #1. Plant glycolysis exists in the plastid and cytosol, with the parallel reactions catalyzed by distinct nuclear-encoded isozymes (Plaxton and Podestá, 2006). The prime functions of glycolysis in darkened chloroplasts and non-photosynthetic plastids are to participate in the breakdown of starch as well as to generate carbon skeletons, reductants and ATP for anabolic pathways such as fatty acid synthesis and N-assimilation.

Remarkable Attribute #2. An extraordinary feature of plant carbohydrate metabolism is that the cytosolic glycolytic pathway in itself is a complex network containing parallel enzymatic reactions at the level of sucrose, fructose-6-phosphate, glyceraldehyde-3-phosphate and phosphoenolpyruvate (PEP) metabolism. Each bypass reaction of plant cytosolic glycolysis circumvents a classical glycolytic reaction that is dependent upon an adenine nucleotide or inorganic orthophosphate (P_i) as a cosubstrate. As discussed below, this flexibility allows the preferential utilization of inorganic pyrophosphate (PP_i) as an alternate energy donor, particularly when cellular ATP pools become diminished during stresses such as anoxia and nutritional P_i starvation.

Remarkable Attribute #3. Respiratory O_2 consumption by the plant miETC can be mediated by the ATP-producing cytochrome pathway or by a non-energy conserving alternative pathway that involves the rotenone-insensitive NAD(P)H dehydrogenase bypasses to Complex I and the cyanide-resistant alternative oxidase (AOX) bypass to Complex III and IV.

Remarkable Attribute #4. A number of bypass reactions to the plant citric acid cycle exist and include: (i) an $NADP^+$-specific isocitrate dehydrogenase within the mitochondrial matrix that can circumvent the NAD^+-specific isocitrate dehydrogenase of the conventional citric acid cycle, (ii) the glyoxylate cycle of germinating oil seeds which necessitates two key enzymes (isocitrate lyase and malate synthase) that collectively bypass both decarboxylating reactions of the citric acid cycle, thereby allowing gluconeogenesis from stored fats to occur, and (iii) the gamma-amino butyric acid (GABA) shunt which may function as an alternative, NAD^+-independent, mechanism for glutamate entry into the citric acid cycle. GABA, a four carbon non-protein amino acid, may represent a significant proportion of the free amino acid pool in plant cells subjected to abiotic or biotic stresses (Plaxton and Podestá, 2006). Various stresses initiate a signal transduction pathway in which increased cytosolic Ca^{2+} activates a calmodulin-dependent glutamate decarboxylase leading to a marked elevation in intracellular GABA levels. After the stress is relieved, GABA can be transported into the mitochondrion and enter the TCA cycle via its conversion to succinate through the sequential action of GABA transaminase and succinic semialdehyde dehydrogenase.

The alternative reactions of glycolysis, the citric acid cycle, and miETC are believed to endow plants with crucial metabolic flexibility that facilitates their development and acclimation to unavoidable abiotic stresses. An ongoing and challenging problem has been to elucidate the respective function(s), control, and relative importance of the many alternative reactions of plant respiration.

Metabolic adjustments during stressing conditions: Osmolyte accumulation

A common defensive mechanism activated in plants exposed to stressing conditions is the production and accumulation of compatible solutes. The chemical nature of these small molecular weight organic osmoprotectants is diverse, these molecules include amino acids (asparagine, proline, serine), amines (polyamines and glycinebetaine), and γ-amino-N-butyric acid (GABA). Furthermore, carbohydrates, including fructose, sucrose, trehalose, raffinose, and polyols (myo-inositol, D-pinitol) (Krasensky and Jonak, 2012; Banu *et al.* 2010), as well as pools of anti-oxidants such as glutathione (GSH) and ascorbate (El-Shabrawi *et al.* 2010; Phang *et al.* 2008), accumulate in response to osmotic stress. Common characteristics of these diverse solutes are a high level of solubility in the cellular milieu and lack of inhibition of enzyme activities even at high concentrations. Accumulation of compatible solutes in response to stress is not only observed in plants, it is a defense mechanism triggered in animal cells, bacteria, and marine algae, indicative of an evolutionarily conserved trait (Empadinhas and da Costa, 2008). Scavenging of reactive oxygen species (ROS) to restore redox metabolism, preservation of cellular turgor by restitution of osmotic balance, and associated protection and stabilization of proteins and cellular structures are among the multiple protective functions of compatible osmoprotectants during environmental stress (Yancey, 2005; Mittler, 2006).

Research has been done on the beneficial effects of compatible solutes on plant tolerance to environmental stress. Correlation between amino acid accumulation (mainly proline) and stress tolerance was described in the mid-1960s in Bermuda grass during water stress (Barnett and Naylor, 1966). Since then, extensive work has proven that proline serves as an osmoprotectant, a cryoprotectant, a signaling molecule, a protein structure stabilizer, and an ROS scavenger in response to stresses that cause dehydration, including salinity, freezing, heavy metals, and drought (low water potential) (Verslues *et al.* 2006; Verbruggen and Hermans, 2008). Proline oxidation may also provide energy to sustain metabolically demanding programs of plant reproduction, once the stress has passed (Mattioli *et al.* 2006).

Proline metabolism and its regulation are processes well characterized in plants. Proline is synthesized from glutamate in the cytoplasm or chloroplasts:

Δ-1-pyrroline-5-carboxylate synthetase (P5CS) reduces glutamate to glutamate semialdehyde (GSA). Then GSA spontaneously cyclizes into pyrroline-5-carboxylate (P5C), which is further reduced by P5C reductase (P5CR) to proline. Conversely, proline is catabolized within the mitochondrial matrix by action of proline dehydrogenase (ProDH) and P5C dehydrogenase (P5CDH) to glutamate. In an alternative pathway, proline can be synthesized from ornithine in a pathway involving ornithine ä-aminotransferase (OAT). Core enzymes P5CS, P5C, P5CR, ProDH, and OAT are responsible for maintaining the balance between biosynthesis and catabolism of proline. Regulation comes at transcriptional level of genes encoding the key enzymes. Transcriptional up-regulation of genes for P5CS and P5C to increase proline synthesis from glutamate and down-regulation of genes for P5CR and ProDH to arrest proline catabolism is observed during dehydration/osmotic stress (Verslues and Sharma, 2010). Also, post-translational regulation of core enzymes is closely associated with proline levels and environmental signals. For example, the *Arabidopsis* P5CS1 enzyme is subjected to feedback inhibition by proline, controlling the carbon influx into the biosynthetic pathway (Hong *et al.* 2000). Considering that proline accumulation is associated with stress tolerance, that core enzymes regulate proline biosynthesis, and that these core enzymes are likely rate-limiting steps for its accumulation, logic dictates that overexpression of biosynthetic proline enzymes might increase the levels of the compatible solute and thus improve the tolerance in plants against abiotic stress. Several studies have tested this by overexpressing genes for P5CS or P5C enzymes in different plant species, reporting the expected rise in proline levels and the associated resistance to dehydration, salinity, or freezing (Su and Wu, 2004; Gleeson *et al.* 2005; Yamada *et al.* 2005; Vendruscolo *et al.* 2007). Furthermore, deletion of genes coding ProDH (Nanjo *et al.* 1999) or P5CDH (Deuschle *et al.* 2004; Borosani *et al.* 2005), expression of a feedback-insensitive P5CS (Hong *et al.* 2000), or the overexpression of OAT (Wu *et al.* 2005) increase the cellular levels of proline and osmoprotection to some abiotic stresses.

Comparable extensive work has been done for other compatible solutes such as δ-aminobutyric acid (Akcay *et al.* 2012), glycine betaine (Giri, 2011), trehalose (Fernandez *et al.* 2010), mannitol, and sorbitol (Rathinasabapathi, 2000), these solutes are efficient protectors against some abiotic stressors. Metabolic pathways for biosynthesis and catabolism of compatible solutes, their regulation, participant enzymes, and compartmentalization are well characterized in most important plant species. This knowledge has led to strategies for improvement of plant tolerance involving the accumulation of those protective osmolytes in plants by expression of core biosynthetic enzymes or their improved derivatives, expression of related transporters, and deletion of osmolyte-consuming enzymes. These numerous studies have provided evidence that enhanced accumulation of compatible solutes correlates with reinforcement of plant resistance to adverse growth conditions.

Plant metabolomics and applications

The traditional approach of enhancing the accumulation of a specific compounds in response to a determined stimulus, as done with compatible solutes, have resulted in some degree of tolerance in plants, and also demonstrates that the ability to redirect nutrients to imperative processes and the induction of adequate metabolic adjustments are crucial for plant survival during conditions of stress. However, this is a sectioned view of how plants regulate their entire metabolism in response to stressing conditions. In order to achieve a more comprehensive understanding, we must consider that plant metabolism is an intricate network of interconnected reactions. Plants have a high degree of subcellular compartmentation, a vast repertory of metabolites, and developmental stage strongly influences metabolism. Therefore, metabolic responses are complex and dynamic and involve the modification of more than one metabolite. Also, accumulation of a specific compound is not an absolute requirement indicative of a tolerance trait, adjustment of the flux through a certain metabolic pathway might be enough to contribute to stress tolerance (Allen *et al.* 2009). Recently, it has been reported that plants modulate stoichiometry and metabolism in a flexible manner in order to maintain optimal fitness in mechanisms of storage, defense, and reproduction under varying conditions of temperature and water availability (Rivas-Ubach *et al.* 2012). Furthermore, time-series experiments in *Arabidopsis thaliana* plants subjected to temperature and/or light alterations revealed that time-resolved metabolic activities respond more quickly than transcriptional activities do (Caldana *et al.* 2011).

Traditional molecular approaches for tracing metabolic phenotypes in plants responding to abiotic stress have identified and manipulated specific genes or groups of genes in plant models. These have primarily been genes involved in early responses or in down-stream assembly of the response reaction. With the application of new powerful tools of molecular biology and bioinformatics, large collections of genes have been subjected to complete analysis. To arrive at a complete and comprehensive knowledge of physiology in the plant response to abiotic stress, researchers are embracing ionomic profiling, transcriptomic, proteomic and metabolomic analysis. A deep dissection of the biochemical pathways in plants facing stressing conditions requires integrative and comprehensive analyses in order to identify all the simultaneous metabolic responses and, more importantly, to be able to link these responses to specific abiotic stress. In this sense, metabolomics could contribute significantly to the study of metabolic responses to stress in plants by identifying diverse metabolites, such as the by-products of stress metabolism, stress signal transduction molecules, and molecules that are part of the acclimation response (Ruan *et al.* 2011).

The metabolome is the entirety of small molecules present in an organism and can be regarded as the ultimate expression of its genotype in response to environmental changes. Metabolomics is gaining importance in plant research in both basic and applied contexts. Metabolomic studies have already shown how detailed information gained from chemical composition can help us to understand the various physiological and biochemical changes occurring in the plants and their influence on the phenotype. The analytical measurement of several hundreds to thousands of metabolites is becoming a standard laboratory technique with the advent of "hyphenated" analytical platforms of separation methods and various detection systems. Separation methods include gas chromatography (GC), liquid chromatography (LC), and capillary electrophoresis (CE). Different types of mass spectrometry (MS), nuclear magnetic resonance (NMR), and ultraviolet light spectroscopy (UV/VIS) devices are utilized for detection. Fourier transform ion cyclotron resonance mass spectrometry (FT-ICR-MS) is a specialized technique often used in direct infusion (DI) mode for metabolomics analyses, as its high mass accuracy allows a separation solely based on this parameter. Each methodology offers advantages and disadvantages, and the method of choice will depend on the type of sample and metabolites to be determined, and the combination of analytical platforms (Allwood *et al.* 2011).

GC and MS were the first pair of techniques to be combined, delivering high robustness and reproducibility. GC-MS remains one of the most widely used methods for obtaining metabolomic data because of its ease of use, excellent separation power, and its reproducibility. The main drawback of GC-MS is that only thermally stable volatile metabolites, or non-volatile compounds that can be chemically altered to make them volatile, can be detected (Han *et al.* 2009; Kueger *et al.* 2012). NMR spectroscopy is a fingerprinting technique that offers several advantages over high-throughput metabolite analyses, such as relatively simple sample preparation and the non-destructive analysis of samples. NMR can detect different classes of metabolites in a sample, regardless of their size, charge, volatility, or stability with excellent resolution and reproducibility (Kim *et al.* 2011). Labeling of metabolites with isotopes and subsequent NMR analysis is also useful for metabolic flux analysis and fluxomics as it allows tracking the selective signal enhancement of isotopologues (O'Grady *et al.* 2012). Recent advances with high-throughput approaches using ultra-high-field FT-ICR-MS alone or in combination with other tools of 'first pass' metabolome analysis as electrospray ionization mass spectrometry (ESI-MS) are expected to make inventory of the entire metabolome in a single sample possible in the near future (Han *et al.* 2008; Draper *et al.* 2013).

In metabolomics, the implicit objective is to identify and quantify all possible metabolites in a cellular system under defined states of stress conditions (biotic or abiotic) over a particular time scale in order to characterize accurately the metabolic profile (Johnson *et al.* 2003). But metabolome studies have some analytical limitations. It is important to have in mind that from the total amount of metabolites in a sample, only an informative portion can be reliably identified and quantified. In addition, metabolic networks in multicellular eukaryotes, specifically in plants, are challenging because of the large size of the metabolome, extensive secondary metabolism, and the considerable variation in tissue-specific metabolic activity (Mintz-Oron *et al.* 2012). Therefore, experimental design and sample preparation need to be done with great care because environmental and experimental variations confer noticeable impact on the resulting metabolic profiles. This has been demonstrated in legumes in which a high proportion of nutritional and metabolic changes depend on non-controllable environmental variables (Sanchez *et al.* 2010).

Metabolomic analyses have been applied to the functional identification of unknown genes through metabolic profiling of plants in which some genes are up- or down-regulated, the discovery of biomarkers associated with disease phenotypes, the safety assessment of genetically modified organisms (GMOs), the characterization of plant metabolites of nutritional importance and significance in human health, and the discovery of compounds involved in plant resistance to biotic and abiotic stresses (Okazaki and Saito, 2012). Metabolic profiles can be used as signatures for assessing the genetic variation among different cultivars or species of the same genotype at different growth stages and environments. The metabolite profile represents phenotypic information, this means that qualitative and quantitative metabolic measurements can be related to the genotypes of the plants to differentiate closely related individuals (Wahyuni *et al.* 2012). Once the identification of individual metabolites is available, connections among metabolites can be established, and then metabolic profiles can be used to infer mechanisms of defense. Metabolic profiles will guide tailoring of genotypes for acceptable performance under adverse growth conditions and will be of help in design and development of crop plant cultivars best suited to sustainable agriculture (Reguera *et al.* 2012; Nadella *et al.* 2012). Metabolomics tools have been used to evaluate the impact of the genotype and the environment on the quality of plant growth in the study of interpecific hybrids between *Jacobaea aquatica* and *J. vulgaris* (common weeds native to Northern Eurasia). An NMR-based metabolomics profiling approach was used to correlate the expression of high and low concentrations of particular compounds, including phenylpropanoids and sugars, with results of quantification of genetically controlled differences between major primary and secondary metabolites (Kirk *et al.* 2012). In melon (*Cucumis melo* L.), metabolomic and elemental profiling of fruit quality were found to be affected by genotype and environment (Bernillon *et al.* 2013).

Plant metabolomics and drought stress

The variable and often insufficient rainfalls in extended areas of rain-fed agriculture, the unsustainable groundwater use for irrigated agriculture worldwide, and the fast-growing demands for urban water are putting extreme pressure on global food crop production. The demand for water to sustain the agriculture systems in many countries will continue to increase as a result of growing populations (FAO, 2007). This progressively worsening water scarcity is imposing hydric stress on both rain-fed and irrigated crops. Water deficiency stress induces a wide range of physiological and biochemical alterations in plants, arrestment of cell growth and photosynthesis and enhanced respiration are among the early affects. Genome expression is extensively remodeled, activating and repressing a variety of genes with diverse functions (Shinozaki *et al.* 2007). Sensing water deficit and activation of defense mechanisms comes through chemical signals in which abscisic acid (ABA) plays a central role. ABA accumulates in tissues of plants subjected to hydric stress and promotes transpiration reduction via stomatal closure. Through this mechanism, plants minimize water losses and diminish stress injury. ABA regulates expression of many stress-responsive genes, including the late embryogenesis abundant (LEA) proteins, leading to a reinforcement of drought stress tolerance in plants (Aroca *et al.* 2008). Many questions remain unresolved concerning hydric stress-plant metabolic response: How does drought stress perturb metabolism in crop plants? How does hydric stress affect the metabolism of wild plants? What modern strategies of "omics" could be exploited to support future programs of crop breeding to lead to a more sustainable agriculture?

One of the main mechanisms by which plants cope with water deficits is osmotic adjustment. These adjustments maintain a positive cell turgor via the active accumulation of compatible solutes. Traditionally, the analysis of metabolic responses to drought stress was limited to analysis of one or two classes of compounds considered as "role players" in the development of tolerance. Application of metabolomic approaches is providing a less biased perspective of metabolic profiles of response and also is aiding in the discovery of novel metabolic phenotypes. Unbiased GC-MS metabolomic profiling in *Eucalyptus* showed that drought stress alters a larger number of leaf metabolites than the previously reported in targeted analysis. Accumulation of shikimic acid and two cyclohexanepentol stereoisomers in response to drought stress was described for the first time in *Eucalyptus*. Also, the magnitude of metabolic adjustments in response to water stress correlates with the sensitivity/tolerant phenotype observed, drought affected around 30-40% of measured metabolites in *Eucalyptus dumosa* (a drought-sensitive specie) compared to 10-15% in *Eucalyptus pauciflora* (a drought-tolerant specie) (Warren *et al.* 2012). Similarly, critical differences in the metabolic responses were observed when drought-tolerant (NA5009RG) and drought-sensitive (DM50048) soybean cultivars were analyzed by ^{1}H NMR-based metabolomics. Interestingly, no enhanced accumulation of the traditional osmoprotectants, such as proline, soluble sugars as sucrose or myo-inositol, organic acids or other amino acids (except for aspartate), were detected in the leaves of either genotype during water stress. In contrast, levels of 2-oxoglutaric acid, pinitol, and allantoin were affected differentially in the genotypes when drought was imposed, suggesting possible roles as osmoprotectants (Silvente *et al.* 2012). In contrast to soybean, levels of amino acids, including proline, tryptophan, leucine, isoleucine, and valine, were increased under drought stress in three different cultivars of wheat (*Triticum aestivum*) analyzed for 103 metabolites in a targeted GC-MS approach (Bowne *et al.* 2012). Metabolic adjustments in response to adverse conditions are transient and depend on the severity of the stress. In a 17-day time course experiment in maize (*Zea mays*) subjected to drought stress, GC-MS metabolic analysis revealed changes in concentrations of 28 metabolites. Accumulation of soluble carbohydrates, proline and eight other amino acids, shikimate, serine, glycine, and aconitase, was accompanied by the decrement of leaf starch, malate, fumarate, 2-oxoglutarate, and seven amino acids during the drought treatment course. However, as the water potential became more negative, between the 8th and 10th days, the changes in some metabolites were more dramatic, demonstrating their dependence on stress severity (Sicher and Barnaby, 2012).

Accumulation of compatible solutes is an evolutionary conserved trait in bacteria, plants, animal cells, and marine algae. A recent GC-MS metabolomic analysis confirmed that the moss *Physcomitrella patens* also triggers compatible solute accumulation in response to drought stress. After two weeks of physiological drought stress, 26 metabolites were differentially affected in gametophores, including altrose, maltitol, L-proline, maltose, isomaltose, and butyric acid, comparable to metabolic adjustments previously reported in stressed *Arabidopsis* leaves. More interesting is the recent report of a new compound, annotated as EITTMS_N12C_ATHR_2988.6_1135EC44, with no previously mass spectra matching record, accumulated specifically in response to drought stress in this moss (Erxleben *et al.* 2012).

Plant metabolomics and salinity stress

A current problem for crop plants worldwide, which will become more critical in the future, is salt stress imposed by salinity in soils due to poor practices in irrigation and over-fertilization, among other causes. Salt stress induces abscisic acid synthesis, abscisic acid transported to guard cells closes stomata, resulting in decreased photosynthesis, photo-inhibition, and oxidative stress. This causes an immediate inhibition of cell expansion, visible as general plant growth inhibition, accelerated development, and senescence (Chinnusamy *et al.* 2006). To cope with salt stress plants implement strategies that include lowering of rates of photosynthesis, stomatal conductance, and transpiration (Huang *et al.* 2012). Sodium ion, by its similar chemical nature to potassium ion, competes with and inhibits the potassium uptake by the root. Potassium deficiency results in growth inhibition because this ion is involved in the capacitance of a plethora of enzyme activities in addition to its participation in maintaining membrane potential and cell turgor (Chinnusamy *et al.* 2006).

The metabolic perturbation in plants exposed to salinity involves a broad spectrum of metabolic pathways and both primary and secondary metabolism. For example, in a proteomic study in foxtail millet, 29 proteins were significantly up or down-regulated due to NaCl stress, with great impact on primary metabolism. These proteins were classified into nine functional categories, cell wall biogenesis (lignin biosynthesis), among these were caffeic acid 3-O-methyltransferase and caffeoyl CoA 3-O-methyltransferase, photosynthesis and energy metabolism, which included proteins like cytochrome P450 71D9, phytochrome 1, photosystem I reaction center subunit IV B, and ATP synthase F1 sector subunit beta, among others, nitrogen metabolism, proteins like glutamine synthetase root isozyme 4, ferredoxin-dependent glutamate synthase, chloroplast precursor (Fd-GOGAT), and urease, carbohydrate metabolism, proteins such as UDP-glucose 4-epimerase GEPI42 (galactowaldenase) and beta-amylase, and lipid metabolism including isovaleryl-CoA dehydrogenase 2 and aldehyde dehydrogenase (Veeranagamallaiah *et al.* 2008).

Studies using metabolomic tools in plant models and plant crops have shown that the physiology in salt stress courses through a complex metabolic response including different systematic mechanisms, time-course changes, and salt-dose dependence. The biochemical changes involve metabolic pathways that fulfill crucial functions in the plant adaptation to salt stressing conditions. Time-course metabolite profiling in cell cultures of *A. thaliana* exposed to salt stress demonstrates that glycerol and inositol are abundant 24 h after salt stress exposure, whereas lactate and sucrose accumulate 48 h later. The methylation cycle, the phenylpropanoid pathway, and glycine betaine biosynthesis exhibit induction as a short-term response to salinity stress, whereas glycolysis and sucrose metabolism and reduction in methylation are long-term responses. Long-term salt exposure also causes a reduction in the metabolites that were initially responsive (Kim *et al.* 2007). In tobacco plants treated with various doses of salt, 1 day of treatment with 50 mM NaCl induced accumulation of sucrose, and to a lesser extent glucose and fructose, through gluconeogenesis. Further stress (500 mM NaCl for another day) led to elevation of proline and even higher elevation in sucrose levels compared to the lower dose, at the same time, glucose and fructose levels decreased as transamination-related metabolites (asparagine, glutamine, and GABA) did. These data suggest that sugar and proline biosynthesis pathways are metabolic mechanisms for control of salt stress over one- to two-day periods (short-term). Proline continues to be observed at high levels at later stages (3 to 7 days under highly stressing concentrations of 500 mM NaCl) and sucrose decreases (although it remains at high levels compared to control). There are also significant elevations in levels of asparagine, valine, isoleucine, tryptophan, myo-inositol, uracil, and allantoin, and reductions in glucose, fructose, glutamine, GABA, malate, fumarate, choline, uridine, hypoxantine, nicotine, N-methylnicotina-mide, and formate (Zhang *et al.* 2011a). Similarly, in maize plants stressed with salt solutions ranging in concentration from 50 to 150 mM NaCl, the metabolic profile of the shoot extracts changes most dramatically compared to controls in the plants exposed to the highest salt concentration (Gavaghan *et al.* 2011).

Another complexity in the metabolic perturbations in salt-stressed plants consists of tissue-specific response differences. In maize plants exposed to 50-150 mM NaCl saline solution, levels of sucrose and alanine were increased and levels of glucose decreased in roots and shoots. Other osmoprotectants exhibited differentiated behavior: GABA, malic acid, and succinate levels increased in roots, while glutamate, asparagine and glycine betaine were at higher concentrations in shoots. There were decreased levels of acetoacetate in roots and of malic acid and *trans*-aconitic acid in shoots. A progressive metabolic response was more evident in shoots than in roots (Gavaghan *et al.* 2011).

In comparative ionomics and metabolite profiling of related *Lotus* species (*Lotus corniculatus*, *L. tenuis*, and *L. creticus*) under salt stress, the extremophile *L. creticus* (adapted to highly saline coastal regions) exhibits better survival after long-term exposure to salinity and is more efficient at excluding Cl^- from shoot tissue than the two cultivated glycophytes *L. corniculatus* and *L. tenuis* (grassland forage species). Sodium ion levels are higher in the extremophile than the cultivars under both control conditions and salt stress. In *L. creticus*, a differential homeostasis of Cl^-, Na^+, and K^+ is accompanied by distinct nutritional changes compared to the glycophytes *L. corniculatus* and

L. tenuis. Magnesium and iron levels increase in *L. creticus* after salt treatment, but levels of potassium, manganese, zinc, and calcium do not. In non-stressed control plants, 41 metabolites are found at lower levels in *L. creticus*than in the two glycophytes, and 10 metabolites are at higher levels in *L. creticus*. These data demonstrate that each of these species has a distinct basal metabolic profile and that these profiles do not show a concordance with salt stress or salt tolerance. In salt stress conditions, 48 metabolites show similar changes in all species, either increasing or decreasing, with increased levels the amino acids proline, serine, threonine, glycine, and phenylalanine, the sugars sucrose and fructose, myo-inositol and other unidentified metabolites, and with decreased levels of organic acids such as citric, succinic, fumaric, erythronic, glycolic, and aconitic acid, including ethanolamine and putrescine, among others. Of note is that more than half of the metabolites affected by salt treatment are common among the three species, and only one-third of responsive metabolites in *L. creticus* are not shared with the glycophytes. Interestingly, the changes in the pool sizes of these metabolites are only marginal (Sanchez *et al.* 2011a). A few changes in the metabolic profile are extremophile-specific, but most salt-elicited changes in metabolism are similar. Other studies in glycophytes under salt stress indicate that organic acids and intermediates of the citric acid cycle tend to decrease (Sanchez *et al.* 2008). Also in genus *Lotus*, model species (*L. japonicus*, *L. filicaulis*, and *L. burttii*) and cultivated species (*L. corniculatus*, *L. glaber*, and *L. uliginosus*) exhibit consistent negative correlation in the Cl^- levels in the shoots and tolerance to salinity, but metabolic profiles diverge amongst genotypes, asparagine levels are higher in the more tolerant genotypes. These results support the conclusion that Cl^-exclusion from the shoots represents a key physiological mechanism for salt tolerance in legumes, moreover, an increased level of the osmoprotectant asparagine is typical (Sanchez *et al.* 2011*)*. In *L. japonicus*, which has a robust metabolic response to salt stress, levels of proline and serine, polyolsononitol and pinitol, and myo-inositol increase (Sanchez *et al.* 2010).

All these studies demonstrate that the metabolic plant response to salinity stress is variable depending on the genus and species and even the cultivar under consideration. Differential metabolic rearrangements are in intimate correlation with genetic backgrounds. Furthermore, the plant physiology in salt stress with time proceeds through a complex metabolic response including different systematic mechanisms and changes. Inside a salt-stressed plant as a biological unit, different tissues respond differentially and in some cases the responses are even contrasting. From comparative ionomics studies, it is evident also that under salinity stress, differential homeostasis of ions as Cl^-, Na^+, and $K^+$is correlated with distinct nutritional changes in extremophile and glycophyte species, even inside the same genus. Noticeable differences exist between plant species in the way they react to surpass the osmotic pressure imposed by high soil salt content through mechanisms such as tolerance, efficiency in salt exclusion, changes in nutrient homeostasis, and osmotic adjustment. From the aforementioned studies, metabolic markers in the response to high salinity in plants include glycine betaine, sucrose, asparagine, GABA, malic acid, aspartic acid, and *trans*-aconitic acid. In legumes, increases in levels of the amino acids asparagine, proline, and serine are notable as are increases in polyolsononitol, pinitol, and myo-inositol (Sanchez *et al.* 2010).

Plant metabolomics and oxidative stress

An increase in intracellular levels of ROS is a common consequence of adverse growth conditions. An imbalance between ROS synthesis and scavenging is caused in a manner independent of the nature of the stress, it is induced by both biotic and abiotic types of stress. Toxic concentrations of ROS cause severe damage to protein structures, inhibit the activity of multiple enzymes of important metabolic pathways, and result in oxidation of macromolecules including lipids and DNA. All these adverse events compromise cellular integrity and may lead to cell death (Gill and Tuteja, 2010; Kar, 2011). Normal cellular metabolic activity also results in ROS generation under regular growth conditions. Thus, cells sense uncontrolled elevation of ROS and use them as a signaling mechanism to activate protective responses (Moller and Sweetlove, 2010). In this context plants have developed efficient mechanisms for removal of toxic concentrations of ROS. The antioxidant system is composed of protective enzymes (*e.g.*, superoxide dismutase, catalase, peroxidase, reductase, and redoxin) and radical scavenger metabolites (mainly GSH and ascorbate). GSH is an essential component of the antioxidant system that donates an electron to unstable molecules such as ROS to make them less reactive and also can acts as a redox buffer in the recycling of ascorbic acid from its oxidized form to its reduced form by the enzyme dehydroascorbate reductase (Jozefckzak *et al.* 2012). Organized remodeling of metabolic networks is a crucial response that gives the cells the best chance of surviving the oxidative challenge.

In *Arabidopsis thaliana*, oxidative treatment with methyl viologen causes the down-regulation of photosynthesis-related genes and concomitant cessation of starch and sucrose synthesis pathways, meanwhile catabolic pathways are activated. These metabolic adjustments avoid the waste of energy used in non-defensive processes and mobilize carbon reserves towards actions of emergency relief such as the accumulation of maltose, a protein structure-stabilizer molecule (Scarpeci and Valle, 2008). A GC-MS metabolomic study, together with an analysis of key metabolic fluxes of cell cultures and roots of *A. thaliana* treated with the oxidative stressor menadione, revealed the similarities and divergences in the metabolic adjustments triggered in both culture systems. Inhibition of the tricarboxylic acid cycle (TCA) by accumulation of pyruvate and citrate is accompanied by a decrement of malate, succinate, and fumarate pools. This early (0.5 h) response was observed in both systems. Inhibition of TCA cycle concomitantly causes a decrement in the pools of glutamate and aspartate due to the inhibition of the synthesis of TCA-linked precursors 2-oxoglutarate and oxaloacetate, respectively. Another mutual early metabolic redistribution is the redirection of the carbon flux from glycolysis to the oxidative pentose phosphate (OPP) pathway. This is also reflected by the decrement in the glycolytic pools of glucose-6 phosphate and fructose 6-P, and the increment in the OPP pathway intermediates ribulose 5-phosphate and ribose 5-phosphate. Increased carbon flux through the OPP pathway might supply reducing power (via nicotinamide adenine dinucleotide phosphate, NADPH) for antioxidant activity, since oxidative stress decreases the levels of the reductants GSH, ascorbate, and NADPH. After 2 and 6 h of stress progression, metabolic adjustments in response to oxidative stress are different in roots than in cell suspension cultures. In roots, pools of TCA cycle intermediates and amino acids are recovered. In contrast, in cell cultures, the concentrations of these metabolites remains depressed throughout the time course, indicating higher basal levels of oxidative stress in cell cultures. At the end of the treatment time (6 h), 39 metabolites, including GABA, aromatic amino acids (tryptophan, phenylalanine, and tyrosine), proline, and other amino acids, were significantly altered in roots. These results showed the broad spectrum of metabolic modifications elicited in response to oxidative stress and the influence of the biological system analyzed (Lehmann *et al.* 2009).

Redirection of carbon flux from glycolysis through the OPP pathway and subsequent increase in the levels of NADPH was also reported in rice cell cultures treated with menadione. CE-MS analysis of these rice cultures showed the depletion of most sugar phosphates resulting from glycolysis (pyruvate, 3-phosphoglyceric acid, dihydroxyacetone phosphate, fructose-6-phosphate, glucose-1-phosphate (G1P), G6P, G3P, phosphoenolpyruvate) and TCA-organic acids (2-oxoglutarate, aconitate, citrate, fumarate, isocitrate, malate, succinate) and increases in the levels of OPP pathway intermediates (6-phosphogluconate, ribose 5-phosphate, ribulose 5-phosphate). Incremental increases in the biosynthesis of GSH and intermediates (*O*-acetyl-L-serine, cysteine, and γ-glutamyl-L-cysteine) are also observed in the menadione-treated rice cell cultures (Ishikawa *et al.* 2010).

1.12. Metabolic Compounds in Defense to Stresses

Antioxidants, is designing chemicals, when added in small quantities to a materials, react rapidly with the free radical intermediates of an autooxidation chain and stop it from progressing. An excellent example of this type of inhibitor is the synthetic hindered phenol 2,6-di-tert-butyl-4 methyl phenols often called BHT which react with mol- of peroxy radical and converts them to much less active products. It has been recognized for some time that naturally occurring substances including those found in higher plants, also have antioxidant activity. Recently, there has been increasing interest in oxygen-containing free radicals in biological systems and their implied roles as causative agents in the etiology of a variety of chronic disorders. Accordingly attention is being focused on the protective biochemical functions of naturally occurring antioxidants in the cells of the organisms containing them, and on the mechanisms of their action.

It has also been reported that plants with high levels of antioxidants, whether constitutive or induced have a greater resistance to such oxidative damage (Foyer *et al.* 1994; Scandalios, 1993; Mullineaux and Creissen, 1997). The primary components of this antioxidant system include carotenoids, ascorbate, glutathione, vitamin E (α-tocopherols), flavonoids, phenolic acids, other phenols, alkaloids, polyamines, chlorophyll derivatives, amino acids and amines and miscellaneous compounds. A number of studies indicated that the degree of oxidative cellular damage in plants exposed to a biotic stress is controlled by the capacity of antioxidative systems (Zhang and Kirkham, 1994; Noctor and Foyer, 1998; Zhu and Scandalios, 1994).

Phenolic compounds

Phenolics are diverse secondary metabolites (flavonoids, tannins, hydroxycinnamate esters and lignin) abundant in plant tissues (Grace and Logan, 2000). Polyphenols possess ideal structural chemistry for free radical scavenging activity, and they have been shown to be more effective antioxidants *in vitro* than tocopherols and ascorbate. Antioxidative properties of polyphenols arise from their high reactivity as hydrogen or electron donors, and from the ability of the polyphenol-derived radical to stabilize and delocalize the unpaired electron (chain-breaking function), and from their ability to chelate transition metal ions (termination of the Fenton reaction) (Rice-Evans *et al.* 1997). Another mechanism underlying the antioxidative properties of phenolics is the ability of flavonoids to alter peroxidation kinetics by modification of the lipid packing order and to decrease fluidity of the membranes (Arora *et al.* 2000). These changes could sterically hinder diffusion of free radicals and restrict peroxidative reactions. Moreover, it has been shown recently that phenolic compounds can be involved in the hydrogen peroxide scavenging cascade in plant cells (Takahama and Oniki, 1997). According to our unpublished results the content of condensed tannins (flavonols), as measured by high performance liquid chromatography, was 100 times higher in *Iris pseudacorus* rhizomes than in those of *Iris germanica.* The effect of anoxia on the flavonol content (a decrease after 35 d of treatment) suggests their participation in the antioxidative defense in *Iris pseudacorus* rhizomes.

Vitamin E (α-tocopherol)

Tocopherols and tocotrienols are essential components of biological membranes where they have both antioxidant and non-antioxidant functions (Kagan *et al.* 2000). There are four tocopherol and tocotrienol isomers (α-, β-, γ-, δ-) which structurally consist of a chroman head group and a phytyl side chain giving vitamin E compounds amphipathic character (Kamal-Eldin and Appelqvist, 1996). Relative antioxidant activity of the tocopherol isomers *in vivo* is $\alpha > \beta > \gamma > \delta$ which is due to the methylation pattern and the amount of methy I groups attached to the phenolic ring of the polar head structure. Hence, α-tocopherol with its three methyl substituents has the highest antioxidant activity of tocopherols (Kamal-Eldin and Appelqvist, 1996). Though antioxidant activity of tocotrienols *vs.* tocopherols is far less studied, α-tocotrienol is proven to be a better antioxidant than α-tocopherol in a membrane environment (Packer *et al.* 2001). Tocopherols, synthesized only by plants and algae, are found in all parts of plants (Janiszowska and Pennock, 1976). Chloroplast membranes of higher plants contain α-tocopherol as the predominant tocopherol isomer, and are hence well protected against photooxidative damage (Fryer, 1992). There is also evidence that α-tocopherol quinone, existing solely in chloroplast membranes, shows antioxidant properties similar to those of α-tocopherol (Kruk *et al.* 1997). Vitamin E is a chain-breaking antioxidant, *i.e.* it is able to repair oxidizing radicals directly, preventing the chain propagation step during lipid autoxidation (Serbinova and Packer, 1994). It reacts with alkoxy radicals (LO'), lipid peroxyl radicals (LOO') and with alkyl radicals (L'), derived from PUPA oxidation (Kamal-Eldin and Appelqvist, 1996; Buettner, 1993). The reaction between vitamin E and lipid radical occurs in the membrane-water interphase where vitamin E donates a hydrogen ion to lipid radical with consequent tocopheroxyl radical (TOH') formation (Buettner, 1993). Regeneration of the TOH' back to its reduced form can be achieved by vitamin C (ascorbate), reduced glutathione (Fryer, 1992) or coenzyme Q (Kagan *et al.* 2000). In addition, tocopherols act as chemical scavengers of oxygen radicals, especially singlet oxygen (*via* irreversible oxidation of tocopherol), and as physical deactivators of singlet oxygen by charge transfer mechanism (Fryer, 1992).

TOH' formation sustains prooxidant action of tocopherol. At high concentration tocopherols act as prooxidant synergists with transition metal ions, lipid peroxides or other oxidizing agents (Kamal-Eldin and Appelqvist, 1996). It has been clearly shown, that prooxidant function of tocopherol on low density lipoprotein was clearly inhibited *in vitro* by antioxidants (ascorbate or ubiquinol) (Upston *et al.* 1999).

In addition to antioxidant functions vitamin E has several non-antioxidant functions in membranes. Tocopherols have been suggested to stabilize membrane structures. Earlier studies have shown that α-tocopherol modulates membrane fluidity in a similar manner to cholesterol, and also membrane permeability to small ions and molecules (Fryer, 1992). In recent studies α-tocopherol has been shown to decrease the permeability of digalactosyldiacyl-glycerol vesicles for glucose and protons (Berglund *et al.* 1999). There is also recent evidence of interaction between PS II with α-tocopherol and α-tocopherol quinone (Kruk *et al.* 2000). Complexation of tocopherol with free fatty acids and lysophospholipids protects membrane structures against their deleterious effects. The process is of great physiological relevance, since phospholipid hydrolysis products are characteristics of pathological events such as hypoxia, ischemia or stress damage (Kagan, 1989). In addition, several other non-antioxidant functions of a-tocopherol have been described such as protein kinase C inhibition, inhibition of cell proliferation, *etc.* as reviewed by Azzi and Stocker (2000).

Naturally occurring compounds with vitamin E activity are the tocopherols, a group of closely related phenolic benzochroman derivatives having extensive ring alkylation. These compounds occur not only in plant but also in mammalian tissues. Among the latter, antioxidant compounds of low molecular weight such as α-tocopherols, play an important role in protecting chloroplastic membranes from the deleterious effects of lipid peroxy radicals and singlet oxygen (Fryer, 1992; Munne-Bosch and Alegre, 2001a; Munne-Bosch *et al.* 2001).

The α-tocopherols are usually recycled back by ascorbic acid or reduced glutathione following oxidation by lipid peroxy radicals. However, it can be irreversible converted to the corresponding quinone and quinone epoxide after reacting with singlet oxygen (Fryer, 1992; Munne-Bosch *et al.* 1999). Besides, an increased synthesis of antioxidant such as α-tocopherols has been correlated with a higher tolerance to drought (Munne-Bosch and Alegre, 2001b) and other environmental stresses (Fryer *et al.* 1998). Munne-Bosch *et al.* (2001) reported that α-tocopherols progressively decreased in sage during the drought. Therefore, the leaves contained smaller pools of antioxidant defenses to counteract oxygen toxicity during the drought and this explain among other biochemical and structural feature, the susceptibility of this species to stress.

The most biologically active of the four major tocopherols is α-tocopherols. The peroxy radical derived from α-tocopherols is also stabilized because the unpaired electrons of the chroman ring oxygen are held nearly perpendicular to the plane the phenyl ring calculations suggest the stabilization is on the order of 3 kcal/mol (Munne-Bosch *et al.* 2000a). Vitamin E is also one of the best quenchers for $1O_2$ yet test, with a quenching rate constant of approximately 6x 108 (in methanol) and it also appears to react with O^{-2} to gave a phenoxy radical 15 mg g^{-1} fresh weight of α-tocopherols (Evans *et al.* 1999). This compound may serve to protect symbiosome membranes and other nodule membranes against lipid peroxidation.

Flavonoids

It has been recognized that several classes of flavonoid showed antioxidant activity toward a variety of easily oxidizable compounds. Flavanoids occur widely in the plant kingdom, and are especially common in leaves flowering tissues, and pollens. They are also abundant in woody parts such as stems and barks. Flavanoids are usually accumulated in the plant vacuole as glycosides. The concentration of flavonoid in plant cells often exceeds 1 mM, with concentrations 3 to 10 mM being reported in the epidermal cells of *Vicia faba* (Vierstra *et al.* 1982). The synthesis of many flavonoids and other phenolic compounds is greatly affected by light, for example tobacco plants grown under supplemental levels of UV contained about twice the concentration of total soluble phenolic compounds compared to the control plants. Plants grown in full sun have also been shown to contain higher levels of flavonoids than shade grown plants (Takahama and Oniki, 1997).

Flavonoids are not only accumulated in the plant vacuole as glycosides but also found as exudates on the leaves and other aerial surfaces of some species of plants. Their physiological functions have long remained unknown except for a rot as a protective filter against harmful UV radiation. Additional physiological functions of flavonoids have been discovered, for example, their role as antioxidants (Rice-Evans and Miller, 1998) as factor inducing pollen germination and pollen-tube elongation (Yistra *et al.* 1992; Mo *et al.* 1992) and as antifungus agents (phytoalexins). Phenylpropanoid compounds might alleviate oxidative stress by shading visible light or UV. Furthermore, some flavonoids and anthocyanins have been shown to work as antioxidants *in vitro* (Kubo *et al.* 1999). Flavonoids and other phenolics are abundant in nodules, where, despite their obvious role as signal molecules during nodule initiation-they can inhibit lipid peroxidation by intercepting the peroxyl radicals formed in nodule membranes (Moran *et al.* 1997). Several flavonoids were shown to be potent inhibitors of the enzymes lipoxygenase and prostaglandin synthetase, which convert polyunsaturated fatty acids to oxygen-containing derivatives. Highest activity against both enzymes was shown by luteolin (5,7,3,4-tetrahydroxyflavone) and 3,4-dihydroxy flavone, at about 3x 10^{-5}, they inhibited 50% of lipoxygenase activity. The author did not speculate on possible mechanisms for the inhibition of these enzymes by these particular flavonoids. Caldwell *et al.* (1983) have poineered a concept that flavonoids, with strong absorption in the 300-400nm UV region, are acting as internal light filters the protection of chloroplasts and other organelles from UV damage .The light-filtering ability of these compounds may reinforce their powerful antioxidant effects to provide a high level of protection against damaging oxidants generate either thermally or by light. Phenolic compounds and flavonoids are among the most influential and wide distributed secondary products in the plant kingdom. Many of these play important physiological and ecological roles, being involved in resistance to different of stress (Rice-Evans and Miller, 1998; Ayaz *et al.* 2000).

Phenolic acids

Acidic compounds incorporating phenolic groups have been repeatedly implicated as active antioxidants (Taiz and Zeiger, 1991). Caffeic acid, chlorogenic acid and its isomers including 4-Ocaffeoylquinic acid were isolated from sweet potatoes. Chlorogenic acid was found to be the most abundant phenolic acid in the plant extract and also the most active antioxidant, a 1.2 x 10^5 M solution inhibited over 80% of peroxide formation in a linoleic acid test system (Taiz and Zeiger, 1991). Esters of caffeic acid with sterols and triterpene alcohols have been isolate from the seed of the grass *Phalaris canariensis.* The fatty acids of the seed were predominantly unsaturated, suggesting that the esters were acting to protect then from oxidation (Fengel and Wegener, 1984). The lipid soluble esters were effective antioxidants in tests with lard or sardine oil heated at 60°C. In the tests, the esters were added as mixtures but at least some components appeared to have activity approaching or exceeding that of BHT2, 6-di-tert-buty,-4-methylphenol. Plant phenolics have often been referred to secondary metabolites and many of these compounds play an essential role in the regulation of plant growth development and interaction with other organisms. In higher pants, most secondary phenolics are derived at least in part from phenylalanine, a product of the shikimic acid pathway. The shikimic acid pathway begins from simple carbohydrates and proceeds to amino acids such phenylalanine and tyrosine (Taiz and Zeiger, 1991; Fengel and Wegener, 1984). Most researches showed that the levels of total and free amino acids increase remarkably during water stress. Gzik (1996) reported that the total of 18 amino acids including phenylalanine and tyrosine increased in sugar beet leaf during water stress. In addition, a remarkable increase in free amino acid content in some plants subjected to water stress was also reported (Handa *et al.* 1983; Svenningson *et al.* 1990). Ayaz *et al.* (2000) reported that, an increase in the content of phenolic acids in rolled leaves could be related with increasing level of amino acids synthesis induced during water stress. However, the increasing levels of amino acids (mainly phenylalanine and tyrosine) may trigger the production of phenolic acids (cinnamic acid pathway) leading to lignin biosynthesis. Lignin is such a key component of water transporting tissue, the ability to make lignin most have been one of the most important adaptation permitting primitive plants to colonize dry land (Taiz and Zeiger, 1991). So, the increase of some phenolic acids in rolled leaves of Cotenant may be explained by lignin biosynthesis in cell wall for preventing water loss (Ayaz *et al.* 2000). On the other hand, most of the increased phenolic acids analyzed in this work are insoluble and semisoluble state and play a role as cell solute as well as sugars for osmotic adjustment (Ayaz *et al.* 2000). The increase of phenolic acids content may be linked to the lignifications of cell walls and, in part, the synthesis of certain amino acids maintaining osmotic adjustment in cell.

Other phenols

Rosmaridiphenol, a diterpene derivative with adjacent OH groups was isolated from *Rosmarinus officinalis* (rosemary). Its antioxidant activity in heated lard exceeded that of BHA and approached that of BHT. Related phenolic diterpenes with antioxidant activity have also been isolated from this plant (Munne-Bosch and Alegre, 2000a, 2001a).

Carnosic acid is a diterpene that displays high antioxidant activity and which protects biological membranes from lipid peroxidation (Aruoma *et al.*,1992; Haraguchi *et al.* 1995). Previous studies, it has been shown that carnosic acid may protect chloroplasts from oxidative stress, but this mechanism has only been tested in rosemary, a drought- tolerant species. A group of potent antioxidant for the air oxidation linoleic acid was isolated from the methanol extract of the rhizome of *Curcuma langal* (turmeric). The abundant and most active constituent of the extract was the orange pigment, curcumin. Its 50% inhibitory concentration for the linoleic acid test system was about 5 x 10^{-4} M, it was more active then vitamin E in the procedure used by the authors and the synthetic antioxidants, BHA and BHT. The mechanism of curcumin activity may include metal ion chelation by central B-diketone group. The lignin isolated from sesame (*Sesamum indicum*) seed, significantly inhibited the autooxidation of linoleic acid at 40°C when added at 5.8 x 10^{5} M. Polyhydroxylated chalcones such as butein which are biosynthetic intermediates between cinnarnic acids and flavonoids, also show considerable antioxidant activity for lard (Aruoma *et al.*1992; Haraguchi *et al.* 1995).

Nitrogen compounds

Alkaloids

Increasingly evidence from a variety of sources is indicating that the basic nitrogen compounds of higher plants include many representatives that are potent inhibitors of various oxidative induced by radioactive cobalt irradiation (in soybean lecithin liposomes) was inhibited by the bisbenzylisoquinoline alkaloid cepharanthine (Vaneil *et al.* 1980). Caffeine, from the leaves of tea (*Thea sinensis*) and coffee (*Coffea arabica*) was shown to have antioxidative activity (in a linoleic acid oxidation test) comparable to that of BHA and BHT. Several alkaloids of various structural types have been found to be potent inhibitors of $1O_2$. Particularly effective are indole alkaloids such as strychnine and brucine that have a basic nitrogen atom in a rigid, cage like structure. Such alkaloids appear to be strictly physical quenchers and are not destroyed chemically by the process of quenching. Thus, in principle, they could inactivate many molecules of singlet oxygen per molecule of alkaloid. Polyamines such as spermine are related simple cyclic alkaloids found in legumes. This compound was shown in electron spin resonance experiments to scavenge the superoxide radical at rather high spermine concentration (0.01 - 0.03 M). This finding may have some bearing on the observed effects of polyamines as membrane stabilizing substance (Vaneil *et al.* 1980).

Alkaloids of quinolizidine type, for example sparteine have been found to be stored principally in the epidermal cells of four plants in the genus lipinus. Ahmed *et al.* (1989) and Ali (2000) suggested that this storage pattern was consistent with a phytochemical role for these substances as antifeedant chemical defense compounds, but it would also be consistent with an antioxidant role. It is unlikely that these alkaloids are acting as UV light filtering agents because their absorption in the solar UV range would be minimal. Alkaloids are commonly accumulated in the tissue of plants subjected to different types of stress (Ahmed *et al.* 1989; Ali, 2000).

Polyamines

Polyamines (spermidine and spermine) play a variety of physiological roles in plant growth and development (Ali, 2000). They are also potent ROS scavengers and inhibitors of lipid peroxidation (Droiet *et al.* 1986). Furthermore, exogenous application of polyamines has been shown to protect against various stress conditions such as cold, wilting, pollution and salinity (Ali and Abbas, 2003). Among the common polyamines putrescine appears to be the most sensitive external stress. Accumulation of putrescine has been observed in response to low pH, high salt concentration, potassium and magnesium deficiency, NH_4 treatment exposure to SO_2 and ozone, water stress, heat stress, chilling and lake of oxygen (Ali, 2000). The protection of plant against ozone damage (Ormord and Beckerson, 1986) by an exogenous supply of polyamines is believed to be cause by the free radical scavenge property of the polyamines (Droiet *et al.* 1986). Also, the protection of plant against stress damage by an exogenous supply of polyamines is believed to be cause by the free radical scavenge of the polyamines.

Chlorophyll derivatives

Although both chlorophyll (Smirnoff, 1993) and pheophytin (Hendry, 1993) promote the oxidation of lipids in the light, they are inhibitors of autooxidation under dark conditions. At 30°C, 2 x 10^{-5} M chlorophyll A was superior by a factor of Ca two to BHT (Rousos *et al.* 1986).The compounds appear to be unreactive toward lipid hydroperoxides, but do react with peroxy radicals, electron spin resonance data indicate the presence of tetrapyrrole radical cation (Munne-Bosch and Alegre, 2000a).

Amino acids and amines

Many amino acids have been tested for their antioxidant activity especially in food, based system. Among the amino acids for which antioxidant activity has been claimed are arginine, histidine, cysteine, tryptophane, lysine, methionine and threonine (Riisom *et al.* 1980). The literature reports are often very confusing with data suggesting that some amino acids may exhibit antioxidant potential under some conditions of temperature or pH or oxygen concentration but have no effect or actually promote oxidation in others. For example, alanine and histidine were reported to inhibit the oxidation of linoleic acid at pH 9.5 and to promote it at pH 7.5 (Riisom *et al.* 1980).

Carotenoids

Carotenoids fulfill two major roles in photosynthetic organisms. Their first role is to act as light harvesting pigments, extending the range of the light spectrum available for use in the photosynthetic process. They absorb light in the region from 450-570 nm, where the chlorophyll molecules do not, and pass the captured energy on the chlorophylls. Secondly, carotenoids provide photosynthetic systems with methods of photo-protection. O^{-2} has been detected in chloroplasts of water stressed wheat (Price *et al.* 1989). Singlet oxygen is an extremely powerful oxidant, powerful enough to cause the death of the organism in question. Carotenoids prevent the formation of singlet oxygen by quenching the triplet state of the chlorophyll molecules as they arise (Fyfe *et al.* 1995).

Other compounds

Vitamin C (Ascorbic acid)

Ascorbic acid (AA) has been proposed for a long time as a biological antioxidant. It exists in rather high concentrations in many cellular environments, such as the stroma of chloroplasts where its level is 2.3 x 10^{-3} M. Ascorbate has been demonstrated in many qualitative studies to possess significant antioxidant activity (Smirnoff, 1996). For example 10^3 M ascorbate inhibited the photooxidation of a kampferol by illuminated spinach chloroplasts. Ascorbate reduces two equivalents of O^{-2} produce H_2O_2 and triketo derivative dehydroascorbic acid. Ascorbate also reacts with $1O_2$ at a relatively fast rate (Noctor and Foyer, 1998). Ascorbic acid is one of the most studied and powerful antioxidants (Arrigoni and De Tullio, 2000; Horemans *et al.* 2000). It has been detected in the majority of plant cell types, organelles and in the apoplast. Under physiological conditions ascorbic acid exists mostly in the reduced form (90% of the ascorbate pool) in leaves and chloroplasts (Smirnoff, 1996), and its intracellular concentration can build up to millimolar range (*e.g.* 20 mM in the cytosol and 20-300 mM) in the chloroplast stroma (Foyer and Lelandais, 1996). The ability to donate electrons in a wide range of enzymatic and non-enzymatic reactions makes ascorbic acid the main ROS detoxifying compound in the aqueous phase. Ascorbic acid can directly scavenge superoxide, hydroxyl radicals and singlet oxygen and reduce H_2O_2 to water via ascorbate peroxidase reaction (Noctor and Foyer, 1998). In chloroplasts, ascorbic acid acts as a cofactor of violaxantin de-epoxidase thus sustaining dissipation of excess exitation energy (Smirnoff, 2000). Ascorbic acid regenerates tocopherol from tocopheroxyl radical providing membrane protection (Thomas *et al.* 1992). In addition, ascorbic acid carries out a number of non-antioxidant functions in the cell. Ascorbic acid has been implicated in the regulation of the cell division, cell cycle progression from G1 to S phase (Liso *et al.* 1988) and cell elongation (De Tullio *et al.* 1999).

Vitamin C is a universal reductant and antioxidant of plants. It is found at concentration of 1-2 mM in legume nodules (Matamoros *et al.* 1999) and is positively correlated with nodule effectiveness (Dalton *et al.* 1993). It is an essential metabolite for the operation of the ASC- GSH pathways, but it also has beneficial effects that do not require the presence of APX. ASC can directly scavenge ROS and reduce ferric Lb and LbIV (Moreau *et al.* 1995). It is also involved in hydroxylation of proline, regulation of the cell cycle and numerous fundamental processes of plant growth and development (Noctor and Foyer, 1998). Some reports concerning the exogenous application of the vitamins to salinized plants and their role in stimulation of their growth are scarce (Oertii,

1987). These compounds were also scarcely tried to counteract some of the adverse effects of salinity stress (Shaddad, 1990). Thus exogenous addition of such substances to the test organism could lead to growth stimulation through the activation of some enzymatic reactions (Makled, 1995). Shalate and Neumann (2001) reported that salt stress increased the accumulation in roots, stems and leaves of lipid peroxidation products produced by interactions with damaging active oxygen species. Additional ascorbic acid partially inhibited this response but did not significantly reduce sodium uptake or plasma membrane leakiness. Ascorbate is a key soluble antioxidant (Smirnoff, 1996). Isolation of an *Arabidopsis* mutant containing 30% of the wild type ascorbate concentration has provided the first genetic evidence for its importance in stress resistance. It was selected by hypersensitivity to ozone and is also hypersensitive to UV-B and sulfur dioxide (Conkin *et al.* 1996). The mutation has no pleiotropic effects on other parts of the antioxidant system with the exception of reduced APX activity, possibly because ascorbate is required to stabilize APX (Conkin *et al.* 1996). Reduction of cytosolic APX activity by expression of APX antisense mRNA also causes increased sensitivity to ozone damage, suggesting that intracellular, as well as extracellular, ozone detoxification by ascorbate is required (Orvar and Ellis, 1997). Overexpression of APX in tobacco chloroplasts had no effect on ozone resistance (Torsethaugen, *et al.* 1997), either it is in the wrong subcellular compartment or it is not a limiting factor.

Lack of oxygen defense

Mechanisms for the generation of ROS in biological systems are represented by both non-enzymatic and enzymatic reactions. The partition between these two pathways under oxygen deprivation stress can be regulated by the oxygen concentration in the system. Nonenzymatic one electron O_2 reduction can occur at about 10^{-4} M and higher oxygen concentrations (Torsethaugen *et al.* 1997), while in very low O_2 concentrations plant terminal oxidases *(Km* 10^{-6} M for oxygen) and the formation of ROS via mitochondrial ETC still remain functional. Among enzymatic sources of ROS, xanthine oxidase (XO), an enzyme responsible for the initial activation of dioxygen should be mentioned. As electron donors XO can use xanthine, hypoxanthine or acetaldehyde (Bolwell and Wojtaszek, 1997). The latter has been shown to accumulate under oxygen deprivation (Pfister-Sieber and Braendle, 1994) and can represent a possible source for hypoxiastimulated ROS production. The next enzymatic step is the dismutation of the superoxide anion by superoxide dismutase (SOD, EC.1.15.1.1) to yield H_2O_2. Due to its relative stability the level of H_2O_2 is regulated enzymatically by an array of catalases (CAT) and peroxidases localized in almost all compartments of the plant cell. Peroxidases, besides their main function in H_2O_2 elimination, can also catalyse O^{2-} and H_2O_2 formation by a complex reaction in which NADH is oxidized using trace amounts of H_2O_2 first produced by the non-enzymatic breakdown of NADH. Next, the NAD' radical formed reduces O_2 to $O^{2.-}$, some of which dismutates to H_2O_2 and O_2 (Lamb and Dixon, 1997). Thus, peroxidases and catalases play an important role in the fine regulation of ROS concentration in the cell through activation and deactivation of H_2O_2 (Elstner, 1987). Lipoxygenase (LOX, linoleate:oxygen oxidoreductase, EC.1.13.11.12) reaction is another possible source of ROS and other radicals. It catalyses the hydroperoxidation of polyunsaturated fatty acids (PUPA) (Rosahl, 1996).

The hydroperoxy derivatives of PUF A can undergo autocatalytic degradation, producing radicals and thus initiating the chain reaction of lipid peroxidation (LP). In addition LOX-mediated formation of singlet oxygen (Berglund *et al.* 1999) or superoxide (Lynch and Thompson, 1984) has been shown. A specific LOX activity increase and its positive correlation with the duration of anoxia have been detected in potato cells (Pavelic *et al.* 2000). Several apoplastic enzymes may also lead to ROS production under normal and stress conditions. Other oxidases, responsible for the two-electron transfer to dioxygen (amino acid oxidases and glucose oxidase) can contribute to H_2O_2 accumulation. Also an extracellular germin-like oxalate oxidase catalyses the formation of H_2O_2 and CO_2 from oxalate in the presence of oxygen (Bolwell and Wojtaszek, 1997). Amine oxidases catalyse the oxidation of biogenic amines to the corresponding aldehyde with a release of NH_3 and H_2O_2. Data on polyamine (putrescine) accumulation under anoxia in rice and wheat shoots (Reggiani and Bertani, 1989) and predominant localization of amine oxidase in the apoplast, suggest amine oxidase participation in H_2O_2 production under oxygen deprivation.

ROS can be also formed as by-products in the electron transport chains of chloroplasts (Asada, 1999), mitochondria and the plasma membrane (cytochrome b-mediated electron transfer) (Elstner, 1987). Plant mitochondrial ETC, with its redox-active electron carriers, is considered as the most probable candidate for intracellular ROS formation. Mitochondria have been shown to produce ROS (superoxide anion $O^{2.-}$ and the succeeding H_2O_2) due to the electron leakage at the ubiquinone site-the ubiquinone:cytochrome *b* region (Gille and Nohl, 2001) and at the matrix side of complex I (NADH dehydrogenase) (Moller, 2001). Hydrogen peroxide generation by higher plant mitochondria and its regulation by uncoupling of ETC and oxidative phosphorylation have been demonstrated by Braidot *et al.* (1999).

Lipid peroxidation is a natural metabolic process under normal aerobic conditionsand it is one of the most investigated consequences of ROS action on membrane structure andfunction. PUFA, the main components of membrane

Differential response of the SOD to oxygen deprivation stress

Plants	Site	Stress	SOD activity	Reference
Iris pseudaconus L.	Rhizome	Anoxia + Reaeration	Increase	Monk *et al.* 1987
Lotus (*Nelumbo nucifera)*	Seedling	Hypoxia + Reoxygenation	Increase	Ushimaru *et al.* 2001
Rice (*Oryza sativa* L.)	Roots	Anoxia	Decline	Chirkova *et al.* 1998
Rice (*Oryza sativa* L.)	Seedlings	Hypoxia	Plastid SOD Decline	Ushimura *et al.* 1999
		Submerged plants	Mitochondria SOD Decline	
Iris germanica L.	Rhizomes	Anoxia + reaeration	Deline	Monk *et al.* 1987
Soybean (*Glycine max* L.)	Seedling	Anoxia	Increase	van Toai and Bolles, 1994.
Barley (*Hordeum vulgare* L.	Root	Hypoxia	Increase	Kalasnikov *et al.* 1994.
Narrow leaved Lupin		Water logging	Fe SOD and CuZn Sod	Yu and Renegel, 1999.
(*Lupinus angustifolius* L.)		Reoxygenation	Increase Mn SOD	
Wheat (*Triticum arstivum* L.)	Root	Hypoxia	Unaffected	Biemelt *et al.* 2000.
		Anoxia	Increase	
Wheat (*Triticum aestivum* L.)	Roots	Anoxia	Decline	Chirkova *et al.* 1991a
Maize (*Zea mays* L.)		Hypoxia	Decline	Yan *et al.* 1996
Potato (*Solanum tuberosum* L.)	Cell culture	Anoxia + Reoxygenation	Decline	Pvelic *et al.* 2000

lipids, is susceptible to peroxidation. The initiation phase of LP includes activation of O_2 which is rate limiting. Hydroxyl radicals and singlet oxygen can react with the methylene groups of PUFA forming conjugated dienes, lipidperoxy radicals and hydroperoxides (Smirnoff, 1995):

PUFA-H + X. → PUFA. + X-H

PUFA. + O2 → PUFA-OO.

The peroxyl radical formed is highly reactive and is able to propagate the chain reaction:

PUFA-OO. + PUFA-H → PUFA-OOH + PUFA

The formation of conjugated dienes occurs when free radicals attack the hydrogens of methylene groups separating double bonds and leading to a rearrangement of the bonds (Recknagel and Glende, 1984). The lipid hydroperoxides produced (PUFA-OOH) can undergo reductive cleavage by reduced metals, such as Fe^{2+}, according to the following equation:

Fe^{2+} complex + PUFA-OOH → Fe^{3+} complex + OH- + PUFA-O.

The lipid alkoxy I radical produced, PVFA-O, can initiate additional chain reactions (Buettner, 1993):

PUFA-O + PUFA-H → PUFA-OH + PUFA'

The multi-stage character of the process, i.e. branching of chain reactions, allows several ways of regulation (Shewfelt and Purvis, 1995). Among the regulated properties are the structure of the membranes: composition and organization of lipids inside the bilayer in a way which prevents LP (Merzlyak, 1989), the degree of PUFA unsaturation, mobility of lipids within the bilayer, localization of the peroxidative process in a particular membrane and the preventive antioxidant system (ROS scavenging and LP product detoxification). The idea of LP as a solely destructive process has changed during the last decade. It has been shown that lipid hydroperoxides and oxygenated products of lipid degradation as well as LP initiators (*i.e.* ROS) can participate in the signal transduction cascade (Tarchevskii, 1992).

Lipid and membrane integrity during oxygen deprivation are among the key factors in the survival of plants. Under anoxia a decrease in membrane integrity is a symptom of injury, and it can be measured as changes in the lipid content and composition (Chirkova *et al.* 1989), as activation of lipid peroxidation (Blokhina, 1999, Chirkova *et al.* 1998), as enhanced electrolyte leakage (Chirkova *et al.* 1991a, b) and as a decrease in adenylate energy charge (Hanhijiirvi and Fagerstedt, 1994, 1995). Since *de novo* lipid synthesis is energy dependent, and could hardly occur under anoxia, the preservation of membrane lipids is the most efficient way to maintain functional membranes. In previous studies it has been shown that anoxia-tolerant plant species such as *Acorus calamus* and *Schoenoplectus lacustris* are able to preserve their polar lipids during anoxia and in post-anoxia, while in anoxia-sensitive plants (*e.g. Iris germanica)* a significant decrease in polar lipids and a simultaneous increase in free fatty acids (FFA) occur during anoxic stress with markedly enhanced lipid peroxidation during reoxygenation (Henzi and Braendle, 1993).

A decrease in unsaturated to saturated fatty acid ratio under anoxia may represent a result of LP and, at the same time sets limits for substrates of LP, the PUFA. This is the case in the anoxia-tolerant *Acorus calamus,* where a decrease in linolenic acid (18:3) is compensated by linoleic (18:2) and oleic (18:0) acids under oxygen deprivation. The original lipid composition is recovered during 2 d of reaeration (Pfister-Sieber and Braendle 1994). Similar results have been obtained for the anoxia-tolerant and -intolerant cereals rice and wheat, respectively (Chirkova *et al.* 1989). On the other hand, no significant qualitative and quantitative changes have been detected in the composition of fatty acids in anaerobically treated rice seedlings (Generosova *et al.* 1998). In that study it was postulated that the reduction of unsaturated fatty acids esterified in lipids was of no significance as a mechanism of plant adaptation to anaerobic conditions. The key role in survival was assigned to energy metabolism (Generosova *et al.* 1998). Indeed, a correlation exists between the leakage of electrolytes (*i.e.* membrane damage) under low ATP and a release of FFA from anoxic tissue (Crawford and Braendle, 1996). The role of ATP in the maintenance of membrane lipid integrity under anoxia has been confirmed by Rawyler *et al.* (1999) in potato cell culture. It has been shown that, when the rate of ATP synthesis falls below 10 μmol g^{-1} fresh weight h^{-1}, the integrity of membranes cannot be preserved and FFA are liberated via lipolytic acyl hydrolase (Rawyler *et al.* 1999). In general, lipids of anoxia-tolerant plants are more preserved during oxygen deprivation in respect to the composition and the degree of unsaturation. During recent years evidence has accumulated on the importance of lipid metabolism, and especially on unsaturated fatty acids, in the induction of defense reactions under biotic and abiotic stresses. Linolenic acid (18:3) has been shown to be a precursor of jasmonic acid, a signal transducer in defense reactions in plant-pathogen interactions (Rickauer *et al.* 1997). FFA, liberated during membrane breakdown under stress conditions, is not only the substrates for LP, but can act also as uncouplers in mitochondrial ETC (Skulachev, 1998). Lipid hydroperoxides, formed as a result of LP, can affect membrane properties, *i.e.* increase hydrophilicity of the internal side of the bilayer (Frenkel, 1991). This phenomenon is very important for the termination of LP, since increased hydrophilicity of the membrane favours the regeneration of tocopherol by ascorbate.

Reoxygenation injury is a well-documented fact for both animal and plant tissues. Indeed, under anoxia-saturated electron transport components, the highly reduced intracellular environment (including transition metal ions), and low energy supply are factors favourable for ROS generation (Pavelic *et al.* 2000). Formation of free radicals within minutes after restoration of the oxygen supply has been shown by electron paramagnetic resonance (EPR) spectroscopy in the rhizodermis of the anoxia-intolerant *I. germanica,* while in the tolerant *I. pseudacorus* no signal was detected (Crawford *et al.* 1994). An investigation on the dynamics of LP [changes in conjugated dienes, trienes and thiobarbituric acid reactive susbstances (TBARS)] in the same plant species confirmed this observation: neither dienes nor TBARS production was detected in the anoxia-tolerant *I. pseudacorus,* with the exception of a 45-d anoxic treatment (Blokhina *et al.* 1999). Accumulation of various LP products as a result of reoxygenation has been observed in the roots of the anoxia-intolerant wheat and anoxia-tolerant rice, the latter showing higher membrane stability and lower level of LP after several days of anoxia (Blokhina *et al.* 1999, Chirkova *et al.* 1998). The length of the anoxic/ hypoxic treatment has been shown to affect the intensity of LP in post-anoxia. Cultured potato cells are known to exhibit a two-phase response to anoxia in respect to lipid hydrolysis: no FFA release has been detected up to 12 h under oxygen deprivation, while after 12 h intensive liberation of FF A sustained by lipolytic acid hydrolase has been observed. This behavior was mirrored by post-anoxic LP: negligible after short-term anoxia and elevated after lipid hydrolysis had occurred (Pavelic *et al.* 2000).

The existence of anoxia-inducible changes in plant metabolism implies that plant cells sense anoxic conditions and respond to them quickly by glycolytic production of ATP and the regeneration of $NAD(P)^+$ (Richard *et al.* 1994). Impairment of membrane structure and function under anoxia contribute to ROS-induced post-anoxic injury. This causes peroxidation of lipid membranes, depletion of reduced glutathione, an increase in cytosolic Ca^{2+} concentration, oxidation of protein thiol groups and membrane depolarization. Hypoxic pretreatment of plants prior to anoxia leads to increased survival (Drew, 1997; Vartapetian and Jackson, 1997). The minimal duration of hypoxia required for the acclimation has been estimated at 2-4 h for the root tips of maize seedlings (Chang *et al.* 2000). The biochemical and physiological features induced by this pretreatment suggest the involvement of several systems for increased stress tolerance. Of these, one is aimed at the maintenance of energy resources through the support of sugar utilization and ATP formation via the glycolytic pathway, while avoiding lactate accumulation and cytoplasmic acidosis. The majority of the genes induced codes for enzymes involved in starch and glucose mobilization, glycolysis and ethanol fermentation (Russel and Sachs, 1991). For example, anaerobic induction of enolase (2-phospho-D-glycerate hydratase, EC 4.2.1.11), an integral enzyme in glycolysis, which catalyses the interconversion of 2-phosphoglycerate to phosphoenolpyruvic acid (PEP), has been reported in maize (Lal *et al.* 1998). Some other glycolytic and fermentation pathway enzymes, such as alcohol dehydrogenase (ADH, EC 1.1.1.1), glucose phosphate isomerase, pyruvate decarboxylase (pDC, EC 4.1.1.1), and sucrose synthase have been characterized as hypoxic ally induced in maize. ADH and PDC, enzymes of ethanolic fermentation, were induced by hypoxic pretreatment in rice cultivars with different tolerance to anoxia (Ellis and Setter, 1999, Ellis *et al.* 1999). Interestingly, both abscisic acid (ABA) and hypoxic pretreatment of *Lactuca sativa* L. seedlings have resulted in increased survival of roots and elevated ADH activity (Kato-Noguchi, 2000). However, endogenous ABA level did not respond to hypoxic pretreatment, suggesting that ABA was not involved in hypoxia-induced anoxia tolerance. The crucial role of protein synthesis under hypoxic conditions, but not under anoxia, has been shown in root tips of maize seedlings (Chang *et al.* 2000). Among 46 individual proteins analysed, four anaerobic proteins have been identified: ADHl, enolase, glyceraldehyde-3-phosphate dehydrogenase and PDC. The rate of their synthesis under hypoxia was enhanced (or comparable) under normoxic conditions. As expected, cycloheximide treatment during hypoxic acclimation (but not under anoxia) resulted in decreased anoxia tolerance (Chang *et al.* 2000). Interestingly, low oxygen (5 %) treatment of *Arabidopsis* plants resulted in higher tolerance to hypoxia (0.1% O_2) but not anoxia (Ellis and Setter, 1999; Ellis *et al.* 1999). In these experiments differential response of shoots and roots was observed. In conclusion, early (hypoxic) induction of the ethanolic fermentation pathway and sugar utilization allows the maintenance of the energy status through regeneration of NADH and, hence, improves anoxia tolerance. Under natural conditions oxygen concentration would decrease gradually, and hence anoxia is always preceeded by hypoxia. Another metabolic feature that has been shown to be upregulated (though not always) under lack of oxygen is the antioxidant system. In an investigation on SOD activity and expression under hypoxia, anoxia and subsequent reaeration, the appearance of additional isozymes has been shown under anoxia. Judged by a cycloheximide treatment, this activity could not be attributed to *de novo* synthesis (Biemelt *et al.* 2000). It has been shown also that anoxic pretreatment protected soybean cells from H_2O_2 induced cell death. Such resistance was associated with up-regulation of peroxidases and alternative oxidase (Amor *et al.* 2000). The beneficial effect of alternative oxidase protein accumulation under anoxia is due to electron flow bifurcation and reduced probability of ROS formation under subsequent reoxygenation. It has been known for a long time that the main damage caused by anoxic stress occurs during readmission of oxygen. Some ROS formation can take place in hypoxic tissues as a result of over reduction of redox chains. Hence, anoxic stress is always accompanied to some extent by oxidative stress (generation of ROS) and its consequences. Induction of some components of the antioxidant system by hypoxic pretreatment can be due to such ROS accumulation and signalling (Semenza, 1999).

Less information is available on tocopherol status under oxygen deprivation. Since oxidative stress is non-specific and many diverse environmental stress factors, *e.g.* light, drought, chilling temperature and flooding, affect plant tissues enhancing production of ROS in chloroplasts and inducing photo-oxidation of thylakoid membranes (Elstner and Osswald, 1994), the response of tocopherols to other abiotic stresses will be discussed. In an experiment where isolated spinach thylakoids and thylakoids with an exogenously added high concentration of α-tocopherol were exposed either to photosynthetically active radiation (PAR) or to UV-B light, lipid peroxidation occurred only in normal thylakoids while no peroxidation was detected in membranes with high amounts of α-tocopherol. According to the results, there was no decrease in endogenous α-tocopherol in normal thylakoids, while in artificially treated thylakoids α-tocopherol contents decreased though no significant lipid peroxidation could be detected (DeLong and Steffen, 1998). The latter results contradict previous studies on lipid peroxidation since increased peroxidation of membranes has been described to occur only after significant amounts of membrane α-tocopherol have been depleted (Shewfelt and Purvis, 1995). During drought, plants show a general response to stress by increasing tocopherol and carotenoid contents in photosynthetic tissues (Munne-Bosch and Alegre, 2000a) which is accompanied by a similar sized rise in total glutathione pool and a depletion of ascorbate at least in many grass species (Price and Hendry, 1989). Substantial increases in a-tocopherol during water-stress have been detected in leaves of *Rosmarinus officinalis* L. (Munne-Bosch and Alegre, 2000a), *Melissa officinalis* L. (Munne-Bosch and Alegre, 2000b) and *Fagus sylvatica* L. (Garcia-Plazaola and Becerril, 2000). Enhanced activity of the xanthophyll cycle measured as increases in de-epoxidized xanthophylls (antheraxanthin and zeaxanthin) during drought is also a feature shared in water-stressed plant species. Rosemary plants also have species-specific antioxidants, abietane diterpenes, known for their function in inhibiting lipid peroxidation and superoxide generation in chloroplasts and micro somes, which are consumed during drought-stress in scavenging oxygen radicals (Munne-Bosch *et al.* 1999).

There is evidence that the chilling-tolerance of plants is correlated with increasing amounts of antioxidants and increasing activity of radical scavenging enzymes. A chilling tolerant maize genotype has been shown to contain higher amounts of both α-tocopherol and glutathione and higher GR activity than a chilling-sensitive maize genotype (Leipner *et al.* 1999). It is known that ascorbate regenerates tocopherols from their radical forms (Buettner, 1993). However, artificially increased ascorbate content in maize leaves did not improve the preservation of endogenous tocopherol during high light and chilling stress, but the high ascorbate content increased the usage of glutathione (Leipner, *et al.* 2000). Studies on vitamin E in the underground parts of plants during stress are scarce,which might in part be a consequence of the fact that generally the predominating tocopherol isomer of plant tissues, α-tocopherol, is mainly localized in chloroplasts (Kamal-Eldin and Appelqvist, 1996), and tocopherol synthesis is described to take place only in chloroplasts and chromoplasts (Schultz *et al.* 1991). During long term anoxic stress vitamin E contents in the rhizomes of two iris species, highly anoxiatolerant *Iris pseudacorus* and anoxia-sensitive *I. germanica,* have been determined. Tocopherols (α- and β-) were identified in both iris species, 13-tocopherol being the predominant tocopherol isomer especially in rhizomes of *I. germanica* which also possessed markedly higher total tocopherol content than *I. pseudacorus.* Anoxia caused a decrease in tocopherol isomers in both iris species (Blokhina *et al.* 2000).

The vitamin E composition in rhizomes of the iris species is unique since there are no previous reports of plant species having β-tocopherol as the main tocopherol isomer in vegetative tissues. In addition, according to mass spectrometry the identified isomer is β-dehydro-tocopherol with one double bond in its phytyl side chain (Blokhina *et al.* 2000), while tocopherols have saturated phytyl chains. Dehydrotocopherols have been found previously in etiolated shoots of maize and barley (Threlfall and Whistance, 1977). There is evidence that tocopherol isomers differ from each other in their functional properties. When the effectiveness of tocopherol isomers in quenching of singlet oxygen was studied, α- and β-tocopherols were equally effective in quenching singlet oxygen physically, but β-tocopherol showed almost no chemical reactivity with singlet oxygen, while α-tocopherol had the highest chemical reactivity of tocopherol isomers (Kaiser *et al.* 1990). Inhibition of protein kinase C activity and cell proliferation is a specific non-antioxidant function of α-tocopherol in animal cells. β-tocopherol lacks this ability but when the two isomers are present together β-tocopherol prevents the inhibitory effect of α -tocopherol (Azzi and Stocker, 2000). Though the total tocopherol content was higher in *I. germanica* than in the more anoxia-tolerant *I. pseudacorus* (Blokhina *et al.* 2000), this could not prevent the massive lipid degradation in *I. germanica* during anoxia found by Henzi and Braendle (1993). There are also earlier reports suggesting that anoxia causes more pronounced lipid peroxidation in the rhizomes of

I.germanica than in *I. pseudacorus* during reaeration (Blokhina *et al.* 1999). In anaerobically germinated rice seedlings a three-fold increase in tocopherol and low TBARS formation have been observed (Ushimaru *et al.* 1994). However, an anoxia-induced elevation in the tocopherol level observed in the anoxiaintolerant wheat and oat seedlings could not be detected in rice seedlings subjected to anoxia (Chirkova *et al.* 1998). There are *in vitro* studies suggesting that under anaerobic conditions a radical initiated reaction between linoleic acid hydroperoxide or methyl linoleate hydroperoxide and α-tocopherol occurs, forming an addition compound of the two reactants, the reaction being terminated in the presence of air (Gardner *et al.* 1972). Carbon-centred radicals are also formed during anaerobic conditions, tending to add to the oxygen of tocopheroxyl radicals forming 6-0-lipid alkyl-chromanol adducts (Kamal-Eldin and Appelqvist, 1996).

1.13. Abiotic Stress and Secondary Metabolites

Plant secondary metabolites are often referred to as compounds that have no fundamental role in the maintenance of life processes in the plants, but they are important for the plant to interact with its environment for adaptation and defense. However, we are beginning to understand the crucial role played by them in plant growth and development. In higher plants a wide variety of secondary metabolites are synthesized from primary metabolites (*e.g.*, carbohydrates, lipids and amino acids). They are needed in plant defense against herbivores and pathogens. Often they may confer protection against environmental stresses (Seigler, 1998). Secondary metabolites also contribute to the specific odours, tastes and colors in plants (Bennett and Wallsgrove, 1994). Plant secondary metabolites are unique sources for food additives, flavors, pharmaceuticals, and industrially important pharmaceuticals (Ravishankar and Rao, 2000). Chemicals include calcium, abscisic acid (ABA), salicylic acid (SA), polyamines and Jasmonates (JA), nitric oxide are involved in stress responses in plants (Tuteja and Sopory, 2008). Accumulation of metabolites often occurs in plants subjected to stresses including various elicitors or signal molecules. Secondary metabolites have significant practical applications in medicinal, nutritive and cosmetic purposes, besides, importance in plant stress physiology for adaptation (Seigler, 1998). The production of these compounds is often low (less than 1% dry weight) and depends greatly on the physiological and developmental stage of the plant GA (Rao and Ravishankar, 2002). Some of the plant derived natural products include drugs such as morphine, codeine, cocaine, quinine *etc.* Catharanthus alkaloids, belladonna alkaloids, colchicines, phytostigminine, pilocarpine, reserpine and steroids like diosgenin, digoxin and digitoxin, flavonoids, phenolics *etc.*

A wide range of environmental stresses (high and low temperature, drought, alkalinity, salinity, UV stress and pathogen infection) are potentially harmful to the plants (Seigler, 1998). Elicitation has been widely used to increase the production or to induce de novo synthesis of secondary metabolites in in vitro plant cell cultures (Dicosmo and Misawa, 1985). A number of researchers have applied various elicitors for enhancement of secondary metabolite production in cultures of plant cell, tissue and organ (Sudha and Ravishankar, 2003, Karuppusamy, 2009). Environmental stresses, such as pathogen attack, UV-irradiation, high light, wounding, nutrient deficiencies, temperature and herbicide treatment often increase the accumulation of phenylpropanoids (Dixon and Paiva, 1995). Nutrient stress also has a marked effect on phenolic levels in plant tissues (Chalker-Scott and Fnchigami, 1989). Concentrations of various secondary plant products are strongly dependent on the growing conditions and have impact on the metabolic pathways responsible for the accumulation of the related natural products. Exposure to drought or salt stress causes many common reactions in plants. Both stresses lead to cellular dehydration, which causes osmotic stress and removal of water from the cytoplasm to vacuoles.

Influence of various abiotic signals on secondary metabolites in plants

Deficiencies in nitrogen and phosphate directly influence the accumulation of phenylpropanoids (Dixon and Paiva, 1995). Potassium, sulfur and magnesium deficiency are also reported to increase phenolic concentrations. Low iron level can cause increased release of phenolic acids from roots (Chalker-Scott and Fnchigami, 1989). Calcium levels have been implicated in plant response to many abiotic stresses including cold, drought and salinity. Salt stress in soil or water is one of the major stresses especially in arid and semi-arid regions and can severely limit plant growth and productivity (Mahajan and Tuteja, 2005). Bryant *et al.* (1983) have hypothesized that when plants are stressed, an exchange occurs between carbon to biomass production or formation of defensive secondary compounds. A stress response is induced when plants recognizes stress at the cellular level. Secondary metabolites are involved in protective functions in response to both biotic and abiotic stress conditions. Formation of phenyl amides and dramatic accumulation of polyamines in bean and tobacco under the influence of abiotic stresses were reported, suggesting antioxidant role of these secondary metabolites (Edreva *et al.* 2000). Similarly, anthocyanin accumulation is stimulated by various environmental stresses, such as UV, blue light, high intensity light, wounding, pathogen attack, drought, sugar and nutrient deficiency (Winkel-Shirley, 2001).

Salt stress

Salt environment leads to cellular dehydration, which causes osmotic stress and removal of water from the cytoplasm resulting in a reduction of the cytosolic and vacuolar volumes. Salt stress often creates both ionic as well as osmotic stress in plants, resulting in accumulation or decrease of specific secondary metabolites in plants (Mahajan and Tuteja, 2005). Anthocyanins are reported to increase in response to salt stress (Parida and Das, 2005). In contrast to this, salt stress decreased anthocyanin level in the salt-sensitive species (Daneshmand *et al.* 2010). Petrusa and Winicov (1997) demonstrated that salt tolerant alfalfa plants rapidly doubled their proline content in roots, whereas in salt sensitive plants the increase was slow. However, Aziz *et al.* (1998) reported a correlation between proline accumulation and salt tolerance in Lycopersicon esculentum and Aegiceras corniculatum respectively. In tomato cultivars under salt stress endogenous JA was found to accumulate (Pedrazani *et al.* 2003). Polyphenol synthesis and accumulation is generally stimulated in response to biotic or abiotic stresses (Muthukumarasamy *et al.* 2000). Increase in polyphenol content in different tissues under increasing salinity has also been reported in a number of plants (Parida and Das, 2005). Navarro *et al.* (2006) showed increased total phenolics content with moderately saline level in red peppers. Plant polyamines have been shown to be involved in plant response to salinity. Salinity-induced changes of free and bound polyamine levels in sunflower (*Helianthus annuus* L.) roots was reported (Mutlu and Bozcuk, 2007).

Drought stress

Drought stress is one of the most significant abiotic stresses that affect plant growth and development (Xu *et al.* 2010). Drought stress occurs when the available water in the soil is reduced to such critical levels and atmospheric conditions add to continuous loss of water. Drought stress tolerance is seen in all plants but its extent varies from species to species. The drought stress arises due to the water deficit, usually accompanied by high temperatures and solar radiation (Xu *et al.* 2010). Water deficit and salt stress are global issues to ensure survival of agricultural crops and sustainable food production (Gosal *et al.* 2010). Drought often causes oxidative stress and was reported to show increase in the amounts of flavonoids and phenolic acids in willow leaves (Larson, 1988). Drought stress influenced changes in the ratio of chlorophyll a and b and carotenoids (Anjum *et al.* 2003). A reduction in chlorophyll content was reported in cotton under drought stresse (Massacci *et al.* 2008) and Catharanthus roseus (Soliz-Guerrero *et al.* 2002). Drought conditions decreased the content of saponins in Chenopodium quinoa from 0.46% dry weight (dw) in plants growing under low water deficit conditions to 0.38% in high water deficit plants (Soliz-

Guerrero *et al.* 2002). Anthocyanins are reported to accumulate under drought stress and at cold temperatures. Plant tissues containing anthocyanins are usually rather resistant to drought (Chalker-Scott, 1999). For example, a purple cultivar of chilli resists water stress better than a green cultivar (Bahler *et al.* 1991). Flavonoids have protective functions during drought stress. Flavonoids are implicated to provide protection to plants growing in soils that are rich in toxic metals such as aluminum (Winkel-Shirley, 2001).

Influence of heavy metal stress on secondary metabolites

Metal ions (Lanthanum, Europium, Silver and Cadmium), and oxalate also influenced secondary metabolite production (Marschne, 1995). The trace metal Nickel (Ni) is essential component of urease enzyme, is needed for plant development (Marschne, 1995). However, elevated Ni concentrations reduce plant growth (Hawrylak *et al.* 2007). The significant decrease in anthocyanin levels due to Ni stress has been reported by Hawrylak *et al.* (2007). Moreover, Ni has been shown to inhibit accumulation of anthocyanins (Krupa et al. 1996). Trace metals obviously limit anthocyanin biosynthesis by inhibiting activity of l-phenylalanine ammonia-lyase (PAL) (Krupa *et al.* 1996). Effective accumulation of metals (chromium, iron, zinc, and manganese) also produced an increase of oil content up to 35% in Brassica juncea (Singh and Sinha, 2005). Cu^{2+} and Cd^{2+} have been shown to induce higher yields of secondary metabolites such as shikonin (Mizukami *et al.* 1977) and also on the production of digitalin (Ohlsson and Berglund, 1989). Cu^{2+} also stimulated the production of betalains in *Beta vulgaris* (Trejo-Tapia *et al.* 2001). Co^{2+} and Cu^{2+} have been shown to have the stimulatory effect on the production of secondary metabolites (Trejo-Tapia *et al.* 2001). In an attempt to enhance betalaines production, the hairy roots were exposed to metal ions (Trejo-Tapia *et al.* 2001). Obrenovic (1990) has demonstrated stimulatory effects of Cu^{2+} on the accumulation of betacyanins in callus cultures of *Amaranthus caudatus*. Addition of Zn^{2+} (900 μM) enhanced the yield of lepidine in cultures of *Lepidium sativum* (Obrenovic, 1990). However, Cu proved more effective than Zn in enhancing the yield (Saba *et al.* 2000). $AgNO_3$ or $CdCl_2$ elicited overproduction of two tropane alkaloids, scopolamine and hyoscyamine, by in hairy root cultures of *Brugmansia candida* (Angelova *et al.* 2006). Rare-earth metal (Lanthanum) had influence on production of taxol in cell culture of *Taxus* sp (Pitta-Alvarez *et al.* 2000). Oat and bean plants treated with cadmium and copper significantly increased putrescine (Put) content (Weinstein *et al.* 1986). However, a decrease in Put level in Cd^{2+} or Cu^{2+} treated sunflower leaf disks has been reported. Sunflower leaf disks showed a significant decreased in spermidine (Spd) content and no variation in spermine (Spm) level when they were treated with Cd^{2+} or Cu^{2+} respectively (Groppa *et al.* 2001). However, Jacobsen *et al.* (1992) reported no changes in Spd or Spm content in chromium-exposed leaves of barley and rape plants, but Put accumulated with increasing chromium concentrations or exposure time. Lin and Kao (1999) reported that copper treatment increased but, but a decrease in Spm concentration in rice leaves.

Influence of cold stress on secondary metabolites

Low temperature is one of the most harmful abiotic stresses affecting temperate plants. These species have adapted to variations in temperature by adjusting their metabolism during autumn, increasing their content of a range of cryo-protective compounds to maximize their cold tolerance (Janska *et al.* 2010). In the cryopreservation process, environmental changes including osmotic injury, desiccation, and low temperature can impose a series of stresses on plants (Janska *et al.* 2010). During over wintering, temperate plant metabolism is redirected toward synthesis of cryoprotectant molecules such as sugar alcohols (sorbitol, ribitol, inositol) soluble sugars (saccharose, raffinose, stachyose, trehalose), and low-molecular weight nitrogenous compounds (proline, glycine betaine) (Janska *et al.* 2010). Cold stress increases phenolic production and their subsequent incorporation into the cell wall either as suberin or lignin (Griffith and Yaish, 2004). In addition, apple tree adaptation to cold climate was found to be associated with a high level of chlorogenic acid (Perez-Ilzarbe *et al.* 1997). Lignification and suberin deposition are also shown to increase resistance to cold temperatures. A mechanism by which suberin and lignin may protect plants from freeze damage (Griffith and Yaish, 2004). Christie *et al.* (1994) reported the accumulation of anthocyanins during cold stress. Pedranzani *et al.* (2003) reported that cold and water stresses produce changes in endogenous jasmonates in *Pinus pinaster*. Lei *et al.* (2004) reported that melatonin protect against cold-induced apoptosis in carrot suspension cells by upregulation of polyamines (putrescine and spermine). Moreover, Melatonin applied to cucumber (*Cucumis sativus* L.) seeds improves germination during chilling stress (Lei *et al.* 2004). Zhao *et al.* (2011) reported that melatonin improves the survival of cryopreserved callus of *Rhodiola crenulata*. The survival rate of the cryopreserved callus increased when the callus was pretreated with 0.1 μM melatonin.

Effect of cold stress on polyamime accumulation was reported (Kovács *et al.* 2011). When leaves of wheat (*Triticum aestivum* L.) are exposed to a cold temperature, accumulation of putrescine (6–9 times), spermidine accumulates to a lesser extent and, spermine decreases slightly. Moreover, alfalfa (*Medicago sativa* L.) also accumulates putrescine under low temperature stress (Nadeau *et al.* 1987). Hummel *et al.* (2004) reported that cold tolerance was associated with increased levels of polyamines (agmatine and putrescine) and their levels could be a significant marker of chilling tolerance in seedlings of *P. antiscorbutica*.

Influence of light on secondary metabolite production

It is well known that light is a physical factor which can affect the metabolite production. Light can stimulate such secondary metabolites include gingerol and zingiberene production in *Z. officinale* callus culture (Anasori and Asghari, 2008). A positive correlation between increasing light intensity and levels of phenolics has been reported (Chalker-Scott and Fnchigami, 1989). Larsson *et al.* (1986) reported decreases in foliar tannin and phenolic glycosides in shaded willow foliage. Arakawa (1993) studied the effect of UV light on anthocyanin accumulation in light colored sweet cherry. In apples, UV light from 280–320 nm synergistically stimulate anthocyanin synthesis when it was combined with red light (Arakawa *et al.* 1985). Effect of light irradiation on anthocyanin production in cell suspension cultures of *Perilla frutescens* was reported (Zhong *et al.* 1993). Chan *et al.* (2010) investigated the effects of different environmental factors, such as light intensity, irradiance (continuous irradiance or continuous darkness), on cell biomass yield and anthocyanin production in cultures of *Melastoma malabathricum*. Moderate light intensity (301-600 lx) induced higher accumulation of anthocyanins, the cultures exposed to 10-d continuous darkness showed the lowest pigment content, while the cultures exposed to 10-d continuous irradiance showed the highest pigment content. Light irradiation exhibited significant influence on the accumulation of anthocyanins by cell cultures of strawberry (Sato *et al.* 1996), *Daucus carota (Zhang et al. 1997),* and *Centaurea cyanus* (Kakegawa *et al.* 1991).

UV-B have been seen to increase in flavonoids in barley (Liu *et al.* 1995), and in polyamines in cucumber (Kramer *et al.* 1991). Hagimori *et al.* (1982) reported the effect of light and plant growth regulators on digitoxin formation in *Digitalis purpurea* L. Moreover, effect of light irradiation influenced artemisinin biosynthesis in hairy roots of *Artemisia annua* (Liu *et al.* 2002). Fett-Neto *et al.* (1995) reported the effect of white light on taxol and baccatin III accumulation in cell cultures of *Taxus cuspidate.* UV-B irradiation enhanced the concentration of flavonols in Norway spruce (*Picea abies*) (Fischbach *et al.* 1999). *Catharanthus roseus* plants, exposed to UV-B light show significant increases in the production of vinblastine and vincristine, which have proven effective in the treatment of leukemia and lymphoma (Bernard *et al.* 2009). UV-B radiation could increase in flavonoid content and phenylalanine ammonia-lyase (PAL) activity, associated with a decrease in chlorophyll content (Liang *et al.* 2006). UV (300–400 nm) increased flavonoids in the roots of pea plants (Shiozaki *et al.* 1999). UV-B was also shown to induce the production of flavonols in silver birch and grape leaves (Tegelberg *et al.* 2004). Moreover, under six different daily doses of UV-radiation (UV-A and UV-B), photosynthetic pigments, condensed tannins were accumulated where as, its precursor, (+)-catechin decreased significantly (Lavola *et al.* 2003). Our recent report suggests that photoperiod regimes influence endogenous indoleamines (serotonin and melatonin) in cultured green algae *Dunaliella bardawil (*Ramakrishna *et al.* 2011).

Influence of climate change on secondary metabolites

Climate change is the major threat to biodiversity and one of the main factors affecting human health and well-being over the coming decades (Pimm, 2009). Cold weather crops like rye, oats, wheat, and apples are expected to decline their productivity by about 15% in the next 50 y and strawberries will drop as much as 32% simply because of projected climate changes (Pimm, 2009). Plants are extremely sensitive to such changes, and do not generally adapt quickly. Ozone exposure has been shown to increase conifer phenolic concentrations (Rosemann *et al.* 1991), but low ozone exposure had no effect on monoterpene and resin acid concentrations (Kainulainen *et al.* 1998). Changes in crop quality due to ozone exposure have been studied in a limited number of crops. For example, in wheat, ozone reduced yield but increased grain protein concentration (Pleijel *et al.* 1999). Moreover, ozone was found to have positive effects on the quality of potato tubers by reducing sugars and increasing the vitamin C content (Piikki *et al.* 2003). In contrast, O_3 has been found to reduce the oil, protein, and carbohydrate contents in rape seeds (Ollerenshaw and Lyons, 1999). Moreover, in leaves of *Ginkgo biloba* ozone fumigation increased the concentrations of terpenes, decreased the concentrations of phenolics (He *et al.* 2009).

Plants grown at high CO_2 levels exhibit significant changes in their chemical composition (Idso and Idso, 2000). A prominent example of a CO_2 effect is the decrease of the nitrogen concentration in vegetative plant parts as well as in seeds and grains resulting in the decrease of the protein levels (Idso and Idso, 2000). Studies have shown that elevated CO_2 increases phenolics and condensed tannins in the leaves. In conifers, elevated CO_2 influenced a decrease/increase in concentrations of some individual monoterpenes (Williams *et al.* 1994) and an increase in total phenolics have been reported (Idso and Idso, 2000). Increased concentration of the monoterpene a-pinene was noticed in elevated CO_2 composition (Idso and Idso, 2000). In contrast to this, Williams *et al.* (1994) found decreased concentrations of b-pinene in needles under elevated CO_2. Several studies have been reported the effect of temperature on secondary metabolite production in plants. Secondary metabolites increase in response to elevated temperatures (Hummel *et al.* 2004; Jochum *et al.* 2007; Gera *et al.* 2007). In contrast to this, Snow *et al.* (2003) reported that high temperature decreases monoterpene levels in Douglas fir (*Pseudotsuga menziesii*).

Perspectives

Metabolome analysis has become an invaluable tool in the study of plant metabolic changes that occur in response to abiotic stresses. Despite progress achieved, metabolomics is a developing methodology with room for improvement. From a technical perspective, further developments are required to improve sensitivity for identification of previously uncharacterized molecules and for quantification of cellular metabolites and their fluxes at much higher resolution. This will allow the identification of novel metabolites and pathways and will allow linkage to responses to specific stresses, and, therefore, increase our level of knowledge of the elegant regulation and precise adjustments of plant metabolic networks in response to stress.

Another challenging task is the integration of metabolic data with data from experiments profiling the transcriptome, proteome, and genetic variations obtained from the same tissue, cell type, or plant species in response to a determined environmental condition. Integrated information can be used to map the loci underlying various metabolites and to link these loci to crop phenotypes, to understand the mechanisms underlying the inheritance of important traits, and to understand biochemical pathways and global relationships among metabolic systems. Elucidation of the regulatory networks involved in the activation/repression of key genes related to metabolic phenotypes in response to determined abiotic stress is becoming possible. Transcription factors (TFs) are central player in the signal transduction network, connecting the processes of stress signal sensing and expression of stress-responsive genes. Thus engineered TFs have emerged as powerful tools to manipulate complex metabolic pathways in plants and generate more robust metabolic phenotypes (Hussain *et al.* 2011; Agarwal *et al.* 2013).

Metabolic networks are highly dynamic, and changes with time are influenced by stress severity, plant developmental stage, and cellular compartmentalization. Since metabolic profiling only reveals the steady-state level of metabolites, detailed kinetics and flux analyses will support a better understanding of metabolic fluctuations in response to stress (Kruger *et al.* 2012). Genome-scale models (GSM) are in *silico metabolic* flux models derived from genome annotation that contain stoichiometry of all known metabolic reactions of an organism of interest. Construction of detailed GSMs applied to plant metabolism will provide information about distribution of metabolic fluxes at a specific genotype, a determined developmental stage, or a particular environmental condition. This detailed knowledge of the metabolic and physiological status of the cell can be used to design rational metabolic engineering strategies and to predict required genetic modifications to obtain a desired metabolic phenotype such as optimized biomass production, increased accumulation of a valuable metabolite, accumulation of a metabolite of response towards abiotic stress, or modification of metabolic flux through a specific pathway of significance (Collakova *et al.* 2012). Recently advances have been made in this field. For example, in rice, by using four complementary analytical platforms based on high-coverage metabolomics, molecular backgrounds of quality traits and metabolite profiles were correlated with overall population structure and genetic diversity, demonstrating that quality traits could be predicted from the metabolome composition, and that traits can be linked with metabolomics data. Results like these are opening the doors to modern plant breeding programs (Redestig *et al.* 2011).

Once a metabotype (metabolic phenotype) is confirmed to strengthen the tolerance to a particular abiotic stressor, the next challenge will be the transfer of this metabolic trait to a non-adapted plant species of interest. Engineering of more tolerant plants will then require the efficient integration and expression of one to several transgenes in order to modify an existent metabolic pathway or reconstruct a new complete one. Development and optimization of protocols for robust transformation of nucleus, mitochondria, and chloroplasts must be made available for higher plants including economically important crops, this will open new opportunities for plant metabolic engineering (Krichevsky *et al.* 2012). Future research progress on these topics will lead to novel strategies for plant breeding and elevating the health and performance of crops under adverse growth conditions to keep up with the ever-increasing needs for food and feed worldwide.

1.14. Conclusions

Metabolomics is the comprehensive and quantitative analysis of the entirety of small molecules present in an organism that can be regarded as the ultimate expression of its genotype in response to environmental changes, often characterized by several simultaneous abiotic and biotic stresses. Results obtained from a number of metabolomic studies in plants in response to different abiotic stresses have shown detailed relevant information about chemical composition, including specific osmoprotectants, directly related to physiological and biochemical changes, and have shed light on how these changes reflect the plant phenotype. Metabolomic studies are impacting both basic and applied research. Metabolomic studies will generate knowledge regarding how plant metabolism is differentially adjusted in relation to a specific stress and whether metabolic adjustments are stress specific or common to different types of stress. These studies will also reveal how metabolic pathways coordinate their fluxes and enzymes activities in order to strength their cellular energy requirements under stressing conditions. In an applied context, metabolomic approaches are providing a broader, deeper, and an integral perspective of metabolic profiles in the acclimation plant response to stressing environments. This information will reveal metabotypes with potential to be transferred to sensitive, economically important crops and will allow design of strategies to improve the adaptation of plants towards adverse conditions. Ultimately, design strategies will consider plant metabolism as a whole set of interconnected biochemical networks and not as sections of reactions that lead to the accumulation of a final metabolite. The task is challenging as it must take into account that reactions to stress course through a complex metabolic response, including different systematic mechanisms, time-course changes, and stress-dose dependences. Moreover, there are differences among plant tissues, and, as expected, marked differences between plants at the genus and species levels, exposing intimate correlation with genetic backgrounds. Nevertheless, the application of more advanced metabolomics tools will lead to new knowledge that will accelerate the design and the improvement of plant breeding projects, that surely will lead to the next generation of crops for specific applications in particular circumstances to cope with abiotic and biotic stress on agricultural crops worldwide.

Much research has focused on the molecular biology of plant stress responses, and this has revealed the ability of plants to respond to stress by altered gene expression leading to the synthesis of various stress-inducible proteins. Somewhat less attention has been devoted to the organization and control of intermediary metabolism as pertains to the acclimation of plants to stress. However, studies of plant metabolic responses to extreme environments have revealed some remarkably adaptive mechanisms that may serve to limit the deleterious consequences of stress. Although these adaptations are by no means identical in all plants, certain aspects are conserved in a wide variety of species from very different environments. A better understanding of the extent to which changes in flux through alternative enzymes and pathways influences plant stress tolerance is of significant practical interest. This knowledge is relevant to the applied efforts of agricultural biotechnologists to engineer transgenic crops exhibiting an improved resistance to abiotic stress, including nutritional P_i deprivation.

Abiotic stress factors influence growth and secondary metabolite production in higher plants. The influences are well marked. Infact, productivities depend on the changed ecosystem also. For example influence of climate change on bees, butterflies, soil microflora, etc. also effect plant antogeny, adaptation and productivities. Such holistic studies are lacking. Most importantly, climate change drastically influence water availability, salinity and several adverse soil conditions which will have direct bearing on original yields. The major advantage of the cell cultures includes synthesis of bioactive secondary metabolites, independently from climate and soil conditions. The use of *in vitro* plant cell culture for the production of chemicals and pharmaceuticals has made great strides. The use of genetic tools and the structure and regulation of pathways for secondary metabolism will provide the basis for the commercial production of secondary metabolites. The increased level of natural products for medicinal purposes coupled with the low product yields and supply concerns of plant harvest has renewed interest in large-scale plant cell culture technology. Hence, knowledge gained through isolated studies need to be translated into comprehensive investigations on the effect of climate change in a multi-pronged approach. Remedial measures are difficult to advocate but are needed in addition to preventive approach at global level since climate change do not have political or geographical boundaries.

References

Abebe, T., Guenzi, A.C., Martin. B. and Cushman, J.C. 2003. Tolerance of mannitol-accumulating transgenic wheat to water stress and salinity. *Plant Physiol.*, **131:** 1748–1755.

Agarwal, P.K., Shukla, P.S., Gupta, K. and Jha, B. 2013. Bioengineering for salinity tolerance in plants: Stae of the art. *Mol Biotechnol.*, **54(1):** 102-123.

Agboma, P. C., Jones, M. G. K., Peltonen-Saino, P., Rita, H. and Pehu, H. 1997a. Exogenous glycinebetaine enhances grain yield of maize, sorghum and wheat grown under two supplementary watering regimes. *J Agron Crop Sci.*, **178:** 29-37.

Agboma, P. C., Sinclair, T. R., Jokinen, K., Peltonrn-Sainio, P. and Pehu, E. 1997b. An evaluation of the effect of exogenous glycinebetaine on the growth and yield of soybean: timing *of* application, watering regimes and cultivars. *Field Crops Res.*, **54:** 51-64.

Agboma, P. C., Peltonen-Sainlo, P., Hinkkanen, R. and Pehu, E. 1997c. Effect of foliar application of glycinebetaine on yield components of drought-stressed tobacco plants. *Expt. Agric.*, **33:** 345-52.

Ahmed, A. M., Heikal, M. D. and Ali, A. 1989. Changes in amino acids and alkaloids in *Hyoscyamus muticus* and *Datura stramonium* in response to salinization. *Phyton*, **29:** 137-147.

Ahmad, M., Jarillo, J. A., Klimczak, U., Landry, L. G., Peng, T., Last, R. L. and Cashmore, A. R. 1997. An enzyme similar to animal type II photolyase mediates photoreactivation in *Arabidopsis. Plant Cell*, **9:** 199-207.

Ahmad, P., Jaleel, C.A., Salem, M.A., Nabi, G. and Sharma, S. 2010. Roles of enzymatic and non-enzymatic antioxidants in plants during abiotic stress. *Crit. Rev. Biotechno.*, **30(3):** 161-175.

Akcay, N., Borm, M., Karabudak, T., Ozdemir, F. and Turkan, I. 2012. Contribution of gama amino acid (GABA) to salt stress responses of *Nicotiana sylvestris* CMSII mutant and wild type plants. *J. Pl. Physiol.*, **169(5):** 452-458.

Alcázar, R., Cuevas, J.C., Patrón, M., Altabella, T. and Tiburcio, A.F. 2006a. Abscisic acid modulates polyamine metabolism under water stress in *Arabidopsis thaliana. Physiol Plant,* **128:** 448–455.

Alcázar, R., Marco, F., Cuevas, J.C., Patrón, M., Ferrando, A., Carrasco, P., Tiburcio, A.F., Altabella, T.2006b. Involvement of polyamines in plant response to abiotic stress. *Biotechnol Lett,* **28:**1867–1876.

Alcázar, R., Planas, J., Saxena, T., Zarza, X., Bortolotti, C., Cuevas, J.C., Bitrián, M., Tiburcio, A.F. and Altabella, T. 2010a. Putrescine accumulation confers drought tolerance in transgenic *Arabidopsis* plants overexpressing the homologous *Arginine decarboxylase 2* gene. *Plant Physiol Biochem.*, **48:** 547-552.

Alcázar, R., Altabella, T., Marco, F., Bortolotti, C., Reymond, M., Koncz, C., Carrasco, P. and Tiburcio, A.F. 2010b.Polyamines: molecules with regulatory functions in plant abiotic stress tolerance. *Planta*, **231:** 1237–1249.

Alcazar, R., Cuevas, J.C., Planas, J., Zarza, X., Bortolotti, C., Carrasco, P., Salinas, J., Tiburcio, A.F. and Altabella, T. 2011. Integration of polyamines in the cold acclimation response. *Plant Science,* **180:** 31-38.

Allen, D.K., Libourel, I.G. and Shachar-Hill, Y. 2009. Metabolic flux analysis in plants: coping with complecity. *Pl. Cell Environ.*, **32(9):** 1241-1257.

Ali, A. 2000. Role of putrescine in salt tolerance of *Atropa belladonna* plant. *Plant Science*, **152:**173-179.

Ali, A. and Abbas, H. M. 2003. Response of salt stressed barley seedling to phenylurea. *Plant Soil Environ*, **49:** 158-162.

Ali, A.A. and Alqurainy, F. 2006. Activities of antioxidants in plants under environmental stress. In,"*The Lutein-prevention and treatment for diseases*", Ed. N. Motohashi, Transworld Research Network, India, pp 187–256.

Allwood, J.W., de Vos, R.C., Moing, A., deborde, C. Erban, A., Kopka, J., Goodacre, R. and Hall, R.D. 2011. Plant metabolomics and its potential for systems biology research concepts, technology, and methodology. *Methods of Enzymology*, **500:** 299-336.

Anasori, P. and Asghari, G. 2008. Effects of light and differentiation on gingerol and zingiberene production in callus culture of *Zingiber officinale Rosc. Res. Pharm Sci.* **3:** 59–63.

Angelova, Z., Georgiev, S. and Roos, W. 2006. Elicitation of plants. *Biotechnol, Biotechnol Equip.*, **20:** 72–83.

Alscher RG, Donahue J. I., Cramer CL. 1997. Reactive oxygen species and antioxidants: relationships in green cells. *Physiol. Plant*, **100:** 224-233.

Altabella, T., Tiburcio, A.F. and Ferrando, A. 2009. Plant with resistance to low temperature and method of production thereof. Spanish patent application WO2010/004070.

An, Z.F., Jing, W., Liu, Y.L. and Zhang, W.H. 2008. Hydrogen peroxide generated by copper amine oxidase is involved in abscisic acid-induced stomatal closure in *Vicia faba. J Exp Bot,* **59:** 815–825.

Anjum, F., Yaseen, M., Rasul, E., Wahid, A. and Anjum, S. 2003. Water stress in barley (*Hordeum vulgare* L.). II. Effect on chemical composition and chlorophyll contents. *Pak J Agric Sci.*, **40:** 45–49.

Apel, K. and Hirt, H. 2004. Reactive oxygen species: metabolism, oxidative stress, and signal transduction. *Annual Review of Plant Biology*, **55:** 373-399.

Arrigoni, O., and De Tullio, M. C. 2000. The role of ascorbic acid in cell metabolism: between gene-directed functions and unpredictable chemical reactions. *Journal of Plant Physiology*, **157:** 481-488.

Arakawa, O., Hori, Y. and Ogata, R. 1985. Relative effectiveness and interaction of ultraviolet-B, red and blue light in anthocyanin synthesis of apple fruit. *Physiol. Plant.*, **64:** 323–327.

Arakawa, O. 1993. Effect of ultraviolet light on anthocyanin synthesis in light-colored sweet cherry, cv.Sato Nishiki. *J. Japan Soc. Hort. Sci.*, **62:** 543–546.

Arora, A., Byrem, T. M., Nair, M. G., and Strasburg, G. M. 2000. Modulation of liposomal membrane fluidity by flavonoids and isoflavonoids. *Archives of Biochemistry and Biophysics*, **373:** 102-109.

Arora, A., Sairam, R.K, and Srivastava, G.C. 2002. Oxidative stress and antioxidative system in plants. *Current Science*, **82(10):** 1227-1238.

Aroca, R., Vernierim, P. and Ruiz-Lozano, J.M. 2008. Mycorhizal and non-mycorhizal *Lactuca sativa* plants exhibit contrasting responses to exogenous ABA during drought stress and recovery. *J. Expt. Bot.*, **59(8):** 2029-2041.

Amor, Y., Chevion, M. and Levine, A. 2000. Anoxia pretreatment protects soybean cells against H_2O_2-induced cell death: possible involvement of peroxidases and of alternative oxidase. *FEBS Letters*, **477:** 175180.

Aruoma, O. I., Halliwell, B., Aeschbach, R. and Loliger, J. 1992. Antioxidant and prooxidant properties of actve rosemary constituents, carnosol and carnosic acid. *Xenobiotica*, **22:** 257-268.

Asada, K. 1980. Formation and scavenging of superoxide in chloroplasts, with relation to injury by sulfur dioxide in studies on the effects of air pollutants on plants and mechanisms of phototoxicity. *Res. Rep. Natt. Inst-Environ. Stud,* **11:** 165-179.

Asada, K. 1997. The role of ascorbate peroxidase and monodehydrpascorbate reductase in H_2O_2 scavenging in plants. In,"*Oxidative stress and the molecular biology of antioxidant defenses*", Ëd. J.G. Scandolios, Cold Spring Harbor Laboratory Press Plainview, pp. 715 -735.

Asada, K. 1999. The water-water cycle in chloroplasts: Scavenging of active oxygen and issipation of excess photons. *Annu. Rev. Plant Physiol. Plant Mol. Biol.*, **50:** 601-639.

Ashraf, M. and Foolad, M.R. 2007. Roles of glycine betaine and proline in improving plant abiotic stress resistance. *Environ. Exp. Bot.* **59:** 206–216.

Athwal, G.S. and Huber, S.C. 2002. Divalent cations and polyamines bind to loop 8 of 14-3-3 proteins, modulating their interaction with phosphorylated nitrate reductase. *Plant J,* **29:** 119–129.

Atkinson, N.J. and Urwin, P.E. 2012. The interaction of plant biotic and abiotic stresses from genes to the field. *J. Exp.Bot.*, **63(10)**: 3623-3643.

Avonce, N., Leyman, B., José, O., Mascorro-Gallardo, van Dijck, P., M. J., Thevelein and Iturriaga, G. 2004. The *Arabidopsis* Trehalose-6-P Synthase *AtTPS1* Gene Is a Regulator of Glucose, Abscisic Acid, and Stress Signaling. *Plant Physiology,* **136(3):**3649-3659.

Ayaz, F. A., Kadioglu, A. and Turgul, R. 2000. Water stress effects on the content of low molecular weight carbohydrates and phenolic acids in *Ctenanthe setasa* (Rosc.) Eichler. *Can. J. Plant Sci.*, **80:** 373-378.

Aziz, A., Martin-Tanguy, J. and Larher, F. 1998. Stress-induced changes in polyamine and tyramine levels can regulate proline accumulation in tomato leaf discs treated with sodium chloride. *Physiol Plant.*, **104:** 195–202.

Azzi, A., and Stocker, A. 2000. Vitamin E: non-antioxidant roles. *Progress in Lipid Research*, **39:** 231-255.

Baena-Gonzalez, E. and Sheen, J. 2008. Convergent energy and stress signaling. *Trends. Plant Sci.*, **13(9):** 474-482.

Baena-Gonzalez, E. 2010. Energy signaling in the regulation of gene expression during stress. *Mol. Plant,* **3(2):** 300-313.

Bahler, B.D., Steffen. K.L. and Orzolek. M.D. 1991. Morphological and biochemical comparison of a purple-leafed and a green-leafed pepper cultivar. *Hort. Science.*, **26:** 736.

Banu, M.N., Hoque, M.A., Watanabe-Sugimoto, M., Islam, M.M., Uraji, M., Matsuoka, K., Nakamura, Y., Murata, Y. 2010. Proline and glycinebetaine ameliorated NaCl stress via scavenging of hydrogen peroxide and methylglyoxal but not superoxide or nitric oxide in tobacco cultured cells. *Bioscience, biotechnology, and biochemistry.*, **74(10):** 2043-2049.

Bansal, K.C., Goel, D., Singh, A.K., Yadav, V., Babbar, S.B. and Murata, N. 2011. Transformation of tomato with a bacterial codA gene enhances tolerance to salt and water stresses. *Journal of Plant Physiology,* **168:** 1286-1294.

Barnett, N.M. and Naylor, A.W. 1966. Amino acid and protein metabolism in Bermuda grass during water stress. *Plant Physiol.*, **41(7):** 1222-1230.

Bartels, D. and Sunkar, R. 2005. Drought and salt tolerance in plants. *Crit Rev Plant Sci.*, **24:** 23–58.

Bartosz, G. 1997. Oxidative stress in plants. *Acta Physiol. Plant*, **19:** 47-64.

Basu, P.S., Ali, M. and Chaturvedi, S.K. 2007. Osmotic adjustment increases water uptake, remobilization of assimilates and maintains photosynthesis in chickpea under drought. *Indian Journal of Experimental Biology*, **45:** 261-267.

Bernard, Y. K. Binder., Christie, A. M. Peebles., Jacqueline, V. Shanks., and, San, Ka-Yiu. 2009. The effects of UV-B stress on the production of terpenoid indole alkaloids in *Catharanthus roseus* hairy roots. *Biotechnol. Prog.*,**25:** 861-865.

Bennett, R.N and Wallsgrove, R.M. 1994. Secondary metabolites in plant defence mechanisms. *New Phytol*, **127:** 617–33.

Berglund, A. H., Nilsson, R., and Liljenberg, C. 1999. Permeability of large unilamellar digalacto-syldiacylglycerol vesicles for protons and glucose-influence of α-tocopherol, β-carotene, zeaxanthin and cholesterol. *Plant Physiology and Biochemistry*, **37:** 179-186.

Bernillon, S., Biais, B., Deborde, C., Maucourt, M., Cabasson, C., Gibon, Y., Hansen, T.H., Husted, S., de Vos, C.H., Mumm, R., Jonker, H., Ward, J.L., Miller, S.J., Baker, J.M., Burger, J., Tadmor, Y., Beale, M.H., Schjoerring, J.K., Schäffer, A.A., Rolin, D., Hall, R.D. and Moing, A . 2013. Metabolomic and elemental profiling of melon fruit quality/as affected by genotype and environments. *Metabolmics*, **9:** 57-77.

Bianchi, G., Gamba, A., Limiroli, R., Pozzi, N., Elster, R., Salamini, F. and Bartels, D. 1993. The unusual sugar composition in leaves of the resurrection plant *Myrothamnus flabellifolia. Physiologia Plantarum,* **87:** 223-226.

Biehler, K. and Fock, H. 1996. Evidence for the contribution of the Mehler-peroxidase reaction in dissipating excess electrons in drought stressed wheat. *Plant Physiol.*, **112:** 265-272.

Biemelt, S., Keetman, U., Mock, H. and Grimm, B. 2000. Expression and activity of isoenzymes of superoxide dismutase in wheat roots in response to hypoxia and anoxia. *Plant Cell and Environment*, **23:** 135-144.

Blokhina, O. B., Fagerstedt, K. V. and Chirkova, T. V. 1999. Relationships between lipid peroxidation and anoxia tolerance in a range of species during post-anoxic reaeration. *Physiologia Plantarum*, **105:** 625-632.

Blokhina, O. B., Virolainen, E., Fagerstedt, K. V., Hoikkala, A., Wiihiilii, K. and Chirkova, T.V. 2000. Antioxidant status of anoxia-tolerant and -intolerant plant species under anoxia and reaeration. *Physiologia Plantarum*, **109:** 396-403.

Blokhina, O. B., Chirkova, T. V. and Fagerstedt, K. V. 2001. Anoxic stress leads to hydrogen peroxide formation in plant cells. *Journal of Experimental Botany*, **52:** 1-12.

Blokhina, O., Virolainen, E. and Fagerstedt, K.V. 2003. Antioxidants, oxidative damage and oxygen deprivation stress: a review. *Ann Bot.*, **91:** 179–194.

Bohnert, H.J. and Jensen, R.G. 1996. Strategies for engineering water-stress tolerance in plants. *Trends Biotechnol.*, **14:** 89–97.

Bolwell, G. P., and Wojtaszek, P. 1997. Mechanisms for generation of reactive oxygen species in plant defense-a broad perspective. *Physiological and Molecular Plant Pathology*, **51:** 347-366.

Borosani, O., Zhu, J., Verslues, P.E., Sunkar, R. and Zhu, J.K. 2005. Endogenous siRNA derived from a pair of natural cis-antisense transcripts regulate salt tolerance in *Arabidopsis. Cell*, **123(7):** 1279-1291.

Bouche, N., Fait, A., Bouchez, D., Moller, S.G. and Fromm, H. 2003. Mitochondrial succinic-semialdehyde dehydrogenase of the gamma-aminobutyrate shunt is required to restrict levels of reactive oxygen intermediates in plants. *Proceedings of the National Academy of Sciences*, USA **100:** 6843-6848.

Bouche, N. and Fromm, H. 2004. GABA in plants: just a metabolite? *Trends in Plant Science*, **9:** 110-115.

Bowler, C., Van Montagu, M., and Inze, D. 1992. Superoxide dismutase and stress tolerance. *Annu. Rev. Plant Physiol. Plant Mol. Biol.*, **43:** 83-116.

Bowne, J.B., Erwin, T.A., Juttner, J., Schnurbusch, T., Langridge, P. and Roessner, U. 2012. Drought responses of leaf tissues from wheat cultivars of differing drought tolerance at the metabolite level. *Mol. Plant,* **5(2):** 418-429..

Braidot, E., Petrussa, E., Vianello, A., and Macri, F. 1999. Hydrogen peroxide generation by higher plant mitochondria oxidizing complex I of complex II substrates. *FEBS Letters*, **451:** 347-350.

Bray, E.A. 1997, Plant responses to water deficit. *Trends Plant Sci*, **2:** 48–54.

Bray, E.A., Bailey-Serres, J. and Weretilnyk, E. 2000. Responses to abiotic stresses. In: "*Biochemistry and molecular biology of plants* ,"Eds. B.B. Buchanan, W. Gruissem, and R.L. Jones, American Society of Plant Physiologists, Rockville, pp 1158–1203.

Bray, E.A. 2002. Classification of genes differentially expressed during water deficit in *Arabidopsis thaliana*: An analysis using microarray and differential expression data. *Annals of Botany*, **89:** 803-811.

Bright, J., Desikan, R., Hancock, J.T., Weir, I.S. and Neill, S.J. 2006. ABA-induced NO generation and stomatal closure in Arabidopsis are dependent on H_2O_2 synthesis. *Plant J*, **45:** 113–122.

Bryant, J.P., Chapin. F.S.I. and Klein, D.R. 1983. Carbon/nutrient balance of boreal plants in relation to vertebrate herbivory. *Oikos*, **40:** 357–68.

Buettner, G. R. 1993. The pecking order of free radicals and antioxidants: lipid reroxidation, α-tocopherol, and ascorbate. *Archives of Biochemistry and Biophysics*, **300:** 535-543.

Caasi-Lit, M., Whitecross, M. I., Nayudu, M., Tanner, G. J. 1997. UV-B irradiation induces differential leaf damage ultrastructural changes and accumulation of specific phenolic compounds in rice cultivars. *Aust. J. Plant Physiol.*, **24:** 261–274.

Cai, H.M., Zhou, Y., Xiao, J.H., Li, X.H., Zhang, Q.F. and Lian, X.M. 2009. Overexpressed glutamine synthetase gene modifies nitrogen metabolism and abiotic stress responses in rice, *Plant Cell Reports*, **28(3):** 527–537.

Caldana, C., Degenkolbe, T., Cuadros-Inostroza, A., Klie, S., Sulpice, R., Leisse, A., Steinhauser, D., Fernie, A.R., Willmitzer, L. and Hannah, M.A. 2011. High density kinetic analysis of the metabolic and transcriptomic responses of *Arabidopsis* to eight environmental conditions. *Plant J.*, **67(5):** 869-884.

Capell, T., Bassie, L., Christou, P. 2004. Modulation of the polyamine biosynthetic pathway in transgenic rice confers tolerance to drought stress. *Proceedings of the National Academy of Sciences, USA,* **101:** 9909-9914.

Chaitanya, K.V., Rasineni, G.K. and Reddy, A.R. 2009. Biochemical responses to drought stress in mulberry (*Morus alba* L.): Evaluation of proline, glycine betaine and abscisic acid accumulation in five cultivars. *Acta Physiol. Plant.* **31:** 437–443.

Chamnongpol, S., Willekens, H., Langebartels, C., Van Montagu, M., Inze, D. and Van Camp, W. 1997. Transgenic tobacco with 8 reduced catalase activity develops necrotic lesions and induces pathogenesis-related proteins under high light. *Plant J.*, **10:** 491-503.

Chan, L.K., Koay, S.S., Boey, P.L. and Bhatt, A. 2010. Effects of abiotic stress on biomass and anthocyanin production in cell cultures of *Melastoma malabathricum.* Biol. Res., **43:** 127–135.

Chang, W. W. P., Huang, L., Shen, M., Webster, C., Burlingame, A. L., and Roberts, J. K. M. 2000. Patterns of protein synthesis and tolerance of anoxia in root tips of maize seedlings acclimated to a low-oxygen environment, and identification of proteins by mass spectrometry. *Plant Physiology*, **122:** 295-317.

Chalker-Scott, L. and Fnchigami, L.H. 1989. The role of phenolic compounds in plant stress responses, In, "*Low temperature stress physiology in crops*" Ed. H.L. Paul, CRC Press Inc., Boca Raton, Florida, pp. 40.

Chalker-Scott, L. 1999. Environmental significance of anthocyanins in plant stress responses. *Photochem Photobiol.* **70:** 1–9.

Chakraborty, U. and Tongden, C. 2005. Evaluation of heat acclimation and salicylic acid treatments as potent inducers of thermotolerance in *Cicer arietinum* L. *Curr Sci.*, **89:** 384–389.

Chaturvedi, V., Bartiss, A. and Wong, B. 1997. Expression of bacterial *mtlD* in *Saccharomyces cerevisiae* results in mannitol synthesis and protects a glycerol-defective mutant from high-salt and oxidative stress. *J. Bacterial*, **179:** 157-162.

Chen, T.H.H. and Murata, N. 2002. Enhancement of tolerance of abiotic stress by metabolic engineering of betaines and other compatible solutes. *Current Opinion in Plant Biology*, **5:** 250-257.

Chen, X.M., Hu, L., Lu, H., Liu, Q.L. and Jiang, X.N. 2005. Overexpression of *mtlD* gene in transgenic *Populus tomentosa* improves salt tolerance through accumulation of mannitol. *Tree Physiology,* **25:** 1273-1281.

Chen, T.H. and Murata, N. 2008. Glycinebetaine: an effective protectant against abiotic stress in plants. *Trends Plant Sci.*, **13:** 499–505.

Chen, S. and Polle, A. 2010. Salinity tolerance of Populus. *Plant Biology (Stuttgart)*, **12(2):** 317-333.

Chen, T.H. and Murata, N. 2011. Glycinebetaine protects plants against abiotic stress: mechanisms and biotechnological applications. *Plant Cell Environ.*, **34:** 1–20.

Chinnusamy, V., Zhu, J. and Zhu, J.K. 2006. Salt stress signaling and mechanism of plant salt tolerance. *Genetic Engineering*, **24:** 141-177.

Chinnusamy, V., Zhu, J., and Zhu, J.K. 2007. Cold stress regulation of gene expression in plants. *Trends in Plant Science*, **12:** 444-451.

Chirkova, T. V. 1988. Plant adaptation to hypoxia and anoxia. Leningrad: Leningrad State University Press.

Chirkova, T. V., Sinyutina, N. F., Blyudzin, Y. A., Barsky, I. E., and Smetannikova, S.V. 1989. Phospho-lipid fatty acids of root mitochondria and microsomes from rice and wheat seedlings exposed to aeration or anaerobiosis. *Russian Journal of Plant Physiology*, **36:** 126-134.

Chirkova, T. V., Zhukova, T. M., and Goncharova, N. N. 1991a. Method of determination of plant resistance to oxygen shortage. *Russian Journal of Plant Physiology*, **38:** 359-364.

Chirkova, T. V., Zhukova, T. M., and Goncharova, N. N. 1991b. Specificity of membrane permeability in wheat and rice seedlings under anoxia. *Fiziologiai Biokhimia KulturnuxRastenii,* **23:** 541-546.

Chirkova, T. V., Novitskaya, L. O., and Blokhina, O. B. 1998. Lipid peroxidation and antioxidant systems under anoxia in plants differing in their tolerance to oxygen deficiency. *Russian Journal of Plant Physiology*, **45:** 55-62.

Chirkova, T. V., and Voitzekovskaya, S. A. 1999. Plant protein synthesis under hypoxia and anoxia. *Uspekhi Sovremennoi. Biologii*, **119:** 178-189.

Cho, Y.H., Hong, J.W., Kim, E.C. and Yoo, S.D. 2012. Regulatory functions of SNF1 in stress-responsive gene expression and in plant growth and development. *Plant Physiol.*, **158(4):** 1955-1964.

Choudhary, N.L., Sairam, R.K. and Tyagi, A. 2005. Expression of Ä1-pyrroline-5-carboxylate synthetase gene during drought in rice (Oryza sativa L.), *Indian Journal of Biochemistry and Biophysics*, **42(6):** 366–370.

Christie, P.J., Alfenito, M.R. and Walbot, V. 1994. Impact of low-temperature stress on general phenylpropanoid and anthocyanin pathways: Enhancement of transcript abundance and anthocyanin pigmentation in maize seedlings. *Planta*, **194:** 541–549.

Choudhary, N.L., Sairam. R.K. and Tyagi, A. 2005. Expression of $Ä^1$-pyrroline-5-carboxylate synthetase gene during drought in rice (*Oryza sativa* L.) *Indian Journal of Biochemistry and Biophysics,* **42(6):**366–370.

Chu, T.M., Aspinall, D. and Paleg, L.G. 1974. Stress metabolism. VI. Temperature stress and the accumulation of proline in barley and radish. *Aust J Plant Physiol.*, **1:** 87–97.

Caldwell, M.M., Robberecht, R. and Flint, S.D. 1983. Protection against damaging oxidants using Flavanoids. *Physiol. Plant*, **58:** 445-453.

Close, T. J. 1997. Dehydrins: a commonality in the response of plants to dehydration and low temperature. *Physiol. Plant*, **100:** 291.296.

Collakova, E., Yen, J.Y. and Senger, R.S. 2012. Are we ready for genome scale modeling in plants? *Plant Sci.*, **191:** 53-70.

Cona, A., Rea, G., Angelini, R., Federico, R. and Tavladoraki, P. 2006. Functions of amine oxidases in plant development and defence. *Trends Plant Sci*, **11:** 80–88.

Conkin, P.L., Williams, E.H. and Last, R.L. 1996. Environmental stress sensitivity of an ascorbic acid-deficient *Arabidopsis* mutant. *Proc Natl Acad Sci U S A.* **93(18):** 9970–9974.

Conklin, P. L., Pallanca, J. E., Last, R. L. and Smirnoff, N. 1997. L-ascorbic acid metabolism in the ascorbate-deficient *Arabidopais* mutant *vtc1. Plant Physiol.*, **15:** 1277-1285.

Conley, T. R., Sharp, R. E., Walker, J. C. 1997. Water deficit stimulates the activity of a protein kinase in the elongation zone of the maize primary root. *Plant Physiol.*,**113:** 219-226.

Convello, P. S, Chang, A, Dumbroff, E. B. and Thompson, J. E. 1989. Inhibition of photosystem II precedes thylakoid membrane lipid peroxidation in bisulfite treated leaves of *Phaseolus vulgaris. Plant Physiol.*, **90:** 1492-1497.

Cramer, G.R., Ergül, A., Grimplet, J., Tillett, R.L., Tattersall, E.A., Bohlman, M.C., Vincent, D., Sonderegger, J., Evans, J., Osborne, C., Quilici, D., Schlauch, K.A., Schooley, D.A. and Cushman, J.C. 2007. Water and salinity stress in grapevines: early and late changes in transcript and metabolite profiles. *Functional and Integrative Genomics*, **7:** 111-134.

Crawford, R.M.M., Walton, J.C. and Wollenweber Ratzer, B.1994. Similarities between post ischaemic injury to animal tissues and post anoxic injury in plants. *Proceedings of the Royal Society of Edinburgh,* **102B:** 325–332.

Crawford, R. M. M., and Braendle, R. 1996. Oxygen deprivation stress in a changing environment. *Journal of Experimental Botany*, **47:** 145-159.

Cromwell, B.T. and Rennie, S.D. 1953. The biosynthesis and metabolism of betaines in plants. 1. The estimation and distribution of glycinebetaine (betaine) in *Beta vulgaris* L. and other plants. *Biochemical Journal*, **55:** 189-192.

Cuevas, J.C., Lopez-Cobollo, R., Alcázar, R., Zarza, X., Koncz, C., Altabella, T., Salinas, J., Tiburcio, A.F. and Ferrando, A. 2008. Putrescine is involved in Arabidopsis freezing tolerance and cold acclimation by regulating abscisic acid levels in response to low temperature. *Plant Physiol*, **148:** 1094–1105.

Cuevas, J.C., Lopez-Cobollo, R., Alcazar, R., Zarza, X., Koncz, C., Altabella, T., Salinas, J., Tiburcio, A.F. and Ferrando, A. 2009. Putrescine as a signal to modulate the indispensable ABA increase under cold stress. *Plant Signal Behav*, **4:** 219–220.

Dalton, D. A. Langeberg, L. and Treneman, N. 1993. Correlations between the ascorbate glutathione pathway and effectiveness in legume root nodules. *Physiol. Plant*, **87:** 365-370.

Davies, K. J. A. 1987. Protein damage and degradation by oxygen radicals. 1. General aspects. *J. Biol. Chem.*, **262:** 9895-9901.

DeTullio, M. C., Paciolla, C., Dalla Vecchia, F., Rascio, N., D'Emerico, S., De Gara, L., Liso, R. and Arrigoni, O. 1999. Changes in onion root development induced by the inhibition of peptidyl-prolyl hydroxylase and influence of the ascorbate system on cell division and elongation. *Planta*, **209:**424-434.

Deeb, F., van der Weele, C.M. and Wolniak, S.M. 2010. Spermidine is a morphogenetic determinant for cell fate specification in the male gametophyte of the water fern *Marsilea vestita. The Plant Cell,* **22:** 3678-3691.

Delano-Frier, J.P., Aviles-Arnaut, H., Casarrubias-Castillo, K., Casique-Arrovo, G., Castrillon-Arbelaez, P.A., Herrera-Estrella, L., Massange-Sanchez, J., Martinez-Gallardo, N.A., Parra-Cota, F.I., Vargas-Ortiz, E. and Estrada-Hernandez, M.G. 2011. Transcriptomic analysis of grain amaranth (*Amaranthus hypochondriacus*) using 454 pyrosequencing comparison with *A. tuberculatus* expression profiling in stems and in response to biotic and abiotic stress. *BMC Genomics*, **12:** 363.

Daneshmand, F., Arvin, M.J. and Kalantari, K.M. 2010. Physiological responses to NaCl stress in three wild species of potato *in vitro*. *Acta Physiol Plant.*, **32:** 91–101.

DeLong, J. M., and Steffen, K. L. 1998. Lipid peroxidation and α-tocopherol content in α-tocopherol-supplemented thylakoid membranes during UV-B exposure. *Environmental and Experimental Botany*, **39:** 177-185.

Demple, B. 1991. Regulation of bacterial oxidalive stress genes. *Ann. Rev. Genet.*, **25:** 315.337.

Deshnium, P., Gombos, Z., Nishiyama, Y. and Murata, N. 1997. The action of glycine betaine in enhancement of tolerance of *Synechococcus* sp. Strain PCC 7942 to low temperature. *J. Bacteriol*, **179:** 339-344.

Desikan, R., Cheung, M.K., Bright, J., Henson, D., Hancock, J.T. and Neill, S.J. 2004. ABA, hydrogen peroxide and nitric oxide signalling in stomatal guard cells. *J Exp Bot*, **55:** 205-212.

Dixon, R.A. and Paiva, N. 1995. Stressed induced phenyl propanoid metabolism. *Plant Cell.* **7:** 1085–97.

Dixon, J., Hull, M., Cobb, A. and Sanders, G. 1996. Phenmedipham-ozone pollution interactions in sugarbeet *(Beta vulgaris* L cv. Saxon): a physiological study. *Pestic. Sci.*,**46:** 381-390.

Deuschle, K., Funck, D., Forlani, G., Stransky, H., Biehl, A., Leister, D., van der Graff, E., Kunze, R. and Frommer, W.B. 2004. The role of (Delta)1-pyrroline-5-carboxylate dehydrogenase in proline degradation. *Plant Cell*, **161(2):** 3413-3250.

Dhaubhadel, S., Browning, K.S., Gallie, D.R. and Krishna, P. 2002. Brassin steroid functions to protect the translational machinery and heat shock protein synthesis following thermal stress. *Plant J.*, **29(6):** 681-691.

Dicosmo, F. and Misawa, M. 1985 Eliciting secondary metabolism in plant cell cultures. *Trends Biotechnol.* **3:** 318–22.

Dobra, J., Motyka, V., Dobrev, P., Malbeck, J., Prasil, I.T., Haisel, D., Gaudinova, A., Havlova, M., Gubis, J. and Vankova, R. 2010. Comparison of hormonal responses to heat, drought and combined stress in tobacco plants with elevated proline content. *Journal of Plant Physiology*, **167:** 1360-1370.

Dodge, A. D. 1971. The mode of action of the bipyridylium herbicides, paraquat and diqual. *Endeavour*, **30:** 130-135.

Doke, N., Miura, Y., Sanchez, L. and Kawakita, K. 1994. Involvement of superoxide in signal transduction responses to attack by pathogens, physical and chemical shocks and UV irradiation. In, "*Causes of photooxidative stress and amelioration of defense systems in plants*", Eds. C.I.I. Foyer, and P.M. Muliineaux, CRC press, Boca Raton, FL, ISBN 0-8493-5443-9. pp.. 1

Dong, H. L., and Chin, B. L. 2000. Chilling stress-induced changes of antioxidant enzymes in the leaves of cucumber: in gel enzyme activity assays. *Plant Science*, **159:** 75–85.

Draper, J., Lloyed, A.J., Goodcare, R. and Beckman, M. 2013. Flow infusion electrospray ionization mass spectrometry from high throughput non-targeted metabolite fingerprinting: a review. *Metabolomics*, **9(1):** 4-29.

Drew, M. C. 1997. Oxygen deficiency and root metabolism: injury and acclimation under hypoxia and anoxia. *Annual Review of Plant Physiology and Plant Molecular Biology*, **48:** 223-250.

Droiet, G., Dumbroff, E. B., Legge, R. L. and Thkopson, J. E. 1986. Radical scavenging properties of polyamines. *Phyochemistry,* **25:** 367-371.

Du, H.M., Wang, Z.L., Yu, W.J., Liu, Y.M., Huang, B.R. 2011. Differential metabolic responses of perennial grass *Cynodon transvaalensis* × *Cynodon dactylon* (C_4) and *Poa pratensis*(C_3) to heat stress. *Physiol. Plant.* **141:** 251–264.

Eastmond, P.J., van Dijken, A.J., Spielman, M., Kerr, A., Tissier, A.F., Dickinson, H.G., Jones, J.D., Smeekens, S.C. and Graham, I.A. 2002. Trehalose-6-phosphate synthase 1, which catalyses the first step in trehalose synthesis, is essential for Arabidopsis embryo maturation. *The Plant Journal*, **29:**225-235.

Edreva, A.M., Velikova. V. and Tsonev, T. 2000. Phenylamides in plants. *Russ J Plant Physiol.*, **54:** 287–301.

Egneus, H., Heber, U., Mathiesen , U., and Kirk, M. 1975. Reduction of oxygen by electron transport chain of chloroplasts during assimilation of carbon dioxide. *Biochem. Biophys Acta.*, **408:** 252-268.

Ellis, M. H., and Setter, T. L. 1999. Hypoxia induces anoxia tolerance in completely submerged rice seedlings. *Journal of Plant Physiology*, **154:** 219-230.

Ellis, M. H., Dennis, E. S, and Peacock, W. L. 1999 *Arabidopsis* roots and shoots have different mechanisms for hypoxic stress tolerance. *Plant Physiology*, **119:** 57-64.

El-Shabrawi, H., Kumar, B., Kaul, T., Reddy, M.K., Singla-Pareek, S.L. and Sopory, S.K. 2010. Redox homeostatis antioxidant defense and methylglyoxal detoxification as markers for salt tolerance in Pokkali rice. *Protoplasma*, **245(1-4):** 85-96.

Elstner, E. F. 1987. Metabolism of activated oxygen species. In, "*Biochemistry of plants*", Vol. 11, Ed. D.D. Davies, London: Academic Press, PP. 253-315.

Elstner, E. F., Wagner, G. A. and Schutz, W. 1988. Activated oxygen in green plants in relation to stress situations. *Curr. Topics Plant Biochem. Physiol.*, **7:** 159-187.

Elstner, E. F. 1991. Mechanisms of oxygen activation in different compartments of plant cells. In, "*Active oxygen: oxidative stress and plant metabolism*", Eds. E.J. Pell, and K.L. Steffen, *Am. Soc. Plant Physiol., Rockville*, MD, pp. 13–25.

Elstner, E. F., and Osswald, W. 1994. Mechanisms of oxygen activation during plant stress. *Proceedings of the Royal Society of Edinburgh*, **102B:** 131-154.

Empadinhas, N. and da Costa, M.S. 2008. Osmoadaptation mechanism in prokaryotes: distribution of compatible solutes. *Int. Microbiol.*, **11(3):** 151-161.

Ericson, J., Freudenberger, M and Boege, E. 1999. Population dynamics, migration, and the future of the Calakmul Biosphere Reserve. American Association for the Advancement of Science, Washington.

Escoubas, J. M., Lomas, M., LaRoche, J. and Falkowski, P. G. 1995. Light intensity regulation of *cab* gene expression is signaled by the redox state of the plastoquinone pool. *Proc. Nat. Acad. Sci., USA*, **92:** 10237-10241.

Espinoza, C., Degenkolbe, T., Caldana, C., Zuther, E., Leisse, A., Willmitzer, L., Hincha, D.K., and Hannah, M.A. 2010. Interaction with diurnal and circadian regulation results in dynamic metabolic and transcriptional changes during cold acclimation in *Arabidopsis. PLoS One,* **5:** e14101.

Erxleben, A., Gessler, A., Vervliet-Scheebaum, M. and Reski, R. 2012. Metabolite profiling of the moss *Physcomitrella* patterns reveals evolutionary conservation of osmoprotective substances. *Plant Cell Rep.*, **31(2):** 427-436.

Evans, P. J., Gallesi , D., Mathieu, C., Hernan der, M. J. De Felipe, M. R., Helliwell, B. and Puppo, A. 1999. Oxidative stress occurs during soybean nodule senescence. *Planta*, **208:** 73-79.

Fait, A., Fromm, H., Walter, D., Galili, G. and Fernie, A.R. 2008. Highway or byway: the metabolic role of the GABA shunt in plants. *Trends in Plant Science*, **13:** 14-19.

F.A.O. 2005. Management of irrigated-induced salt-affected soils. FAO, Rome.

F.A.O. 2007. Adaptation to climate change in agriculture, forestry and fisheries: Prospective frame work and priorities. Rome, FAO, IDWG on Climate Change.

F.A.O. 2011. The stae of the world's land and water resources for food and agriculture (SOLAW)–Managing systems at risk. FAO, Rome and Earthscan, London.

F.A.O. 2012. Water quality: agriculture as water polutatnt. FAO, Rome.

Fengel, D. and Wegener, G. 1984. Wood chemistry ultrastructure, reactions. Water de Gruyter, New York, NY. PP. 133-135.

Fernandez, O., Bethencourt, L., Quero, A., Sangwan, R.S. and Clement, C. 2010. Trehalose and plant stress responses friend or foe? *Trends Plant Science,* **15(7):** 409-417.

Fett-Neto, A.G., Pennington, J.J. and Di Cosmo, F. 1995. Effect of white light on taxol and baccatin III accumulation in cell cultures of *Taxus cuspidata* and *Zucc. J. Plant Physiol.*, **146:** 584–590.

Fischbach, R.J., Kossmann, B., Panten, H., Steinbrecher, R., Heller, W., Seidlitz, H.K., H. Sandermann, H., Hertkorn, N. and Schnitzler, J.-P. 1999. Seasonal accumulation of ultraviolet-B screening pigments in needles of Norway spruce (*Picea abies* (L.) Karst). *Plant Cell Environ.*, **22:** 27–37.

Flores, H.E. and Galston, A.W. 1982. Polyamines and plant stress—activation of putrescine biosynthesis by osmotic shock. *Science*, **217:**1259–1261.

Foyer, C. H., Furbank, R. T., Harbinson, J. and Horton, P. 1990. The mechanisms contributing to photosynthetic control of electron transport by carbon assimilation in leaves. *Photosynth. Res.*, **25:** 63-100.

Foyer, C. H., Desco Urvieres, P. and Kunert, K. J. 1994. Protection against oxygen radicals, an important defense mechanism studied in transgenic plants. *Cell and Environment*, **17:** 507-523.

Foyer, C. H., and Lelandais, M. A. 1996. A comparison of the relative rates of transport of ascorbate and glucose across the thylakoid, chloroplast and plasmalemma membranes of pea leaves mesophyll cells. *Journal of Plant Physiology*, **148:** 391-398.

Foyer, C. H., Lopez-Delgardo, H., Dat, J. F. and Scott, I. M. 1997. Hydrogen peroxideand glutathione- associated mechanisms of acclimatory stress tolerance and signalling, *Physiol. Plant*, **100:** 241-254.

Frenkel, C. 1991. Disruption of macromolecular hydration-a possible origin of chilling destabilisation of biopolymers. *Trends in Food Science and Technology*, **2:** 39-41.

Fridovich, I. 1986. Biological effects of the superoxide radical. *Arch. Biochem-Biophys.*, **247:** 1-11.

Fry, J., Huang, B. 2004. Applied turfgrass science and physiology. Wiley, Hoboken, NJ. Fryer, M. J. 1992. The antioxidant effects of thylakoid vitamin E (α- tocopherol). *Plant Cell & Environment*, **15:** 381-392.

Funck, D., Stadelhofer, B. and Koch, W. 2008. Ornithine-delta-aminotransferase is essential for arginine catabolism but not for proline biosynthesis. *BMC Plant Biology*, **8:** 40.

Fyfe, P., Cogdell, R. J., Hunter, C. N. and Jones, M. R. 1995. Study of the carotenoid binding pocket of the photosynthetic reaction center from the purple bacterium *Rhodobacter sphaeroides*. In, "*Photosynthesis from light to Biosphere*", Ed. P. Mathis, Vol. IV, *Proceeding of 10th International Photosynthesis Congress*, Montplleir, France, 20-25 August. pp. 47-50

Gagneul, D., Ainouche, A., Duhaze, C., Lugan, R., Larher, F.R. and Bouchereau, A. 2007. A reassessment of the function of the so-called compatible solutes in the halophytic Plumbaginaceae *Limonium latifolium*. *Plant Physiology*, **144:** 1598-1611.

Galliard, T. and Chan, H. W. S. 1980. In Biochemistry of plants. Vol., 4. Academic Press. New York. pp. 131.

Gao, M., Tao, R., Miura, K., Dandekar, A.M. and Sugiura, A. 2001. Transformation of Japanese persimmon (*Diospyros kaki* Thunb.) with apple cDNA encoding NADP-dependent sorbitol-6-phosphate dehydrogenase. *Plant Science*, **160:** 837-845.

Garcia, A. B., de Almeida Engler, J., Iyer, S., Gerats, T., Van Montagu, M. and Caplan, A. B. 1997. Effects of osmoproetectants upon NaCl stress in rice. *Plant Physiol.*, **115:** 159-169.

Garcia-Plazaola, J. I., and Becerril, J. M. 2000. Effects of drought on photoprotective mechanisms in European beech (*Fagus sylvatica* L.) seedlings from different provenances. *Trees*, **14:** 485-490.

Gracia, G.O., Ferreira, P.A., Miranda, G.V., Neves, J.C.L., Moraes, W.B. and Santos, D.B. 2007. Leaf content of cationinc macronutrients and their relationships with sodium in maize plants under saline stress. *Idesia*, **25(3):** 93-106.

Gardner, H. W., Eskins, K., Grams, G. W., and Inglett, G. E. 1972. Radical addition of linoleic hydroperoxides to α-tocopherol or the analogous hydroxychroman. *Lipids*, **7:** 324-334.

Gavaghan, C.L., Li, J.V., Hadfield, S.T., Hole, S., Nicholson, J.K., Wilson, I.D., Howe, P.W., Stanley, P.D. and Holmes, E. 2011. Application of NMR based Metabolomics to the investigation of salt stress in maize (*Zea mays*). *Phytochem. Ana.*, **22(3):** 214-224.

Ge, L.F., Chao, D.Y., Shi, M., Zhu, M.Z., Gao. J,P. and Lin, H.X. 2008. Overexpression of thc trehalose-6-phosphate phosphatase gene *OsTPP1* confers stress tolerance in rice and results in the activation of stress responsive genes. *Planta*, **228:** 191-201.

Geissler, N., Hussin, S. and Koyro, H.W. 2010. Elevated atmospheric CO_2 concentration enhances salinity tolerance in *Aster tripolium* L. *Planta*, **231:** 583–594.

Generosova, I. P., Krasavina, M. S., Polyakova, L. I., Burmistrova, N. A., Lyubomilova, M. Y. and Vartapetian, B.B. 1998. On some molecular aspects of adaptation of *Oryza sativa* seedlings to anoxia. *Russian Journal of Plant Physiology*, **45:** 227-233.

Gera, M.J., Kenneth, W.M. and Richard, B.T. 2007. Elevated temperatures increase leaf senescence and Root secondary metabolite concentrations in the understory herb *Panax quinquefolius* (Araliaceae). *Am. J Bot.*, **94:** 819-826,

Germino, M.J., Smith, W.K., 2000. High resistance to low-temperature photoinhibition in two alpine, snowbank species. *Physiol. Plant.*, **110:** 89–95.

Ghillebert, R., Swinnen, E., Wen, J., Vandesteene, L., Ramon, M., Noroa, K., Rolland, F. and Winderickx, J. 2011. The AMPK/SNF1 fuel gauge and energy regulator: structure, function and regulation. *FEEBS J.*, **278(21):** 3978-3990.

Gibon, Y., Bessieres, M. A. and Larher, F. 1997. Is glycine betaine a noncompatible solute in higher plants that do not accumulate it? *Plant Cell Env.*, **20:** 329-340.

Gille, L., and Nohl, H. 2001. The ubiquinollbc] redox couple regulates mitochondrial oxygen radical formation. *Archives of Biochemistry and Biophysics*, **388:** 34-38.

Gill, S.S. and Tuteja, N. 2010. Reactive oxygen species and antioxidant mechinary in abiotic stress tolerance in crop plants. *Plant Physiol. Biochem.*, **48(12):** 909-930.

GIPB – The Global Partnership Initiative for Plant Breeding Capacity Building, 2012. Available from http://km.fao.org/gipb/

Giri, J. 2011. Glycine betaine and abiotic stress tolerance in plants. *Pl. Signal Behav.*, **6(11):** 1746-1751.

Gleeson, D., Lelu-Walter, M-A. and Parkinson, M. 2005. Overproduction of proline in transgenic hybrid larch (*Lartix leptoeuropaea* (Dengler)) cultures renders them tolerant to cold, salt and frost. *Mol.Breed.*, **15(1):** 21-29.

Gleick, P.H. 2000. The World's Water 2000–2001. The Biennial Report on Freshwater Recources. Island Press, Washington.

Godde, D. and Buchhold, J. 1992. Effect of long term fumigation with ozone on the turnover of the D-1 reaction center polypeptide of photosystem II in spruce (*Picea abiers*). *Physiol. Plant*, **86:** 568-574.

Gong, Q., Li, P., Ma, S., Rupassara, S.I. and Bohnert, H.J. 2005. Salinity stress adaptation competence in the extremophile *Thellungiella halophila* in comparison with its relative *Arabidopsis thaliana*. *Plant Journal*, **44(5):** 826–839.

Gosal, S.S., Wani, S.H. and Kang, M.S. 2010. Water and Agricultural sustainability strategies. Ed. M.S. Kang. CRC Press, pp. .259.

Gout, E., Boisson, A. M., Aubert, S., Douce, R., and Bligny, R. 2001. Origin of the cytoplasmic pH changes during anaerobic stress in higher plant cells. Carbon[13] and phosphorus[31] nuclear magnetic resonance studies. *Plant Physiology*, **125:** 912-925.

Grace, S., and Logan, B. A. 2000. Energy dissipation and radical scavenging by the plant phenylpropanoid pathway. *Philosophical Transactions of the Royal Society of London B*, **355:** 1499-1510.

Grativol, C., Hemerly, A.S. and Ferreira, P.C. 2012. Genetic and epigenetic regulation of stress responses in natural plant populationa. *Biochim. Biophys. Acta.,* **1819(12):** 176-186.

Griffith, M. and Yaish, M.W.F. 2004. Antifreeze proteins in overwintering plants: a tale of two activities. *Trends Plant Sci.,* **9:** 399–405.

Grimes, H. D., Perkins, K. K. and Boss, W. F. 1983. Ozone degrades into hydroxyl radical under physiological conditions. A spin trapping study. *Plant Physiol.*, **72:** 1016-1020.

Groppa, M.D., Tomaro, M.L. and Benavides, M.P. 2001. Polyamines as protectors against cadmium or copper-induced oxidative damage in sunflower leaf discs. *Plant Sci.*, **161:** 481–488.

Groppa, M.D. and Benavides, M.P. 2008. Polyamines and abiotic stress: recent advances. *Amino Acids*, **34:** 35–45.

Guo, Y., Zhang, L., Ziao, G., Cao, S.Y., Gu, D. M., Tian, W. Z. and Chen, S. Y. 1997. Expression of betaine aldehyde deydrogenase gene and salinity tolerance in rice transgenic plants. *Sci. China Ser.*, **40:** 496-501.

Gupta, R., Huang, Y.F., Kieber, J. and Luan, S. 1998. Identification of a dualspecificity protein phosphatase that inactivates a MAP kinase from *Arabidopsis. Plant J*, **16:** 581–589.

Gupta, A.K. and Kaur, N. 2005. Sugar signalling and gene expression in relation to carbohydrate metabolism under abiotic stresses in plants. *Journal of Biosciences*, **30(5):** 761-776.

Guy, C., Kaplan, F., Kopka, J., Selbig, J. and Hincha, D.K. 2008. Metabolomics of temperature stress. *Physiologia Plantarum*, **132:** 220-235.

Gzik, A. 1996. Accumulation of proline and pattern of a amino acids in sugars beet plants in response to osmotic water and salt stress. *Environ. Exp. Bot.*, **36:** 29-38.

Hagimori, M., Matsumoto, T. and Obi, Y. 1982. Studies on the production of *Digitalis cardenolides* by plant tissue culture III. Effects of nutrients on digitoxin formation by shoot-forming cultures of *Digitalis purpurea* L. Grown in Liquid Media. *Plant Cell Physiol.*, **23:** 1205–1211.

Halliwell, B. 1987. Oxidative damage, lipid peroxidation and antioxidant protection in chloroplasts. *Chem. Phys. Lipids*, **44:** 327-340.

Hamerlynck, E.P., Huxman, T.E., Loik, M.E., Smith, S.D. 2000. Effects of extreme high temperature, drought and elevated CO_2 on photosynthesis of the Mojave Desert evergreen shrub, *Larrea tridentata*. *Plant Ecol.* **148:** 183–193.

Han, J., Danell, R.M., Patel, J.R., Gumerov, D.R., Scarlett, C.O., Speir J.P., Parker, C.E., Rusyn, I., Zeisel, S. and Borchers, C.H. 2008. Towards high-throughput ,etabolomics using ultra-high field Fourier transform ion cyclotron resonance mass spectrometry. *Metabolomics*, **4(2):** 128-140.

Han, J., Delta, R., Chan, S. and Borchers, C.H. 2009. Mass spectrometry based technologies for high throughput metabolomics. *Bioanalysis*, **1(9):** 1665-1684.

Handa, A. K., Bressan, R. A., Handa, A. K., Carpita, N.C. and Hasegawa, R.M. 1983. Solutes contributing to osmotic adjustment in cultured plant cells adapted to water stress. *Plant Physiol.*, **73:** 834-843.

Hanhijiirvi, A. M., and Fagerstedt, K. V. 1994. Comparison of the effect of natural and experimental anoxia on carbohydrate and energy metabolism in *Iris pseudacorus* rhizomes. *Physiologia Plantarum*, **90:** 437-444.

Hanhijiirvi, A. M., and Fagerstedt, K. V. 1995. Comparison of carbohydrate utilization and energy charge in the yellow flag iris *(Iris pseudacorus)* and garden iris *(Iris germanica)* under anoxia. *Physiologia Plantarum*, **93:** 493-497.

Hannah, M.A., Wiese, D., Freund, S., Fiehn, O., Heyer, A.G. and Hincha, D.K. 2006. Natural genetic variation of freezing tolerance in *Arabidopsis. Plant Physiology*, **142:** 98-112.

Haraguchi, H., Saito, T., Okamura, N. and Yagi, A. 1995. Inhibition of lipid peroxidation and superoxide generation by diterpenoids from *Rosmarinus officinalis*. *Planta Med.*, **61:** 333-336.

Hawrylak, B., Matraszek, R. and Szymanska, M. 2007. Response of lettuce (*Lactuca sativa* L.) to selenium in nutrient solution contaminated with nickel. *Veg Crop Res. Bull.*, **67:** 63.

Hauser, M.T., Aufsatz, W., Jonak, C. and Luschnig, C. 2011. Transgenerational epigenetic inheritance in plants. *Biochimica Biophysica Acta*, **1809:** 459-468.

Hay, S., Mayerhofer, H., Halford, N.G. and Dickinson, J.R. 2007. SNA sequences from *Arabidopsis* which encode protein kinase and function as upstream regulators of Snf1 in yeast. *J. Biol. Chem.*, **282(14):** 10472-10479.

Hayashi. H., Alia, Mustardy, L., Deshnium, P., Ida, M. and Murata, N. 1997. Transformation of *Arabidopsis thaliana* with the codA gene for choline oxidase: accumulation of glycine betaine and enhanced tolerance to salt and cold stress. *Plant J.*,**12:** 133-142.

He, X., Huang, W., Chen, W., Dong, T., Liu, C., Chen Z., Xu, S. and Ruan, Y. 2009. Changes of main secondary metabolites in leaves of *Ginkgo biloba* in response to ozone fumigation. *J. Environ. Sci.* (China), **21:** 199–203.

Heagle, A. S. 1989. Ozone and crop yield. *Annu. Rev. Phytopathol.*, **27:** 397-423.

Heath, R. L. 1987. The biochemistry of ozone attack on plasma membrane of plant cells. *Adv. Phytochem.*, **21:** 29-54.

Henzi, T., and Braendle, R. 1993. Long term survival of rhizomatous species under oxygen deprivation. In, "*Interacting stresses on plants in a changing climate*", Vol. 116, Ëds. M. Jackson and C.R. Black, NATO ASI Series, Berlin, Heidelberg: Springer- Verlag, pp. 305-314.

Hendry, G.A.F. 1993. Evolutionary origins and natural functions of fructans: a climatological, biogeogra-phy and mechanistic appraisal. *New Phytologist*, **123:** 3-14.

Hernandez, J. A., Corpas, F. J., Gomez, M. del Rio, L. A. and Sevilla, F. 1993. Saltinduced oxidative stress mediated by activated oxygen species in pea leaf mitochondria. *Physiol. Plant*, **89:** 103-10.

Hernandez, J. A., Ohnos, E., Corpas, F. J., Sevilla, L. and del Rio, L. A. 1995. Salt induced oxidative stress in chloroplasts of pea plants. *Plant Sci.*, **105:** 151-167.

Hernández, J.A., Jiménez, A., Mullineaux, P. and Sevilla, F. 2000. Tolerance of pea (*Pisum sativum* L.) to long-term salt stress is associated with induction of antioxidant defenses. *Plant, Cell and Environment*, **23:** 853-862.

Herrera-Rodríguez, M.B., Pérez-Vicente, R. and Maldonado, J.M. 2007. Expression of asparagine synthetase genes in sunflower (*Helianthus annuus*) under various environmental stresses, *Plant Physiology and Biochemistry*, **45(1):** 33–38.

Hidema, J., Kumagai, T., Sutherland, J. C. and Sutherland, B. M. 1997. Ultraviolet Bsensitive rice cultlvar deficient in cyclobutyl pyrimidine dimer repair. *Plant Physiol.*, **113:** 39-44.

Hoekstra, F.A. 2002. Essay 11.5: Role of respiration in desiccation tolerance. In, "Plant Physiology", 13thEd, Sinauer Associates Inc., Sunderland, MA, pp. 690.

Holcik, M. and Sonenberg, N. 2005. Translational control in stress and apoptosis. *Nat. Rev. Mol. Cell Biol.*, **6(4):** 318-327.

Hong, Z., Lekkineni, K., Zhang, Z. and Verma, D.P. 2000. Removal of feedback inhibition of delta(1)-pyrroline-5-carboxylate synthatase results in increased proline accumulation and protection of plants from osmotic stress. *Plat Physiiol.*, **122(4):** 1129-1136.

Hong-Sheng, L., Feng-min, L. and Xao, X. 2004. Deficiency of water can enhance root respiration rate of drought-sensitive but not in drought-tolerant spring wheat. *Agriculture Water Management: An International J.,* **64(1):** 41-48.

Hong, S.W. and Vierling, E. 2000. Mutants of Arabidopsis thaliana defective in the acquisition of tolerance to high temperature stress. *Proc Natl Acad Sci USA*, **97:** 4392–4397.

Horemans, N., Foyer, C. H., Potters, G., and Asard, H. 2000b. Ascorbate function and associated transport systems in plants. *Plant Physiology and Biochemistry*, **38(7-8):** 531-540.

Hua, X.J., Van de Cotte, B., Van Montagu, M. and Verbruggen, N. 2001. The 5# untranslated region of the At-P5R gene is involved in both transcriptional and post-transcriptional regulation. *Plant J.*, **26:** 157–169

Huang, Z., Long, X., Wang, L., Kang, J., Zhang, Z., Zed, R. and Liu, Z. 2012. Growth, photosynthesis and H^+-ATPase activity in two *Jerusalem artichoke* varieties under NaCl-induced stress. *Process Biochem.*, **47(4):** 591-596.

Huchzermeyer, B. and Koyro, H.W. 2005. Salt and drought stress effects on photosynthesis. In,"*Handbook of photosynthesis*", Ed. M. Pessarakli, 2nd edn. CRC, Boca Raton, pp 751–777.

Hummel, I., EI-Amrani, A., Gouesbet, G., Hennion, F. and Couee, I. 2004. Involvement of polyamines in the interacting effects of low temperature and mineral supply on *Pringlea antiscorbutica* (Kerguelen cabbage) seedlings. *J. Exp. Bot.*, **399:** 1125–1134.

Hussain, S.S., Kayani, M.A. and Amjad, M. 2011. Transcription factors as tools to engineer enhanced drought stress tolerance in plants. *Biotehnol Prog.*, **27(2):** 297-306.

Idso, C.D. and Idso, K.E. 2000. Forecasting world food supplies: The impact of rising atmospheric CO_2 concentration. *Technology*, **7:** 33–55.

Intergovernmental Panel on Climate Change. 2007. Climate change: Fourth assessment report. Cambridge University Press, London, UK.

Iordachescu, M. and Imai, R. 2008. Trehalose biosynthesis in response to abiotic stresses. *Journal of Integrative Plant Biology,* **50:** 1223-1229.

Ishikawa, T., Takahara, K. Hirabayashi, T. Matsumura, H., Fujisawa, S. Terauchi, R., Uchimiya, H. and Kawai-Yamada, M. 2010. Metablone analysis of response to oxidative stress in rice suspension cells over-expressing cell death suppressor Bax inhibitor. *Planr Cell Physiol.*, **51(1):** 9-20.

Jacobsen, S., Hauschild. M. and Rasmussen. U. 1992. Induction by chromium ion of chitinases and polyamines in barley (*Hordeum vulgare* L.) and rape (*Brassica napus* L ssp. oleifera) *Plant Sci.*, **84:** 119–128.

Jagtap, V. and Bhargava, S. 1995. Variation in the antioxidant metabolism of drought tolerant and drought susceptible varieties of *Sorghum bicolor* (L.) Moench. exposed to high light, low water and high temperature stress. *J. Plant Physiol.*, **145:**195-197.

Janiszowska, W., and Pennock, J. F. 1976. The biochemistry of vitamin E in plants. *Vitamins and Hormones*, **34:** 77-105.

Janska, A,, Marsik, P., Zelenkova, S., and Ovesna, J. 2010. Cold stress and acclimation: what is important for metabolic adjustment? *Plant Biology*, **12:** 395-405.

Janz, D., Behnke, K., Schnitzler, J.P., Kanawati, B., Schmitt-Kopplin, P. and Polle A. 2010. Pathway analysis of the transcriptome and metabolome of salt sensitive and tolerant poplar species reveals evolutionary adaption of stress tolerance mechanisms. *BMC Plant Bioogy,* **10:**150.

Jiang, M. and Zhang, J. 2002. Water stress-induced abscisic acid accumulation triggers the increased generation of reactive oxygen species ad up-regulates the activities of antioxidant enzymes in maize leaves. *Journal of Experimental Botany*, **53(379):** 2401-2410.

Jochum, G.M., Mudge, K.W. and Thomas, R.B. 2007. Elevated temperatures increase leaf senescence and root secondary metabolite concentration in the understory herb *Panax quinquefolius* (Araliaceae). *Am. J. Bot.*, **94:** 819-826,

Johnson, H.E., Broadhurst, D., Goodcare, R. and Smith, A.R. 2003. Metabolic fingerprinting of salt-stressed tomatoed. *Phytochem.*, **62(6):** 919-928.

Jones, H.G and Jones, M.B. 1989. Introduction: some terminology and common mechanisms. In "*Plants Under Stress*" Eds. H.G. Jones, T.J. Flowers, and M.B. Jones, Cambridge university Press, Cambridge, pp. 1–10.

Jozefckzak, M., Remans, T., Vangronsveld, J. and Cuypers, A. 2012. Glutathion is a key player in induced oxidative stress defences. *Int. J. Mol. Sci.*, **13(3):** 3145-3175.

Kagan, V. E. 1989. Tocopherol stabilizes membrane against phospholipase A, free fatty acids, and lysophospholipids. In, "*Vitamin E: biochemistry and health implications*", Vol. 570, Eds. A.T. Diplock, J. Machlin, L. Packer, W. Pryor, New York, NY: Annals of the New York Academy of Sciences, pp. 121-135.

Kagan, V. E., Fabisiak, J. P., and Quinn, P. J. 2000. Coenzyme Q and vitamin E need each other as antioxidants. *Protopslasma*, **214:** 11-18.

Kainulainen, P., Holopainen, J.K. and Holopainen, T. 1998. The infuence of elevated CO_2 and O_3 concentrations on Scots pine needles: changes in starch and secondary metabolites over three exposure years. *Oecologia*, **114:** 45560.

Kaiser, S., Di Mascio, P., Murphy, M.E. and Sies, H.1990. Physical and chemical scavenging of singlet molecular oxygen by tocopherols. *Archives of Biochemistry and Biophysics*, **277:** 101–108.

Kakegawa, K., Hattori, E., Koike, K. and Takeda, K. 1991. Induction of anthocyanin synthesis and related enzyme activities in cell cultures of *Centaurea cyanus* by UV-light irradiation. *Phytochemistry*, **30:** 2271–2273.

Kalashnikov, E., Balakhnina, T. I., and Zakrzhevsky, D. A. 1994. Effect of soil hypoxia on activation of oxygen and the system of protection from oxidative destruction in roots and leaves of *Hordeum vulgare. Russian Journal of Plant Physiology*, **41:** 583-588.

Kaplan, F., Kopka, J., Haskell, D.W., Zhao, W., Schiller, K.C., Gatzke, N., Sung, D.Y., Guy, C.L. 2004. Exploring the temperature-stress metabolome of *Arabidopsis*. *Plant Physiology*, **136:** 4159-4168.

Kaplan, F. and Guy, C.L. 2004. beta-Amylase induction and the protective role of maltose during temperature shock. *Plant Physiology*, **135:** 1674-1684.

Kamal-Eldin, A., and Appelqvist, L. 1996. The chemistry and antioxidant properties of tocopherols and tocotrienols. *Lipids*, **31:** 671-701.

Kar, R.K. 2011. Plant responses to water stress: role of reactive oxygen species. *Plant Signal Beha.*, **6(11):** 1741-1745.

Karpinski, S., Escobar, C., Karpinska, B., Creissen, G. and Mullineaux, P. M. 1997. Photosynthetic electron transport regulates the expression of cytosolic ascorbate peroxidase genes in *Arabidopsis* during excess light. *Plant Cell*, **9:** 627-640.

Karuppusamy, S. 2009. A review on trends in production of secondary metabolites from higher plants by in vitro tissue, organ and cell cultures. *J Med Plants Res*., **3:** 1222–1239.

Kasinathan, V. and Wingler, A. 2004. EVect of reduced arginine decarboxylase activity on salt tolerance and on polyamine formation during salt stress in *Arabidopsis thaliana. Physiol Plant*, **121:** 101–107.

Kasukabe, Y., He, L., Nada, K., Misawa, S., Ihara, I., Tachibana, S. 2004. Overexpression of spermidine synthase enhances tolerance to multiple environmental stresses and up-regulates the expression of various stress-regulated genes in transgenic *Arabidopsis thaliana. Plant and Cell Physiology*, **45:** 712-722.

Kato-Noguchi, H. 2000. Abscisic acid and hypoxic induction of anoxia tolerance in roots of lettuce seedlings. *Journal of Experimental Botany*, **51:** 1939-1944.

Kawakami, A., Sato, Y. and Yoshida M. 2008. Genetic engineering of rice capable of synthesizing fructans and enhancing chilling tolerance. *Journal of Experimental Botany*, **59:** 793-802.

Kempa, S., Krasensky, J., Dal Santo, S., Kopka, J. and Jonak, C. 2008. A central role of abscisic acid in stress-regulated carbohydrate metabolism. *Anals of* Botany, **97:** 479-495.

Khraiwesh, B., Zhu, J.K. and Zhu, J. 2011. Role of miRNAs and siRNAs in biotic and abiotic stress responses of plants. *Biochim Biophys Acta*, **1819:** 137-148.

Kishor, P.B.K., Hong, Z.L., Miao, G.H., Hu, C.A.A. and Verma, D.P.S. 1995. Overexpression of delta-pyrroline-5-carboxylate synthetase increases proline production and confers osmotolerance in transgenic plants. *Plant Physiol*., **108:** 1387–1394.

Kishor, P.B.K., Sangam, S., Amrutha, R.N., Laxmi, P.S., Naidu, K.R., Rao, K.R.S.S., Rao, S., Reddy, K.J., Theriappan, P.and Sreenivasulu, N. 2005. Regulation of proline biosynthesis, degradation, uptake and transport in higher plants: Its implications in plant growth and abiotic stress tolerance. *Curr. Sci.* **88:** 424–438.

Kim, J.K., Bamba, T., Harada, K., Fukusaki, E. and Kobayashi, A. 2007. Time-course metabolic priling in *Arabidopsisthaliana* cell cultures after salt stress treatment. *J. Expt. Bot*., **58(3):** 415-424.

Kim, H.K., Choi, Y.H. and Verpoorte, R. 2011. NMR-based plant metabolomics where do we stand, where do we go? *Trends Biotechnol,* **29(6):** 267-275.

Kirk, H., Cheng, D., Choi, Y., Vrieling, K. and Kinkhamer, P.G.L. 2012. Transgressive segregation of primary and secondary metabolites in F2 hybrids between *Jacobaea aquatic* and *J. vulgaris. Metabolomics*, **8:** 211-219.

Kirkham, M.B. 2011. Elevated carbon dioxide: Impact on soil and plant water relations. CRC Press, Boca Raton, FL Kirschbaum, M.U.F. 2004. Direct and indirect climate change effects on photosynthesis and transpiration. *Plant Biol*., **6:** 242–253.

Korn, M., Gartner, T., Erban, A., Kopka, J., Selbig, J. and Hincha, D.K. 2010. Predicting *Arabidopsis* freezing tolerance and heterosis in freezing tolerance from metabolite composition. *Molecular Plant,* **3:** 224-235.

Kotting, O., Kossmann, J., Zeeman, S.C. and Lloyd, J.R. 2010. Regulation of starch metabolism: the age of enlightenment? *Current Opinion in Plant Biology*, **13:** 321-329.

Kovács, Z., Simon-Sarkadi, L. Vashegyi, I. and Kocsy, G. 2011. Different accumulation of free amino acids during short- and long-term osmotic stress in wheat. *The Scientific World Journal*, 2012: Article ID 216521

Koyro, H.W. 2006. Effect of salinity on growth, photosynthesis, water relations and solute composition of the potential cash crop halophyte *Plantago coronopus* (L.). *Environ Exp Bot*., **56:** 136–146.

Kramer, G.F., Norman, H.A., Krizek, D.T. and Mirecki, R.M. 1991. Influence of UV-B radiation on polyamines, lipid peroxidation and membrane lipids in cucumber. *Phytochemistry*, **30:** 2101–2108.

Krasensky, J and Jonak, C. 2012. Drought, salt and temperature stress-induced metabolic rearrange ments and regulatory network. *J. Exp. Bot.*, **63(4):** 1593-1608.

Kreeb, K.H. 1996. Salzstreß. In, *Streß bei Pfl anzen*", Ëds. C. Brunold, A. Rüegsegger, and R Brändle, UTB, Bern, pp 149–172.

Krichevsky, A., Zaltsman, A., King, L. and Citovsky, V. 2012. Expression of complete metabolic pathways in transgenic plants. *Biotechnol Genet Eng Rev.,* **28:** 13.

Kruger, N.J., Masakapalli, S.K. and Ratcliff, R.G. 2012. Strategies for investigating the plant metabolic network with steady state metabolic flux analysis from an *Arabidopsis* cell culture and other systems. *J Expt Bot.,* **63(6):** 2309-2323.

Kruk, J., Schmid, G. H., and Strzalka, K. 2000. Interaction of á-tocopherol quinone, á-tocopherol and other prenyllipids with photosystem. *Plant Physiology and Biochemistry*, **38:** 271-279.

Krupa, Z., Baranowska, M. and Orzol, D. 1996. Can anthocyanins be considered as heavy metal stress indicator in higher plants? *Acta Physiol. Plant*, **18:** 147–151.

Kubo, A., Aono, M., Nakajima, N., Saji, H., Tanaka, K. and Kondo, N. 1999. Differential responses in activity of antioxidant enzymes to different environmental stresses in *Arabidopsis thaliana. J. Plant Res.*, **112:** 279-390.

Kueger, S., Steinhauser, D., Willmitzer, L. and Giayalisco, P. 2012. High-resoluion plant metabolomics from mass spectal features to metabolites and from whole-cell analysis to subcellular metabolite distribution. *Plant J.*, **70(1):** 39-50.

Kumar, V., Shriram ,V., Kishor, P.B.K., Jawali, N. and Shitole, M.G. 2010. Enhanced proline accumulation and salt stress tolerance of transgenic indica rice by over-expressing P5CSF129A gene. *Plant Biotechnol Rep.*, **4:** 37–48.

Kumriaa, R. and Rajam, M.V. 2002. Ornithine decarboxylase transgene in tobacco affects poly amines, *in vitro*-morphogenesis and response to salt stress. *Journal of Plant Physiology,* **159:** 933-990.

Kuppusamy, K., Walcher, C. and Nemhauser, J. 2009. Cross-regulatory mechanisms in hormone signaling. *Plant Mol Biol*, **69:** 375–381.

Kruk, J., Jemiola Rzeminska, M., Strzalka, K.1997. Plastoquinol and á tocopherol quinol are more active than ubiquinol and á tocopherol in inhibition of lipid peroxidation. *Chemistry and Physics of Lipids*, **87:** 73–80.

Kusano, T., Yamaguchi, K., Berberich, T., Takahashi, Y. 2007. The polyamine spermine rescues *Arabidopsis* from salinity and drought stresses. *Plant Signal Behav*, **2:** 2.

Kusano, T., Berberich, T., Tateda, C., Takahashi, Y. 2008. Polyamines:essential factors for growth and survival. *Planta*, **228:** 367–381.

Kuznetsov, V.V. and Shevyakova, N.I. 1997. Stress responses of tobacco cells to high temperature and salinity: proline accumulation and phosphorylation of polypeptides. *Physiol Plant.*, **100:** 320–326.

Kwon, S.Y., Jeong, Y.J., Lee, H.S., Kim, J.S., Cho, K.Y., Allen, R.D., Kwak, S.S. 2002. Enhanced tolerances of transgenic tobacco plants expressing both superoxide dismutase and ascorbate peroxidase in chloroplasts against methyl viologen-mediated oxidative stress. *Plant, Cell and Environment*, **25:** 873-882.

Laisk, A., Pfanz, V., Schramm, M. J. and Heber, U. 1988a. Sulfur dioxide fluxes into different cellular compartments of leaves photosynthesizing in a polluted atmosphere .I. Computer analysis. *Planta*, **173:** 230-240.

Laisk, A., Pfanz, H. and Heber, U. 1988b. Sulfur dioxide fluxes into different cellular compartments of leaves photosynthesizing in a polluted atmosphere. II. Consequences of SO_2 uptake as revealed by computer analysis. *Planta*, **173:** 241-252.

Lal, S. K., Lee, C., and Sachs, M. M. 1998. Differential regulation of enolase during anaerobiosis in maize. *Plant Physiology*, **118:** 1285-1293.

Lamb, C., and Dixon, R. A. 1997. The oxidative burst in plant disease resistance. *Annual Review of Plant Physiology and Plant Molecular Biology*, **48:** 251-275.

Lamont, B. B. and Lamont, H. C. 2000. Utilizable water in leaves of 8 arid species as derived from pressure–volume curves and chlorophyll fluorescence. *Physiol. Plant.*, **110:** 64–67.

Landry, L. G. and Pell, E. J. 1993. Modification of Rubisco and altered proteolytic activity in O_3 stressed hybrid popler (*Populus maximowizil trichocarpa*). *Plant Physiol.*, **101:** 1355-1362.

Landry, L. G., Stapleton, A. E., Um, J., Hoftman, P., Hays, J. B., Walbot, V. and Last, R. L. 1997. An Arabidopsis photolyase mutant is hypersensitive to ultraviolet-B radiation. *Proc. Nat. Acad. Sci., USA,* **94:** 328-332.

Larcher, W. 2001. Ökophysiologie der Pflanzen: Leben, Leistung und Stressbewältigung der Pflanzen in ihrer Umwelt. Ulmer, Stuttgart.

Larkindale, J., Hall, J.D., Knight, M.R. and Vierling, E. 2005. Heat stress phenotypes of *Arabidopsis* mutants implicate multiple signaling pathways in the acquisition of thermotolerance. *Plant Physiol.*, **138:** 882–897.

Larsson, S., Wiren, A., Ericsson, T. and Lundgren, L. 1986. Effects of light and nutrient stress on defensive chemistry and susceptibility to *Galerucella lineola* (*Coleoptera*, *Chrysomelidae*) in two Salix species. *Oikos*, **47:** 205–210.

Larson, R. A. 1988. The antioxidants of higher plants. *Phytochemistry*, **27:** 969-978.

Lavola, A., Aphalo, P.J., Lahti, M. and Julkunen-Tiitto, R. 2003. Nutrient availability and the effect of increasing UV-B radiation on secondary plant compounds in Scots pine. *Environ. Exp. Bot.*, **49:** 49–60.

Leakey, A.D.B., Uribelarrea, M., Ainsworth, E.A., Naidu, S.L., Rogers, A., Ort, D.R., Long, S.P. 2006. Photosynthesis, productivity, and yield of maize are not affected by open-air elevation of CO_2 concentration in the absence of drought. *Plant Physiol*, **140:** 779–790.

Lei, X.Y., Zhu, R.Y., Zhang, G.Y. and Dai, Y.R. 2004. Attenuation of cold induced apoptosis by exogenous melatonin in carrot suspension cells: the possible involvement of polyamines. *J. Pineal Res.*, **36:**126–131.

Leipner, J., Fracheboud, Y., and Stamp, P. 1999. Effect of growing season on the photosynthetic apparatus and leaf antioxidative defenses in two maize genotypes of different chilling tolerance. *Environmental and Experimental Botany*, **42:** 129-139.

Leipner, J., Stamp, P., and Fracheboud, Y. 2000. Artificially increased ascorbate content affects zeaxanthin formation but not thermal energy dissipation or degradation of antioxidants during coldinduced photooxidative stress in maize leaves. *Planta*, **210:** 964-969.

Lehmann, M., Schwarzländer, M., Obata, T., Sirikantaramas, S., Burow, M., Olsen, C.E., Tohge, T., Fricker, M.D., Møller, B.L., Fernie, A.R., Sweetlove, L.J., Laxa, M. 2009. The metabolic responses of *Arabidopsis* roots to oxidative stress is distint from that of heterotrophic cells in culture and highlight a complex relationship between levels of transcripts metabolites, and flux. *Mol. Pl.*, **2(3):** 390-406.

Levitt, J. 1980. Responses of plants to environment stress: chilling, freezing and high temperature stress, 2nd ed., Academic Press, NY.

Li, H.J., Yang, A.F., Zhang, X.C., Gao, F. and Zhang, J.R. 2007. Improving freezing tolerance o transgenic tobacco expressing sucrose: sucrose 1-fructosyltransferase gene from *Lactuca sativa. Cell Tisssue and Organ Culture*, **89:** 37-48

Liang, B,, Huang, X., Zhang, G., Zhang, F. and Zhou, Q. 2006. Effect of lanthanum on plants under supplementary ultraviolet-B radiation: Effect of lanthanum on flavonoid contents in Soybean seedlings exposed to supplementary ultraviolet-B radiation. *J. Rare Earths*, **24:** 613–616.

Loiacono, F.V. and DeTullio, M.C. 2012. Why we should stop interfering simple correlations between antioxidants and plant stress resistance: towards the antioxidomic era. *OMICS,* **16(4):** 160-167.

Lugan, R., Niogret, M.F., Leport, L., Guegan, J.P., Larher, F.R., Savoure, A., Kopka, J. and Bouchereau A. 2010. Metabolome and water homeostasis analysis of *Thellungiella salsuginea* suggests that dehydration tolerance is a key response to osmotic stress in this halophyte. *The Plant Journal*, **64:** 215-229.

Li, H.J., Yang, A.F., Zhang, X.C., Gao, F. and Zhang, J.R. 2007. Improving freezing tolerance of transgenic tobacco expressing sucrose: sucrose 1 fructosyltransferase gene from *Lactuca sativa. Cell Tisssue and Organ Culture*, **89:** 37-48.

Li, H.W., Zang, B.S., Deng, X.W. and Wang, X.P. 2011. Over-expression of the trehalose-6-phosphate synthase gene OsTPS1 enhances abiotic stress tolerance in rice. *Planta*, **234:** 1007-1018.

Lin, C.C. and Kao, C.H. 1999. Excess copper induces an accumulation of putrescine in rice leaves. *Bot Bull Acad Sin.*, **40:** 213–218.

Liso, R., Innocenti, A. M., Bitonti, M. B. and Arrigoni, O. 1988. Ascorbic acid induced progression of quiescent centre cells from G1 to S phase. *New Phytologist*, **110:** 469-471.

Liu, L., Dennis, C., Gitz, I,I.I., Jerry, W. and McClure. 1995. Effects of UV-B on flavonoids, ferulic acid, growth and photosynthesis in barley primary leaves. *Physiol. Plant.*, **93:** 734–738.

Liu, K., Fu, H., Bei, Q. and Luan, S. 2000. Inward potassium channel in guard cells as a target for polyamine regulation of stomatal movements. *Plant Physiol*, **124:** 1315–1326.

Liu, C., Guo, C., Wang, Y. and Ouyang, F. 2002. Effect of light irradiation on hairy root growth and artemisinin biosynthesis of *Artemisia annua* L. *Process Biochem.*, **38:** 581–585.

Liu, C., Zhao, L. and Yu, G. 2011. The dominant glutamic acid metabolic flux to produce gamma-amino butyric acid over proline in *Nicotiana tabacum* leaves under water stress relates to its significant role in antioxidant activity. *Journal of Integrative Plant Biology*, **53:** 608-618.

Livingston, D.P., Hincha, D.K., Heyer, A.G. 2009. Fructan and its relationship to abiotic stress tolerance in plants. *Cellular and Molecular Life Sciences*, **66:** 2007-2023.

Lutz, C., Steiger, A. and Godde, D. 1992. Influence of air pollutants and nutrient deficiency on D-1 protein content and photosynthesis in young spruce trees. *Physiol. Plant.*, **85:** 611-617.

Lv, W.T., Lin, B., Zhang, M. and Hua, X.J. 2011. Proline accumulation is inhibitory to *Arabidopsis* seedlings during heat stress. *Plant Physiology*, **156:** 1921-1933.

Lynch, D. V., and Thompson, J. E. 1984. Lipoxygenase mediated production of superoxide anion in senescing plant tissue. *FEBS Lett.*, **173:** 251-254.

Mackay, E., Senaratna, T., McKersie, B. and Fletcher, R. 1987. Ozone induced injury to cellular membranes in *Triticum aestivum* L. and protection by the triazole S-3307. *Plant Cell Physiol.*, **28:**1271-1278.

Mahajan, S. and Tuteja, N. 2005. Cold, salinity and drought stresses: An overview. *Arch Biochem. Biophys.*, **444:** 139–58.

Majee, M., Maitra, S., Dastidar, K.G., Pattnaik, S., Chatterjee, A., Hait, N.C., Das. K,P. and Majumdar, A. L. 2004. A novel salt-tolerant L-myo-inositol-1-phosphate synthase from *Porteresia coarctata* (Roxb.) Tateoka, a halophytic wild rice: molecular cloning, bacterial overexpression, characterization, and functional introgression into tobacco-conferring salt tolerance phenotype. *Journal of Biological Chemistry*, **279:** 28539-28552.

Makled, M. M. 1995. Physiological studies on some species of unicellular green algae with special reference to protein production. M. Sc. Thesis, Menoufia Univ., Menoufia, Egypt.

Marschner, H. 1995. Mineral nutrition of higher plants, Academic Press, London. Pp. 889.

Martinelli, T., Whittaker, A., Bochicchio, A., Vazzana, C., Suzuki, A. and Masclaux-Daubresse, C. 2007. Amino acid pattern and glutamate metabolism during dehydration stress in the "resurrection" plant *Sporobolus stapfianus*: a comparison between desiccation-sensitive and desiccation-tolerant leaves, *Journal of Experimental Botany*, **58(11):** 3037–3046.

Massacci, A.M., Nabiev, L., Pietrosanti, S.K., Nematov. T.N., Chernikova, K. and Thor Leipner. J. 2008. Response of the photosynthetic apparatus of cotton (*Gossypium hirsutum*) to the onset of drought stress under field conditions studied by gas-exchange analysis and chlorophyll fluorescence imaging. *Plant Physiol Biochem.*, **46:** 189–195.

Matamoros, M. A., Baird, L. M., Escuredo, P. R., Dalton, D. A., Minchin, F. R., Iturbe-Ormaetxe, I., Rubio, M., Moran, J., Gordon, A. and Becana, M. 1999. Stress-induced legume root nodule senescence. Physiological biochemical, and structural alterations. *Plant Physiol.*, **121:** 97-111.

Mattioli, R., Marchese, D., D'Angeli, S., Altamura, M.M., Costantino, P. and Trovato, M. 2006. Modulation of intracellular proline levels affects flowering time and inflorescence architecture in *Arabidopsis. Plant Molecular Biology*, **66:** 277-288.

Mattioli, R., Costantino, P. and Trovato, M. 2008. Proline accumulation in plants: not only stress. *Plant Signal Behav.*, **4(11):** 1016-1018.

Mattioli, R., Falasca, G., Sabatini, S., Altamura, M.M., Costantino, P. and Trovato, M. 2009. The proline biosynthetic genes *P5CS1* and *P5CS2* play overlapping roles in Arabidopsis flower transition but not in embryo development. *Physiologia Plantarum*, **137:** 72-85.

Maurel, C. 1997. Aquaporins and water permeability of plant membranes. *Ann. Rev. Plant Physiol. Plant Mol. Biol.*, **48:** 399-429.

Mayer, R.R., Cherry, J.H., Rhodes, D. 1990. Effects of heat shock on amino acid metabolism of cowpea cells. *Plant Physiol.* **94:** 796–810.

McAinsh, M. R., Clayton, H., Mansfield, T. A., Hetherington A. M. 1996. Changes in stomatal behavior and guard cell cytosolic free calcium in response to oxidative stress. *Plant Physiol.*, **111:** 1031-1042.

Mehlhorn, H. 1990. Ethylene promoted ascorbate peroxidase activity protects plants against hydrogen peroxide, ozone and paraquat. *Plant Cell Environ.*, **13:** 971-976.

Merewitz, E.B., Du, H., Yu, W., Liu, Y., Gianfagna, T., Huang, B. 2012. Elevated cytokinin content in *ipt* transgenic creeping bentgrass promotes drought tolerance through regulating metabolite accumulation. *J. Expt. Bot.* **63:**1315–1328.

Mercer, P. and Ruddock, A. 1998. Evaluation of azoxystrobin and range of conventional fungicides on yield, *Septaria tritici* and senescence in winter wheat. *Test-Agrochem-Cultiv.*, **19:** 24-25.

Merzlyak, M. N. 1989. Activated oxygen and oxidative processes in plant cell membranes. Itogi Nauki Tekhhiki, Ser. *Plant Physiology*, **6:** 11-67.

Miller, G., Honig, A., Stein, H., Suzuki, N., Mittler, R. and Zilberstein, A. 2009. Unraveling delta1-pyrroline-5-carboxylate-proline cycle in plants by uncoupled expression of proline oxidation enzymes. *J Biol Chem.*, **284:** 26482–26492.

Mintz-Oron, S., Meir, S., Malitsky, S., Ruppin, E., Aharoni, A. and Shlomi, T. 2012. Reconstruction of *Arabidopsis* metabolic network accounting subcellular compartmentization and tissue specificity. *Proc. Nat. Acad. Sci. USA.*, **109(1):** 339-344.

Mittler, R. 2006. Abiotc stress: the field environment and stress combination. *Trends Plant Sci.*, **11(1):** 15-19.

Mizukami, H., Konoshima, M. and Tabata, M. 1977. Effect of nutritional factors on shikonin derivative formation in Lithospermum callus cultures. *Phytochemistry*, **16:** 1183–1186.

Mo, Y., Nagel, C. and Taylor, L. P. 1992. Biochemical complementation of chalcone synthetase mutants defines a role for flavonols in functional pollen. *Proc. Natt. Acad. Sci.*, **89:** 7213-7218.

Moench, M. 2002. Water and the potential for social instability: livelihoods, migration and the building of society. *Nat Resour Forum*, **26:** 195–204.

Moller, M. 2001. Plant mitochondria and oxidative stress: electron transport, NADPH turnover, and metabolism of reactive oxygen species. *Annual Review of Plant Physiology and Plant Molecular Biology*, **52:** 561-591.

Moller, I.M. and Sweetive, L.J. 2010. ROS signaling-specificity is required. *Trends Plant Sci.*, **15(7):** 370-374.

Molinari, H.B.C., Marur, C.J., Daros, E., de Campos, M.K.F., de Carvalho, J.F.R.P., Filho JCB, Pereira, L.F.P. and Vicira, L.G.E. 2007. Evaluation of the stress-inducible production of proline in transgenic sugarcane (*Saccharum spp.*): osmotic adjustment, chlorophyll fluorescence and oxidative stress. *Physiol Plant.*, **130:** 218–229.

Monk, L. S., Fagerstedt, K. V., and Crawford, R. 1987. Superoxide dismutase as an anaerobic polypeptide-a key factor in recovery from oxygen deprivation in *Iris pseudacorus? Plant Physiology*, **85:** 1016-1020.

Moran, J. F., Klucas, R. V., Grayer, R. J., Abian, J. and Becana, M. 1997. Complexes of iron with phenolic compounds from soybean nodules and other legume tissues: Prooxidant and antioxidant properties. *Free Radic. Biol. Med.* **22:** 861-870.

Moreau, S., Puppo, A. and Davies, M. J. 1995. The reactivity of ascorbate with different redox states of leg hemoglobin. *Phytochemistry*, **39:** 1281-1286.

Mullineaux, P. M. and Creissen, G. P. 1997. Glutathione reductase regulation and role in oxidative stress. In, "*Oxidative stress and the molecular biology of antioxidant defenses*", Ed. J.G. Scandalios, Cold Spring Harbor. Laboratory Press, NY.

Munne-Bosch, S., Schwarz, K. and Alegre, L. 1999. Enhanced formation of α-tocopherol and highly oxidized a bietane diterpenes in water stressed rosemary plants. *Plants Physiol.*, **121:** 1047-1052.

Munne-Bosch, S., and Alegre, L. 2000a. Changes in carotenoids, tocopherols and diterpenes during drought and recovery, and the biological significance of chlorophyll loss in *Rosmarinus officinalis* plants. *Planta*, **210:** 925-931.

Munne-Bosch, S., and Alegre, L. 2000b. The significance of β-carotene, α-tocopherol and the xanthophyll cycle in droughted *Melissa officinalis* plants. *Australian Journal of Plant Physiology*, **27:** 139-146.

Munne-Bosch, S., Muller, M., Schwarz, K. and Alegre, L. 2001. Diterpenes and antioxidative protection in drought-stressed *Saliva officinalis* plants. *J. Plant Physiol.*, 158: 1431-1437.

Munne-Bosch, S. and Alegre, L. 2001a. Function of tocopherols and tocotrienols in plants. *Crit. Rev. Plant Sci.*, **20:** 2332-2339.

Munne-Bosch, S., and Alegre, L. 2001b. Subcellular compartmentation of the diterpene carnosic acid and its derivatives in the leaves of rosemary. *Plant Physiology*, **125:** 1094-1102.

Munns, R., and Tester, M. 2008. Mechanisms of salinity tolerance. *Annual Review of Plant Biology*, **59:** 651-681.

Muthukumarasamy, M., Gupta, S.D. and Pannerselvam, R. 2000. Enhancement of peroxidase, olyphenol oxidase and superoxide dismutase activities by tridimefon in NaCl stressed *Raphanus sativus* L. Biol. Plant, **43:** 317–320.

Mutlu, F. and Bozcuk, S. 2007. Salinity-induced changes of free and bound polyamine levels in sunflower (*Helianthus annuus* l.) roots differing in salt tolerance. *Pak J Bot.* **39:** 1097–1102.

Nadeau, P., Delaney, S. and Chouinard, L. 1987. Effects of cold hardening on the regulation of polyamine levels in wheat (*Triticum aestivum* L.) and Alfalfa (*Medicago sativa* L.) *Plant Physiol.* **84:** 73–77

Nadella, K.D., Maria, S.S. and Kumar, P.A. 2012. Metabolomics in agriculture. *OMICS*, **16(4):** 149-159.

Nakamura, T., Yokata, S., Muramoto, Y., Tsutsui, K., Oguri, Y., Fukui, K. and Takabe, T. 1997. Expression of a betaine aldehyde dehydrogenase gene in rice, a glycinebetaine nonaccumulator, and possible localization of its protein in peroxisomes. *Plant J.*, **11:** 1115-1120.

Nanjo, T., Kobayashi, M., Yoshiba, Y., Kakubari, Y., Yamaguchi-Shinozaki, K. and Shinozaki, K. 1999a. Antisense suppression of proline degradation improves tolerance to freezing and salinity in *Arabidopsis thaliana. FEBS Lett.*, **461:** 205–210.

Nanjo, T., Kobayashi, M., Yoshiba, Y., Sanada, Y., Wada, K., Tsukaya, H., Kakubari, Y., Yamaguchi-Shinozaki, K. and Shinozaki, K. 1999b. Biological functions of proline in morphogenesis and osmotolerance revealed in antisense transgenic *Arabidopsis thaliana. Plant J.*, **18:** 185–193.

Navari-Izzo, F., Quratacci, M. F., Izzo, R. and Pinzino, C. 1992. Degradation of membrane lipid components and antioxidant levels in *Horedeum vulgare* exposed to long -term fumigation with SO_2. *Physiol. Plant,* **84:** 73-79.

Navarro, J.M., Flores, P., Garrido, C. and Martinez, V. 2006. Changes in the contents of antioxidant compounds in pepper fruits at ripening stages, as affected by salinity. *Food Chem.*, **96:** 66–73.

Neill, S., Desikan, R. and Hancock, J. 2002. Hydrogen peroxide signaling. *Current Opinion in Plant Biology*, **5(5):** 388-395.

Netto, L.E.S. 2001. Oxidative stress response in sugarcane. *Genetics and Molecular Biology.* **24(1-4):** 93-102.

Nishizawa, A., Yabuta, Y. and Shigeoka, S. 2008. Galactinol and raffinose constitute a novel function to protect plants from oxidative damage. *Plant Physiology*, **147:** 1251-1263.

Noctor, G. and Foyer, C.H. 1998. Ascorbate and glutathione: Keeping active oxygen under control. *Annual Review of Plant Physiology and Plant Molecular Biology*, **49:** 249-279.

Oertii, J. J. 1987. Exogenous application of vitamins as regulators for growth and development of plants, a review. *Z. Pflanzenemahr Bodenk,* **150:** 375-391.

Ogawa, A. and Yamauchi, A. 2006. Root osmotic adjustment under osmotic stress in maize seedlings 2. Mode of accumulation of several solutes for osmotic adjustment in the root, *Plant Production Science*, **9(1):** 39–46.

Ohlsson, A.B. and Berglund, T. 1989. Effect of high $MnSO_4$ levels on cardenolide accumulation by *Digitalis lanata* tissue cultures in light and darkness. *J. Plant Physiol.*, **135:** 505–507.

Obrenovic, S. 1990. Effect of Cu (11) D-penicillanine on phytochrome mediated betacyanin formation in*Amaranthus caudatus* seedlings. *Plant Physiol. Biochem.*, **28:** 639–646.

O'Grady, J., Schwender, J., Shachar-Hill, Y. and Morgan, J.A. 2012. Metabolic cartographic: experimental quantification of metabolic fluxes from isotopic labeling studies. *J Expt Bot.*, **63(6):** 2293-2308.

O'Kane, D., Gill, V., Boyd, A. and Burdon, R. 1996. Chilling, oxidative stress and antioxidant responses in *Arabidopsis thaliana* callus. *Planta*, **198:** 371–377.

Okazaki, Y. and Saito, K. 2012. Recent advances of metabolomics in plant biotechnology. *Plant Biotechnol Rep.*, **6(1):** 1-15.

Olien, C.R. and Clark, J.L. 1995. Freeze-induced changes in carbohydrates associated with hardiness of barley and rye. *Crop Science*, **35(2):** 496-502.

Olga, B., Eija, V. and Kurt, V. F. 2003. Antioxidants, oxidative damage and oxygen deprivation stress: a review. *Annals of Botany*, **91:** 179-194.

Ollerenshaw, J.H. and Lyons, T. 1999. Impacts of ozone on the growth and yield of field grown winter wheat. *Environ Pollut.,* **106:** 67–72.

Oncel, I., Keles, Y., Ustun, A. S., 2000. Interactive effects of temperature and heavy metal stress on the growth and some biochemical compounds in wheat seedlings. *Environ. Pollut.*, **107:** 315–320.

Oncel, I., Yurdakulol, E., Keles, Y., Kurt, L., and Yyldyz, A. 2004. Role of antioxidant defense system and biochemical adaptation on stress tolerance of high mountain and steppe plants. *Acta Oecologica*, **26:** 211-218.

Ormord, D. P. and Beckerson, A. 1986. Polyamines as antioxidants for tomato. *Hort. Sci.*, **21:** 1070-1071.

Orthen, B., Popp, M. and Smirnoff, N. 1994. Hydroxyl radical scavenging properties of cyclitols. *Proc. Royal Soc. Edin.*, **1026:** 269-272.

Ostnrk, Z.N., Talamì V., Deyholos, M., Michalowski, C.B., Galbraith, D.W., Gozukirmizi, N., Tuberosa, R. and Bohnert, H.J. 2002. Monitoring large-scale in transcript abundance in drought-and-salt-stressed barley. *Pl. Mol. Biol.*,**48:** 551-573.

Orvar, B. L. and Ellis, B. E. 1997. Transgenic tobacco plants expressing antisense RNA for cytosolic ascorbate peroxidase show increased susceptibility to ozone injury. *Plant J.*, **11:** 1297-1305.

Packer, L., Weber, S., and Rimbach, G. 2001. Molecular aspects of á-tocotrienol antioxidant action and cell signalling. *Journal of Nutrition*, **131:** 369S-373S

Palanivelu, R., Brass, L., Edlund, A.F. and Preuss, D. 2003. Pollen tube growth and guidance is regulated by POP2, an *Arabidopsis* gene that controls GABA levels. *Cell,* **114(1):** 47-59.

Palta, Jiwan P. and Farag, Karim. 2006. "Methods for enhancing plant health, protecting plants from biotic and abiotic stress related injuries and enhancing the recovery of plants injured as a result of such stresses." United States Patent 7101828, September 2006.

Pan, Z., Zhao, Y., Zheng, Y., Liu, J., Jiang, X. and Guo, Y. 2012. A high-throughput method for screening *Arabidopsis* mutants with disordered abiotic stress-induced calcium signal. *J. Genet.. Genomics.*, **39(5):** 225-236.

Pang, X., Wang, D.H., Xing, X.Y., Peng, A.and Zhang, F.S. 2002. Effect of La^{+3} on the activities ofantioxidant enzymes in wheat seedlings under lead stress in solution culture. *Chemosphere*, **47:** 1033-1039.

Panikulangara, T.J., Eggers Schumacher, G., Wunderlich, M., Stransky, H. and Schoffl, F. 2004. Galactinol synthase1. A novel heat shock factor target gene responsible for heat-induced synthesis of raffinose family oligosaccharides in *Arabidopsis. Plant Physiology*, **136:** 3148-3158.

Parida, A.K. and Das, A.B. 2005. Salt tolerance and salinity effects on plants: a review. *Ecotoxicol Environ Saf.*, **60:** 324–349.

Park, E.J., Jeknic, Z., Sakamoto, A., DeNoma, J., Yuwansiri, R., Murata, N. and Chen, T.H. 2004. Genetic engineering of glycinebetaine synthesis in tomato protects seeds, plants, and flowers from chilling damage. *The Plant Journal*, **40:** 474-487.

Park, E.J., Jeknic, Z., Pino, M.T., Murata. N. and Chen, T.H. 2007. Glycinebetaine accumulation is more effective in chloroplasts than in the cytosol for protecting transgenic tomato plants against abiotic stress. *Plant Cell Environ.*, **30:** 994–1005.

Patra, B., Ray, S., Richter, A. and Majumder, A.L. 2010. Enhanced salt tolerance of transgenic tobacco plants by co-expression of PcINO1 and McIMT1 is accompanied by increased level of myo-inositol and methylated inositol. *Protoplasma*, **245:** 143-152.

Paul, M.J., Primavesi, L.F., Jhurreea, D. and Zhang, Y.H. 2008. Trehalose metabolism and signaling. *Annual Review of Plant Biology*, **59:** 417-441.

Pavelic, D., Arpagaus, S., Rawyler, A., and Braendle, R. 2000. Impact of postanoxia stress on membrane lipids of anoxia pretreated potato cells. A re-appraisal. *Plant Physiology*, **24:** 1285-1292.

Pedranzani, H., Sierra-de-Grado, R., Vigliocco, A., Miersch, O. and Abdala, G. 2003. Cold and water stresses produce changes in endogenous jasmonates in two populations of *Pinus pinaster* Ait. *Plant Growth Regul.,* **52:** 111–116.

Pedrazani, H., Racagni. G., Alemano. S., Miersch. O., Ramirez. I., Pena-Cortes. H., Taleisnik, E., Machado-Domenech, E. and Abdala, G. 2003. Salt tolerant tomato plants show increased levels of jasmonic acid. *Plant Growth Regul.*, **412:** 149–58.

Peiser, G. and Yang, S. F. 1979. Ethhylene production sulfur dioxide injured plants. *Plant Physiol.* **63:**1142-1145.

Peiser, G. and Yang, S. F. 1985. Biochemical and physiological effects of SQ2 on nonphotosynthetic processes in plants. In, "*Sulfur Dioxide and vegetation*", Eds. W.E. Winner, I.A. Monney, and R.A. Goldstein, Stanford University Press. Standford, C.A. ISBN 0-8047-1234-4. Pp. 148-161.

Peleg, L., Zer, H. and Chevion, M. 1992. Paraquat toxicity in *Pisum sativum*. Effects on soluble and membrane bound proteins. *Physiol. Plant.*, **86:** 131-135.

Peleg, Z., Apse, M.P. and Blumwald, E. 2011. Engineering salinity and water stress tolerance in crop plants: Getting closer to field. *Adv. Bot. Res.*, **57:** 405-403.

Perez-Ilzarbe, J., Hernandez, T., Estrella, I. and Vendrell, M. 1997. Cold storage of apples (cv. Granny Smith) and changes in phenolic compounds. *Z Lebensm Unters Forsch.*, **204:** 52-55.

Perez-Amador, M.A., Leon, J., Green, P.J. and Carbonell, J. 2002. Induction of the arginine decarboxylase *ADC2* gene provides evidence for the involvement of polyamines in the wound response in *Arabidopsis. Plant Physiol*, **130:** 1454-1463.

Peterbauer, T. and Richter, A. 2001. Biochemistry and physiology of raffinose family oligosaccharides and galactosyl cyclitols in seeds. *Seed Science Research*, **11:** 185-197.

Petrusa, L.M. and Winicov, I. 1997. Proline status in salt tolerant and salt sensitive alfalfa cell lines and plants in response to NaCl. *Plant Physiol Biochem.*, **35:** 303–310.

Pfister-Sieber, M., and Braendle R. 1994. Aspects of plant behavior under anoxia and post-anoxia. *Proceedings of the Royal Society of Edinburgh,* **102B:** 313-324.

Pimm, S.L. 2009. Climate disruption and biodiversity. *Curr. Biol.*, **19:** 595–601.

Pitta-Alvarez, S.I., Spollansky, T.C. and Giullietti, A.M. 2000. The influence of different biotic and abiotic elicitors on the production and profile of tropane alkaloids in hairy root cultures of *Brugmansia candida. Enzyme Microb Technol.* **26:** 252–258.

Phang, T.H., Shao, G. and Lam, H.M. 2008. Salt tolerance in soybean. *Journal of Intergrative Plant Biology*, **50(10):** 1196-2120.

Piikki, K., Vome, V., Ojanpera, K. and Pleijel, H. 2003. Potato tuber sugars, starch and organic acids in relation to ozone exposure. *Potato Res.,* **46:** 67–79.

Plaxton, W. C., and Podestá, F. E. 2006. The functional organization and control of plant respiration. *Crit. Rev. Plant Sci.* **25:** 159–198.

Pleijel, H., Mortensen, L., Fuhrer, J., Ojanpera, K. and Danielsson, H. 1999. Grain protein accumulation in relation to grain yield of spring wheat (*Triticum aestivum* L.) grown in open-top chambers with different concentrations of ozone, carbon dioxide and water availability. *Agric Ecosyst Environ.,* **72:** 265–270.

Polle, A., Baumbusch, L. O., Oschinski, C., Eiblmeier, M., Kuhlenkamp, V., Vollrath, B., Scholz, F. and Rennenberg, H. 1999. Growth and protection against oxidative stress in young clones and mature spruce trees (*Picea abies* L.) at high altitudes. *Oecologia,* **121:** 149–156.

Price, A.H., and Hendry, G. 1989. Stress and the role of activated oxygen scavengers and protective enzymes in plants subjected to drought. *Biochemical Society Transactions*, 17: 493-494.

Price, A. H., Athertan, N. M. and Hendry, G. A. F. 1989. Plants under drought-stress generate activated oxygen. *Free Radical Res.Commun.*, **8:** 61-66.

Price, A. H., Taylor, A., Ripley, H. J., Graffiths, A., Trewaves, A. J., Knight, M. R. 1994. Oxidative signals increase cytosolic calcium. *Plant Cell*, **6:** 1301-1310.

Qaderi, M.M., Kurepin, L.V., Reid, D.M. 2006. Growth and physiological responses of canola (*Brassica napus*) to three components of global climate change: Temperature, carbon dioxide and drought. *Physiol. Plant.* **128:** 710–721.

Qu, L.J., Wu, L.Q., Fan, Z.M., Guo, L., Li, Y.Q. and Chen, Z.L. 2005. Over-expression of the bacterial *nhaA* gene in rice enhances salt and drought tolerance. *Plant Science*, **168:** 297-302.

Qu, C., Liu, C., Ze, Y., Gong, X., Hong, M., Wang, L. and Hong, F. 2011. Inhibition of nitrogen and photosynthetic carbon assimilation of maize seedlings by exposure to a combination if salt stress and potassium difiencient stress. *Biol. Trace Elem. Res.*, **144(1-3):** 11569-11574.

Quinet, M., Ndayiragije, A., Lefevre, I., Lambillotte, B., Dupont-Gillain, C.C. and Lutts S. 2010. Putrescine differently influences the effect of salt stress on polyamine metabolism and ethylene synthesis in rice cultivars differing in salt resistance. *Journal of Experimental Botany*, **61:** 2719-2733.

Quinn, P. J. 1988. Effects of temperature on cell membrane, *Symp. Soc. Expt. Biol.*, **42:** 237–258.

Rabbani, M.A., Maruyama, K., Abe, H., Khan, A.M., Katsura, K.I., Yoshimara, K., Seki, M., Shinozaki, K. and Yamaguchi-Shinozaki, K. 2003. Monitoring expression profiles of rice genes under cold, drought, and high-salinity stresses and abscisic acid application using cDNA microarray and RNA gel-blot analyses. *Plant Physiol.*, **133:** 1755-1767.

Rai, V.K. 2002. Role of amino acids in plant responses to stresses, *Biologia Plantarum*, **45(4):** 481–487.

Ramakrishna, A., Dayananda, C., Giridhar, P., Rajasekaran, T. and Ravishankar, G.A. 2011. Photoperiod influences endogenous indoleamines in cultured green alga *Dunaliella bardawil. Indian J. Exp. Biol.*, **49:** 234–240.

Rao, S.R. and Ravishankar, G.A. 2002. Plant cell cultures: chemical factories of secondary metabolites. *Biotechnol Adv.,* **20:** 101–53.

Rathinasabapathi, B., Burnet, M., Russell, B. L., Gage, D. A., Liao, P. C., Nye, G. J., Scolt, P., Goldbeck, J. H. and Hanson, A. D. 1997. Choline monooxygenase, an unusual iron-sulfur enzyme catalyzing the first step of glycine betaine synthesis in plants: prosthetic group characterization and cDNA cloning. *Proc. Nat. Acad. Sci., USA*, **94:** 3454-3458.

Rathinasabapathi, B. 2000. Metabolic engineering for salt tolerance: Installing osmoprotectants synthesis pathways. *Analas Bot.*, **86(4):** 709-716.

Ravishankar, G.A. and Rao, S.R. 2000. Biotechnological production of phytopharmaceuticals. *J. Biochem Mol. Biol. Biophys.,* **4:** 73–102.

Rawyler, A., Pavelic, D., Gianiazzi, C., Oberson, J., and Braendle, R. 1999. Membrane lipid integrity relies on a threshold of ATP production rate in potato cell cultures submitted to anoxia. *Plant Physiology,* **120:** 293-300.

Recknagel, R. O., and Glende, E. A., Jr. 1984. Spectrophotometric detection of lipid conjugated dienes. *Methods in Enzymology*, **105:** 331-337.

Reddy, A.R., Rasineni, G.K., Raghavendra, A.S. 2010. The impact of global elevated CO_2 concentration on photosynthesis and plant productivity. *Curr. Sci.* **99:** 46–57.

Reddy, A.S., Ali, G.S., Celesnik, H. and Day, I.S. 2011. Cropping with stresses: roles of calcium and calcium/calmodulin-regulated gene expression. *Plant Cell.*, **23(6):** 2010-2032.

Redestig, H., Kusano, M., Ebana, K., Kobayashi, M., Oikawa, A., Okazaki, Y., Matsuda, F., Arita, M., Fujita, N. and Saaito, K. 2011. Exploring molecular backgrounds of quality traits in rice by predictive models based on high coverage metabolomics. *BMC Syst Biol.,* **5:** 176.

Reggiani, R., and Bertani, A. 1989. Effect of decreasing oxygen concentration on polyamine metabolism in rice and wheat shoots. *Journal of Plant Physiology*, **135:** 375-377.

Renault, H., Roussel, V., El Amrani, A., Arzel, M., Renault, D., Bouchereau, A., Deleu, C. 2010. The *Arabidopsis* pop2-1 mutant reveals the involvement of GABA transaminase in salt stress tolerance. *BMC Plant Biology*, **10:** 20.

Rhodes, D. and Hanson, A.D. 1993. Quaternary ammonium and tertiary sulfonium compounds in higher plants. *Ann. Rev. Plant Physiol.* **44:** 357–384.

Rice-Evans, C. A., Miller, N. J., and Paganga, G. 1997. Antioxidant properties of phenolic compounds. *Trends in Plant Sciences*, **2:** 152-159.

Rice-Evans, C. A. and Miller, N. J. 1998. Structure-antioxidant activity relationships of flavonoids and isoflavonoides. In, "*Flavonoids in health and disease*", Eds. C.A. RiceEvans, and L. Packer, Marcel Dekker Inc., New York. PP. 199-220.

Richards, F.J. and Coleman, R.G. 1952. Occurrence of putrescine in potassium-deficient barley. *Nature*, **170:** 460.

Richard, B., Couce, I., Raymond, P., Saglio, P., Saint-Ges, V., and Pradet, A. 1994. Plant metabolism under hypoxia and anoxia. *Plant Physiology and Biochemistry*, **32:** 1-10.

Rickauer, M., Brodschehn, W., Bottin, A., Veronesi, C., Grima, H., and Esquerretugaye, M. 1997. The jasmonate pathway is involved differentially in the regulation of different defense responses in tobacco cells. *Planta*, **202:** 155-162.

Riisom, T., Sims, R. J and Fioriti, J. A. 1980. Antioxidant activity in food. *J. Am. Oil Chem. Soc.*, **57:** 354-361.

Rivas-Ubach, A., Sardans, J., Perez-Truji,,o. M., Estiarte, M. and Penuelas, J. 2012. Strong relationship between elemental stochiometry and metablome in plants. *Proc. Natl. Acad. Sci., USA.*, **109(11):** 4181-4186.

Rizhsky, L., Liang, H.J., Shuman, J., Shulaev, V., Davletova, S. and Mittler, R. 2004. When defense pathways collide. The response of *Arabidopsis* to a combination of drought and heat stress. *PlantPhysiology*, **134:** 1683-1696.

Reguera, M., Peleg, Z. and Blumwald, E. 2012. Targeting metabolic pathways for genetic engineering abiotic stress-tolerance in crops. *Biochim Biophys Acta.*, **18(1):** 186-194.

Rocha, I.M.A., Vitorello, V.A., Silva, J.S., Ferreira-Silva, S.L., Viégas, R.A., Silva, E.N. and Silveira, J.A.G. 2012. Exogenous ornithine is an effective precursor and the ä-ornithine amino transferase pathway contributes to proline accumulation under high N recycling in salt-stressed cashew leaves. *J. Plant Physiol,* **169:** 41–49.

Rodrigues, F.A., Laia, M.A. and Zingaretti, S.M. 2009. Analysis of gene expression profile under water stress in tolerant and sensitive sugarcane plant. *Pl. Sci.*, **176(2):** 286-302.

Rolf D. Vinebrooke, Kathryn L. Cottingham, Jon Norberg, Marten Scheffer, Stanley I. Dodson, Stephen C.Maberly and Ulrich Sommer. 2004. Impacts of multiple stressors on biodiversity and ecosystem functioning: the role of species co-tolerance. *Oikos,* **104 (3):** 451– 457.

Rogers, A., Allen, D.J., Davey, P.A., Morgan, P.B., Ainsworth, E.A., Bernacchi, C.J., Cornic, G., Dermody, O., Dohleman, F.G., Heaton, E.A., Mahoney, J., Zhu, X.G., Delucia, E.H., Ort, D.R., Long, S.P. 2004. Leaf photosynthesis and carbohydrate dynamics of soybeans grown throughout their life-cycle under Free-Air Carbon Dioxide Enrichment. *Plant Cell Environ.*, **27:** 449–458.

Rosahl, S. 1996. Lipoxygenases in plants-their role in development and stress response. *Zeitschrift fur Natuiforschung*, **51c:** 123-138.

Rosemann, D., Heller, W. and Sandermann, H. 1991. Biochemical plant responses to ozone. II. Induction of stilbene biosynthesis in Scots pine (*Pinus sylvestris* L.) seedlings. *Plant Physiol.*, **97:** 1280–1286.

Rousos, P. A., Harrison, H. C., Palta, J. P. 1986. Reduction of chlorophyll in cabbage leaf disks following Cu^2 exposure. *Hort. Science*, **21:** 499-501.

Ruan, C.J., Teixeira and da Silva, J.A. 2011. Metabolomics: creating new potentials for unveiling the mechanism in response to salt and drought stress and for the biotechnological improvement of xerophytes. *Crit. Rev. Biotechniol.*, **31(2):** 153-169.

Russel, D., and Sachs, M. 1991. The maize cytosolic glyceraldehyde-3phosphate dehydrogenase gene family: organ specific expression and genetic analysis. *Molecular and General Genetics*, **229:** 219-228.

Saba, P.D., Iqbal, M, and Srivastava, P.S. 2000. Effect of ZnSO4 and $CuSO_4$ on regeneration and lepidine content in Lepidium sativum. *Biol. Plant.*, **43:** 253–256.

Sairam, R.K., Srivastava, G.C., Saxena, D.C. 2000. Increased antioxidant activity under elevated temperatures: A mechanism of heat stress tolerance in wheat genotypes. *Biol. Plant.* **43:** 245–251.

Sairam, R.K. and Tyagi, A. 2004. Physiology and molecular biology of salinity stress tolerance in plants. *Current Science*, **86(3):** 407-421.

Sanchez, D.H., Siahoosh, M.R., Roessner, U., Udvardi, M. and Kooka, J. 2008. Plant metabolismics reveales conserved and divergent metabolic responses to salinity. *Physiology Planta*, **132(2):** 209-219.

Sanchez, D.H., Szymanski, J., Erban, A., Udvardi, M.K. and Kopka, J.2010. Mining for robust transcriptional and metabolic responses to long-term salt stress: a case study on the legume *Lotus japonicas*. *Plant Cell & Evnironment,* **33(4):** 468-480.

Sanchez, D.H., Pieckenstain, F.L., Escaray, F., Erban, A., Kraemer, U., Udvardi, M.K., Kopka, J. 2011a. Comparative ionomics and metabolomics in extremophils and glycophytic lotusspecies under salt stress challenge the metabolic pre-preadaptation hypothesis. *Plant Cell & Evnironment,* **34(4):** 605-617.

Sanchez, D.H., Pieckenstain, F.L., Szymanski, J., Erban, A., Bromke, M., Hannah, M.A., Kraemer, U., Kopka, J., Udvardi, M.K. 2011b. Comparative functional genomics of salt stress in related model and cultivated plants identifies and overcomes limitations to translational genomics. *PloS one*, **6(2):** e17094.

Sandmann, G. and Gonzalea, H. G. 1989. Peroxidative processes induced in bean leaves by fumigation with sulphur dioxide. *Environ. Pollut.*, **56:** 145-154.

Sato, K., Nakayama, M., Shigeta, J. 1996. Culturing conditions affecting the production of anthocyanin in suspended cell cultures of strawberry. *Plant Sci.*, **113:** 91–98.

Satoh-Nagasawa, N., Nagasawa, N., Malcomber, S., Sakai, H. and Jackson, D. 2006. A trehalose metabolic enzyme controls inflorescence architecture in maize. *Nature*, **441:** 227-230.

Scandalios, J. G. 1993. Oxygen stress and superoxide dismutase. *Plant Physiology*, **101**: 7-12.

Sawahel, W.A. and Hassan, A.H. 2002. Generation of transgenic wheat plants producing high level of osmoprotectant proline. *Biotechnology Letter*, **24(9):** 721-725.

Scarpeci, T.E. and Valle, E.M. 2008. Rearrangement of carbon metabolism in *Arabidopsis thaliana* subjected to oxidative stress condition: an emergency survival strategy. *Plant Growth Regul.*, **54:**133–142.

Schoner, S. and Krause, G. H., 1990. Protective systems against active oxygen species in spinach: response to cold acclimation in excess light. *Planta*, **180:** 383–389.

Schmidhuber, J. and Tubiello, F.N. 2007. Global food security under climate change. *Proc Natl Acad Sci U S A.*, **104(50):** 19703-19708.

Schulze, E.D., Beck, E., and Müller-Hohenstein, K. 2002. Pfl anzenökologie. Spektrum Akademischer, Heidelberg.

Seigler, D.S. 1998. Plant Secondary Metabolism. Chapman and Hall, Kluwer Academic Publishers, Boston, MA, pp. 711.

Sengupta, S., Patra, B., Ray, S. and Majumder, A.L. 2008. Inositol methyl tranferase from a halophytic wild rice, *Porteresia coarctata* Roxb. (Tateoka): regulation of pinitol synthesis under abiotic stress. *Plant, Cell and Environment*, **31:** 1442-1459.

Semenza, G. L. 1999. Perspectives on oxygen sensing. *Cell*, **98:** 281-284.

Serbinova, E. A., and Packer, L. 1994. Antioxidant properties of α-tocopherol and á-tocotrienol. *Methods in Enzymology*, **234:** 354-366.

Shaddad, M. A., Radi, A. F., Abdel-Rahaman, A. M. and Azooz, A. A. 1990. Response of seed of *Lupimis termis* and *Vicia faba* to the interactive effect of salinity and ascorbic acid or pyridoxine (B6). *Plant and Soil*, **122:** 177-183.

Shalata, A. and Neumann, P. M. 2001. Exogenous ascorbic acid (vitamin- C) increases resistance to salt stress and reduces lipid peroxidation. *J. Exp. Botany*, **52:** 2207-2211.

She, X.P., Song, X.G., He, J.M. 2004. Role and relationship of nitric oxide and hydrogen peroxide in light/dark-regulated stomatal movement in *Vicia faba*. *Acta Bot Sin*, **46:** 1292–1300.

Shelp, B.J., Bown, A.W. and McLean, M.D. 1999. Metabolism and functions of gamma-aminobutyric acid. *Trends in Plant Science*, **4:** 446-452.

Shen, B., Jensen, G. and Bohnert, H. J. 1997a. Increased resistance to oxidative stress in transgenic plants by targetting mannitol biosynthesis to chloroplast. *Plant Physiol.*, **113:** 1177-1183.

Shen, B., Jensen, R.G. and Bohnert, H. J. 1997b. Mannitol protects against oxidation by hydroxyl radicals. *Plant Physiol.*,**115:** 527-532.

Shewfelt, R. L., and Purvis, A. C. 1995. Toward a comprehensive model for lipid peroxidation in plant tissue disorders. *Hort. Science*, **30:** 213-218.

Shiozaki, N., Hattori, I., Gojo, R. and Tezuka, T. 1999. Activation of growth and nodulation in symbiotic system between pea plants and leguminous bacteria by near UV radiation. *J Phtochem Phtoboi B*. Biology, **50:** 33–37.

Shinozaki, K. and Yamaguchi-Shinozaki, K. 1997. Gene expression and signal transduction in water-stress response. *Plant Physiol.*,**115:** 327-334.

Shimazaki, K. and Sugahara, K. 1980. Inhibition site of the electron transport system in lettuce chloroplasts by fumigation of leaves with SO_2. *Plant Cell Physiol.*, **12:** 303-312.

Shinozaki, K. and Yamaguchi-Shinozaki K. 2007. Gene networks involved in drought stress response and tolerance. *Journal of Experimental Botany*, **58:** 221-227.

Sicher, R.C. and Barnaby, J.Y. 2012. Impact of carbon dioxide enrichment on the response of maize leaf transcripts and metabolites to water stress. *Physiol. Plant*, **144(3):** 238-253.

Silva-Ortega, C.O., Ochoa-Alfaro, A.E., Reyes-Agüero, J.A., Aguado-Santacruz, G.A. and Jiménez-Bremont, J.F. 2008. Salt stress increases the expression of p5cs gene and induces proline accumulation in cactus pear. *Plant Physiol. Biochem.* **46:** 82–92.

Silvente, S., Sobolev, A.P., Lara, M. 2012. Metabolite adjustments in drought tolerant and sensitive soybean genotypes in response to water stress. *PLoS One*, **7(6):** e3855

Singh, S, and Sinha, S. 2005. Accumulation of metals and its effects in *Brassica juncea* (L.) Czern. (cv. Rohini) grown on various amendments of tannery waste. *Ecotoxicol Environ. Saf.*, **62(1):** 118–127.

Siripornadulsil, S., Traina, S., Verma, D.P. and Sayre, R.T. 2002. Molecular mechanisms of proline-mediated tolerance to toxic heavy metals in transgenic microalgae. *Plant Cell,* **14:** 2837–2847.

Skulachev, V. P. 1998. Cytochrome c in the apoptotic and antioxidant cascades. *FEBS Letters*, **423:** 275-280.

Smirnoff, N. and Cumbes, Q. J. 1989. Hydroxyl radical scavenging activity of compatible solutes. *Phytochem.*, **28:** 1057-1060.

Smirnoff, N. 1993. The role of active oxygen in the response of plants to water deficit and desiccation. *New Phytol.*, **125:** 27-58.

Smirnoff, N. 1995. Antioxidant systems and plant response to the environment. In, *"Environment and plant metabolism*" Ed. N. Smirnoff, Bios Scientific Publishers. pp. 217-243.

Smirnoff, N. 1996. The function and metabolism of ascorbic acid in plants. *Ann. Bot.*, **78:** 661-669.

Smirnoff, N. 2000. Ascorbic acid: metabolism and functions of a multifacetted molecule. *Current Opinion in Plant Biology*, **3:** 229-235.

Smith, T.A. and Richards, F.J. 1964. The biosynthesis of putrescine in higher plants and its relation to potassium nutrition. *Biochem J*, **84:** 292–294.

Snow, M.D., Bard, R.R., Olszyk, D.M., Minster, L.M., Hager, A.N. and Tingey, D. 2003. Monoterpene levels in needles of *Douglas firexposed* to elevated CO_2 and temperature. *Physiol. Plant,* **117:** 352–358.

Soares-Cavalcanti, N.M., Belarmino, L.C., Kido, E.A., Pandolf, V., Marcelino-Guimarães, F.C., Rodrioques, F.A., Pereira, G.A. and Benko-Iseppon, A.M. 2012. Overall picture of expressed heat shock factors in *Glycine max*, *Lotus japonicas* and *Medicago truncatula. Genet. Mol. Biol.*, **35(1-suppl):** 247-259.

Soliz-Guerrero, J.B., de Rodriguez, D.J., Rodriguez-Garcia, R., Angulo-Sanchez, J.L. and Mendez-Padilla, G. 2002. Quinoasaponins: concentration and composition analysis. In: "*Trends in new crops and new uses*" Eds. J. Janick, A. Whipkey, ASHS Press, Alexandria, pp. 110.

Somashekaraiah, B. V., Padmaja, K. and Prasad, A. R. K. 1992. Phytotoxicity of cadmium ions on germinating seedlings of mung bean (*Phaseolus vulgairs*): Involvement of lipid peroxides in chlorophyll degradation. *Physiol. Plant*, **851:** 85-89.

Song, H.M., Xu, X.B., Wang, H., Wang, H.Z. and Tao, Y.Z. 2010. Exogenous gamma-aminobutyric acid alleviates oxidative damage caused by aluminium and proton stresses on barley seedlings. *Journal of the Science of Food and Agriculture*, **90:** 1410-

Soyka, S. and Heyer, A.G. 1999. Arabidopsis knockout mutation of ADC2 gene reveals inducibility by osmotic stress. *FEBS Lett*, **458:** 219–223.

Spollen, W.G. and Nelson, C.J. 1994. Response of fructan to water-deficit in growing leaves of tall fescue. *Plant Physiology*, **106:** 329-336.

Stein, H., Honig, A., Miller, G., Erster, O., Eilenberg, H., Csonka, L.N., Szabados, L., Koncz, C. and Zilberstein, A. 2011. Elevation of free proline and proline-rich protein levels by simultaneous manipulations of proline biosynthesis and degradation in plants. *Plant Sci.*, 181: 140–150 / *Int. J. Mol. Sci.*, 2012, **13:** 8642.

Stoop, J. M. H., Williamson, J. D. and Pharr, D. M. 1996. Mannitol metabolism In plants:a method for coping with stress. *Trends Plant Sci.*, **1:** 139-155.

Streb, P., Michael-Knauf, A. and Feierabend, J. 1993. Preferential photoinactivation of catalase and photoinhibition of photosystem II are common early symptoms under various osmotic and chemical stress conditions. *Physiol. Plant.*, **88:** 590-598.

Streb, P., Telor, E. and Feierabend, J. 1997. Light stress effects and alternative protection in two desert plants. *Funct. Ecol.*, **11:** 416–424.

Streb, P., Shang,W., Feierabend, J. and Bligny, R. 1998. Divergent strategies of photoprotection in high-mountain plants. *Planta*, **207:** 313–324.

Strid, P. 1993 Alteration an expression of defense genes in *Pisum sativum* after exposure to supplementary ultraviolet-B radiation. *Plant Cell Physiol.*, **34:** 949 953.

Su, J. and Wu, R. 2004. Stress-inducible synthesis of proline in transgenic rice confers faster growth under stress conditions than that with constitutive synthesis. *Plant Sci.*, **166(4):** 941-948.

Sudha, G. and Ravishankar, G.A. 2003. Elicitation of anthocyanin production in callus cultures of *Daucus carota* and involvement of calcium channel modulators. *Curr Sci.*, **84:** 775–779.

Suzuki, N., Bajad, S., Shuman, J., Shulaev, V. and Mittler, R. 2008. The transcriptional co-activator MBF1c is a key regulator of thermotolerance in Arabidopsis thaliana. *Journal of Biological Chemistry*, **283:** 9269-9275.

Svenningson, H., Sundin, P. and Lilljenberg, G. 1990. Lipids carbohydrates and amino acids exuded from the axenic roots of rape seedlings exposed to water deficit stress. *Plant. Cell. Environ.*, **13:** 155-162.

Szabados, L. and Savoure, A. 2010. Proline: a multifunctional amino acid. *Trends in Plant Science*, **15:** 89-97.

Szekely, G., Abraham, E., Cselo, A., Rigo, G., Zsigmond, L., Csiszar, J., Ayaydin, F., Strizhov, N., Jasik, J., Schmelzer, E., Koncz, C. and Szabados, L. 2008. Duplicated *P5CS* genes of *Arabidopsis* play distinct roles in stress regulation and developmental control of proline biosynthesis. *Plant Journal,* **53:** 11–28.

Tadege, M., Dupuis, I., and KuWemeier, C. 1999. Ethanolic fermentation: new functions for an old pathway. *Trends in Plant Sciences*, **4:** 320-325.

Taiz, L. and Zeiger, E. 1991. Plant physiology. The Benjamin/Cummings Comp., Inc., Red wood City, CA.USA.

Taji T, Ohsumi C, Iuchi S, Seki M, Kasuga M, Kobayashi M, Yamaguchi-Shinozaki, K. and Shinozaki, K. 2002. Important roles of drought- and cold-inducible genes for galactinol synthase in stress tolerance in *Arabidopsis thaliana. Plant J.*, **29:** 417–426.

Takahama, U. and Oniki, T. 1997. A peroxidase / phenolics/ ascorbate system can scavenge hydrogen peroxide in plant cells. *Physiologia Plant.*, **101:** 845-852.

Takahashi, T. and Kakehi, J.-I. 2009. Polyamines: ubiquitous polycations with unique roles in growth and stress responses. *Ann Bot*, **105:** 1–6.

Tanaka, K., Kondo, N. and Sugahara, K. 1982. Accumulation of hydrogen peroxide in chloroplasts of SO_2 fumigated spinach leaves. *Plant Cell Physiol.*, **23:** 999-1000.

Tarchevskii, I. A. 1992. Regulatory role of degradation of biopolymers and lipids. *Fiziologiya Rastenii*, **39:**1215-1223.

Tarczynski, M.C., Jensen, R.G. and Bohnert, H.J. 1993. Stress protection of transgenic tobacco by production of the osmolyte mannitol. *Science*, **259(5094):** 508-510.

Tegelberg, R., Julkunen-Tiitto, R., Aphalo, P.J. 2004. Red: far-red light ratio and UV-B radiation: their effects on leaf phenolics and growth of silver birch seedlings. *Plant Cell Environ.*, **27:** 1005–1013.

Tetlow, I.J., Morell, M.K. and Emes, M.J. 2004. Recent developments in understanding the regulation of starch metabolism in higher plants. *Journal of Experimental Botany*, **55:** 2131-2145.

Thomas, C. E., McLean, L. R., Parker, R. A., and Ohlweiler, D. F. 1992. Ascorbate and phenolic antioxidant interactions in prevention of liposomal oxidation. *Lipids*, **27:** 543-550.

Threlfall, D. R., and Whistance, G. R. 1977. Dehydrophylloquinone, (á-hydrotocopherolquinone) and dehydrotocopherols from etiolated maize and barley shoots. *Phytochemistry*, **16:** 1903-1907.

Torsethaugen, G., Pitcher, L. H., Zilinskas, B.A. and Pell, E. J. 1997. Overproduction of ascorbate peroxidase in the tobacco chloroplast does not provide protection against ozone. *Plant Physiol.*, **114:** 529-537.

Trejo-Tapia, G., Jimenez-Aparicio, A., Rodriguez-Monroy, M., De Jesus-Sanchez, A. and Gutierrez-Lopez, G. 2001. Influence of cobalt and other microelements on the production of betalains and the growth of suspension cultures of *Beta vulgaris. Plant Cell Tissue Organ Cult.*, **67:** 19–23.

Trossat, C., Rathinasabapathi, B. and Hanson, A. D. 1997. Transgenically expressed belaine aldehyde dehydrogenase efficiently catalyses oxidation of dimethylsulfonioproprion-aldehyde and omega-aminoaldehydes. *Plant Physiol.*, **113:** 1457-1461.

Tun, N.N., Santa-Catarina, C., Begum, T., Silveira, V., Handro, W., Floh, E.I.S. and Scherer, G.F.E. 2006. Polyamines induce rapid biosynthesis of nitric oxide (NO) in *Arabidopsis thaliana* seedlings. *Plant Cell Physiol*, **47:** 346–354.

Türkan, .I and Demiral, T. 2009. Recent developments in understanding salinity tolerance. *Environ Exp Bot.*, **67:** 2–9.

Tuteja, N. and Sopory, S.K. 2008. Chemical signaling under abiotic stress environment in plants. *Plant Signal Behav.*, **3:** 525–336.

Umezawa, T., Fujita, M., Fujita, Y., Yamaguchi-Shinozaki, K. and Shinozaki, K. 2006. Engineering drought tolerance in plants: discovering and tailoring genes to unlock the future. *Curr Opin Biotechnol.*, **17:** 113–22.

Upston, J., Terentis, A., and Stocker, R. 1999. Tocopherol-mediated peroxidation of lipoproteins: implications for vitamin E as a potential antiatherogenic supplement. *FASEB Journal*, **13:** 977-979.

Urano, K., Yoshiba, Y., Nanjo, T., Igarashi, Y., Seki, M., Sekiguchi, F., Yamaguchi-Shinozaki, K. and Shinozaki, K. 2003. Characterization of *Arabidopsis* genes involved in biosynthesis of polyamines in abiotic stress responses and developmental stages. *Plant Cell Environ*, **26:** 1917-1926.

Urano, K., Yoshiba, Y., Nanjo, T., Ito, T., Yamaguchi-Shinozaki, K. and Shinozaki, K. 2004. *Arabidopsis* stress-inducible gene for arginine decarboxylase *AtADC2* is required for accumulation of putrescine in salt tolerance. *Biochem Biophys Res Commun*, **313:** 369-375.

Urano, K., Hobo, T. and Shinozaki, K. 2005. *Arabidopsis* ADC genes involved in polyamine biosynthesis are essential for seed development. *FEBS Lett,* **579:** 1557-1564.

Urano, K,, Maruyama, K., Ogata, Y., Morishita, Y., Takeda, M., Sakurai, N., Suzuki, H., Saito, K., Shibata, D., Kobayashi, M., Yamaguchi-Shinozaki, K. and Shinozaki, K. 2009. Characterization of the ABA-regulated global responses to dehydration in *Arabidopsis* by metabolomics. *Plant J*, **57:**1065–1078.

Urban, O. 2003. Physiological impacts of elevated CO_2 concentration ranging from molecular to whole plant responses. *Photosynthetica*, **41:** 9–20.

Usadel, B., Blasing, O.E., Gibon, Y., Poree, F., Hohne, M., Gunter, M., Trethewey, R., Kamlage, B., Poorter, H. and Stitt M. 2008. Multilevel genomic analysis of the response of transcripts, enzyme activities and metabolites in *Arabidopsis* rosettes to a progressive decrease of temperature in the non-freezing range. *Plant, Cell and Environment,* **31:** 518-547.

Ushimaru, T., Shibazaka, M., and Tsuji, H. 1994. Resistance to oxidative injury in submerged rice seedlings after exposure to air. *Plant Cell Physiology*, **35:** 211-218.

Ushimaru, T., Kanematsu, S., Shibasaka, M., and Tsuji, H. 1999. Effect of hypoxia on the antioxidative enzymes in aerobically grown rice *(Oryza sativa)* seedlings. *Physiologia Plantarum*, **107:** 181-187.

Ushimaru, T., Kanematsu, S., Katayama, M., and Tsuji, H. 2001. Antioxidative enzymes in seedlings of *Nelumbo nucifera* germinated under water. *Physiologia Plantarum*, **112:** 39-46.

Valerio, C., Costa, A., Marri, L., Issakidis-Bourguet, E., Pupillo, P., Trost, P. and Sparla, F. 2011. Thioredoxin regulated beta-amylase (BAM1) triggers diurnal starch degradation in guard cells, and in mesophyll cells under osmotic stress. *Journal of Experimental Botany*, **62:** 545-555.

Valliyodan, B. and Nguyen, H.T. 2006. Understanding regulatory networks and engineering for enhanced drought tolerance in plants. *Current Opinion in Plant Biology*, **9:** 189-195.

Valluru, R., Van den Ende, W. 2008. Plant fructans in stress environments: emerging concepts and future prospects. *Journal of Experimental Botany*, **59:** 2905-2916.

van Assche, F. and Clijsters, H. 1990. Effects of metals on enzyme activity in plants. *Plant Cell Environ.*, **13:** 195-200.

van Dijken, A.J., Schluepmann, H. and Smeekens, S.C. 2004. *Arabidopsis* trehalose-6-phosphate synthase 1 is essential for normal vegetative growth and transition to flowering. *Plant Physiology*, **135:** 969-977.

van Toai, T. T., and Bolles, C. S. 1991. Postanoxic injury in soybean *(Glycine max)* seedlings. *Plant Physiology*, **97:** 588-592.

Vaneil, A., Pinturo, R., Geremia, E., Tiriolo, P., Durso, G., Rizza, V., Disilvestro, L., Grimaldi, R. and Brai, M. 1980. *JRCS Med. Sci. Biochem.*, **8:** 940-947.

Vartapetian, B. B., and Jackson, M. B. 1997. Plant adaptations to anaerobic stress. *Annals of Botany*, **79:** (Suppl. A), 3-20.

Veeranagamallaiah, G., Jyothsnakumari, G., Thippeswamy, M., Reddy, P.C.O., Surabhi, G.K.,Sriranganayakulu, G., Mahesh, Y., Rajasekhar, B., Madhurarekha, C., Sudhakar, C. 2008. Proteomic analysis of salt stress responses in foxtail millet (*Setaria italica* L. cv. Prasad) seedlings. *Plant Sci.,* **175:** 631–641.

Veljovic-Jovanovic, S., Bilger, W. and Heber, U. 1993. Inhibition of photosynthesis, stimulation of zeaxanthin formation and acidification in leaves by SO_2 and reversal of these effects. *Plant.*, **191:** 365-376.

Vendruscolo, E.C., Schuster, I., Pileggi, M., Scapim, C.A., Molinari, H.B., Marur, C.J., Vieira, L.G. 2007. Stress-induced synthesis of proline confers tplerance to water deficit in transgenic wheat. *J. Plant Physiol.*, **164(10):** 1367-1376.

Vener, A. V., VanKam, P. J. M., Rich, P. R., Ohad, I. and Andersson, B. 1997. Plastoquinol at the quinol oxidation site of reduced cytochrome of mediates signal transduction between light and protein phosphorylation: thylakoid protein kinase deactivation by a single turnover flash. *Proc. Nat. Acad. Sci., USA,* **94:** 1585-1590.

Verbruggen, N. and Hermans, C. 2008. Proline accumulation in plants: a review. *Amino acids*, **35(4):** 753-759.

Verdoy, D., Coba de la Peña, T., Redondo, F.J., Lucas, M.M. and Pueyo, J.J. 2006. Transgenic *Medicago truncatula* plants that accumulate proline display nitrogen-fixing activity with enhanced tolerance to osmotic stress. *Plant, Cell and Environment*, **29(10):** 1913–1923.

Verslues, P.E., Agrawal, M., Katuyar-Agrawal, S., Zhu, J. and Zhu, J.K. 2006. Methods and concepts in quantifying resistance to drought, salt and freezing abiotic stresses that effect plant water status. *Plant J.*, **45(4):** 523-539.

Verslues, P.E., and Sharma, S. 2010. Proline metabolism and its implications for plant-environment interaction. *Arabidopsis* book, Edited by Georg Jander.

Vierstra, R. D., John, T. R. and Proff, K. L. 1982. Kaempferol 3-O-galactoside, 7-O-rhamnoside is the major green fluorescing compound in the epidermis of *Vicia faba. Plant Physiology*, **69:** 522–525.

Vinocur, B. and Altman, A. 2005. Cellular basis of salinity tolerance in plants. *Environ Exp Bot.,* **52:** 113–122.

Vojtechova, M., Hanson, A. D., Munrioz-Clares, R. A. 1997. Betaine-aldehyde dehydrogenase from amaranth leaves efficiently catalyses the NAD-dependent oxidation of dimethylsulfonioproprion-aldehyde to dimethylsulfonioproprionate. *Arch Biochem. Biophys.*, **337:** 888-901.

Waditee, R., Bhuiyan, M.N., Rai, V., Aoki, K., Tanaka, Y., Hibino, T., Suzuki, S., Takano, J, Jagendrof, A.T. and Takabe T. 2005. Genes for direct methylation of glycine provide high levels of glycinebetaine and abiotic-stress tolerance in *Synechococcus* and *Arabidopsis. Proceedings of the National Academy of Sciences, USA*, **102:** 1318-1323.

Wahyuni, Y., Ballester, A.R., Tikunov, Y., de Vos, R.C., Pelgrom, K.T. and Maharijaya, A. Tarczynski, M. C., Jensen, R. G. and Bohnert, H. J. 1993. Stress protection of transgenic tobacco by production of the osmolyte mannitol. *Science*, **259:** 508-5

Wang, W., Vinocur, B. and Altman, A. 2003. Plant responses to drought, salinity and extreme temperatures: towards genetic engineering for stress tolerance. *Planta*, **218:**1–14.

Wang, H., Liu, D., Sun, J. and Zhang, A. 2005. Asparagine synthetase gene TaASN1 from wheat is up-regulated by salt stress, osmotic stress and ABA, *Journal of Plant Physiology*, **162(1):** 81–89.

Wang, Z., Zhu, Y., Wang, L., Liu, X., Liu, Y., Phillips, J. and Deng, X. and Wrky, A. 2009a. A transcription factor participates in dehydration tolerance in *Boea hygrometrica* by binding to the W-box elements of the galactinol synthase (BhGolS1) promoter. *Planta*, **230:** 1155-1166.

Wang, X.Q., Yang, P.F., Liu, Z., Liu, W.Z., Hu, Y., Chen, H., Kuang, T.Y., Pei, Z.M., Shen, S.H. and He, Y.K. 2009b. Exploring mechanis of Physcomitrella paterns desiccation tolerance through a proteomic strategy. *Plant Physiol.*, **149(4):** 1739-1750.

Warren, C., Ananda, I. and Cano, F. 2012. Metabolomics demonstrates divergent responses of two *Eucalyptus* species to water stress. *Metabolomics*, **8(2):** 186-200.

Weinstein, L.H., Kaur-Sawhney, R., Venkat Rajam, M., Wettlaufer. S.H. and Galston, A.W. 1986.Cadmium-induced accumulation of putrescine in oat and bean leaves. *Plant Physiol.*, **82:** 641–645.

Wi, S.J. and Park, K.Y. 2002. Antisense expression of carnation cDNA.encoding ACC synthase or ACC oxidase enhances polyamine.content and abiotic stress tolerance in transgenic tobacco plants. *Mol Cells*, **13:** 209–220.

Wi, S.J., Kim, W.T. and Park, K.Y. 2006. Overexpression of carnation *S-adenosylmethionine decarboxylase* gene generates a broad-spectrum tolerance to abiotic stresses in transgenic tobacco plants. *Plant Cell Rep*, **25:** 1111–1121.

Widodo, Patterson, J.H., Newbigin, E., Tester, M., Bacic, A. and Roessner, U. 2009. Metabolic responses to salt stress of barley (*Hordeum vulgar*e L.) cultivars, Sahara and Clipper, which differ in salinity tolerance. *Journal of Experimental Botany*, **60:** 4089-4103.

Wildi, B. and Lütz, C., 1996. Antioxidant composition of selected high alpine plant species from different altitudes. *Plant Cell Environ.*, **19:** 138–146.

Winkel-Shirley, B. 2001. Flavonoid biosynthesis, A colorful model for genetics, biochemistry, cell biology, and biotechnology. Plant Physiol., **126:** 485 493.

Williams, R.S., Lincoln, D.E. and Thomas, R.B. 1994. Loblolly pine grown under elevated CO_2 affects early instar pine sawfly performance. *Oecologia*, **98:** 64–71.

Witkowski, D. A. and Hailing, B. P. 1989. Inhibition of plant protoporphyrinogen oxidase by the herbicide acifluorfenmethyl. *Plant Physiol.*, **90:** 1239-1242.

Wu, L., Fan, Z., Guo, L., Chen, Z-L. and Qu, L-J. 2005. Over expression of the bacterial nhaA gene in rice enhances salt and drought tolerance. *Plant Sci.*, **168(2):** 297-302.

Wullschleger, S.D., Tschaplinski, T.J. and Norby, R.J. 2002. Plant water relations at elevated CO_2– implications for water-limited environments. *Plant Cell Environ.*, **25:** 319–331.

Yamada, M., Morishita, H., Urano, K., Shiozaki, N., Yamaguchi-Shinozaki, K., Shinozaki, K. and Yoshiba, Y. 2005. Effect of free accumulation in petunias under drought stress. *J. Exp. Bot.*, **56(417):** 1975-1981.

Xiong, L., Schumaker, K.S. and Zhu, J-K. 2002. Cell signaling during cold, drought, and salt stress. *Plant Cell*, **8:** 165-183.

Xu, D. P., Duan, X. L., Wang, B. Y., Hong, B. M., Ho, T. H. O. and Wu, R. 1996. Expression of a late embryogenesis abundant protein gene, HVA1, from barley confers tolerance to water-deficit and salt stress in transgenic rice. *Plant Physiol.*, **110:** 249-257.

Xu, Z., Zhou, G. and Shimizu, H. 2010. Plant responses to drought and rewatering. *Plant Signal Behav.* **5:** 649–54.

Xu, S., Zhou, J., Liu, L. and Chen, J. 2011. Arginine: a novel compatible solute to protect *Candidaglabrata* against hyperosmotic stress, *Process Biochemistry*, **46(6):** 1230–1235.

Yamaguchi, K., Takahashi, Y., Berberich, T., Imai, A., Miyazaki, A., Takahashi, T., Michael, A., and Kusano, T. 2006. The polyamine spermine protects against high salt stress in *Arabidopsis thaliana. FEBS Letters*, **580:** 6783-6788.

Yamaguchi, K., Takahashi, Y., Berberich, T., Imai, A., Takahashi, T., Michael, A.J. and Kusano, T. 2007. A protective role for the polyamine spermine against drought stress in *Arabidopsis. Biochem Biophys Res Commun*, **352:** 486–490.

Yamasaki, H. and Cohen, M.F. 2006. NO signal at the crossroads: polyamine-induced nitric oxide synthesis in plants? *Trends Plant Sci*, **11:** 522–524.

Yan, B., Dai, Q., Liu, X., Huang, S., and Wang, Z. 1996. Flooding-induced membrane damage, lipid oxidation and activated oxygen generation in corn leaves. *Plant and Soil*, **179:** 261-268.

Yancey, P.H. 2005. Organic osmolytes as compatible metabolic and counteracting cytoprotactants in high osmolarity and other stress. *J. Expt.Biol.*, **208(15):** 2819-2830.

Yang, G., Zhou, R., Tang, T., Chen, X., Ouyang, J., He, L., Li, W., Chen, S., Guo, M., Li, X., Zhong, C. and Shi, S. 2011. Gene expression profiles in response to salt stress in *Hibiscus tiliaceus*. *Plant. Mol. Biol. Rep.,* **29:** 609–617.

Yano, R., Nakamura, M., Yoneyama, T. and Nishida I. 2005. Starch-related alpha-glucan / water ikinase is involved in the cold-induced development of freezing tolerance in *Arabidopsis*. *Plant Physiology*, **138:** 837-846.

Yistra, B., Touraev, A., Benito, M. Moreno, R. M., Stoger, E., Van-Tunen A. J., Vicente, O., Moi, J. N. M. and Heberie-Bors, E. 1992. Flavonols stimulate development, germination and tube growth of tobacco pollen. *Plant Physiol.*, **100:** 902-907.

Yoshiba, Y., Kiyosue, T., Nakashima, K., Yamaguchi-Shinozaki, K. and Shinozaki, K. 1997. Regulation of levels of proline as an osmolyte in plants under water stress, *Plant and Cell Physiology*, **38(10):** 1095–1102

Yu, Q., and Rengel, Z. 1999. Drought and salinity differentially influence activities of superoxide dismutase in narrow-leafed lupins. *Plant Science*, **142:** 1-11.

Yu, J., Chen, L., Xu, M., Huang, B. 2012. Effects of elevated CO_2 on physiological responses of tall fescue (*Festuca arundinacea*) to elevated temperature, drought stress, and the combined stresses. *Crop Sci.*, **52:** 1848-1858.

Zhang, J. and Kirkham, M. B. 1994. Drought-stress induced changes in activities of superoxide dismutase, catalase and peroxidase in wheat species. *Plant Cell Physiol.*, **35:** 785-791.

Zhang, J. and Kirkham, M. B. 1996. Enzymatic responses of the ascorbate glutathione cycle to drought in sorghum and sunflower plants. *Plant. Sci.*, **113:** 139-147.

Zhang, W.., Seki, M. and Furusaki, S. 1997. Effect of temperature and its shift on growth and anthocyanin production in suspension cultures of strawberry cells. *Plant Sci.*, **127:** 207-214.

Zhang, J., Zhang, Y., Du, Y., Chen, S. and Tang, H. 2011a. Dynamic metabolomics responses of tobacco (*Nicotiana tabacum*) plants to slat stress. *J. Proteome. Res.*, **10:** 1904-1914.

Zhang, Y., Wu, R., Qin, G., Chen, Z., Gu, H. and Qu, L.J. 2011b. Over-expression of WOX1 leads to defects in meristem development and polyamine homeostasis in Arabidopsis. *Journal of Integrative Plant Biology*, **53:** 493-506.

Zhao, Y., Qi, L.. Wei- Ming Wang, W., Saxena, P.K. and Chun-Zhao Liu, C. 2011. Melatonin improves the survival of cryopreserved callus of *Rhodiola crenulata*. *J. Pineal Res.*, **50:** 83–88.

Zhong, J.J.T., Seki, T., Kinoshita, S.I. and Yoshida, T. 1993. Effect of light irradiation on anthocyanin production by suspended culture of *Perilla frutescens*. *Biotechnol Bioeng*, **38:** 653–658.

Zhu, D. and Scandalios, J. G. 1994. Differential accumulation of manganese-superocide dismutase transcripts in maize in response to abscisic acid and high osmoticum. *Plant Physiol.*, **106:** 173-178.

Zingaretti, S.M., Rodrigues, F.A., Graca, J.P., Perrira, L.M. and Lourenco, M.V. 2011. Sugarcane responses to water deficit conditions. In "*Plant Stress*" Ed. M.D. Ismail, M. Rahman, H. Hasegawa, Intech, Croatia, pp. 255-276.

Zobayed, S.M.A., Afreen, F., Kozai, T. 2005. Temperature stress can alter the photosynthetic efficiency and secondary metabolite concentrations in st. john's wort. *Plant Physiol. Biochem.* **43:** 977–984.

Zuther, E., Buchel, K., Hundertmark, M., Stitt, M., Hincha, D.K. and Heyer, A.G. 2004. The role of raffinose in the cold acclimation response of *Arabidopsis thaliana. FEBS Letters,* **576:**169-173.

Zuther, E., Koehl, K. and Kopka, J. 2007. Comparative metabolome analysis of the salt response in breeding cultivars of rice. In,"*Advances in molecular breeding toward drought and salt tolerant crops*", Eds. M.A. Jenks, P.M. Hasegawa, and S.M. Jain, Netherlands: Springer, pp. 285-315.

2

Abiotic Stress and Secondary Metabolites in Plants

A. Bhattacharya

Climate change drastically influence water availability, salinity and several adverse soil conditions which will have direct bearing on original yields. Abiotic stress factors influence growth and secondary metabolite production in higher plants. The influences are well marked. In fact, productivities depend on the changed ecosystem also. Secondary metabolites play a major role in the adaptation of plants to the changing environment and in overcoming stress constraints. This flows from the large complexity of chemical types and interactions underlying various functions: structure stabilizing, determined by polymerisation and condensation of phenols and quinones, or by electrostatic interactions of polyamines with negatively charged loci in cell components, photoprotective, related to absorbance of visible light and UV radiation due to the presence of conjugated double bonds, antioxidant and antiradical, governed by the availability of –OH, $-NH_2$, and –SH groupings, as well as aromatic nuclei and unsaturated aliphatic chains, signal transducing. Several plant-abiotic stress stimuli systems evidenced the multiplicity of biochemical mechanisms involved in the protective role of secondary metabolites: condensation of chlorogenoquinone with proteins yielding brown pigments limiting the spread of stress-induced tissue damage in tobacco, accumulation of polyamines and formation of phenylamides in tobacco and bean subjected to water stress and heat shock, respectively, with phenylamides performing ROS-scavenging ability, accumulation of anthocyanins in leaves of cotton suffering Na/K imbalance, and shift from mono- to orthodihydroxy substitution in the B-ring of anthocyanin aglycone, with this conferring a higher ROS-scavenging capacity, relation of drought tolerance in cotton to the level of ROS-scavenging polyphenol compounds. The protective effect of a gaseous secondary metabolite, isoprene, against ozone fumigation and heat shock was shown, and the ability of isoprene to scavenge singlet oxygen was demonstrated. Altogether, the data provided evidence that secondary metabolites through their diversity of functions can be involved in the non-enzymatic plant defense strategy. Abiotic factors which *influence secondary metabolite production have a bearing on enhancing the potential to over produce useful phytochemicals for varied applications. Moreover, molecular understanding of stress response will be useful in plant improvement with enhanced adaptation and efficacy.*

The term "secondary" introduced by A. Kossel in 1891 implies that while primary metabolites are present in every living cell capable of dividing, the secondary metabolites are present only incidentally and are not of paramount significance for plant life. In the last decade secondary metabolites, low molecular compounds occurring in all living organisms while largely distributed in plants, became a subject of dramatically increasing interest relevant to their significant practical implication for medicinal, nutritive and cosmetic purposes, as well as to their indisputable importance in plant stress physiology. Plant organisms being devoid of motility and immune system, have elaborated alternative defense strategies, involving the huge variety of secondary metabolites as tools to overcome stress constraints, adapt to the changing environment and survive. The large diversity of chemical types and interactions displayed by the secondary metabolites can underlie the impressive multiplicity of protective functions ranging from toxicity and light/UV shielding to signal transduction (Yang *et al.* 1997, Grassmann *et al.* 2002, Hadacek, 2002, Osbourn *et al.* 2003, Vasconsuelo and Boland, 2007). Revealing the putative structure-function relationships of secondary metabolites contributes to the better understanding of their stress-protective role in plants, and can constitute a rationale for their extensive exploitment as pharmaceuticals, food additives and cosmetic products (Edreva *et al.* 2008).

The secondary metabolites from plants, which are distinguished from primary metabolites such as nucleic acids, amino acids, carbohydrate, fat, *etc*. (Weinberg, 1971) are extremely diverse; thousands of them have been identified in several classes. Each plant family, genus, and species produce a characteristic mix of these chemicals, and they can sometimes be used as taxonomic characters in classifying plants (Thrane, 2001). Many scientific sources state that their role is not crucial for living cells in normal growth, development, and reproduction (Fraenkel 1959), but they act in defense purposes to protect a plant from any possible harm in the ecological environment (Stamp, 2003) and other interspecies protection (Samuni-Blank *et al.* 2012). Therefore, they are usually synthesized in plants for particular needs, while the primary metabolites have generally the shared biological purposes across all species. Secondary metabolites may often be created by modified synthetic pathways from primary metabolite, or share substrates of primary metabolite origin. Plants have been evolving to adapt the environment with genetic encoding of useful and diverse synthases for secondary metabolites (Waterman, 1992). In human life, these compounds are used as medicines, flavorings, or relaxing drugs, especially essential oils. In most references, it is stated that the secondary metabolites extracted from plants are subdivided in three major classes: terpenoids, alkaloids and phenolics. They contain numerous natural products with interesting pharmacology activities (Verpoorte, 1998; Savithramma *et al.* 2011).

Specimens of same plant species growing under different environmental conditions show significant differences in the production and accumulation of the primary and secondary metabolites (Pavarini *et al.* 2012, Ramakrishna and Ravishankar, 2011, Gutbrodt *et al.* 2012, Edreva *et al.* 2008, Oh *et al.* 2009, Theis and Lerdau, 2003, Bennett and Wallsgrove, 1994, Wink, 1988). Influenced by environmental factors, respective group of secondary metabolites act as a chemical interface between the plant and its environment. The chemical interaction between plants and their environment is mediated mainly by the biosynthesis of secondary metabolites, which exert their biological roles, as a plastic adaptive response to their environment. Such chemical interaction often includes variations in the production of plant metabolites (Pavarini *et al.* 2012, Ramakrishna and Ravishankar, 2011, Gutbrodt *et al.* 2012, Szakiel *et al.* 2010, Miranda *et al.* 2015, Treutter, 2005, Winkel-Shirley, 2002). Therefore, the study of these variations is very useful in the chemical characterization of plants of the same species which are collected from different regions and this is when the different geographical origin of a plant material is taken into account (Stashenko *et al.* 2010, Lukas *et al.* 2009, Vilela *et al.* (2013) O. Some processes can be the main sources of variation in the levels of metabolites for individual plant species. These processes involve long term acclimation or local adaptation, seasonal differences related to phenology or environmental changes in the biotic and abiotic factors, geographical differences involving different populations (genetic differences within a plant species), or different environmental conditions of the growth location of the species individuals, especially when they have genetic homogeneity (*i.e.* cultivars and/or clones) (Telascrea *et al.* 2007, Rahimmalek *et al.* 2009)..

2.1. Plant Seconadary Metabolites

Plant secondary metabolism has been investigated for decades as a dynamic phenomenon (Schultz, 2002) with fixed and predictable results. What guided us to such a paradigm? Possibly it was cases of plants in which compounds of pleasant odors, and therefore easily noticeable, display an accumulation of major compounds, such as eugenol in cloves (*Syzygium aromaticum*) or nutmeg (*Myristica fragrans*) and 1,8-cineole in *Eucalyptus* spp. However, the boundaries of science were pushed toward consideration of the chemical (Reddy and Guerrero, 2004), ecological (Baldwin *et al.* 2006) and evolutionary (Conrath *et al.* 2006) roles of secondary metabolites. Evolved mechanisms of plant defense such as tri-trophic herbivory protection (Degenhardt, 2008) by release of volatile compounds, and also the jasmonate induced elicitation of cellular responses against predators (Liechti and Farmer, 2002) were reported. Considered in a larger perspective, these roles are of tremendous importance for safe and reliable use of plants, be it for drug discovery and agrochemical research and development or in animal feed science and technology. Recent reviews dealing with variations in secondary metabolites in medicinal plants showed the influences of abiotic factors and reinforced the importance of chemical control for safe use of such plants in phytomedicine (Gobbo-Neto and Lopes, 2007; Ramakrishna and Ravishankar, 2011; Gouvea *et al.* 2012).

Considering all the evidence raised by the joint efforts of phytochemistry and plant biology it can be assumed that secondary metabolites are a way which plants communicate or respond to external stimuli (Bouwmeester *et al.* 2007; Maffei, 2010; Rasmann and Turlings, 2008; Frost *et al.* 2008). Such bioactive compounds accumulate in plant tissues through biochemistry mechanisms encrypted by other mechanisms of regulation. More specifically, we might conclude that each plant species has evolved up to a singular set of mechanisms for the biosynthetic regulation of secondary metabolites, from prior generations, which successfully expressed these strategies.

The classification of secondary metabolites consists of terpenoids, alkaloids and phenolics (Verpoorte, 1998). Glycosides, tannins and saponins are part of them according their specific structure.

Terpenoids

Terpenoids constitute a large family of phytoconstituents of little functional and structural common ground. Steroids, carotenoids, and gibberelic acid are just some of its members. They are composed by the most important group of active compounds in plants with over than 23,000 known structures. They are polymeric isoprene derivatives and synthesized from acetate via the mevalonic acid pathway. During their formation, the isoprene units are linked in head and tail fashion. The number of units incorporated into a particular terpene serves as a basis for their classification. Many of them have pharmacological activity and are used for diseases treatment both in humans and animals. Beaulieu and Baldwin, (2002) has pointed out that diterpenes tend to be most abundant in Lamiaceae family and have antimicrobial and antiviral properties. Some interesting compounds are extensively used in the industry sector as flavors, fragrance, spices (Styger *et al.* 2011; Gershenzon and Dudareva, 2007). Several thousand different types of molecules from very different plant groups have been isolated and haracterized. Despite their varied structures, all of them are synthesized by only a few pathways. Plants tissues are related to adaptation to both abiotic stressors. However, the high volatility and reactivity of some terponoids may also affect the atmosphere composition (Maffei *et al.* 2011). Volatility of terpenoids provides, for sessile plants, a tool for communication with other organisms such as neighboring plants, pollinators and foes of herbivores, via air-bone infochemicals. From a physiological standpoint, briefly, plant volatiles are involved in three critical processes, namely plant to plant interaction, the signaling between symbiotic organisms, and the attraction of pollinating insects. Their role in these "housekeeping" activities underlies agricultural applications that range from the search for sustainable methods for pest control to the production of flavors and fragrances (Maffei, 22010).

Important molecules of terpenoids

Number of carbon	Name	Example
C5	Hemiterpene	Isoprene, prenol, isovaleric acid
C10	Monoterpene	Limonene, eucalyptol, pinene
C15	Sesquiterpene	ABA (abscisic acid)
C20	Diterpene	Gibberellin
C25	Sesterterpenes	
C30	Triterpene	Brassinosteroids, squalen, lanosterol
C40	Tetraterpene	Carotenoids, lycopen
C>40	Polyterpene	Ubuquinone

After Kogan *et al.* 2006

Alkaloids

The alkaloids present the group of secondary metabolites that contain basic nitrogen atoms. Some related compounds with neutral and weakly acid properties are also included in the alkaloids. In addition to carbone, hydrogen and nitrogen, this group may also contain oxygen, sulfur and rarely other element such as chlorine, bromine and phosphorus (Nicolaou *et al.* 2011). Alkaloids are produced by a large variety of organisms, such as bacteria, fungi, animals but mostly by plants as secondary metabolites. Most of them are toxic to other organisms and can be extracted by acid-base. The boundary between alkaloids and other nitrogen-containing natural compounds is not clear-cut (Giweli *et al.* 2013). Compounds like amino acids, proteins, peptides, nucleotides, nucleic acid, and amines are not usually called alkaloids. Compared with most other classes of secondary metabolites, alkaloids are characterized by a great structural diversity and there is no uniform classification of them (Verpoorte, 1998). First classification was based on the common source because no information about chemical structure was yet available. Recent classification is based on similarity of the carbon skeleton (Savithramma *et al.* 2011).

Phenolics

Phenolic compounds from plants are one of largest group of secondary plants constituents synthesized by fruits, vegetables, teas, cocoa and other plants that possess certain health benefits. They are characterized by the antioxidant, anti-inflammatory, anti-carcinogenic and other biological properties, and may protect from oxidative stress (Park *et al.* 2001). Simple phenolics are bactericidal, antiseptic and anthlemintic. Phenol itself is a standard for other antimicrobial agents (Pengelly, 2004). They are distributed in almost all plants and subject to a great number of chemical, biological, agricultural, and medical studies (Dai *et al.* 2010) .

Phenols are diverse in structure, and present in common the hydroxylated aromatic rings (*e.g*., flavan-3-ols). Most of phenolic compounds are polymerized into larger molecules such as the PA (proanthocyanidins; condensed tannins) and lignans. Furthermore, phenolic acids may occur in food plants as esters or glycosides conjugated with other natural compounds such as flavonoids, alcohols, hydroxyfatty acids, sterols, and glucosides (Dai *et al.* 2010). Hydroxybenzoic and hydroxycinnamic acids present two main phenolic compounds found in plants. In tea, coffee, berries and fruits, the total phenolic comounds could reach up to 103 mg/100 g fresh weigh (Mmanach *et al.* 2004). The approach to classifying plant phenolics are based on: (1) a number of hydroxylic groups. So, they may be divided on 1-, 2- and polyatomic phenols. Phenolic compounds containing more than one OH-group in aromatic ring are polyphenols; (2) chemical composition: mono-, di, oligo- and polyphenols; (3) substitutes in carbon skeleton, a number of aromatic rings and carbon atoms in the side chain. According to the latter principle, phenolic compounds are divided into four main groups: phenolics with one aromatic ring, with two aromatic rings, quinones and polymers.

Phenolic compounds with one aromatic ring: a large number of compounds, among them are simple phenols (C6), phenol with attached one (C6-C1), two (C6-C2) and three (C6-C3) carbon atoms. Phenolic compounds with two aromatic rings: this group includes benzoquinones and xanthones (C6-C1-C6) containing two aromatic rings which are linked by one carbon atom; stylbenes (C6-C2-C6) which are linked by two carbon atoms; and flavonoids, containing three carbon atoms (C6-C3-C6). Flavonoids, depending on the structure of propane unit and an attaching place of side chain B, are divided into flavonoids in strict sense, which are derived from chromane or chromone, isoflavonoids and neoflavonoids. Polyphenolics are more than 8,000 different compounds identified to date. That is why the terminology and classification of polyphenols is complex and confusing. Although all polyphenols have similar chemical structures, there are some distinctive differences. Based on these differences, polyphenols can be subdivided into two classes: flavoinoids and non flavonoids, like tannins (Somasegaran and Hoben, 1994).

Flavonoid

Flavonoids stands as the first class of polyphenols, They are water soluble pigments found in vacuoles of plant cells. Flavonoids can also be divided in three groups: anthocyanins, flavones and flavonols. They are widely distributed in plants, fulfilling many functions such as flower colouration, producing yello and red or blue pigmentation in petals. In higher plants, flavonois are involved in UV filteration, symbiotic nitrogeb fixation and floral pigmentation. They may also act as chemical messengers, physiological regulation, and cell cycle imhibition.

Tannins

Tannins are the phenolic compounds that precipitate proteins. They are composed by a very diverse group of oligomers and polymers. They can form complex with proteins, starch, cellulose and minerals. Tannins are water soluble compounds with exception of some higher molecular weight structures. Tannins are usually subdivided in two groups: hydrolysable tannins (HT) that include gallotannins, elligatannins complex tannins, and proanthocyanidins; condensed tannins, (PA).

Glycosides

Glycosides may be phenol, alcohol or sulfur compounds. They are characterized by a sugar proteins attached by a special bond to one or non-sugar proteins. Many plants store chemicals in the form of inactive glycosides, which can be activated by enzyme hydrolysis (Polt, 1995). For this reason, most glycosides can be classified as prodrug as they remain inactive until they are hydrolysed leading to the release of aglycogen, the active constituent.The classification of glycosides is based on the nature of aglycogen, which can be any of a wide range of molecular types including phenols, quinines, terpenes and steroids. These glycosides are hetergenous in structure and are not easy to learn as specific group. Glycosides are of great significance, since they link monosaccharides together to form oligosacchraides and polysaccharides (Levy and Ting, 1995; Newman *et al.* 2008).

Saponines

Saponines are compounds whose active portions form colloidal solution in water, which produce lather on shaking and precipitate chlosterol. Saponines occur as glycosides whose aglycogen is tripenoid or steroidal structure. The combination of lipophilic sugars at the end gives them the ability to lower surface tension, producing the detergent characteristics (Guclu-ustundag and Mazza, 2007). Saponines are largely distributed in plants having many physiochemical (foaming, emulsification, solubilisation, sweetness and bitterness) and biological properties.

Other researchers have classified secondary metabolites into following, more specific types

Class	Type	Example
Alkaloids	Nitrogen containing	Cocaine, Psilosin, Caffeine, Nicotine, Morphine, Berberine, Vincristine, Reserpine, Galantamine, Atropine, Vincamine, Quinidine, Ephedrine, Quinine
Non protein amino acids	Nitrogen containing	
Amines	Nitrogen containing	
Cyanogenic glycosides	Nitrogen containing	Amygdalin, Dhurrin, Linamarin, Lotaustralin, Prunasin
Glusinolates	Nitrogen containing	
Alkamides	Nitrogen containing	
Lectins, peptides and polypeptides	Nitrogen containing	
Terpenes	Without nitrogen	Azadhirachtin, Artemisinin, Tetrahydrocannabinol
Steroids and saponins	Without nitrogen	These are terpenoids with a particular ring structure. Cycloaretenol
Flavonoids and Tannins	Wthout nitrogen	
Phenylpropanoids, Lignins, Coumarins and Lignans	Without nitrogen	
Polyacetylenes, fatty acids and waxes	Without nitrogen	
Polyketides	Without nitrogen	
Carbohydrates and organic acids	Without nitrogen	

2.2. Chemical Types of Secondary Metabolites

Plants produce a high diversity of natural products or secondary metabolites with a prominent function in the protection against abiotic stress (*e.g.* UV-B exposure) and also important for the communication of the plants with other organisms (Schafer *et al.* 2009), and are insignificant for growth and developmental processes (Rosenthal *et al.* 1991). There are three major groups of secondary metabolites *viz* terpenes, phenolics and N and S containing compounds. Terpenes composed of 5-C isopentanoid units, are toxins and feeding deterrents to many herbivores. Phenolics synthesized primarily from products of the shikimic acid pathway, have several important defensive role in the plants. Members of the third major group *i.e.* N and S containing compounds are synthesized principally from common amino acids (van Etten *et al.* 2001). *In vitro* and *in vivo* experiments using plants whose secondary metabolites expression has been altered by modern molecular methods have to confirm their defensive roles (Mes *et al.* 2000; Mansfield, 2000). Although the situation is still unclear, it is believed that most of the 100,000 known secondary metabolites are to be involved in plant chemical defense systems, which are formed throughout the millions of years during which plants have co-existed with their attackers (Wink, 1999). Although higher concentrations of secondary metabolites might result in a more resistant plant, the production of secondary metabolites is thought to be costly and reduces plant growth and reproduction (Simens *et al.* 2002). The cost of defense has also been invoked to explain why plants have evolved induced defense, where concentrations generally increase only in stress situations (Harvell and Tollrian, 1999). During the last several years, it has been discovered that hundreds of compounds that plants make have significant ecological and chemical defensive roles, opening a new area of scientific endeavour, often called ecological biochemistry (Harborne, 1988, 1989).

In plants an enormous variety of secondary metabolites has been described, their number amounting to more than 100 000 (Hadacek, 2002). The most characteristic feature of secondary metabolites is the large diversity of chemical types, embracing representatives of all main classes of organic compounds: aliphatic, aromatic, hydroaromatic, and heterocyclic, unique carbon skeletons occur along with multiplicity of functional groups (Table 1).

Chemical characters of secondary metabolites as determinants of their interactions and functions

A broad array of protective functions is performed by secondary metabolites in both biotic and abiotic stress situations: antimicrobial, photoprotective, structure stabilizing, signalling. Diverse chemical characters can be determinants of various functions, as shown in Table 2.

Availability of electrical charges

Electrostatic interactions of secondary metabolites can result in stabilizing of cell structures. Thus, positive charges in polyamines containing protonated amino- and imino groups allow electrostatic interactions with negatively charged loci in macromolecules and cellular substructures, with this exerting a stabilizing effect. Evidence is supplied that polyamine interactions with phosphoric acid residues in DNA, uronic acid residues in cell wall matrix, and negative groups on membrane surfaces contribute to the maintaining of their functional and structural integrity (Edreva, 1996 and Edreva *et al.* 2007, Berta *et al.* 1997).

Presence of carboxy-, hydroxy-, amino-, mono- and o-diphenol-, o-quinoid groups

Covalent bondings of the above functional groups underlie polymerization, condensation and complexation events, occurring mainly at cell wall level. Thus, polymerization of phenolic alcohols (sinapyl-, coniferyl- and p-coumaryl-) yields lignins strengthening the wall structure. Feruliuc acid, an unsaturated aromatic carbonic acid, is bound to polysaccharids chains in cell wall matrix by ester bonds, complexation (cross-linking) via diferulate bridges consolidates cell wall structure (Fry, 1986).

The presence of o-diphenol groups allows formation of o-quinoid grouping, *in vitro* covalent interactions between proteins and oxidation products of caffeoylquinic acid were recently reported (Prigent *et al.* 2007). Quinoid grouping can interact with thiol- and amino groups in proteins as follows:

O
R— =O + R'—SH → R— —OH
OH
R'—S

O
R— =O + R'—NH_2 → R— —OH
OH
R'—NH

The interactions of quinoid groupings with microbial proteins can underlie toxicity of secondary metabolites against fungi, bacteria and viruses, i.e. an antimicrobial activity, given that blocking of active sites and corresponding functions in proteins may occur. Similar interactions of quinoid groupings with plant proteins can be a mechanism of the necrotic reaction against pathogens, as well as of the necrobiosis due to abiotic stresses. High molecular condensation brown products of the quinoid form of o-diphenols and proteins were evidenced to arise during both types of necrobiosis and suggested to limit the spread of stress-induced tissue damage (Edreva, 1975). Hypersensitive response (HR) localizing invading pathogens at infection sites appears as a series of similar processes (Heath, 2000). Experimental data point to the involvement of phenolic-storing cells as keys in the expression of programmed cell death (Beckman, 2000).

Presence of conjugated double bonds

The rationale of the photoprotective function displayed by various chemical types of secondary metabolites (flavonoids, including anthocyanins, cinnamic acid derivatives, xanthophylls) can be the presence of conjugated double bonds, *i.e.* delocalized π-electrons. This electronic configuration allows absorbance in the visible light- and UV- spectrum and easy electron and energy transfers (Cockell, 1997). π-electron configuration occurs in ring-closed, short side-chained and long chained-structures (Table 2). Xanthophylls containing numerous conjugated double bonds in a long chain operate in the xanthophyll cycle, performing dissipation of excess light energy as harmless heat (Demmig-Adams, 2003).

Presence of thiol-, hydroxy-, amino-, o- and p-diphenol groups and unsaturated carbon chains, including conjugated double bond-containing ones

Increasing amount of evidences substantiates the functioning of secondary metabolites as antioxidants and antiradicals, assisting the plants to cope with oxidative stress arising in hostile environments (Grace and Logan, 2000, Grassmann *et al.* 2002, Gould *et al.* 2002). The list is ever growing, involving hydroxy- and thiol-group-containing compounds, such as ascorbic acid and lipoic acid, o-dihydroxy group-containing flavonoids, such as quercetin, aliphatic and arylamines, unsaturated fatty acids, carotenoids (Edreva, 2005, Edreva *et al.* 2007). Chelation of transition metals (Fe) by flavonoids such as quercetin interferes with the generation of reactive oxygen species (ROS) via the Fenton reaction, thus contributing to a powerful antioxidant/antiradical performance (Leopoldini *et al.* 2006).

Plants isoprene, a gaseous secondary metabolite containing two conjugated double bonds (Table 1) was shown to have an antioxidant effect in ozone- and heat-induced stress situations (Loreto and Velikova, 2001, Velikova *et al.* 2005a). Evidence in favour of the singlet oxygen quenching ability of isoprene was obtained (Velikova *et al.* 2004). Moreover, for the first time singlet oxygen quenching effect of phenylamides (conjugates of unsaturated aromatic carbonic acids and polyamines) was demonstrated by *in vitro* experiments (Velikova *et al.* 2007). Formation of phenylamides and dramatic accumulation of polyamines in bean and tobacco following abiotic stresses were reported, thus suggesting an antioxidant role of these secondary metabolites (Edreva *et al.* 1995, 1998, 2007).

Ortho-dihydroxy substitution in the B-ring of anthocyanins potentiating the antioxidant capacity was proposed as a protective mechanism in a physiological disorder of cotton (leaf reddening), due to oxidative stress provoked by Na+ / K+ imbalance (Edreva *et al.* 2006). Flavonoids (quercetin glycosides) and cinnamic acid derivatives endowed with high ROS scavenging capacity were suggested as a rationale of drought tolerance in cotton

Other characters and functions

Color and scent performance by secondary metabolites, such as flavonoids, terpenoids, and other volatives can underlie attraction or repelling of insects and herbivores, while toxins can be involved in plant-plant allelopathic interactions (Hadacek, 2002). Water solubility and lack of electrical charge contributes to the compatible solutes being implicated in osmoregulation and protection of hydrophilic cellular sites, while hydrophobic molecules (carotenoids) are the best candidates to protect lipophilic surfaces such as membranes. It is becoming increasingly clear that secondary metabolites may play an important role as signal molecules. The best known examples are ethylene, salicylic acid and jasmonic acid (Table 1) (Dixon *et al.* 2002, Vasconsuelo and Boland, 2007). However, the chemical rationale of their involvement in the signal transduction cascades remains elusive.

Table 1: Diverse chemical types of secondary metabolites

Chemical types	Formulae	Representatives
Aliphatic	$NH_2—CH_2—CH_2—CH_2—CH_2—NH_2$	Polyamines
	$CH_2=CH_2$	Ethylene
	$CH_2=C(CH_3)—CH=CH_2$	Isoprene
Aromatic	HO–(benzene ring)–CH = CH—CH_2OH	Phenolic alcohols
	(benzene ring) COOH, OH	Phenolic acids
	HO, OCH_3 (benzene ring) CH=CH—COOH	Unsaturated aromatic carbonic acids
Hydroaromatic	H, O, O, O, CH_3, CH_3	Terpenoids
	COOH, O	Jasmonic acid
Heterocyclic	OH, O, C, OH, OH, HO, O, OH	Flavonoids
	NH	Indole derivatives

Table 2: Chemical characteristics of secondary metabolites

Chemical characteristics	Interactions	Functions
Availability of electrical charge	Electrostatic	Structure stabilizing
Presence of – COOH, –OH, –NH_2, 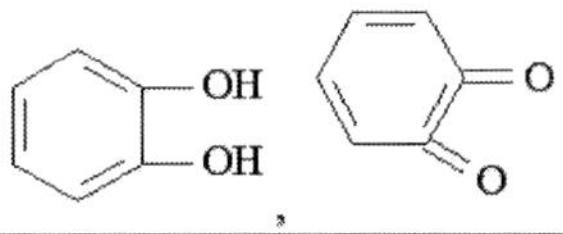 OH OH , O O	Covalent bonding: polymerisation, condensation, complexation	Structure-stabilizing (cell wall-strenghtening) Antimicrobial (blocking of active sites) Hypersensitive response (HR)
Availability of conjugated double bonds (delocalized π-electrons) in:  Ring-closed structures 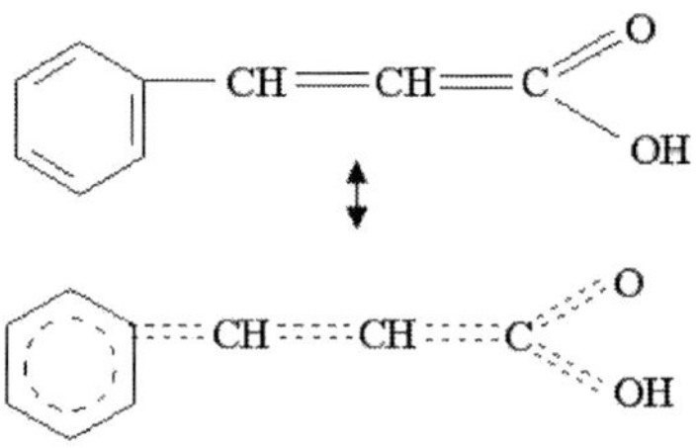 CH=CH=C(=O)OH CH⋯CH⋯C⋯O / OH Ring-closed Short side-chained structures —CH=CH—CH=CH— ⋯CH⋯CH⋯CH⋯CH⋯ Long-chained structures	Light and UV-absorbance Energy dissipation	Photoprotective
Availability of – NH_2, – SH, – OH, 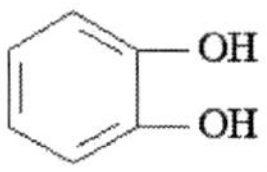 OH OH , – CH – CH = CH – CH = (unsaturated carbon chains)	H and electron transfers	Antioxidant Antiradical

2.3. Drought Stress and Secondary Metabolites

Drought stress occurs when the available water in the soil is reduced to such critical levels and atmospheric conditions adds to continuous loss of water. Drought stress tolerance is seen in all plants but its extent varies from species to species. Drought often causes oxidative stress and was reported to show increase in the amounts of flavonoids and phenolic acids in willow leaves (Larson, 1988). Drought stress influenced changes in the ratio of chlorophyll a and b and carotenoids (Anjum *et al.* 2003). A reduction in chlorophyll content was reported in cotton under drought stresse (Massacci *et al.* 2008) and *Catharanthus roseus* (Soliz-Guerrero *et al.* 2002). Drought conditions decreased the content of saponins in *Chenopodium quinoa* from 0.46% dry weight in plants growing under low water deficit conditions to 0.38% in high water deficit plants (Soliz-Guerrero *et al.* 2002). Anthocyanins are reported to accumulate under drought stress and at cold temperatures. Plant tissues containing anthocyanins are usually rather resistant to drought (Chalker-Scott, 1999). For example, a purple cultivar of chilli resists water stress better than a green cultivar (Bahler *et al.* 1991). Flavonoids have protective functions during drought stress. Flavonoids are implicated to provide protection to plants growing in soils that are rich in toxic metals such as aluminum (Winkel-Shirley, 2001). The influence of drought stress on various secondary metabolites are given in Table 3.

Table 3: Influence of drought stress on various plant secondary metabolites

Plant species	Secondary metabolite	Reference
Scrophularia ningpoensis	Glycoside	Wang *et al.* 2010
Papaver somniferum	Morphine alkaloid	Szabo *et al.* 2003
Glycine max	Trigonelline	Cho *et al.* 2003
Brassica napus	Glucosinolates	Jensen *et al.* 1996
Lupinus angustifolius	Chinolozdin alkaloids	Christiansen *et al.* 1996
Camellia cinemsis	Epicatechins	Hernàndez *et al.* 2006
Hypericum brasiliense	Betulnic acid	Nacif de Abreu and Mazzafera, 2005
Hypericum brasiliense	Rutine	Nacif de Abreu and Mazzafera, 2005
Prisms sativum	Flavonoids	27
Pisum sativum	Anthocyanins	Nogues *et al.* 1998
Halianthus annuum	Chloregenic acid	Del Moral, 1972
Salvia miltiorrhiza	Rosmaninic acid	Liu *et al.* 2011

Adapted from Bartels and Sunkar, 2005

According to estimated data, a 70% increase in crop yield could be achieved if the environmental conditions were close to optimum for a given plant. The main factors that restrain the plant distribution are environmental stresses, such as drought, low or high temperature and excessive salinity. These abiotic stress factors generate secondary stresses, *i.e.* osmotic and oxidative stress, which has negative influence on the plant, causing changes in its normal growth, development and metabolism (Bohenert *et al.* 1995, Kranner *et al.* 2010). Water stress is one of these environmental factors that can considerably limit distribution of crops in the world (Passioura, 2007, Cattivelli *et al.* 2008, Farooq *et al.* 2009).

Immediate drought responses, extent of acclimation and drought tolerance strongly differ among plant species (Ogaya and Peñuelas, 2003, 2006; Blackman *et al.* 2009; Gallé *et al.* 2009; Barigah *et al.* 2013) and populations within species (Marron *et al.* 2003; Costa e Silva *et al.* 2004; Monclus *et al.* 2009; Correia *et al.* 2014; Granda *et al.* 2014), but underlying mechanisms are still not wholly resolved even for classical drought-triggered phenomena such as reductions in stomatal conductance and photosynthetic rate (Niinemets and Way, 2016). Apart from changes in photosynthetic metabolism, drought induces a plethora of additional cellular responses that are manifested in changes in whole plant transcriptome and metabolome (Ramakrishna and Ravishankar, 2011, Arbona *et al.* 2013; Xu *et al.* 2013; Granda *et al.* 2014; Zhang *et al.* 2014), ultimately resulting in major changes in plant chemical composition. Accumulation of osmotica, including ubiquitous chemicals such as sugars and salt ions (Morales *et al.* 2013), and species-specific osmotica such as betaine (Grieve and Maas, 1984), glycinebetaine (Sakamoto and Murata, 2000), proline (Morales *et al.* 2013) and quercitol (Arndt *et al.* 2008) constitutes the most conspicuous chemical modification that importantly enhances plant drought resistance. In addition to osmotic modifications, reprogramming of plant metabolism in droughted plants results in multiple other changes in plant secondary metabolism.

Since stress-related reactions extensively impact the entire metabolism, the synthesis and accumulation of secondary metabolites also should be affected. Unfortunately, in the past, these coherences have not been considered adequately (Selmar, 2008). Kleinwächter and Selmar (2014) compiled the relevant literature in order to get a clearer picture of this issue. Due to water shortage, in combination with high light intensities, stomata are closed. As a result, the uptake of CO_2 is markedly decreased. In consequence, the consumption of reduction equivalents (NADPH + H^+) for CO_2 fixation via Calvin cycle declines considerably, generating a massive oversupply of NADPH + H^+. Accordingly, all metabolic processes are pushed towards the synthesis of highly reduced compounds, such as isoprenoids, phenols, or alkaloids (Selmar and Kleinwächter, 2013a and b).

The growing conditions, *e.g.* temperature, light regime, nutrient supply, strongly influence synthesis and accumulation of secondary plant products (Falk *et al.* 2007). Consequently, much more severe environmental influences, such as various stress situations, which strongly impact on general metabolism

(Bohnert *et al.* 1995), also will influence the metabolic pathways responsible for the accumulation of secondary plant products. Whereas a tremendous lot of information dealing with the impact of biological stress on the synthesis of secondary plant products is available, corresponding information, how secondary metabolism is changed in response to abiotic stress is rare. Mainly, the knowledge on the related biological background is limited (Ramakrishna and Ravishankar, 2011, Selmar and Kleinwächter, 2013a and b: Kleinwächter and Selmar, 2014). Accordingly, numerous studies which had been aimed to investigate the impact of one certain abiotic stress on secondary metabolism lack the distinction to other putative stress factors. Nevertheless, thorough reviewing and evaluating the literature adequately allows sound deductions on the impact of one single factor on the accumulation of natural products.

Secondary metabolic products, which are intensively synthesized under drought, are antioxidants (Nascimento and Fett-Neto, 2010). This group of compounds protects cells against the negative effects of ROS as well as against lipid peroxidation, protein denaturation and DNA damage (Kranner *et al.* 2002, Mittler, 2002, Allakhverdiev *et al.* 2008). In turn, environmental stress can cause a decline (Weidner *et al.* 2007, 2009b) or an increase (Wróbel *et al.* 2005, Weidner *et al.* 2009a) in the content of phenolic compounds in a cell. Phenolic compounds can scavenge ROS (Amarowicz *et al.* 2000, 2004, Negro *et al.* 2003, Caillet *et al.* 2006, Amarowicz and Weidner, 2009), they form complexes with the metals, which catalyse oxygenation reactions and inhibit activity of oxidizing enzymes (Elavarthi and Martin, 2010). Phenolic compounds include many secondary metabolites in plants that exhibit antioxidant properties (Oszmański, 1995). Phenols constitute a large group of compounds that may be divided into five subgroups: coumarins, lignins, flavonoids, phenolic acids and tannins (Gumul *et al.* 2007). Precursors for the synthesis of phenolic compounds are made in the shikimic acid and chorismic acid pathways. The metabolism of chorismic acid leads to L-phenylalanine and L-tyrosine. Tyrosine biosynthesis leads to *p*-coumaric acid and L-phenylalanine synthesis leads to cinnamic acid. As a result of the processes of methylation, hydration and dehydration of cinnamic acid, phenolic acids are produced, as part of the response of the plant to abiotic stresses (Dixon and Paiva, 1995). Some of the phenolic compounds, such as phenolic acids or flavonoids, are widely known and present in most of the plant species (Jwa *et al.* 2006). Phenolic compounds were considered as by-products of metabolic alteration (Solecka, 1997). The results investigations unequivocally suggest that phenols not only play a major role in defensive reactions of plants but also influence humans and animals that consume products enriched with phenolic compounds (Franca *et al.* 2001, Amarowicz and Weidner, 2009).

A large numbers of studies manifested that plants exposed to drought stress accumulate higher concentrations of secondary metabolites than those cultivated under well-watered conditions. There is evidence to suggest that the drought stress-related concentration enhancement is a common feature of all different classes of natural products. Corresponding increases are reported for simple as well as for complex phenols, and also for the various classes of terpenes. In equal measure, the concentrations of nitrogen-containing substances, such as alkaloids, cyanogenic glucosides, and glucosinolates, are positively impacted by drought stress, too. There is no doubt that drought stress consistently enhances the concentration of secondary plant products. In this context, however, we have to consider that the drought stressed plants generally are reduced in their growth. Thus, due to the reduction in biomass—even without any enhancement of the overall amount of natural products—their concentration on dry or fresh weight basis simply could be enhanced. Accordingly, corresponding explanations frequently are reported in the literature. Unfortunately, in most of the studies dealing with the impact of drought on natural product, biosynthesis data on the overall biomass per plant are lacking. One obvious reason for this deficit of information is due the fact that mostly only some certain plant parts or organs, *i.e.* roots, leaves, or seeds, had been in the centre of focus, whilst the overall content of the natural products on a whole plant basis was not of interest. However, in some papers, the total content of secondary plant products per entire plant is given or could be calculated from the data presented (Kleinwachter and Selmar, 2015).

In *Hypericum brasiliense* plants grown under drought stress both concentration and the total amount of the phenolic compounds is drastically enhanced in comparison to the control plants (de Abreu and Mazzafera, 2005) Despite the fact that the stressed *H. brasiliense* plants were quite smaller, the product of biomass and concentration of the related phenolics yield in 10% increase of the total amount of these natural products. In the same manner, in stressed peas (*Pisum sativum*), the overall amount of anthocyanins is about 25% higher as in plants cultivated under standard conditions (Nogués *et al.* 1998): although the biomass of the stressed pea plants is only about one third of the control plants, the massive increase in the concentration of phenolic compounds still resulted in a real increase of anthocyanins in the stressed plants. In the same manner, Jaafar *et al.* (2012) reported that not only the concentration but also the overall production of total phenolics and flavonoids per plant is enhanced in plants suffering drought stress, although the explicit data on biomass per plant are not displayed by the authors. In contrast, the overall yield of flavonoids was nearly the same, when the plants were either grown under drought stress or under well-watered, non-stress conditions. In stressed red sage plants (*Salvia miltiorrhiza*), the overall content of furoquinones is slightly lower in plants grown under water deficiency than that of the well-watered controls, although drought stress caused a significant increase of their concentration (Liu *et al.* 2011).

Focusing on terpenoids, there are only few reports available which soundly document a drought stress-related increase in the total amount of terpenoids per plant. In sage (*Salvia officinalis*), drought stress results in a massive increase in the concentration of mono terpenes, which easily overcompensate the reduction in biomass (Nowak *et al.* 2010). As a result, the entire amount of mono terpenes synthesized in sage plants suffering moderate drought stress is significantly higher than that of the well-watered controls. Yet, in parsley (*Petroselinum crispum*), the drought stress-related concentration enhancement of essential oils in leaves is more or less completely counterweighed by the related loss in biomass, resulting in nearly the same overall contents of essential oils in drought-stressed and well-watered plants (Petropoulos *et al.* 2008). In contrast, in drought-stressed catmint and lemon balm plants, the slight increase in the concentrations of mono terpenes could not compensate the stress-related detriment of growth and biomass. Accordingly, the overall content of terpenoids in the drought-stressed plants of *Melissa officinalis* and *Nepeta cataria* is lower than that in the corresponding well-watered controls (*Manukyan*, 2011).

Unfortunately, most of the reports dealing with the impact of drought stress on nitrogen-containing natural products focus on the concentrations of these compounds in certain organs and no data on the biomass of the entire plants are available. Accordingly, no deductions on the impact of drought on the overall content of natural products could be drawn. The only exception is the work of Xia *et al.* (2007), who analyzed the stress-related influence on accumulation of benzylisoquinoline alkaloids in cork tree seedlings (*Phellodendron amurense*) on concentration as well as on total content basis, whereas the concentration of berberine, jatrorrhizine, and palmatine is strongly enhanced by drought stress. Due to the growth reduction of drought-stressed plants, the overall content of alkaloids is considerably reduced and accounted only for about one fourth of the control. In conclusion, in nearly all plants analyzed, the concentrations of secondary plant products are significantly enhanced under drought stress conditions. However, only in few cases, also a corresponding increase of the total content of the natural compounds per plant is reported. This could be either due to the lack of data on the biomass of the corresponding plants, or to the fact that the stress-related decrease in biomass frequently over compensates the increase in the concentration of relevant natural products.

According to Franz (1983), Palevitch (1987) and Marchese and Figueira (2005) one of the most important factors affecting secondary metabolism is soil water capacity. Usually, limited availability of water has a negative effect on plant growth and development. However, an non-severe water deficit has sometimes proven beneficial for the accumulation of biologicallyactive compounds in medicinal and aromatic plants (Palevitch, 1987). Under high water stress, there is a limit on the translocation of carbon to its sinks, with the remaining carbon accumulates as carbohydrates, which leads to an increase in the carbon pool that would be allocated for secondary metabolism, with little or no competition with growth and development (Herms and Mattson, 1992). Ghershenzon (1984) demonstrated that, in herbaceous plants and shrubs, terpenes tend to increase under stress, mainly under severe water deficit conditions. This type of stress is known to increase the amount of secondary metabolites in a variety of medicinal plants, *e.g.*, artemisinin in *Artemisia annua* L. (Charles *et al.* 1993), ajmalicine in *Catharanthus roseus* (Jaleel *et al.* 2008) and hyperforin in *Hypericum perforatum* (Zobayed *et al.* 2005).

Interaction with other factors

Changing and unpredictable climate have resulted into crop losses leading to food insecurity and poverty in a number countries. One of the solution of this changing climate is the use of improved crop varieties with tolerance to drought and ability to give a decent harvest despite the unexpected changes in climate (Sagoe, 2006). The moisture stress induced responses are due to activation of various metabolic pathways resulting into re-establishment of cellular homeostasis as well as structural protection of membranes (Lokko *et al.* 2007). They are expressed in form of biochemical manifestation like increased enzymes activities and levels of secondary metabolites such as phenolics and tannins, osmotically active solutes such as proline, antioxidant enzymes such as catalase and peroxidase, hormones such as absiscic acid and pigments such as chlorophylls and carotenoids. As a result plants are enabled not only to recover after the stress (Okogbenin *et al.* 2010) but also to regain capacity to restore its normal metabolic activities in a shorter time and produce a decent yield.

Various putative functions for secondary plant products are known, *e.g.* protection against UV light or too high light intensities, action as compatible solutes, radical scavenging, or reduction of the transpiration (Edreva *et al.* 2008, Wink, 2010). Initiated by the tremendous progress in molecular biology, we are aware that the synthesis of the relevant secondary metabolites frequently is induced, modulated, and regulated by numerous environmental impacts and abiotic factors, respectively. Yet, the situation becomes even more complex, if we consider that the biosynthesis of phytoalexins is elicited by pathogen attack (Hahlbrock *et al.* 2003, Saunders and O'Neill, 2004) and various defense compounds effective against herbivores are synthesized as result of very complex induction mechanisms (Ferry *et al.* 2004). In consequence, the actual synthesis and accumulation of a certain natural product frequently is influenced and determined by numerous factors. Differentiation between the related effect on the osmotic potential and the water availability within the cell on the one hand, and the increase in redox potential due to the decline of the CO_2 influx caused by stomata closure on the other hand. With respect to the decrease in water availability, on the first sight, water potential seems to be a reliable parameter. Yet, in response to drought, many plants produce and accumulate osmotic active substances—denoted as compatible solutes—which significantly reduce the water potential without changing the amount of water, the actual water content, seems to be a better option. In addition to the classical gravimetrical methods to determine the water content, a new alternative methodology based on terahertz technology was recently presented (Breitenstein *et al.* 2011).

When focussing on the stress-induced enhancement of redox potential, the most appropriate and reliable markers should either be the increased ratio of NADPH + H^+ to $NADP^+$, or the amount of oxygen radicals generated. Unfortunately, the real *in situ* concentration of both components could not be quantified without inappropriate efforts and expenditures. As alternative, frequently, the enzymes responsible for the detoxification of the reactive oxygen species (ROS) generated are estimated, *i.e.* the superoxide dismutases (SOD) and the ascorbate peroxidases (APX). These enzymes occur in various isoforms and are also part of various signal transduction chains. Accordingly, they do not indicate reliably a specific stress situation and their significance as stress markers is limited. Another candidate as general stress marker is glutathione, which also is part of the antioxidative defence against ROS. Yet, the ratio of oxidized to reduced glutathione is not always reflecting the actual redox state (Tausz *et al.* 2004). Moreover, the stress-induced responses of the glutathione system are multi layered and biphasic: an initial response phase is followed by an acclimation phase. Thus, the use of the glutathione system as general stress marker for routine analysis is limited (Tausz *et al.* 2004). As alternative, the occurrence of characteristic stress metabolites, which are synthesized and accumulated pretty much specifically in response to a particular stress situation, could be estimated to determine and quantify its metabolic impact.

In this context, proline is in the centre of focus, since this amino acid is accumulated as compatible solute in plants suffering drought stress (Rhodes *et al.* 1999). However, the drought stress-induced proline accumulation does not occur in all plant species. A further option might be the quantification of γ-amino butyric acid (GABA), a stress metabolite produced by decarboxylation of glutamic acid (Kinnersley and Turano, 2000). Indeed, GABA is produced to a high extent in response to drought stress, but it also is accumulated under various other stress conditions (Satya Narayan and Nair, 1990, Bown and Shelp, 1997). Thus, alternative markers are required. One of the most promising options is the abundance of dehydrins. It is well established that dehydrins are frequently synthesized in plant cells suffering drought stress (Allagulova *et al.* 2003, Bouché and Fromm, 2004). It is assumed that these small hydrophilic proteins reveal various protective functions in desiccating cells (Hara, 2010). Indeed, these small protective proteins had been first discovered in maturing seeds in the course of late embryogenesis. Yet, it was realized that the occurrence of dehydrins during seed development is related to the appearance of maturation drying (Radwan *et al.* 2014), and also in seed dehydrins are synthesized in response to water deficiency. Accordingly, the expression of dehydrins seems to be the best option to monitor the impact of drought stress. However, we have to consider that any stress situation is really complex and eclectic. This vividly was demonstrated by Kramer *et al.* (2010), who found that in coffee seeds whilst drying the expression of dehydrins and the accumulation of the stress metabolite GABA follow different time patterns. Obviously, in the course of stress responses, several metabolic responses occur in parallel and/ or subsequently. As outlined above, we have to consider that in leaves exposed to drought stress, apart from the impact of the decrease in water availability, also the over-reduction due to stomata closure, entail numerous metabolic responses. Consequently, any comprehensive elucidation of the entire metabolic stress syndrome in medicinal plants requires a combination of several markers. In this context, apart from the accumulation of GABA and the expression of dehydrins, especially markers, which reflect the status of the various energy dissipation systems (*e.g.* the non-photochemical quenching, the xanthophyll cycle, or the photorespiration), have to be determined.

2.4. Temperature Stress and Thermotolerance

Like other stresses, heat stress causes accumulation of secondary metabolites of multifarious nature in plants. However, the specific roles they play in enhancing heat-stress tolerance seem to be different and warrant further elucidation. Most of the secondary metabolites are synthesized from the intermediates of primary carbon metabolism via phenyl propanoid, shikimate, mevalonate or methyl erythritol phosphate (MEP) pathways (Wahid and Gazanfar, 2006). Temperature strongly influences metabolic activity and plant ontology, and high temperatures can induce premature leaf senescence (Morison and Lawlor, 1999). Carotenoids in Brassicaceae, including β-carotene, were found to be slightly decreased after thermal treatments (Morison and lawlor, 1999). Elevated temperatures increase leaf senescence and root secondary metabolite concentrations in the herb *Panax quinquefolius (*Jochum *et al.*

2007). Elevated temperatures by 5°C would reduce photosynthesis and biomass production of *P. quinquefolius*, on the contrary storage ginsenoside is reported to be enhanced (Jochum *et al.* 2007).

High-temperature stress induces production of phenolic compounds such as flavonoids and phenylpropanoids. Phenylalanine ammonia-lyase (PAL) is considered to be the principal enzyme of the phenylpropanoid pathway. Increased activity of PAL in response to thermal stress is considered as the main acclamatory response of cells to heat stress. Thermal stress induces the biosynthesis of phenolics and suppresses their oxidation, which is considered to trigger the acclimation to heat stress for example as in watermelon, *Citrulus vulgaris* (Rivero *et al.* 2001). Carotenoids are widely known to protect cellular structures in various plant species irrespective of the stress type (Havaux, 1998, Wahid and Gazanfar, 2006, Wahid, 2007). For example, the xanthophyll cycle (the reversible interconversion of two particular carotenoids, violaxanthin and zeaxanthin) has evolved to play this essential role in photoprotection. Since zeaxanthin is hydrophobic, it is found mostly at the periphery of the lightharvesting complexes, where it functions to prevent peroxidative damage to the membrane lipids triggered by ROS (Horton, 2002). Recent studies have revealed that carotenoids of the xanthophylls family and some other terpenoids, such as isoprene or α-tocopherol, stabilize and photoprotect the lipid phase of the thylakoid membranes (Havaux, 1998, Sharkey, 2005, Velikova *et al.* 2005b). When plants are exposed to potentially harmful environmental conditions, such as elevated temperatures, the xanthophylls including violaxanthin, antheraxanthin and zeaxanthin partition between the light-harvesting complexes and the lipid phase of the thylakoid membranes. The resulting interaction of the xanthophyll molecules and the membrane lipids brings about a decreased fluidity (thermostability) of membrane and a lowered susceptibility to lipid peroxidation under high temperatures (Havaux, 1998).

Phenolics, including flavonoids, anthocyanins, lignins, *etc.,* are the most important class of secondary metabolites in plants and play a variety of roles including tolerance to abiotic stresses (Chalker-Scott, 2002, Wahid and Ghazanfar, 2006, Wahid, 2007). Studies suggest that accumulation of soluble phenolics under heat stress was accompanied with increased phenyl ammonia lyase (PAL) and decreased peroxidase and polyphenol lyase activities (Rivero *et al.* 2001). Anthocyanins, a subclass of flavonoid compounds, are greatly modulated in plant tissues by prevailing high temperature, low temperature increases and elevated temperature decreases their concentration in buds and fruits (Sachray *et al.* 2002). For example, high temperature decreases synthesis of anthocyanins in reproductive parts of red apples (Tomana and Yamada, 1988), chrysanthemums (Shibata *et al.* 1988) and asters (Sachray *et al.* 2002). One of the causes of low anthocyanin concentration in plants at high temperatures is a decreased rate of its synthesis and stability (Sachray *et al.* 2002). On the other hand, vegetative tissues under high temperature stress show an accumulation of anthocyanins including rose and sugarcane leaves (Wahid and Ghazanfar, 2006). It has been suggested that in addition to their role as UV screen, anthocyanins serve to decrease leaf osmotic potential, which is linked to increased uptake and reduced transpirational loss of water under environmental stresses including high temperature (Chalker-Scott, 2002). These properties may enable the leaves to respond quickly to changing environmental conditions. Isoprenoids, another class of plant secondary products, are synthesized via mevalonate pathway (Taiz and Zeiger, 2006). Being of low molecular weight and volatile in nature, their emission from leaves has been reported to confer heat-stress tolerance to photosynthesis apparatus in different plants (Loreto *et al.* 1998). Studies have revealed that their biosynthesis is cost effective. While deriving considerable amount of photosynthates, they show compensatory benefits as to heat tolerance (Funk *et al.* 2004). Plants capable of emitting greater amounts of isoprene generally display better photosynthesis under heat stress, thus there is a relationship between isoprene emission and heat-stress tolerance (Velikova and Loreto, 2005). Sharkey (2005) opined that isoprene production protects the PSII from the damage caused by ROS, including H_2O_2, produced during heat-induced oxygenase action of rubisco, even though the photosynthetic rate approaches zero. It is proposed that endogenous production of isoprene protects the biological membranes from damaging effects by directly reacting with oxygen singlets ($^1O_2$) by means of isoprene-conjugate double bond (Velikova *et al.* 2005b).

It is known that exposure of plants to abiotic factors including high and low temperatures causes oxidative stress in which increased production of reactive oxygen species (ROS) is evident (Aroca *et al.* 2001). ROS reacts with lipids, proteins and DNA, thus, causing immense cellular damage. Therefore, plants contain an array of enzymatic and non-enzymatic mechanisms to protect themselves from increased accumulation of ROS (Inze and van Montagu, 1995). Plants combat oxidative stress with antioxidant substrates and protective enzymes (Polle, 1997, Noctore and Foyer, 1998). Superoxide radicals are detoxified by superoxide dismutase resulting in H_2O_2 formation (Bowler *et el.* 1992). The removal of H_2O_2 is achieved by ascorbate peroxidase, which oxidize ascorbate to monodehydro ascorbate (Asada, 1999). These radicals can dismutate spontaneously to ascorbate and dehydro ascorbate. In addition, monodehydro-ascorbate radicals are reduced to ascorbate by monodehydro ascorbate reductase Dehydroascorbate is reduced to ascorbate with glutathion as reductant. Glutathion (reduced form) ascorbate with glutathione as reductant.

Glutathione (reduced form), which is oxidized in this process to glutathione disulfide, is recycled by glutathione reductase consuming NADPH (Foyer and Halliwell, 1976). In plant cell, tissue and organ culture, the effects of the temperature on the regulation of different physiological processes occurring *in vitro* culture, particularly during somatic embryogenesis, is not fully understood (Shohael *et al.* 2006). It has been shown that the optimal temperature treatment is required even in suspension cultures for accumulation of biomass and production of metabolites (Zhong and Yoshida, 1993).

Elevated temperature prompts strain within stressed cells that lowers quality and yield of various crop species (Maestri *et al.* 2002, Anderson-Teixeira *et al.* 2012). In general, heat stress induces the production of reactive oxygen species (ROS) in plants. Overproduction of ROS such as superoxide (O_2^-), hydrogen peroxide (H_2O_2), hydroxyl radical (OH^-) and singlet oxygen that damage the membrane lipids, nucleic acids, structural proteins, photosynthetic pigments and enzymes (Ozkur *et al.* 2009, Nawaz *et al.* 2013), and consequently disturb the normal metabolic activities of plants. Plant metabolism contributes towards heat stress tolerance by providing the energy and metabolites essential for cellular homeostasis, or via production of protective osmolytes like sucrose, fructose and myo-inositol in excess. The strength and recovery of these systems under heat stress significantly rely upon the plant metabolism and its products (Gu *et al.* 2012, Bartwal *et al.* 2014). As a strategic adaptation, to cope with the environmental challenges, plants modulate the production and accumulation of a number of enzymatic (Xu *et al.* 2011, Ahmad *et al.* 2013) and non-enzymatic (Gu *et al.* 2012, Mahmood *et al.* 2012) antioxidants. These antioxidants scavenge the heat stress triggered ROS. Plants overproduce secondary metabolites under heat stress as one of the adaptive stratageis (Jochum *et al.* 2007, Tanveer *et al.* 2012). They have diverse protective functions ranging from toxicity and light / UV shielding to signal transduction. Their importance in plant stress physiology is mediated by structural and functional modifications, diversity of chemical types and their interactions. They possibly allow the resumption of normal cellular and physiological activities (Vasconsuelo and Boland, 2007). Secondary metabolites interact with phosphoric acid residues in DNA and uronic acid residues in cell wall matrix. They possibly stabilize the plant cell structures with the help of their electrostatic interactions, thereby mediating the defensive adaptations in heat stressed plants (Edreva *et al.* 2008).

Heat stress affects the whole life cycle of plants as it exaggerates the effects of other abiotic stresses. A minor variation in ambient temperature directly upsets the plant growth and development. It considerably alters the physiological settings of the plants, particularly at seedling stage. There are a number of studies that assessed the physiological, molecular and genetic basis of temperature tolerance in different crops (Wang *et al.* 2003). However, the effects of heat stress upon secondary plant metabolism are still not fully unveiled. Furthermore, the comparative data with respect to thermotolerance and thermosensitivity mediated by secondary metabolites in maize seedlings is scarce.

Low temperature stress

Low temperatures are another important plant abiotic stress. Lower temperatures alter the membrane structure, also affecting the activity of membrane-bound enzymes. An excessive production of ROS can be associated with chilling and this has deleterious effects on membranes. Moreover, low temperatures reduce scavenging enzyme activities, impairing the whole antioxidant plant response. In these conditions, some plants can adapt by modifying the membrane composition and activating oxygen-scavenging systems (Oh *et al.* 2009, Lütz *et al.* 2010, Airaki *et al.* 2012). Low temperature conditions also determine the increased production of phenolics, which exert antioxidant activity in chilled tissues. An enhancement of phenylpropanoid metabolism is induced in plant tissues when temperatures decrease below a certain threshold value (Lattanzio *et al.* 2001, 2012) Low, non-freezing temperature stress induces an increase in phenylalanine ammonia-lyase and chalcone synthase activities, as well as the activation of a number of genes involved in phenolic metabolism (Stefanowska *et al.* 2002). Anthocyanins are believed to accumulate in leaves and stems of *Arabidopsis thaliana* in response to low temperatures (Leyva *et al.* 1995, Chalker-Scott, 1999, Solecka *et al.* 1999). Christie *et al.* (1994) show that an increase in anthocyanin and mRNA abundance in the sheaths of maize seedlings are positively related with the severity and duration of the cold.

Temperate plants have adapted to variations in temperature by adjusting their metabolism during autumn, increasing their content of a range of cryo-protective compounds to maximize their cold tolerance (Janska *et al.* 2010). In the cryopreservation process, environmental changes including osmotic injury, desiccation, and low temperature can impose a series of stresses on plants (Janska *et al.* 2010). During over wintering, temperate plant metabolism is redirected toward synthesis of cryoprotectant molecules such as sugar alcohols (sorbitol, ribitol, inositol) soluble sugars (saccharose, raffinose, stachyose, trehalose), and low-molecular weight nitrogenous compounds (proline, glycine betaine) (Janska *et al.* 2010). Cold stress increases phenolic production and their subsequent incorporation into the cell wall either as suberin or lignin (Griffith and Yaish, 2004). In addition, apple tree adaptation to cold climate was found to be associated with a high level of chlorogenic acid (Parez-Ilzarbe *et al.* 1997). Lignification and suberin deposition are also shown to increase resistance to cold temperatures. A mechanism by which suberin and lignin may protect plants from freeze damage (Griffith and Yaish, 2004). Christie *et al.* (1994) reported the accumulation of anthocyanins during cold stress. Pedranzani *et al.* (2003) reported that cold and water stresses produce changes in endogenous jasmonates in *Pinus pinaster*. Lei *et al.* (2004) reported that melatonin protect against cold-induced apoptosis in carrot suspension cells by upregulation of polyamines (putrescine and spermine). Moreover, Melatonin applied to cucumber (*Cucumis sativus* L.) seeds improves germination during chilling stress (Posmyk *et al.* 2009). Zhao *et al.* (2011) reported that melatonin improves the survival of cryopreserved callus of *Rhodiola crenulata.* The survival rate of the cryopreserved callus increased when the callus was pretreated with 0.1 μM melatonin.

The effect of cold stress on polyamime accumulation had been reported (Kovàcs *et al.* 2011). When leaves of wheat (*Triticum aestivum* L.) are exposed to a cold temperature, accumulation of putrescine (6–9 times), spermidine accumulates to a lesser extent and, spermine decreases slightly. Moreover, alfalfa (*Medicago sativa* L.) also accumulates putrescine under low temperature stress (Nadeau *et al.* 1987). Hummel *et al.* (2004) reported that cold tolerance was associated with increased levels of polyamines (agmatine and putrescine) and their levels could be a significant marker of chilling tolerance in seedlings of *P. antiscorbutica*.

Temperature variation stress

Several studies have examined the effects of increased temperatures on secondary metabolite production of plants (Morison and lawlor, 1999). Lower soil temperatures caused an increase in levels of steroidal furostanol and spirostanol saponins (Yu *et al.* 2005). Temperature variations has multiple effects on the metabolic regulation, permeability, rate of intracellular reactions in plant cell cultures (Morison and lawlor, 1999). Changing the culture temperature may change the physiology and metabolism of cultured cells and subsequently affect growth and secondary metabolite production (Morison and lawlor, 1999). Temperature range of 17–25°C is normally used for the induction of callus tissues and growth of cultured cells (Rao and Ravishankar, 2002). Yu *et al.* (2005) reported the influence of temperature and light quality on production of ginsenoside in hairy root culture of *panax ginseng.* Chan *et al.* (2010) reported that *Melastoma malabathricum* cell cultures incubated at a lower temperature range (20 ± 2°C) grew better and had higher anthocyanin production than those grown at 26 ± 2°C and 29 ± 2°C. Optimum temperature (25°C) maximizes the anthocyanin yield as demonstrated in cell cultures of *Perilla frutescens* (Zhong and Yashida, 1993) and strawberry (Zhang *et al.* 1997). Lower temperature favors anthocyanin accumulation, but reduces cell growth. For strawberry cell culture, maximum anthocyanin content was obtained at 15°C and it was about 13-fold higher than that obtained at 35°C (Zhang *et al.* 1997). For suspension cultures of *Perilla frutescens*, anthocyanin production was remarkably reduced at the relatively high temperature of 28°C, whereas 25°C was optimal for the productivity of the pigment (Zhong and Yashida, 1993). Similar observations on optimal productivity of anthocyanin in cell suspension cultures of *Daucus carota* was reported (Narayan *et al.* 2005). Pigment release from hairy root cultures of *Beta vulgaris* under the influence of different temperatures was reported (Thimmaraju *et al.* 2003).

2.5. Soil Salinity Stress and Secondary Metabolites

Salt environment leads to cellular dehydration, which causes osmotic stress and removal of water from the cytoplasm resulting in a reduction of the cytosolic and vacuolar volumes. Salt stress often creates both ionic as well as osmotic stress in plants, resulting in accumulation or decrease of specific secondary metabolites in plants (Mahajan and Tuteja, 2005). Anthocyanins are reported to increase in response to salt stress (Parida and Das, 2005). In contrast to this, salt stress decreased anthocyanin level in the salt-sensitive species (Daneshmand *et al.* 2011). Petrusa and Winicov (1997) demonstrated that salt tolerant alfalfa plants rapidly doubled their proline content in roots, whereas in salt sensitive plants the increase was slow. However, Aziz *et al.* (1998) reported a correlation between proline accumulation and salt tolerance in *Lycopersicon esculentum* and *Aegiceras corniculatum* respectively. In tomato cultivars under salt stress endogenous jasmonic acid was found to accumulate (Pedrazani *et al.* 2003). Polyphenol synthesis and accumulation is generally stimulated in response to biotic or abiotic stresses (Dixon and Paiva, 1995; Muthukumarasami *et al.* 2000). Increase in polyphenol content in different tissues under increasing salinity has also been reported in a number of plants (Parida and Das 1995). Navarro *et al.* (2006) showed increased total phenolics content with moderately saline level in red peppers. Plant polyamines have been shown to be involved in plant response to salinity. Salinity-induced changes of free and bound polyamine levels in sunflower (*Helianthus annuus* L.) roots was reported (Mutlu and Bozcuk, 2006). The influence of salt stress on secondary metabolites in plants are shown in Table 4.

Table 4: Salt stress increases various secondary metabolites in plant.

Secondary metabolites	Plant Species	Reference
Sobitol	*Lycopersicum esculentum*	Tari *et al.* 2010
GABA	*Sesamum indicum* L.	Bor *et al.* 2009
Flavonoids	*Hordeum vulgare*	Ali and Abbas, 2003
Jasmonic acid	*Lycopersicum esculentum*	Pedrazani *et al.* 2003
Polyphenol	*Cakile maritime*	Ksouri *et al.* 2007
Tropane alkaloids	*Datura innoxia*	Brachet and Cosson, 1986
Anthocyanins	*Grevillea* sp.	Parida and Das, 2005
Trigonelline	*Glycine max*	Cho *et al.* 1999
Glycinebetain	*Trifolium repens*	Varshney and Gangwar, 1988
Polyamines	*Oryza sativa*	Krishnamurthy and Bhagwat, 1989
Glycine betaine	*Triticum aestivum*	Krishnamurthy and Bhagwat, 1990
Sucrose and starch	*Cenchrus pennisetiformis*	Ashraf, 1997

Adapted from Parvez and Satyawati, 2008

Soil salinity is a major environmental constraint to plant growth and productivity and is an especially serious problem in agricultural systems that rely heavily on irrigation (Munns, 2002). High salt concentration causes osmotic and ionic stress in plants (Le Rudulier, 2005). Salinity is expressed by a series of morphological, physiological, metabolic and molecular changes that cause delayed germination, poor stand establishment (Almansouri *et al.* 2001), high seedling mortality, stunted growth and lower yields (Allakhverdiev *et al.* 2000, Khan *et al.* 2010, Muhammad and Hussain, 2010). Plants have evolved complex mechanisms for adaptation to osmotic and ionic stresses caused by high salt. These mechanisms include osmotic adjustment by accumulation of compatible solutes, such as proline, glycine, betaine, polyols, sugar alcohols and soluble sugars, and lowering the toxic concentration of ions in the cytoplasm by restriction of Na^+ influx or its sequestration into the vacuole and/or its extrusion (Wuyts *et al.* 2006).

Plants produce a large variety of secondary products that their concentrations are strongly depending on the growing conditions and it is obvious that, especially, stress situations have a strong impact on the metabolic pathways responsible for the accumulation of the related natural products. Due to their capacity to scavenge reactive oxygen species, secondary products like flavonoids and saponins represent important radical scavengers (Haghighi *et al.* 2012).. These aspects of the significance of secondary metabolites could contribute to the understanding of the high plasticity and variability of secondary metabolism. In a series of experimental observations, it could be shown that plants which are exposed to salt stress produce a greater amount of secondary plant products, such as phenols, terpenes as well as nitrogen and sulphur containing substances such as alkaloids (Horborne and Williams, 2000, Stewart *et al.* 2001, Winkel, 2002, Mosaleeyanon *et al.* 2005, Couceiro *et al.* 2006).

Flavonoids have multiple biological activities *in vitro* and *in vivo*, such as antiadherence, antioxidant and antiinflammatory. Their ability to act as antioxidants depends on the reduction potentials of their radicals and accessibility of the radicals (Rice-Evans, 2001, Heim *et al.* 2002). Saponins belong to the secondary metabolites of mixed biosynthesis. They consist of a tri-terpene/steroid nucleus (the aglycone) with mono-or oligosaccharides attached to this core. They have been considered undesirable due to toxicity and their haemolytic activity. There is evidence for saponin regulation of the apoptosis pathways enzymes (AkT, Bcl and ERk1/2), leading to programmed cell death of cancer cells (Ellington *et al.* 2006, Zhu *et al.* 2005).

The earliest plant response of salt stress is a reduction in the rate of leaf surface expansion, followed by cessation of expansion as the stress intensifies (Parida and Das, 2005). The deleterious effect of salinity on plant growth is attributed to the decreased osmotic potential of the growing medium, specific ion toxicity, and nutrient ion deficiency (Luo *et al.* 2005). Saline environments can reduce a wide number of responses in plants, including readjustment of transport and metabolic processes, leading to growth inhibition (Lambers, 1985). Na^+ is the predominant soluble cation in most saline soils and water, particularly in coastal areas. Most crop plants exhibit considerable hypersensitivity to saline environments because inter cellular accumulation of N_a^+ is toxic to cellular metabolism, and for many salt-sensitive plants excess N_a^+ in the soil plays a major role in growth inhibition (Greenway and Munns, 1980). High sodium levels disturb potassium (K^+) nutrition and when accumulated in cytoplasm it inhibits many enzymes (Sharma *et al.* 1987). These effects are also due to a combination of adverse osmotic gradients, and the inhibitory effects of salts and ions on cell metabolism, as well as nutrient imbalance and such secondary stresses as oxidative stress linked to the production of toxic reactive oxygen intermediates (Iqbal *et al.* 2006).

2.6. Mineral Nutrition and Secondary Metabolites

Both the primary and secondary metabolism of higher plants are influenced by mineral nutrition. In most terrestrial ecosystems, plant growth is nitrogen limited, but phosphorus limitation also occurs frequently. In these conditions, high concentrations of phenolics in tissues of low-productivity species growing at infertile sites are observed (Aerts and Chapin, 2000, Scheible *et al.* 2004). Broadly, it was observed that low-productivity species have higher amounts of secondary compounds than high-productivity species. Species growing in nutrient-poor habitats often have traits that lead to high nutrient retention and high levels of secondary metabolites, which have a defense role against herbivores and pathogens (Lillo *et al.* 2008). Deficiencies of essential elements (such as nitrogen, phosphorus and potassium) can increase the amounts of phenolics in plant tissues either as existing pools or by inducing their *de novo* synthesis (Kováèik *et al.* 2007, Glynn *et al.* 2007). Barley plants grown under nitrogen deficiency conditions showed lower biomass, while leaf levels of soluble phenolics increased (Mercure *et al.* 2004). Iron deficiency induced increased amounts of phenolic acids in root exudates of non-graminaceous monocots and dicots (Jin *et al.* 2007, Jin *et al.* 2008, Vigani *et al.* 2013). An increased amount of anthocyanins is recognized as a consequence of phosphorus limitation. Anthocyanin over-accumulation lowers the accumulation of ROS *in vivo* under oxidative and drought stress (Nakabayashi *et al.* 2014). As a consequence of phosphorus limitation, the content of phenylpropanoids and flavonoids resulted in increased *Arabidopsis thaliana* roots and shoots. The overexpression of MYB transcription factors PAP1/MYB75 and/or PAP2/MYB90 led plants to increase the content of anthocyanins and glycosides of quercetin and kaempferol (Tohge *et al.* 2005, Morcuende *et al.* 2007). This indicates that PAP1 and PAP2 have a role in increasing phenolics during pkosphorus limitation (Lillo *et al.* 2008, Pant *et al.* 2015, Rubio *et al.* 2001).

Deficiencies in nitrogen and phosphate directly influence the accumulation of phenylpropanoids (Dixon and Paiva, 1995). Potassium, sulfur and magnesium deficiency are also reported to increase phenolic concentrations. Low iron level can cause increased release of phenolic acids from roots (Chalker-Scott and Fnchigami, 1989). Calcium levels have been implicated in plant response to many abiotic stresses including cold, drought and salinity. Bryant *et al.* (1983) have hypothesized that when plants are stressed, an exchange occurs between carbon to biomass production or formation of defensive secondary compounds. A stress response is induced when plants recognizes stress at the cellular level. Secondary metabolites are involved in protective functions in response to both biotic and abiotic stress conditions. Formation of phenyl amides and dramatic accumulation of polyamines in bean and tobacco under the influence of abiotic stresses were reported, suggesting antioxidant role of these secondary metabolites (Edreva *et al.* 2000). Similarly, anthocyanin accumulation is stimulated by various environmental stresses, such as UV, blue light, high intensity light, wounding, pathogen attack, drought, sugar and nutrient deficiency (Winkel-Shirley, 2001).

When plants are stressed, secondary metabolite production may increase because growth is often inhibited more than photosynthesis, and the carbon fixed is predominantly allocated to secondary metabolites (Seigler, 1998). The *Daucus carota* callus subjected to phosphate stress produced 7.2% dry wt anthocyanin against 5.4% dry weight (DW) in the control (Rajendra *et al.* 1992). Nutrient stress also has a marked effect on phenolic levels in plant tissues

(Chalker-Scott and Fnchigami, 1989). Deficiencies in nitrogen and phosphate lead to the accumulation of phenyl propanoids and lignification (Dixon and Paiva, 1995). In tomato, the 3-fold increase in anthocyanidins level and the simultaneous doubling of quercetin-3-O-glucoside occurs under nutrient stress stress (Bongue-Bartelsman and Phillips, 1995). Zeid (2009) reported that the increased urea concentration in the nutrient solution markedly increased putrescine contents in *Phaseolus vulgaris* cell suspensions. Osmotic stress created by sucrose and other osmatic agents was found to regulate anthocyanin production in *Vitis vinifera* cultures (Tuteja and Mahajan, 2007).

Different hypotheses, such as the carbon-nutrient balance hypothesis and the growth-differentiation balance hypothesis, have been considered in order to explain the influence of nutrient deficiency on secondary phenolic metabolism. These hypotheses affirm that carbon skeletons synthesized by photosynthesis are dynamically used for growth (primary metabolism) or defense (secondary metabolism). Because allocation for plant growth and defense can take place at the same time in plants, these hypotheses suggest that secondary metabolism utilizes extra carbon skeletons when growth is more limited than photosynthesis (*e.g.*, due to mineral element deficiencies) (Stamp, 2003, Liakopoulos and Karabourniotis, 2005, Hamilton*et al.* 2001).

2.7. Secondary Metabolites Under Heavy Metal Stress

Of the implications of human-induced disturbance of natural biogeochemical cycles, accentuated accumulation of heavy metals is a problem of paramount importance for ecological, nutritional, and environmental reasons (Nahajyoti *et al.* 2010, Ali *et al.* 2013). Heavy metals belong to group of nonbiodegradable, persistent inorganic chemical constituents with the atomic mass over 20 and the density higher than 5 $g \cdot cm^{-3}$ that have cytotoxic, genotoxic, and mutagenic effects on humans or animals and plants through influencing and tainting food chains, soil, irrigation or potable water, aquifers, and surrounding atmosphere (Flora *et al.* 2008, Chirlakova, 2009, Rascio and Navari-Izzo, 2011, Wuana and Okieimen, 2011). There are two kinds of metals found in soils, which are referred to as essential micronutrients for normal plant growth (Fe, Mn, Zn, Cu, Mg, Mo, and Ni) and nonessential elements with unknown biological and physiological function (Cd, Sb, Cr, Pb, As, Co, Ag, Se, and Hg) (Rascio and Navari-Izzo, 2011, Schützendübel and Polle, 2002, Tangahu *et al.* 2011, Zhou *et al.* 2014). Both underground and aboveground surfaces of plants are able to receive heavy metals (Patra *et al.* 2004). The essential elements play a pivotal role in the structure of enzymes and proteins. Plants require them in tiny quantities for their growth, metabolism, and development, however, the concentration of both essential and nonessential metals is one single important factor in the growing process of plants so that their presence in excess can lead to the reduction and inhibition of growth in plants (Zengin and Munzuroglu, 2005). Heavy metals at toxic levels hamper normal plant functioning and act as an impediment to metabolic processes in a variety of ways, including disturbance or displacement of building blocks of protein structure, which arises from the formation of bonds between heavy metals and sulfhydryl groups (Hall, 2002), hindering functional groups of important cellular molecules (Hossain *et al.* 2012), superseding or disrupting functionality of essential metals in biomolecules such as pigments or enzymes (Ali *et al.* 2013), and adversely affecting the integrity of the cytoplasmic membrane (Farid *et al.* 2013), resulting in the repression of vital events in plants such as photosynthesis, respiration, and enzymatic activities (Hossain *et al.* 2012). On the other hand, elevated levels of heavy metals are associated with the increased generation of reactive oxygen species (ROS), such as superoxide free radicals ($O_2^{\cdot -}$), hydroxyl free radicals (OH^-), or non-free radical species (molecular forms) such as singlet oxygen (O_2^*) and hydrogen peroxide (H_2O_2) as well as cytotoxic compounds like methylglyoxal (MG), which can cause oxidative stress via disturbing the equilibrium between prooxidant and antioxidant homeostasis within the plant cells (Zengin and Munzuroglu, 2005, Hossain *et al.* 2012, Sytar *et al.* 2013). This condition implicates the causation of multiple deteriorative disorders such as, oxidation of protein and lipids, ion leakage, oxidative DNA attack, redox imbalance, and denature of cell structure and membrane, ultimately resulting in the activation of programmed cell death (PCD) pathways (Nagajyoti *et al.* 2010, Flora *et al.* 2008, Rascio and Navari-Izzo, 2011, Rellán-Álvarez *et al.* 2006, Hatata and Abdel-Aal, 2008, Sharma *et al.* 2012).

Metal ions (lanthanum, europium, silver and cadmium), and oxalate are also exert their influence on secondary metabolite production (Marschner, 1995). The trace metal nickel (Ni) is essential component of urease enzyme, is needed for plant development (Marschner, 1995). However, elevated Ni concentrations reduce plant growth (Hagemeyer, 1999). The significant decrease in anthocyanin levels due to Ni stress has been reported by Hawrylak *et al.* (2007). Moreover, Ni has been shown to inhibit accumulation of anthocyanins (Krupa *et al.* 1996). Trace metals obviously limit anthocyanin biosynthesis by inhibiting activity of L-phenylalanine ammonia-lyase (PAL) (Krupa *et al.* 1996). Effective accumulation of metals (Cr, Fe, Zn, and Mn) also produced an increase of oil content up to 35% in *Brassica juncea* (Singh and Sinha, 2005). Cu^{2+} and Cd^{2+} have been shown to induce higher yields of secondary metabolites such as shikonin (Mizukami *et al.* 1977) and also on the production of digitalin (Ohlsson and Berglund, 1989). Cu^{2+} also stimulated the production of betalains in *Beta vulgaris* (Trejo-Tapia et al. 2001) Co^{2+} and Cu^{2+} having the stimulatory effect on the production of secondary metabolites (Trejo-tapia *et al.* 2001). In an attempt to enhance betalaines production, the hairy roots were exposed to metal ions (Rudrappa *et al.* 2004). Obrenovic (1990) has demonstrated stimulatory effects of Cu^{2+} on the accumulation of betacyanins in callus cultures of *Amaranthus caudatus*. Addition of Zn^{2+} (900 ìM) enhanced the yield of lepidine in cultures of *Lepidium sativum (*Obrenovic, 1990). However, Cu proved more effective than Zn in enhancing the yield (Saba *et al.* 2000). $AgNO_3$ or $CdCl_2$ elicited overproduction of two tropane alkaloids, scopolamine and hyoscyamine, by in hairy root cultures of *Brugmansia candida* (Angelova et al. 2006). Rare-earth metal (lanthanum) had influence on production of taxol in cell culture of *Taxus sp.* (Pitta-Alvarez, *et al.* 2000). Oat and bean plants treated with cadmium and copper significantly increased putrescine (Put) content (Weinstein *et al.* 1986). However, a decrease in Put level in Cd^{2+} or Cu^{2+} treated sunflower leaf disks has been reported. Sunflower leaf disks showed a significant decreased in spermidine (Spd) content and no variation in spermine (Spm) level when they were treated with Cd^{2+} or Cu^{2+} respectively (Groppa *et al.* 2001). However, Jacobsen *et al.* (1992) reported no changes in Spd or Spm content in chromium-exposed leaves of barley and rape plants, but Put accumulated with increasing chromium concentrations or exposure time. Lin and Kao (1999) reported that copper treatment increased Put, but a decrease in Spm concentration in rice leaves.

Plants employ various inherent and extrinsic defense strategies for tolerance or detoxification whenever confronted with the stressful condition caused by the high concentrations of heavy metals. As a first step towards dealing with metal intoxication, plants adopt avoidance strategy to preclude the onset of stress via restricting metal uptake from soil or excluding it, preventing metal entry into plant root (Viehweger, 2014). This can be achieved by some mechanisms such as immobilization of metals by mycorrhizal association, metal sequestration, or complexation by exuding organic compounds from root (Patra *et al.* 2004. Dalvi and Bhalerao, 2013). At next stage, if these strategies fail and heavy metals manage to enter inside plant tissues, tolerance mechanisms for detoxification are activated which include metal sequestration and compartmentalization in various intracellular compartments (*e.g.*, vacuole) (Patra *et al.* 2004), metal ions trafficking, metal binding to cell wall, biosynthesis or accumulation of osmolytes andosmoprotectants, for example, proline, intracellular complexation or chelation of metal ions by releasing several substances, for example, organic acids, polysaccharides, phytochelatins (PCs), and metallothioneins (MTs) (Dalvi and Bhalerao, 2013. Memon *et al.* 2001. Prasad, 2004. John *et al.* 2009), and eventually if all these measures prove futile and plants become overwhelmed with toxicity of heavy metal (HM), activation of antioxidant defense mechanisms is pursued (Manara, 2012).

Large areas of land are contaminated with heavy metals (the main group of inorganic contaminants) resulting from urban activities, agricultural practices and industry (Khan *et al.* 2000, Clemens, 2001). Excessive concentrations of trace elements (Cd, Co, Cr, Hg, Mn, Ni, Pb and Zn) are toxic and lead to growth inhibition, decrease in biomass and death of the plant (Zenk, 1996). Heavy metals inhibit physiological processes such as respiration, photosynthesis, cell elongation, plant-water relationship, N-metabolism and mineral nutrition (Zornoza *et al.* 2002). According to the chemical and physical properties of heavy metals we can divide their harmful action into:

a. generation of ROS (*Reactive oxygen species*) by autooxidation and Fenton reaction,

b. blocking of essential functional groups in biomolecules: proteins (by the inactivation of the SH-groups in enzymes active centers) and polynucleotides (Baranowska –Morek, 2003; Mithöfer *et al.* 2004).

c. substitution of essential metal ions by other incorrect ones (Rai *et al.* 2004).

Several metabolic processes may use ROS in a good way. Some of the ROS are involved in lignin formation in cell walls. They participate in an oxidative burst and act not only as direct protectants against invading pathogens, but also as signals activating further reactions (HR-hypersensitive response or phytoalexin biosynthesis) (Wojtaszek, 1997; Inzé and VanMontagu, 1995).

Generally a plant's cells try to keep the concentration of ROS at the possible low level because they are more reactive than molecular oxygen (O_2) (Wojtaszek, 1997), and they react with almost every organic constituent of the living cell. The high reactivity of ROS is based on the specificity of their electronic configuration.

ROSs are known to damage cellular membranes by inducing lipid peroxidation (Ramadevi and Prasad, 1998). They also can damage DNA, proteins, lipids and chlorophyll (Mittova *et al.* 2000). The most popular ROS are $.O_2^-$ - *superoxide radical*, H_2O_2 *–hydrogen peroxide*, and OH^-*hydroxyl radical* originating from one, two or three electron transfers to dioxygen (O_2). Under physiological conditions O_2^- is not very reactive against the biomolecules of the cell and in aqueous solutions at neutral or slithly acidic pH disproportionate to H_2O_2 and O_2. H_2O_2 is relatively stable and not very reactive, electrically neutral ROS , but is very dangerous because it can pass through cellular membranes and reaches cell compartments far from the site of its formation (Wojtaszek, 1997).

Plants possess a sophisticated and interrelated network of defense strategies to avoid or tolerate heavy metal intoxication. Physical barriers are the first line of defense in plants against metals. Some morphological structures like thick cuticle, biologically active tissues like trichomes, and cell walls as well as mycorrhizal symbiosis can act as barriers when plants are faced with heavy metal stress (Hall, 2002, Wong *et al.* 2004. Harada *et al.* 2010). Trichomes, for instance, can either serve as heavy metal storage site for detoxification purposes or secrete various secondary metabolites to negate hazardous effects of metals (Lee *et al.* 2002. Hauser, 2014). On the other hand, once HMs overcome biophysical barriers and metal ions enter tissues and cells, plants initiate several cellular defense mechanisms to nullify and attenuate the adverse effects of heavy metals. Biosynthesis of diverse cellular biomolecules is the primary way to tolerate or neutralize metal toxicity. This includes the induction of a myriad of low-molecular weight protein metallochaperones or chelators such as nicotianamine, putrescine, spermine, mugineic acids, organic acids, glutathione, phytochelatins, and metallothioneins or cellular exudates such as flavonoid and phenolic compounds, protons, heat shock proteins, and specific amino acids, such as proline and histidine, and hormones such as salicylic acid, jasmonic acid, and ethylene (Viehweger, 2014. Dalvi and Bhalerao, 2013. Sharma and Dietz, 2006). When the above-mentioned strategies are not able to restrain metal poisoning, equilibrium of cellular redox systems in plants is upset, leading to the increased induction of ROS (Mourato *et al.* 2012). To mitigate the harmful effects of free radicals, plant cells have developed antioxidant defense mechanism which is composed of enzymatic antioxidants like superoxide dismutase (SOD), catalase, (CAT), ascorbate peroxidase (APX), guaiacol peroxidase (GPX), and glutathione reductase (GR) and nonenzymatic antioxidants like ascorbate (AsA), glutathione (GSH), carotenoids, alkaloids, tocopherols, proline, and phenolic compounds (flavonoids, tannins, and lignin) that act as the scavengers of free radicals (Sharma *et al.* 2012, Michalak, 2006, Rastgoo *et al.* 2011). Some of the biological molecules involved in cellular metal detoxification can be multifunctional and have antiradical, chelating, or antioxidant activities. Exploitation and upregulation of any of these mechanisms and biomolecules may depend on plant species, the level of their metal tolerance (Solanki and Dhankar, 2011), plant growth stage, and metal type.

Dietz *et al.* (1999) and Sahw *et al.* (2004) have reported that heavy metals induce oxidative stress in cells and tissues in the following ways:

a) they transfer electrons directly in single-electron reactions, which generate free radicals. The so-called transition metals (Fe, Cu, Mn, *etc.*), which have unpaired electrons in their orbitals, accept and donate single electrons, thus promoting monoelectron transfers to O_2 and generally ROS interconversion and oxireduction phenomena,

b) metals disturb metabolic pathways, especially in the thylakoid membrane, which also results in increased formation of free radicals and reactive oxygen species,

c) in addition, heavy metals mainly inactivate the antioxidant enzymes (peroxidases, catalases, superoxide dismutases) responsible for free radical detoxification, although peroxidases also may be activated due to metal stress,

d) finally, heavy metal accumulation results in the depletion of low molecular weight antioxidants, such as glutathione, which is consumed under phytochelate formation.H_2O_2 in the presence of $.O_2^-$ can generate highly reactive. OH *hydroxyl radicals* via the metal-catalyzed Haber-Wiess reaction (scheme 1.) thus the scavenging of H_2O_2 in cells is critical to avoid oxidative damage (Inzé and Van Montagu, 1995, Yamasaki *et al.* 1997). In the presence of redox active transition metals such as Cu^+ and Fe^{2+}, H_2O_2 can be converted to .OH molecule in a metal-catalyzed reaction via the Fenton reaction (Mithöfer *et al.* 2004, Wojtaszek, 1997) (scheme 1.).

Other metals (*e.g.* Hg^{2+}) not belonging to transient metals cannot replace cuprum and iron in Fenton's reaction, but such ions can inhibit the activities of antioxidative enzymes, especially glutathione reductase and lead consequently to accumulation of ROS (Mithöfer *et al.* 2004). Cadmium causes oxidative stress probably through indirect mechanisms such as interaction with the antioxida tive defence, disruption of the electron transport chain or induction of lipid peroxidation. The activation of lipoxygenase, an enzyme that stimulates lipid peroxidation, has been reported after cadmium exposure (Smeets *et al.* 2005).

Scheme 1

Haber-Wiess reaction

$$H_2O_2 + .O_2^- \rightarrow .OH + OH^- + O_2$$

Fenton reaction

$$H_2O_2 + Fe^{2+}(Cu^+) \rightarrow Fe^{3+}(Cu^{2+}) + .OH + OH^-$$

$$O_2 + Fe^{3+}(Cu^{2+}) \rightarrow Fe^{2+}(Cu^+) + O_2$$

The conception of antioxidant action of phenolic compounds is not novel (Bors *et al.* 1990). There have been many reports of induced accumulation of phenolic compounds and peroxidase activity in plants treated with high concentrations of metals. Antioxidant action of phenolic compounds is due to their high tendency to chelate metals. Phenolics possess hydroxyl and carboxyl groups, able to bind particularly iron and copper (Jung *et al.* 2003). The roots of many plants exposed to heavy metals exude high levels of phenolics (Winkel-Shirley, 2002). They may inactivate iron ions by chelating and additionally suppressing the superoxide-driven Fenton reaction, which is believed to be the most important source of ROS (Arora *et al.* 1998. Rice-Evans *et al.* 1997). Tannin-rich plants such as tea, which are tolerant to Mn excess, are protected by the direct chelatation of Mn. Direct chelation, or binding to polyphenols, was observed with methanol extracts of rhizome polyphenols from *Nympheae* for Cr, Pb and Hg (Lavid *et al.* 2001). According to Morgan *et al.* (1997) this general chelating ability of phenolic compounds is probably related to the high nucleophilic character of the aromatic rings rather than to specific chelating groups within the molecule.

There is another mechanism underlying antioxidant ability of phenols. Metal ions decompose lipid hydroperoxide (LOOH) by the hemolytic cleavage of the O-O bond and give lipid alkoxyl radicals, which initiate free radical chain oxidation. Phenolic antioxidants inhibit lipid peroxidation by trapping the lipid alkoxyl radical. This activity depends on the structure of the molecules, and the number and position of the hydroxyl group in the molecules (Millic *et al.* 1997). Arora *et al.* (2000) showed that phenolics (especially flavonoids) are able to alter peroxidation kinetics by modifying the lipid packing order. They stabilize membranes by decreasing membrane fluidity (in a concentration-dependent manner) and hinder the diffusion of free radicals and restrict peroxidative reaction (Blokhina *et al.* 2003. Arora *et al.* 2000). According to Verstraeten *et al.* (2003), in addition to known protein-binding capacity of flavanols and procyanidins, they can interact with membrane phospholipids through hydrogen bonding to the polar head groups of phospholipids. As a consequence, these compounds can be accumulated at the membranes' surface, both outside and inside the cells. Through this kind of interaction, as they suggest, selected flavonoids help maintain membranes' integrity by preventing the access of deleterious molecules to the hydrophobic region of the bilayer, including those that can affect membrane rheology and those that induce oxidative damage to the membrane components. The cross-contamination of human food chain by risk elements (Sharma *et al.* 2009) is one of the most important problems that today's food chemists need to address given the high impact of contaminated crops (*e.g.* carrots, onions and potatoes) which are basic components of our diet. The reasons of heavy metal accumulation are mainly anthro-pogenic: fertilizers, mining and industrial processing (Uprety *et al.* 2009).

Food cross-contamination by heavy metals has been widely studied (Sharma *et al.* 2009; Song *et al.* 2009) since these risk elements do not only affect the nutritional value of food, but also have detri-mental effects on consumers' health. With regard to the levels of heavy metals in food, the trend is that national and international regulations on food safety are getting stricter by lowering the maximum permissible levels of toxic metals. For the case of Cd, according to the EFSA (EFSA, 2009) a lower Provisional Tolerable Weekly Intake (PTWI) is suggested. On the contrary, continuously increasing levels of risk elements in food grown in polluted areas worldwide are being reported (Kirkillis *et al.* 2012), in food grown where the underground water table has been extensively contaminated. These two trends are somehow contradicting and they need to be taken into consideration by policy making and legislative bodies to maximize the "protection net" to Public Health. Plants grown in contaminated soils tend to accumulate heavy metals, which cause changes to their normal metabolic functions (Shankar *et al.* 2005; Biacs *et al.* 1995). A correlation between heavy metals and free radicals production has been indicated (Cheeseman and Slater, 1993).

There has been an increase of interest in the carotenoid role in human nutrition (Calvo, 2005). Apart from their use as natural pigments in food industry, carotenoids are well known for their antioxidant activity and their protective role against human diseases (Bramley, 2005; Cooper *et al.* 1999), such as some types of cancer, UV-induced skin damage, coronary heart disease, cataract and macular degeneration (Clotault *et al.* 2008). Carotenoids are products of secondary metabolism of plants and are synthesized and localized in the plastids of plant cells (Fraser and Bramley, 2004). External factors may interfere and affect their physiological function. Heavy metals, which are known for their ability to replace the soil's essential metals, are one of these factors (Clijsters *et al.* 1999).

2.8. Secondary Metabolites Under CO_2 and Light Intensity Stress

The concentration of plant secondary metabolites has been found to be influenced by environmental conditions such as light intensity, CO_2 levels, temperature, fertilization, and biotic and abiotic factors which can change the concentration of these active constituents (Ibrahim *et al.* 2012; Barz and Koster, 1981). Changes in the accumulation pattern of secondary metabolites due to environmental factors are, however, the result of changes in both the rate of synthesis and the rate of catabolism (Waterman and Mole, 1989; Ibrahim *et al.* 2011). The activity of L-phenylalanine ammonia lyase (PAL, EC 4.3.1.5) determines the extent to which phenylalanine is withdrawn from primary metabolism and enters the general phenylpropanoid pathway (Barz and Koster,

1981). It has been reported that enrichment of Kacip Fatimah (*Labisia pumila* Benth) with high levels of CO_2 increased the secondary metabolite production (phenolics and flavonoids) of this plant (Ibrahim and Jafar, 2011a, b, c, Ibrahim *et al.* 2012; Jafar *et al.* 2012; Ghasemzadeh *et al.* 2010).Similar results were observed in ginger (Zingiber officianale) (Ghasemzadeh *et al.* 2010).

Atmospheric CO_2 concentration has increased by about a third over preindustrial levels and is continuing to increase at around 0.4% per year. By the year 2100, the atmospheric partial pressure of CO_2 is projected to be 700 parts per million (ppm) or higher (Ceulemans and Mousseau, 1994). Increased CO_2 concentration in the air promotes growth and accelerates physiologic processes in young trees because it causes an increase in internal carbon availability and changes partitioning of assimilates among plant parts (roots, stems, and leaves) and primary and secondary metabolites (Bazzaz, 1990). Plant responses to CO_2 will likely be influenced by other environmental factors (Bazzaz and Miao, 1993), including nutrient and water availability (Jhonson and Lincoln, 1991) and climate change (Schneider, 1993). Irradiance also directly affects plant growth and phytochemistry (Schneider, 1993) and is expected to mediate effects of elevated CO_2 on plants. Both irradiance and CO_2 are required resources for photosynthesis and their levels will interactively affect primary and secondary plant metabolism (Schneider, 1993). However, studies on interactive effects between irradiance and CO_2 levels on phytochemistry of herbal plants were limited. Researchers have reported that environmental factors such as irradiance and CO_2 concentration can significantly alter secondary metabolite synthesis and production in plants (Larsson *et al.* 1986; Sfendla *et al.* 2008; Feng *et al.* 2008; Lambreva *et al.* 2006. Bazzaz and Miao, 1993). The synthesis of medicinal components in the plant is affected by CO_2 and light intensity with changes observed in plant morphology and physiological characteristics (Curtis and Wang, 1998; Matsuki, 1996). Both factors were known to adjust not only plant growth and development but also the biosynthesis of primary and secondary metabolites (Palo, 1984).

It had been reported that elevated CO_2 promoted the accumulation of secondary metabolites (flavonoids) progressively to a greater extent as plants became mature. The percentage increase in total wheat leaf flavonoid concentration in going from an atmospheric CO_2 concentration of 400 to 1500 ppm was 22%, 38% and 27% at 14, 21 and 28 days after planting, respectively, while in going from a CO_2 concentration of 400 to 10,000 ppm, the percentage increase in total flavonoid concentration was 38%, 56% and 86%, respectively, at 14, 21 and 28 days after planting. And they found that "both elevated CO_2 levels resulted in an overall 25% increase in biomass over the control plants. In addition, the increased accumulation of secondary metabolites in plants grown under elevated CO_2 may have implications regarding plant-herbivore interactions, decomposition rates for inedible biomass, and potential beneficial effects on plant tolerance to water stress (Idso, 1988) and cold stress (Solecka and Kacperska, 2003) due to their potentials for the scavenging of reactive oxygen species (ROS).

Stutte *et al.* (2008), while studying *Scutellaria* plants - herbaceous perennials that possess numerous medicinal properties - they are rich in physiologically active flavonoids that have a wide spectrum of pharmacological activity. And in regard to this fact, they said that leaf extracts of *Scutellaria barbata* had been found to be limiting to the growth of cell lines associated with lung, liver, prostate, and brain tumors (Yin *et al.* 2004), and that extracts of *S. lateriflora* and the isolated flavonoids from the extracts have been shown to have antioxidant, anticancer, and antiviral properties (Awad *et al.* 2003). Hence, it was only natural for them to wonder how the growth of these important plants - and their significant medicinal properties - might be affected by the ongoing rise in the air's CO_2 content. In an attempt to extend their findings, Stutte *et al.* conducted experiments with both *S. barbata* and *S. lateriflora*, where they measured effects of elevated atmospheric CO_2 (1200 and 3000 ppm vs. a control value of 400 ppm) on plant biomass production and plant concentrations of six bioactive flavonoids - apigenin, baicalin, baicalein, chrysin, scutellarein and wogonin - all of which substances, according to them, have been reported to have anticancer and antiviral properties (Joshee *et al.* 2002, Cole *et al.* 2007). These experiments were conducted in a large step-in controlled-environment chamber that provided a consistent light quality, intensity and photoperiod to six small plant growth chambers that had high-fidelity control of relative humidity, temperature, and CO_2 concentration, each of which chambers was also designed to monitor nutrient solution uptake by six individual plants that they grew from seed for a period of 49 days.

Two theories that best describe plant phyto-chemical responses to elevated CO_2 and different irradiance have received much attention. The carbon/nutrient balance (CNB) theory (Tuomi *et al.* 1988; Högy and Fangmeier, 2008) proposes that allocation to storage and secondary metabolites is determined by the balance between the availability of carbon and nutrients, such that carbon in excess of growth demands is allocated to carbon-based secondary metabolites or storage compounds. The growth-differentiation balance (GDB) theory (Idso *et al.* 2000) is premised upon general patterns for relative responses of net assimilation and growth to resource availability (*e.g.*, water, nutrients, and light), such that conditions limiting growth more than assimilation allow resources to be available for differentiation processes such as secondary metabolism.

There have been several studies that investigated the effects of elevated CO_2 on plant primary and secondary metabolites (Peñuelas and Estiarte, 1998. Estiarte *et al.* 1999; Högy and Fangmeier, 2009; Idso *et al.* 2000), but only a few covered the responses of secondary metabolites under increasing CO_2 and irradiance. Although secondary metabolites constitute a significant sink for assimilated carbon, to date the picture is unclear as to how these compounds respond to different levels of CO_2 and irradiance. However, the interaction of CO_2 enrichment with different irradiance levels is expected to enhance the medicinal value of this plant (Ibrahim *et al.* 2014). Plants grown at high CO_2 levels exhibit significant changes of their chemical composition (Idso and Idso, 2000). A prominent example of a CO_2 effect is the decrease of the nitrogen concentration in vegetative plant parts as well as in seeds and grains resulting in the decrease of the protein levels (Idso and Idso, 2000). Increasing evidence has shown that elevated CO_2 may promote plant growth and mitigate heat stress damage in various plant species, particularly C_3 species (Hamerlynck *et al.* 2000; Krikham *et al.* 2011; Quaderi *et al.* 2006) including perennial turfgrass species such as tall fescue (Yu *et al.* 2012). Various studies report elevated CO_2 promotes shoot growth, root growth, and photosynthesis but inhibits dark respiration and photorespiration rates (Gonzàlez-Meter *et al.* 1996; Leakey *et al.* 2006; Reddy *et al.* 2010). However, the metabolites associated with the improvement in plant growth and changes in photosynthesis and respiration metabolic processes resulting from elevated CO_2, particularly under heat stress, are not well documented (Jinjgin *et al.* 2012).

Earlier studies have shown that elevated CO_2 increases phenolics and condensed tannins in the leaves. In conifers, elevated CO_2 influenced a decrease/increase in concentrations of some individual monoterpenes (Williams *et al.* 1994) and an increase in total phenolics has been reported (Idso and Idso, 2000). Increased concentrations of the monoterpene a-pinene was noticed in elevated CO_2composition (Idso and Idso, 2000). In contrast to this, Williams *et al.* (1994) found decreased concentrations of b-pinene in needles under elevated CO_2. Several studies have been reported the effect of temperature on secondary metabolite production in plants. Secondary metabolites increase in response to elevated temperatures (Hummel *et al.* 2004; Jochum *et al.* 2007). In contrast to this Snow *et al.* (2003) reported that high temperature decreases monoterpene levels in Douglas fir (*Pseudotsuga menziesii*).

Light is a fundamental important environmental signal regulating plant development and gene expression (Chrsitie *et al.* 1996). Elevated UV-B radiation that can be a consequence of ozone depletion has pleiotropic effects on plant life (Frohnmeyer and Staiger, 2003; Hectors, *et al.* 2007). The most common effects are plant growth reductions and increased quantities of phenolic compounds in plant tissues (Cladwell *et al.* 2003). Indeed, as for many abiotic stresses, ROS production is involved in UV radiation stress. A general strategy adopted by plants is to scavenge ROS using both enzymatic and nonenzymatic scavengers such as phenolic compounds (Mittler *et al.* 2004, Gracea and Logan 2000). Phenolic accumulation is induced by UV-B activation of the phenolic biosynthetic pathway. The presence of increased levels of phenolics in the epidermal cells results in hampering UV-B penetration, thus protecting the photosynthetically active tissues (Mazza *et al.* 2000).

Irradiance (photosynthetic photon flux density, PPFD, 400–700 nm) is known to regulate not only plant growth and development, but also the biosynthesis of both primary and secondary metabolites (Briskin *et al.* 2001; Kurata *et al.* 1997). The carbon nutrient balance (CNB) hypothesis predicts that the pattern of allocation to carbon-based secondary metabolites (CBSM) depends on the relative availability of carbon and nutrients, as well as on their relationship with plant growth rate. In accord with the predictions, the stimulation of secondary metabolite production has been demonstrated under high light (Hemm *et al.* 2004) or low light, (Jaakola *et al.* 2004) and by other biotic and abiotic stresses, *e.g.*, pathogen attack, mechanical wounding, ultraviolet (UV) radiation, low temperatures and nutrient deficit (Bernards and Ellis, 1991. Kovacik and Backor, 2007; Jansen *et al.* 2008). Phenylalanine ammonia-lyase enzyme (PAL) catalyzes the conversion of phenylalanine into trans-cinnamic acid, thereby causing the flux of primary metabolites into secondary metabolites in the phenolics pathway. PAL activity and other phenylpropanoid enzymes are mainly regulated via *de novo* synthesis. Hemm *et al.* (2004) showed that PAL genes were highly expressed in the low light-grown Arabidopsis roots, and Fritz *et al.* (2006) reported that PAL gene expression could be induced by nitrate deficiency in *Nicotiana tabacum* plants. In fact, low light intensities were regarded as environmental stimuli for the production of secondary metabolites in some plant species (Coelho *et al.* 2007; Ralphs *et al.* 1998; Zhan *et al.* 2002). Earlier studies report an increase in the concentration of specific secondary metabolites in some medicinal species, such as methylxanthines in *Ilex paraguariensis* (Coelho *et al.* 2007), alkaloid concentration in *Delphinium barbeyi* (Ralphs *et al.* 1998), and aloin (barbaloin) in *Aloe mutabilis* (Paez *et al.* 2000; Chauser-Volfson and Gutterman, 1998), under low light conditions. There were also numerous reports on an increase of secondary metabolites under high light condition and has been reported in tea (Wang *et al.* 2012), red kidney bean (Peng and Bo, 2012) and St. John's wort (Brechner *et al.* 2011). These showed the importance of light in secondary metabolite regulation in medicinal plants.

Polyphenols are very important in the control and prevention of tissue damage done by activated oxygen species due to their antioxidative effects (Rice-Evans and Packer, 1998). The concentration of polyphenols was found to be influenced by environmental conditions such as light intensity, carbon dioxide levels, temperature, fertilization, and biotic and abiotic factors, which can alter the concentration of these active constituents (Fine *at al.* 2006). Irradiance is known to regulate not only plant growth and development, but also the biosynthesis of both primary and secondary metabolites (Hemm *et al.* 2004; Liu *et al.* 2002). Phenolics biosynthesis requires irradiance or is enhanced by irradiance, and flavonoids formation is absolutely irradiance dependent where its biosynthetic rate is related to irradiance intensity and density (Xie and Wang, 2006). Different plants have different responses to irradiance intensity that result in differences in their production of secondary metabolites. In general, the resource availability hypothesis (Bryant *et al.* 1983) predicts a decline in allocation to secondary metabolites when carbon gain is limited relative to nutrient availability, which normally occurs during periods of low irradiance. This prediction was based on the effects of environmental conditions on carbon-nutrient balance (Zuvala and Ravetta, 2001). Secondary metabolites production might increase (Chauser-Volfson and Gutterman, 1998) or decrease (Gershenzon, 1994) under low light intensity conditions, depending on the type of plant. Low light intensity, for example, increased methylxanthine-content in the leaves of *Ilex paraguariensis (*Coelho *et al.* 2007) but decreased resin content in the leaves of *Grindelia chiloensis* (Zuvala and Ravetta, 2001).

Earlier studies have shown that differences in irradiance levels were able to change the production of secondary metabolites in plants (Graham, 1998; Jaafar *et al.* 2008; Gu *et al.* 2010). This simultaneously can affect the medicinal components in these plants, besides changes in the morphology and physiology (Briskin *et al.* 2001; Kurata *et al.* 1997). Both Briskin (2001) and Kurata (1997) observed a positive and significant relationship between production of total phenolics and flavonoids, and antioxidant activities in plants. It seems that different irradiances had a direct effect on antioxidant activities in plants resulting in increased total phenolics and flavonoids contents. It was suggested that increasing phenolic and flavonoid components in shaded plants are related to lower temperatures under this light condition. High temperature increases anthocyanin degradation in grape skin, together with a decrease in expression of flavonoids biosynthesis (Mori *et al.* 2007). On the other hand, low temperature increases anthocyanin production, which previously has been observed in grape berries (Mori *et al.* 2005; Yamane *et al.* 2006).

It has been shown that plants irradiated with high light intensity tend to increase their photosynthetic capacity and enhance the biomass (Abrams and Mostoller, 1995). Many studies have investigated the effects of high light levels on the vegetative and plant primary metabolism but relatively few studies have investigated the response of plant carbon based secondary metabolites (CBSM) to decreasing irradiance (Awale *et al.* 2001), particularly in a herbal plant such as *O. stimaneus*. Previous research by Affendy *et al.* (2010) have shown that *O. stimaneus* are sensitive to low light intensity however, the result was not conclusive because it only measured growth parameters.

In general, plants living in an exposed full sun habitat exhibit different photosynthetic properties and leaf characteristics than those living in the shade. Tropical pioneer species show higher responsiveness to high irradiance than late-successional species (Chazdon *et al.* 2003). However, under high irradiance, plants are predisposed to suffer photoinhibition, which is defined as the slow, reversible decline in photosynthetic efficiency that occurs when the absorbed light is in excess of that required for carbon assimilation (Demmig-Adams and Adams, 1992; Long *et al.* 1994). This phenomenon is very frequent in tropical plants (Krause *et al.* 2001), where the light intensity can reach levels beyond 1800 ìmol/m2/s photosynthetic photon flux density (PPFD) (Laisk *et al.* 2005). The ability to cope with photoinhibition varies between different plant species (Martinez and Maestri, 1995; Kitao *et al.* 2006). Photoinhibition occurs because shade leaves have high concentrations of chlorophyll, thus a higher light-capturing capacity, due to the larger antenna size of the photosystem II (PSII), and lower rates of light-saturated photosynthesis due to lower amounts of photosynthetic enzymes, such as Rubisco. A previous study by Close *et al.* (2003) has shown that photoinhibition can enhanced the production of polyphenolics in *Eucalytus* plants. The result implies that production of plant secondary metabolites can be enhanced by the manipulation of irradiance.

2.9. UV-Radiation Stress and Secondary Metabolites

Ultraviolet (UV) wavelength (400–200 nm) is a small part of the solar radiation reaching the Earth's surface but with significant biological impact on the living organisms, including plants. According to the International Commission on Illumination this wavelength region is divided into UV-A (315 – 400nm), UV-B (280–315nm) and UV-C (200–280nm). The negative effect of UV radiation increases towards the shorter wavelengths. Therefore, due to its highest energy, UV-C quickly provokes high levels of injuries and it is most detrimental for the living organisms (Hollósy, 2002; Häder *et al.* 2007). Both UV-C and UV-B possess enough energy to damage different chemical bonds causing photochemical reactions, which is the main reason for the negative biological effects (Kovács and Keresztez, 2002). The composition of the UV radiation is modified due to its absorption by the atmosphere. Usually the short-wave UV-C radiation is fully absorbed with exception of the high mountain locations (Häder *et al.* 2007), while UV-B radiation is only absorbed by the stratospheric ozone and small part of it reaches the Earth's surface. During the last decades the surface solar UV-B radiation was found to increase which corresponds with depletion of the stratospheric ozone caused by the increased release of anthropogenic pollutants such as chlorofluorocarbons and other gaseous emissions. Along with the atmospheric ozone amount, the spectral irradiance of the environmental UV depends also on the angle at which the solar radiation reaches the Earth's atmosphere, i.e. the "solar zenith angle" (including time of day, season and latitude), altitude, clouds, surface reflection, aerosols and even air pollution (Paul, 2001; McKenzie *et al.* 2007). Since there is not selective absorber for the long-wave UV-A radiation, it is affected mainly by the light scattering, its intensity is much higher than UV-B but it is not so biologically relevant (Vass *et al.* 2005). Additionally, it was found that UV-A irradiation partially protected PSII reaction center from damages caused by UV-B by activating xanthophyll cycle by preserving the level of β-carotene in cluster bean chloroplasts during the steady phase of leaf development (Joshi *et al.* 2007). The report supported previous observations showing that environmentally relevant UV-A doses possess ameliorating effect on UV-B triggered damage (Newsham *et al.* 1998; Bischof *et al.* 2003). In fact, the presence of realistic doses of both UV-A and PAR are highly important in order to obtain environmentally adequate results since they both moderate the negative UV-B effects (Flint *et al.* 2003; Kakani *et al.* 2003; Dolzhenko *et al.* 2010). High doses of UV-B and UV-C radiation affect negatively growth, development, photosynthesis, and other important processes in plants, leading to overproduction of reactive oxygen species (ROS) and development of oxidative stress, acting negatively on macromolecules, may decrease cell viability and cause cell death (Zacchini and de Agazio, 2004; Procházková and Wilhelmová, 2007; Takeuchi *et al.* 2007; Toncheva-Panova *et al.* 2010; Schreiner *et al.* 2012). However, low ROS concentrations were found to play a key role in the signaling processes during plant acclimation (Dat *et al.* 2000).

Along with the nucleic acids (Kucera *et al.* 2003; Takeuchi *et al.* 2007), proteins and lipids, the main target sites of UV radiation are known to be amino acids, membranes, quininones, pigments, photosynthetic machinery, mainly because of the UV absorbing aromatic chemical groups (Hollósy, 2002; Jansen, 2002; Vass *et al.* 2005; Edreva, 2005). UV-induced effects depend also on the plant sensitivity (Lavola *et al.* 2003; Zu *et al.* 2011). Although the role of some secondary metabolites such as anthocyanins is still under question (Hada *et al.* 2003), most authors claim that the production of these compounds (mainly flavonoids and UV-B absorbing metabolites) in plants subjected to low UV-B doses is a major part of the complex plant defense system (Solovchenko and

Schmitz-Eiberger, 2003; Kucera *et al.* 2003; Schmitz-Hoerner and Weissenböck, 2003; Frohnmeyer and Staiger, 2003; Jansen *et al.* 2008). Bashandy *et al.* (2009) also assume that the accumulation of non-pigmented flavonoids in leaves of the double ntra ntrb (lacking NADPH dependent thioredoxin reductases) *Arabidopsis* mutant might lead to the observed UV-C tolerance. The authors strongly support the proposed theory by the fact that UV-C tolerance was lost after crossing the double mutant with the tt4 (mutation in the gene encoding the first enzyme of the flavonoid biosynthesis) showing that production of flavonoids in the ntra ntrb mutant could protect plants against UV-C. In addition, the mRNA level of Chs gene (chalcone synthase gene, involved in flavonoid production) was also induced in UV-C treated plants. The authors suggest that NADPH dependent thioredoxin reductases could be a new negative regulator of flavonoid biosynthesis. In opposite to the high UV-fluency rate, relatively low UV-B or UV-C doses led to an increased production of secondary metabolites (Antognoni *et al.* 2007; Nadeau *et al.* 2012; Schreiner *et al.* 2012). As the pathways for secondary metabolite production are interrelated, the fact that some of these compounds increase and other decrease is not unexpected, and the biosynthesis prevails mostly to compounds possessing higher ROS scavenging activity or UV-shielding properties (Jansen *et al.* 2008). In the current review, we focus on secondary metabolites, which have been reported to alter predominantly after UV-B and/or UV-C treatment.

in vitro cultured plant cells, tissues and calli

UV-C light exposure of sterile cultures of *Scenedesmus quadricauda* (Chlorophyceae) over 1 hour did not influence total soluble phenols and flavonoids (Kováèik *et al.* 2010). Phenolic acids were altered differently by UV-C - vanillic acid increased; gallic, caffeic, chlorogenic and p-coumaric acids were decreased, while protocatechuic and salicylic acids did not change significantly. Selected flavonols (quercetin and kaempferol) were not detected after UV-C treatment. The authors concluded that the exposure time to UV light was not sufficient to stimulate more considerable changes of the phenolic metabolites in *Scenedesmus quadricauda*. Exposure of *Chlamydomonas nivalis* (so-called snow alga) cells to UV-C light resulted in a three-fold increase in free proline occurred within two days after exposure to UV-C, accompanied with a 12–24% increase in phenolics after 7 days of exposure (Duval *et al.* 2000). The authors report that UV-C light exposure can stimulate phenolic-antioxidant production in aplanospores of *C. nivalis* which supports the idea that there is a considerable biotechnological and pharmaceutical potential incorporated within the genome of this UV-tolerant snow alga. Moreover, UV-induced secondary metabolite production, similar to that in *C. nivalis*, may provide a valuable source of pharmacological products targeted for anticancer, anticoagulant, antimicrobial, or antiinflammatory treatments (Duval *et al.* 2000). Additionally, by using blue autofluorescence method Lesniewska *et al.* (2004) also found that UV-C light forced *Vitis vinifera* cells to produce phytoalexins - secondary metabolites with antimicrobial properties. Ghorpade *et al.* (2011) used tissue culture techniques and examined the effect of UV-C on the synthesis of four derivatives of boswellic acid - active metabolite produced in *Boswellia serrata* Roxb. (endangered medicinal plant). The authors found that 5 min of UV-C irradiation was effective for production and accumulation of acetyl-11-keto-β- boswellic acid (10-fold) and β-boswellic acid (7-fold) in the callus culture. The synthesis of stilbenes was extensively investigated in grape calli systems (Liu *et al.* 2010). Katerova *et al.* (2012) reported induced by UV-C *in vitro* production of resveratrols and their glucoside (piceids) in four grape genotypes and three tissue types of each genotype. UV-C irradiated calli accumulated stilbenes which have been already reported to have a number of health-beneficial properties.

Low levels of UV-B radiation can cause distinct changes in the plant's secondary metabolism resulting in the accumulation of a broad range of secondary plant metabolites. Best documented is the UV-B mediated accumulation of flavonoids which have in planta both reactive oxygen species (ROS) scavenging and UV-screening activities. UV-B radiation has frequently been observed to disproportionally enhance the accumulation of the more hydroxylated flavonoids, *i.e.* quercetin to kaempferol ratios increase. Flavonoids with multiple hydroxylgroups have particularly good ROS scavenging activity, and for this reason quercetin and luteolin are considered as better ROS scavengers than kaempferol and apigenin. Another group of metabolites that has been studied extensively in the context of UV-B radiation are the glucosinolates. These sulphur- and nitrogen-containing compounds have been found to be induced in a range of plant species by UV-B exposure under both pre- and postharvest conditions. Specific aliphatic or aromatic glucosinolates accumulate, depending on the species, UV-B dose and developmental stage, and accumulation is either constitutive or transient. Flavonoids and other phenolic compounds. Flavonoids are the main phenolic compounds of plants. They are divers and naturally occur as flavonoid glycosides. Flavonoid compounds have received considerable attention because of their potential health-promoting properties for human consumers. Key flavonoid biosynthesis genes are regulated by UV-B and flavonoids accumulate in a range of cellular compartments, including cell walls, vacuoles, chloroplasts, nucleus and in trichomes. Flavonols, anthocyanidines and hydroxybenzoic acids increased in black currant fruits (Ribes nigrum) exposed to three low, short-term UV-B doses, whereas the hydroxycinnamic acids increased only with the highest dose applied. However, low UV-B can enhance caffeoylquinic acid in apples (Malus domestica). Furthermore, low UV-B as postharvest treatment increased the total phenolic concentration in inflorescences and seeds of nasturtium (*Tropaeolum majus*), but not in the leaves. Indeed, not all flavonoid compounds are similarly induced. Recent studies underline the structure-dependent response of flavonoids and other phenolics to UV-B radiation. In kale (*Brassica oleracea* var. sabellica) exposed to single doses of low UV-B radiation (less than 1 kJ m^{-2} d^{-1} UV-BBE (biologically effective UV-B)), investigated quercetin glycosides decreased under UV-B. In the case of kaempferol glycosides, the amount of sugar moieties and the flavonol glycoside hydroxycinnamic acid residue influenced the response to UV-B. *e.g.* the monoacylated kaempferol tetraglucosides decreased with additional UV-B, whereas the monoacylated kaempferol diglucosides increased strongly with doses of 0.88 kJ m^{-2} d^{-1} UV-BBE. Furthermore, the hydroxycinnamic acid glycosides disinapoyl-gentiobiose and sinapoyl-feruloyl-gentiobiose were enhanced in a dose-dependent manner under UV-B. For flavonoid glycosides, there are structural characteristics regarding the aglycone, the glycosylated sugars, and the acylated phenolic and organic acids. The effect of UV-B is also modified by UV-B dose the structure of the flavonoids and other phenolics and further environmental factors such as photosynthetically active radiation and temperature. Glucosinolates Although the production of flavonoids and related phenolic compounds in response to UV-B in several plant species has been well documented, effects of ecologically relevant UV-B doses on accumulation on other secondary metabolites such as glucosinolates – the characteristic defence compounds in the order Brassicales – received little attention in the past. Glucosinolates are sulfonated thioglycosides sharing a common glycone moiety with a variable aglycone side chain and are, based on this, divided into aliphatic, indolyl, and aromatic glucosinolates. Recent studies have shown that ecologically-relevant low UV-B levels can trigger the induction of glucosinolates in Brassicaceae as demonstrated for *Brassica oleracea* var. italica (broccoli), *Arabidopsis thaliana*, and *Tropaeolum majus*. Low to ambient UV-B doses (0.3 to 1 kJ m–^{2} d^{-1} UV-B) elicited an increase of especially aliphatic methylsulfinylalkyl glucosinolates and the indole 4-methoxy-indol-3ylmethyl glucosinolate in *A. thaliana* and broccoli sprouts and an aromatic glucosinolate in nasturtium. Already 2 hour after UV-B exposure to 0.3–0.6 kJ m^{-2} d^{-1} UV-B, total glucosinolate levels have been increased in broccoli sprouts (Fig. 3). Interestingly, greater doses of up to 0.9 kJ m^{-2} d^{-1} UV-B and multiple exposure times did not elicit a stronger response in glucosinolate accumulation. Due to a lack of studies a comparison is a difficult task. However, Nadeau *et al.* (2012) showed that 4-methoxy-indol-3ylmethyl glucosinolate, 4-hydroxy-indol-3ylmethyl glucosinolate and 4-methylsulfinylbutyl glucosinolate accumulated in broccoli florets in response to exposure to UV-C. Furthermore, it has been shown that the sensitivity of plant organs and response to low levels of UV-B can vary. In nasturtium, unripe green seeds responded to low levels of UV-B with a 6-fold increase in benzyl glucosinolates and mature leaves only with a 3-fold increase

UV radiation can be regarded as a stress factor which is capable of significantly affecting plant growth characteristics. Plant height, leaf area, leaf length have been showed to decrease, whereas leaf thickness was increased in response to UVB radiation (Rozema *et al.* 1997). Plants produce a wide range of flavonoids and related phenolic compounds which tend to accumulate in leaves of higher plants in response to UV radiation (Tevini and Teramura, 1989; Rozema *et al.* 1997). It has been suggested that plants have developed UV-absorbing compounds to protect them from damage to DNA or to physiological processes caused by UV radiation (Stepleton, 1992). These UV-absorbing compounds accumulate in the epidermis, preventing UV radiation from reaching the photosynthetic mesophyll (Stepleton, 1992; Braun and Tevini, 1993). In the red 'Lollo Rosso' lettuce the main phenolic metabolites are phenolic acids (cafeoyltartaric, chlorogenic, dicaffeoylquinic), flavonoids (quercetin-3-glucuronide, quercetin-3-glucoside, quercetin 3-6-malonyglucoside, quercetin-3-6-malonyglucoside 7-glucoside) and the anthocyanin cyanidin 3-malonyglucoside (Ferreres *et al.* 1997).

Plants may produce secondary products to protect them against UV light damage, but these metabolites also play an important role in human health. Phenolics, flavonoids and anthocyanins are responsible for antioxidant activity in fruits and vegetables (Wang *et al.* 1996; Cao *et al.* 1997). Epidemiological studies have shown a positive relationship between fruit and vegetable consumption and reduced incidence of chronic and degenerative diseases (Heinonen *et al.* 1998; Prior *et al.* 1998). These natural antioxidants have scavenging properties against oxygen free radicals, offering protection to lipids, proteins and nucleic acids Heinonen *et al.* 1998). Most published results on the effects of UV radiation on plants are based on the effect of UV light added to either natural or artificial light (using UV lamps) (Rozema *et al.* 1997; Krizek, 2004). Published studies based on UV exclusion from natural radiation are limited (Krizek *et al.* 1997, 1998; Krause *et al.* 1999; Kolb *et al.* 2001, 2003). Krizek (2004). stated that results of experiments using supplementary U Vradiation are difficult to interpret due to unrealistically high UV radiation levels and inappropriate levels of UV-A (320–400 nm) irradiance. It seems that ambient levels of PAR and UVA mitigate damage caused specifically by UV-B radiation. According to Cladwell *et al.* (1994), UV-B radiation caused biomass reduction only when PAR and UV-A levels were reduced to less than half that of sunlight levels. In addition, UV-A proved to be effective in ameliorating UV-B damage when PAR was low.

References

Abrams, M.D. and Mostoller, S.A. 1995. Gas exchange, leaf structure and nitrogen in contrasting successional tree species growing in open and understory sites during a drought. *Tree Physiol.*, **15:** 361-370.

Aerts, R. and Chapin, F.S., III. 2000. The mineral nutrition of wild plants revisited: A re-evaluation of processes and patterns. *Adv. Ecol. Res.,* **30:** 1–67.

Ahmad, I., Basra, S.M.A., Afzal, I., Farooq, M. and Wahid, A. 2013. Growth improvement in spring maize through exogenous application of ascorbic acid, salicylic acid and hydrogen peroxide. *Int. J. Agric. Biol.*, **15:** 95-100.

Affendy, H., Aminuddin, M., Azmy, M., Amini, M.A., Assis, K. and Tamir, A.T. 2010. Effects of light intensity on *Orthosiphon stamineus* Benth treated with different organic fertilizers. *Int. J. Agric. Res.*, **5:** 201-207.

Airaki, M., Leterrier, M., Mateos, R.M., Valderrama, R., Chaki, M., Barroso, J.B., del Rio, L.A., Palma, J.M. 2012. Corpas, F.J. Metabolism of reactive oxygen species and reactive nitrogen species in pepper (*Capsicum annuum* L.) plants under low temperature stress. *Plant Cell Environ.,***35:** 281–295.

Allagulova, C.R., Gimalov, F.R., Shakirova, F.M. and Vakhitov, V.A. 2003. The plant dehydrins: structure and putative functions. *Biochem-US*, **68:** 945–951.

Allakhverdiev,,S.I., Sakamoto, A., Nishiyama, Y., Inaba, M. and Murata, N. 2000. Ionic and osmotic effects of NaCl-induced inactivation of photosystems I and II in *Synechococcus* sp. *Plant Physiol.*, **123:** 1047-1056.

Allakhverdiev, S.I., Kreslavski, V.D., Klimov, V.V., Los, D.A., Carpentier, R. and Mohanty, P. 2008. Heat stress: an overview of molecular responses in photosynthesis. *Photosynth Res*, **98:** 541–550.

Ali, R.M. and Abbas, H.M. 2003. Response of salt stressed barley seedlings to phenylurea. P*lant Soil Ebviron.*, **49:** 158-162.

Ali, H., Khan, E. and Sajad, M.A. 2013. Phytoremediation of heavy metals—concepts and applications. *Chemosphere,* **91(7):** 869–881.

Almansouri, M., Kinet, J.M. and Lutts S. 2001. Effect of salt and osmotic stresses on germination in durum wheat (*Triticum durum* Desf.). *Plant Soil*, **231:** 243-254.

Amarowicz, R., Naczk, M., Zadernowski, R. and Shahid, F. 2000. Antioxid ant activity of condensed tannins of beach pea, canola hulls, evening primrose, and fababeans. *J Food Lipids*, **7:** 199–211.

Amarowicz, R., Pegg, R.B., Rahimi-Moghaddam, P., Barl, B. and Weil, J.A. 2004. Free-radical scavenging capacity and antioxidant activity of selected plant species from the *Canadian prairies*. *Food Chem*, **84:** 551–562.

Amarowicz, R. and Weidner, S. 2009. Biological activity of grapevine phenolic compounds. In,"*Grapevine molecular physiology and biotechnology*, 2nd edn", Ed. K.A. Roubelakis-Angelakis, Springer, New York, pp. 389–405.

Arndt, S.K., Livesley, S.J., Merchant, A., Bleby, T.M. and Grierson, P.F. 2008. Quercitol and osmotic adaptation of field-grown *Eucalyptus* under seasonal drought stress. *Plant Cell Environ*, **31***:* 915–924.

Anderson-Teixeira, K.J., P.K. Snyder, T.E. Twine, S.V. Cuadra, M.H. Costa and E.H. DeLucia, 2012. Climate-regulation services of natural and agricultural ecoregions of the Americas. *Nat. Climate Change,* **2:** 177-181

Angelova, Z., Georgiev, S. and Roos, W. 2006. Elicitation of plants. Biotechnol, *Biotechnol Equip.*, **20:** 72–83.

Anjum, F., Yaseen, M., Rasul, E., Wahid, A. and Anjum, S. 2003. Water stress in barley (*Hordeum vulgare* L.). II. Effct on chemical composition and chlorophyll contents. *Pak J Agric Sci..,* **40:** 45–49.

Antognoni, F., Zheng, S., Pagnucco, C., Baraldi, R., Poli, F. and Biond, S. 2007. Induction of flavonoid production by UV-B radiation in *Passiflora quadrangularis* callus cultures. *Fitoterapia*, **78:** 345–352.

Arbona, V., Manzi, M., de Ollas, C. and Gómez-Cadenas, A. 2013. Metabolomics as a tool to investigate abiotic stress tolerance in plants. *Int J Mol Sci.*, **14:** 4885–4911.

Arora, A., Nair, M.G. and Strasburg, G.M. 1998. Structure-activity relationships for antioxidant activities of a series of plavonoids in a liposomal system. *Free Radic. Biol. Med.*, **24:** 1355.

Arora, A., Byrem T.M., Nari, M.G. and Strasburg, G.M. 2000. Modulation of liposomal membranes fluidity by flavonoids and isoflavonoids. *Archives of Biochemistry and Biophysic,* **373:** 102.

Aroca, R., Irigoyen, J.J. and sanchez-Diaz, M. 2001. Photosynthetic characteristics and protective mechanism against oxidative stress during chilling and subsequent recovery in two maize varieties differing in chilling sensitivity. *Plant Sci.*, **161:** 719-726.

Asada, K. 1999. The water–water cycle in chloroplasts: scavenging of active oxygens and dissipation of excess photons. *Annu. Rev. Plant Physiol. Plant Mol. Biol.*, **50:** 601–639.

Ashraf, M. 1997. Changes in soluble carbohydrates and soluble proteins in three arid-zone grass species under salt stress. *Trop Agric.*, **74:** 234–237.

Awad, R., Arnason, J.T., Trudeau, V., Bergeron, C., Budziinski, J.W., Foster, B.C. and Merali, Z. 2003. Phytochemical and biological analysis of skullcap (*Scutellaria lateriflora* L.): A medicinal plant with anxiolytic properties. *Phytomedicine,* **10**: 640-649.

Awale, S., Tezuka, Y., Banskota, A.H., Kouda, K., Tun, K.M. and Kadota, S. 2001. Five novel highly oxygenated diterpenes of *Orthosiphon stamineus* from Myanmar. *J. Nat. Prod.,* **64:** 592-596.

Aziz, A., Martin-Tanguy, J. and Larher, F. 1998. Stress-induced changes in polyamine and tyramine levels can regulate proline accumulation in tomato leaf discs treated with sodium chloride. *Physiol Plant.*, **104:** 195–202.

Bahler, B.D., Steffen, K.L. and Orzolek, M.D. 1991. Morphological and biochemical comparison of a purple-leafed and a green-leafed pepper cultivar. *HortScience.* **26:** 736.

Baldwin, I.T., Halitschke, R., Paschold, A., Von Dahl, C.C. and Preston, C.A. 2006. Volatile signaling in plant–plant interactions: 'Talking trees' in the genomics era. *Science*, **311:** 812–815.

Baranowska –Morek, A. 2003. Rocelinne mechanizmy tolerancji na toksyczne dzia³anie metali *Kosmos*, **52:** 283.

Bartels, D. and Sunkar, R. 2005. Drought and salt tolerance in plants. *Crit Rev Plant Sci.*, **24:** 23–58.

Bartwal, A., Mall, R., Lohani, P., Guru, S.K. and Arora, S. 2014. Role of secondary metabolites and brassinosteroids in plant defense against environmental stresses. *J. Plant Growth Regul.*, **32:** 216-232.

Barigah, T.S., Charrier, O., Douris, M., Bonhomme, M., Herbette, S., Améglio, T., Fichot, R., Brignolas, F. and Cochard, H. 2013. Water stress-induced xylem hydraulic failure is a causal factor of tree mortality in beech and poplar. *Ann Bot,* **12:** 1431–1437.

Barz, W. and Koster, J. 1981. Turnover and degradation of secondary (natural) products. In, "*The Biochemistry of Plants, Volume 7 Secondary Plant Products*", Academic Press, New York, pp. 35–84.

Bazzaz, F.A. 1990. The response of natural ecosystems to the rising global CO_2 levels. *Annual Review of Ecology and Systematics*, **21(1):** 167–196.

Bazzaz, F.A. and Miao, S.L. 1993. Successional status, seed size, and responses of tree seedlings to CO_2, light, and nutrients. *Ecology*, **74(1):** 104–112.

Bischof, K., Janknegt, P.J., Buma, A.G.J., Rijstenbil, J.W., Peralta, G. and Breeman, A.M. 2003. Oxidative stress and enzymatic scavenging of superoxide radicals induced by solar UV-B radiation in Ulva canopies from southern Spain. *Scientia Marina*, **67(3):** 353–359.

Beaulieu, J.C. and Baldwin, E.A. 2002. Flavor and aroma of fresh-cut fruits and vegetables, Science, Technology, and Market. CRC Press, 391-425.

Beckman, G.H. 2000. Phenolic-storing cells: keys to programmed cell death and periderm formation in wilt disease resistance and in general defence responses in plants? *Physiol. Mol. Plant Pathol.*, **57:** 101-110.

Bernards, M.A. and Ellis, B.E. 1991. Phenylalanine ammonia-lyase from tomato cell cultures inoculated with *Verticillium alboatrum*. *Plant Physiol.*, **97:** 1494–1500.

Bennett, R.N. and Wallsgrove, R.M. 1994. *Secondary Metabolites in plant defence mechanisms. New Phytol.* **127:** 617–633.

Berta, G., Altamura, M.M., Fusconi, A., Cerruti, F., Capitani, F. and Bagni, N. 1997. The plant cell wall is altered by inhibition of polyamine biosynthesis. *New Phytol.*, **137:**, 569-577.

Bashandy, T., Taconnat, L., Renou, J.P., Meyer, Y. and Reichheld, J.P. 2009. Accumulation of flavonoids in an ntra ntrb mutant leads to tolerance to UVC. *Mol. Plant*, **2:** 249–258.

Biacs, P.A., Daood, H.G. and Kadar, I. 1995. Effect of Mo, Se, Zn, and Cr treatments on the yield, element concentration, and carotenoid content of carrot. *Journal of Agricultural and Food Chemistry*, **43(3):** 589-591.

Blackman, C.J., Brodribb, T.J. and Jordan, G.J. 2009. Leaf hydraulics and drought stress: response, recovery and survivorship in four woody temperate plant species. *Plant Cell Environ,* **32***:* 1584–1595.

Blokhina, O., Virolainen, E. and Fagerstedt, K.V. 2003. Antioxidants, oxidative damage and oxygen deprivation stress: a review. *Ann. Bot.*, **91:** 179.

Bohnert, H.J., Nelson, D.E. and Jensen, R.G. 1995. Adaptations to environmental stresses. *Plant Cell*, **7:** 1099–1111.

Bongue-Bartelsman, M. and Phillips, D.A. 1995. Nitrogen stress regulates gene expression of enzymes in the flavonoid biosynthetic pathway of tomato. *Plant Physiol Biochem.,* **33:** 539–546.

Bouché, N. and Fromm, H. 2004. GABA in plants: just a metabolite. *Trends Plant Sci.*, **9:** 100–115.

Bowler, C., van Montagu, M. and Inze,′ D. 1992. Superoxide dismutases and stress tolerance. *Annu. Rev. Plant Physiol. Plant Mol. Biol.*, **43:** 83–116.

Bohenert, H.J., Nelson, D.E. and Jensen, R.G. 1995. Adaptation to environmental stresses. *Plant Cell*, **7:** 1099–1111.

Bor, M., Seckin, B., Ozgur, R., Yilmaz, O., Ozdemir, F. and Turkan, I. 2009. Comparative effects of drought, salt, heavy metal and heat stresses on gamma-aminobutryric acid levels of sesame (*Sesamum indicum* L.). *Acta Physiol Plant*, **31:** 655–659.

Bors, W., Heller, W., Michel, C. and Saran, M. 1990. Flavonoids as antioxidants: determination of radical-scavenging efficiency. *Meth. Enzymol.*, **186:** 343.

Bouwmeester, H.J., Roux, C., Lopez-Raez, J.A. and Becard, G., 2007. Rhizosphere communication of plants, parasitic plants and AM fungi. *Trends Plant Sci.*, **12:** 224–230.

Bown, A.W. and Shelp, B.J. 1997. The metabolism and functions of ã-aminobutyric acid. *Plant Physiol.,* **115:** 1–5.

Bramley, P.M. 2000. Is lycopene beneficial to human health? *Phytochemistry*, **54(3):** 233- 236.

Brachet, J. and Cosson, L. 1986. Changes in the total alkaloid content of *Datura innoxia* Mill. subjected to salt stress. *J Exp Bot.* **37:** 650–656.

Braun, J. and Tevini, M., 1993. Regulation ofUV-protective pigment synthesis in the epidermal layer of rye seedlings (*Secale cereale* L. cv. Kustro). *Photochem. Photobiol.*, **57:** 318–323.

Brechner, M.L., Albright, L.D. and Westa, L.A. 2011. Effect of UVB on secondary metabolites of St John's Wort (*Hypericum perforatum* L.) grown in controlled environment. *Photochem. Photobiol.*, **87:** 680–684.

Breitenstein, B., Scheller, M., Shakfa, M.K., Kinder, T., Müller-Wirts, T., Koch, M. and Selmar, D. 2011. Introducing terahertz technology into plant biology: a novel method to monitor changes in leaf water status. *J Appl Bot Food Qual.*, **82(2):** 158–161.

Briskin, D.P. and Gawienowski, M.C. 2001. Differential effects of light and nitrogen on production of hypericins and leaf glands in *Hypericum perforatum. Plant Physiol.*, **39:** 1075–1081.

Bryant, J.P., Chapin III, F.S. and Klein, D.R. 1983. Carbon/nutrient balance of boreal plants in relation to vertebrate herbivory. *Oikos*, **40(3):** 357–368.

Caillet, S., Salmieri, S. and Lacriox, M. 2006. Evaluation of free radical-scavenging properties of commercial grape phenol extracts by a fast colorimetric method. *Food Chem*, **95:** 1–8.

Cao, G., Sofic, E. and Prior, R.L. 1997. Antioxidant and prooxidaant behavior of flavonoids: structure–activity relationships. *Free Radic. Biol. Med.*, **22:** 749–760.

Ceulemans, R. and Mousseau, M. 1994. Tansley review No.71. Effects of elevated atmospheric CO_2 on woody plants. *New Phytologist*, **127(3):** 425–446.

Calvo, M.M. 2005. Lutein: A valuable ingredient of fruit and vegetables. *Critical Reviews in Food Science and Nutrition*, **45(7):** 671-696.

Cattivelli, L., Rizza, F., Badeck, F.-W., Mazzucotelli, E., Mastrangelo, A.M., Francia, E., Marè, C., Tondelli, A. and Stanca, A.T. 2008. Drought tolerance improvement in crop plants: an integrated view from breeding to genomics. *Field Crop Res.*, **105:** 1–14.

Chalker-Scott, L. 1999. Environmental significance of anthocyanins in plant stress response. *Photochem. Photobiol.,* **70:** 1–9.

Chalker-Scott, L and Fnchigami, L.H. 1989. The role of phenolic compounds in plant stress responses; In, "*Low temperature stress physiology in crops*", Ed. H.L. Paul, CRC Press Inc., Boca Raton, Florida, pp. 40.

Chalker-Scott, L. 2002. Do anthocyanins function as osmoregulators in leaf tissues? *Adv. Bot. Res.* **37:**103–106.

Chan, L.K., Koay, S.S., Boey, P.L. and Bhatt, A. 2010. Effects of abiotic stress on biomass and anthocyanin production in cell cultures of *Melastoma malabathricum. Biol Res.*, **43:** 127–135.

Chauser-Volfson, E. and Gutterman, Y. 1998. Content and distribution of anthrone C-glycosides in the South African arid plant species Aloe mutabilis growing in the direct sunlight and the shade in the Negev Desert of Israel. *J. Arid Environ.,* **40:** 441–451.

Chazdon, R.L., Pearcy, R.W., Lee, D.W. and Fetcher, N. 2003. Photosynthetic response of tropical forest plants to contrasting light environments. In, "*Tropical Forests Plant Ecophysiology*", Eds. S.S. Mulkey, R.L. Chazdon, A.P. Smith, Chapman *and* Hall; New York, NY, USA: pp. 5–55. Cheeseman, K.H. and Slater, T.F. 1993. An introduction to free radical biochemistry. *British Medical Bulletin*, **49(3):** 481-493.

Cho, Y., Njitiv, N., Chen, X.. Lightfood, D.A. and , Wood, A.J. 2003. Trigonelline concentration in field- grown soybean in response to irrigation. *Biol Plant.*, **46:** 405–410.

Christiansen, J.L., Jornsgard, B., Buskov, S. and Olsen, C.E. 1997. Effect of drought stress on content and composition of seed alkaloids in narrow-leafed lupin, *Lupinus angustifolius* L. *Eur J Agron.,* **7:** 307–314.

Christie, P.J., Alfenito, M.R. and Walbot, V. 1994. Impact of low-temperature stress on general phenylpropanoid and anthocyanin pathways: Enhancement of transcript abundance and anthocyanin pigmentation in maize seedlings.*Planta,* **194:** 541–549.

Christie, J.M. and Jenkins, G.I. 1996. Distinct UV-B and UV-A/Blue light signal transduction pathways induce chalcone synthase gene expression in *Arabidopsis* cells. *Plant Cell,* **8:** 1555–1567.

Cirlaková, A. 2009. Heavy metals in the vascular plants of Tatra mountains. *Oecologia Montana*, **18:** 23–26.

Caldwell, M.M., Flint, S.D. and Searles, P.S. 1994. Spectral balance and UV-Bsensitivity of soybean: a field experiment. *Plant Cell Environ.* **17:** 267–276.

Caldwell, M.M., Ballaré, C.L., Bornman, J.F., Flint, S.D., Bjorn, L.O., Teramura, A.H., Kulandaivelu, G. and Tevini, M. 2003. Terrestrial ecosystems, increased solar ultraviolet radiation and interactions with other climatic change factors.*Photochem. Photobiol. Sci.,* **2:** 29–38.

Charles, D.I., Simon, J.E., Shock, C.C., Feibert, E.B.G. and Smith, R.M. 1993. Effect of water stress and post-harvest handling on artemisinin content in the leaves of *Artemisia annua* L. In, "*Proceedings* of the Second International Symposium: New Crops, Exploration, Research and *Commercialization, Indianapolis, USA",* Eds. J. Janick, J.E. Simon, John Wiley and Sons Inc.: New York, NY, USA, pp. 640–643.

Cho, Y., Lightfoot, D.A. and Wood, A.J. 1999. Trigonelline concentrations in salt stressed leaves of cultivated *Glycine max. Phytochemistry*. **52:** 1235–1238.

Clemens, S. 2001. Molecular mechanisms of plant metal tolerance and homeostasis. *Planta* **212:** 475.

Clijsters, H., Cuypers, A. and Vangronsveld, J. 1999. Physiological responses to heavy metals in higher plants, defense against oxidative stress. *Zeitschrift fur Naturfor-schung—A Journal of Biosciences*, **54(9-10):** 730-734.

Close, D., McArthur, C., Paterson, S., Fitzgerald, H., Walsh, A. and Kincade, T. 2003. Photoinhibition: A link between effects of the environment on eucalypt leaf chemistry and herbivory. *Ecology,* **84:** 2952–2966.

Clotault, J.,Peltier, D., Berruyer, R., Thomas, M., Briard, M. and Geoffriau, E. 2008. Expression of carotenoid biosynthesis genes during carrot root development. *Journal of Experimental Botany*, **59(13):** 3563-3573.

Cockell, C.S. 1998. Ultraviolet radiation, evolution and the ð-electron system. *Biol. J. Linnean Soc.*, **63:** 449-457.

Coelho, G.C., Rachwal, M.F.G., Dedecek, R.A., Curcio, G.R., Nietsche, K. and Schenkel, E.P. 2007. Effect of light intensity on methylxanthine contents of *Ilex paraguariensis* A. St. Hil. *Biochem. Syst. Ecol.,* **35:** 75–80.

Cole, I.B., Sacena, P.K. and Murch, S.J. 2007. Medicinal biotechnology in the genus *Scytellaria. In Vitro Cellular and Developmental Biology - Plantm* **43**: 318-327.

Conrath, U., Beckers, G.J.M., Flors, V., Garcia-Agustin, P., Jakab, G., Mauch, F., Newman, M.A., Pieterse, C.M.J., Poinssot, B., Pozo, M.J., Pugin, A., Schaffrath, U., Ton, J., Wendehenne, D., Zimmerli, L. and Mauch-Mani, B. 2006. Priming: getting ready for battle. *Mol. Plant Microbe Interact.*, **19:** 1062–1071.

Cooper, D.A., Eldridge, A.L. and Peters, J.C. 1999. Dietary carotenoids and certain cancers, heart disease, and age-related macular degeneration: A Review of Recent Re-search, *Nutrition Reviews*, **57(7):** 201- 214.

Correia, B., Pintó-Marijuan, M., Neves, L., Brossa, R., Dias, M.C., Costa, A., Castro, B.B., Santos, C. Chaves, M.M. and Pinto, G. 2014. Water stress and recovery in the performance of two *Eucalyptus globulus* clones: physiological and biochemical profiles. *Physiol Plant*, **150:** 580–592.

Costa e Silva, F., Shvaleva, A., Maroco, J.P., Almeida, M.H., Chaves, M.M. and Pereira, J.S. 2004. Responses to water stress in two *Eucalyptus globulus* clones differing in drought tolerance. *Tree Physiol*, **24:** 1165–1172.

Couceiro, M.A., Freen, F.A., Zobayed, S.M.A. and Kozai, T. 2006. Variation in concentrations of major bioactive compounds of St. John's wort: effects of harvesting time, temperature and germplasm. *Plant Sci.*, **170:** 128–134.

Curtis, P.S. and Wang, X. 1998. A meta-analysis of elevated CO_2 effects on woody plant mass, form, and physiology. *Oecologia*, **113(3):** 299–313.

Dai, J., Mumper, and R.J. 2010. "Plant phenolics: extraction, analysis and their antioxidant and anticancer properties. *Molecules*, **15(10):** 7313-7352.

Dalvi A.A. and S. A. Bhalerao, S.A. 2013. Response of plants towards heavy metal toxicity: an overview of avoidance, tolerance and uptake mechanism. *Annals of Plant Sciences*, **2(9):** 362–368.

Daneshmand, F., Arvin, M.J. and Kalantari, K.M. 2010. Physiological responses to NaCl stress in three wild species of potato *in vitro*. *Acta Physiol Plant.*, **32:** 91–101.

Dat J, Vandenabeele, S., Vranová, E., van Montagu, M., Inze, D. and van Breusegem, F. 2000. Dual action of the active oxigen species during plant stress responses. CMLC, *Cell Mol Life Sci*, **57:** 779–795.

de Abreu, I.N. and Mazzafera, P. 2005. Effect of water and temperature stress on the content of active constituents of *Hypericum brasiliense Choisy*. *Plant Physiol Biochem.*, **43:** 241-248.

Degenhardt, J., 2008. Ecological roles of vegetative terpene volatiles chapter 21. In, "*Induced Plant Resistance to Herbivory*", Andreas, J., Schaller, Stuttgart, Germany.

Del Moral, R. 1972. On the variability of chlorogenic acid concentration. *Oecologia.* **9:** 289–300.

Demmig-Adams, B. and Adams, W.W. 1992. Carotenoid composition in sun and shade leaves of plants with different life forms. *Plant Cell Environ.*, **15:** 411–419.

Demmig-Adams, B. 2003. Linking the xanthophyll cycle with thermal energy dissipation. *Photosynth. Res.*, **76:** 73–80.

Dietz, K.J., Baier, M. and Kramer, U. 1999. Free radicals and active oxygen species as mediators of heavy metal toxicity in plants. In, "*Heavy Metal Stress in Plants. From Molecule to Ecosystems*", Eds. M.N.V. Prasad, J. Hagemeyer, Springer, Berlin, pp 73–97.

Dixon, R.A., Achnine, L., Kota, P., Liu, C.J., Srinivasa Reddy, M.S. and Wang, L. 2002. The phenylpropanoid pathway and plant defence - a genomic perspective. *Mol. Plant Pathol.*, **3:** 371-390.

Dixon, R.A. and Paiva, N.L. 1995. Stress-induced phenylpropanoid metabolism. *Plant Cell*, **7:** 1085–1097.

Dolzhenko, Y, Bertea, C.M., Occhipinti, A., Bossi, S. and Maffei, M.E. 2010. UV-B modulates the interplay between terpenoids and flavonoids in peppermint (*Mentha* x *piperita* L.). *Journal of Photochemistry and Photobiology B: Biology*, **100:** 67–75.

Duval, B., Shetty, K. and Thomas, W.H. 2000. Phenolic compounds and antioxidant properties in the snow alga Chlamydomonas nivalis after exposure to UV light. *Journal of Applied Phycology*, **11:** 559–566.

Edreva, A. 1996. Polyamines in plants. *Bulg. J. Plant Physiol.*, **22:** 73-101.

Edreva, A., 2005. Generation and scavenging of reactive oxygen species in chloroplasts: a submolecular approach. *Agric. Ecosyst. Env.*, **106:** 119-133.

Edreva, A., Yordanov, I., Kardjieva, R., Hadjiiska, E. and Gesheva, E. 1995. Expression of phenylamides in abiotic stress conditions. *Bulg. J. Plant Physiol.*, **21:** 15-23.

Edreva, A., Yordanov, I., Kardjieva, R. and Gesheva, E. 1998. Heat shock responses of bean plants: involvement of free radicals, antioxidants and free radical/active oxygen scavenging systems. *Biol. Plant.*, **41:**185-191.

Edreva, A.M., Velikova, V. and Tsonev, T. 2000. Phenylamides in plants. *Russ J Plant Physiol.* **54:** 287–301.

Edreva, A. 2005. The importance of nonphotosynthetic pigments and cinnamic acid derivatives in photoprotection. *Agric Ecosyst Env*, **106(2-3):** 135– 146.

Edreva, A.M., Velikova, V. and Tsonev, T. 2007. Phenylamides in plants. *Russ. J. Plant Physiol.*, **54:** 287-301.

Edreva, A., Velikova, V., Tsonev, T., Dagnon, S., Gürel, A., Akta°, L. and Gesheva, E. 2008. Stress-protective role of secondary metabolites: diversity of functions and mechanisms. *Gen. Appl. Plant Physiology*, **34(1-2):** 67-78

EFSA 2009. Panel on Contaminants in the Food Chain (CONTAM), "Cadmium in Food," 2009. http://www.efsa.europa.eu/en/efsajournal/pub/980.htm

Elavarthi, S. and Martin, B. 2010. Spectrophotometric assays for antioxidant enzymes in plants. *Methods Mol Biol.*, **639:** 273–281.

Ellington, A.A., Berhow. M.A. and Singletary, K.W. 2006. Inhibition of Akt signaling and enhanced ERK1/2 activity are involved in induction of macro autophagy by triterpenoid B-group soya saponins in colon cancer cells. *Carcinogenesis*, **27:** 298-306.

Estiarte, M., Peñuelas, J., Kimball, B.A., Hendrix, D.L., Pinter Jrb, P.J., Wallb, G.W., LaMorteb, R.L. and Hunsakerb, D.J. 1999. Free-air CO_2 enrichment of wheat: leaf flavonoid concentration throughout the growth cycle. *Physiologia Plantarum*, **105(3):** 423–433.

Fajer, E. D., Bowers, M.D. and Bazzaz, F.A. 1992. The effect of nutrients and enriched CO_2 environment on production of carbon-based allelochemicals in Plantago: a test of the carbon/nutrient balance hypothesis. *The American Naturalist*, **140(4):** 707–723.

Falk, K.L., Tokuhisa, J.G. and Gershenzon, J. 2007. The effect of sulfur nutrition on plant glucosinolate content: physiology and molecular mechanisms. *Plant Biol.* **9:** 573–581.

Farid, M., Shakoor, M.B., Ehsan, A., Ali, S., Zubair, M. and Hanif, M.S. 2013. Morphological, physiological and biochemical responses of different plant species to Cd stress. *International Journal of Chemical and Biochemical Sciences*, **3:** 53–60.

Farooq, M., Kobayashi, N., Wahid, A., Ito, O. and Basra, S.M.A. 2009. Strategies for producing more rice with less water. *Adv Agron*, **101:** 352–388.

Feng, Y., Warner, M.E., Zhang, Y. Sun, J., Fu, F.-X., Rose, J.M. and Hutchins, D.A. 2008. Interactive effects of increased pCO_2, temperature and irradiance on the marine coccolithophore *Emiliania huxleyi* (Prymnesiophyceae). *European Journal of Phycology,*, **43(1):** 87–98.

Ferreres, F., Gil, M.I., Casta˜ner, M. and Tom´as-Barber´an, F.A. 1997. Phenolic metabolites in red pigmented lettuce (*Lactuca sativa*). Changes with minimal processing and cold storage. *J. Agric. Food Chem.* ,**45:** 4249–4254.

Ferry, N., Edwards, M.G., Gatehouse, J.A. and Gatehouse, A.M.R. 2004. Plant–insect interactions: molecular approaches to insect resistance. *Curr Opin Biotechnol*, **15:** 155–161.

Fine, P.V.A., Miller, Z.J., Mesones, I., Irazuzta, S., Appel, H.M. and Stevens, M.H.H. 2006. The growth defense trade off and habitat specialization by plants in Amazonian forest. *Ecology*, **87:** 150-162.

Flint, S.D., Ryel, R.J. and Caldwell, M.M. 2003. Ecosystem UV-B experiments in terrestrial communities: a review of recent findings and methodologies. *Agricultural and Forest Meteorology*, **12:** 177–189.

Flora, S.J.S., Mittal, M. and Mehta, A. 2008. Heavy metal induced oxidative stress *and* its possible reversal by chelation therapy. *Indian Journal of Medical Research*, **128(4):** 501–523.

Foyer, C.H. and Halliwell, B. 1976. The presence of glutathione and glutathione reductase in chloroplasts: a proposed role in ascorbate acid metabolism. *Planta*, **133:** 21–25.

Fraenkel, G.S. 1959. The Raison d'Être of secondary plant substances. These Odd Chemicals Arose as a Means of Protecting Plants from Insects and Now Guide Insects to Food. *Science*, **129(3361):** 1466-1470.

Franca, S.C., Roberto, P.G., Marins, M.A., Puga, R.D., Rodrigeuz, A. and Pereira, J.O. 2001. Biosynthesis of secondary metabolites in sugarcane. *Gen Mol Biol.*, **24:** 243–250.

Franz, C. 1983. Nutrient and water management for medicinal and aromatic plants. *Acta Hort.*, **132:** 203–215.

Fraser, P.D. and Bramley, P.M. 2004. The biosynthesis and nutritional uses of carotenoids. *Progress Lipid Research*, **43(3):** 228-265.

Fritz, C., Palacios, N., Fiel, R. and Stitt, M. 2006. Regulation of secondary metabolism by the carbon-nitrogen status in tobacco: Nitrate inhibits large sectors of phenylpropanoids metabolism. *Plant J.*, **46:** 553–548.

Frohnmeyer, H. and Staiger, D. 2003. Ultraviolet-B radiation mediated responses in plants. Balancing damage and protection. *Plant Physiol.,* **133:** 1420–1428.

Frost, C.J., Mescher, M.C., Carlson, J.E. and de Moraes, C.M. 2008. Plant defense priming against herbivores: getting ready for a different battle. *Plant Physiol.* **146:** 818–824.

Fry, S.C., 1986. Cross-linking of matrix polymers in the growing cell walls of angiosperms. *Annu. Rev. Plant Physiol.*, **37:** 165-186.

Funk, J.L., Mak, J.E. and Lerdau, M.T. 2004. Stress-induced changes in carbon sources for isoprene production in *Populus deltoides*. *Plant Cell Environ.*, **27:** 747–755.

Gallé, A., Flórez-Sarasa, I., Tomás, M., Pou, A., Medrano, H., Ribas-Carbo, M. and Flexas, J. 2009. The role of mesophyll conductance during water stress and recovery in tobacco (*Nicotiana sylvestris*): acclimation or limitation? *J Exp Bot*, **60:** 2379–2390.

Giweli, A.A., D•amiæ, A.M., Sokoviæ, M., Ristiæ, M., Janaækoviæ, P. and Marin, P. 2013. The chemical composition, antimicrobial and antioxidant activities of the essential oil of salvia fruticosa growing wild in libya. *Archives of Biological Sciences*, **1(65):** 321-329.

Ghershenzon, J. 1984. Changes in levels of plant secondary metabolites under water and nutrient stress. In, "*Recent Advances in Phytochemistry-Phytochemical Adaptations to Stress*", Eds. B.N. Timmermann, C. Steelin, F.A. Loewus, Plenum Press: New York, NY, USA, pp. 273–220.

Ghorpade, R.P., Chopra, A. and Nikam, T.D. 2011. Influence of biotic and abiotic elicitors on four major isomers of boswellic acid in callus culture of *Boswellia serrata* Roxb. *Plant OMICS*, **4(4):** 169–176.

Gershenzon, J. 1994. Metabolic cost of terpenoid accumulation in higher plants. *J. Chem. Ecol.,* **20:** 1281-2981.

Gershenzon, J., and Dudareva, N. 2007. The function of terpene natural products in the natural world. *Nature Chemical Biology*, **3(7):** 408-414.

Ghasemzadeh, A., Jaafar, H.Z.E. and Rahmat, A. 2010. Elevated carbon dioxide increases contents of flavonoids and phenolic compounds, and antioxidant activities in Malaysian young ginger (*Zingiber officinale* Roscoe.) varieties. *Molecules*, **15(11):** 7907–7922.

Glynn, C., Herms, D.A., Orians, C.M., Hansen, R.C. and Larsson, S. 2007. Testing the growth-differentiation balance hypothesis: Dynamic responses of willows to nutrient availability. *New Phytol.,* **176:** 623–634.

Gonzàlez-Meler, M.A., Ribas-Carbó, M., Siedow, J.N. and Drake, B.G. 1996. Direct inhibition of plant mitochondrial respiration by elevated CO_2. *Plant Physiol.* **112**:1349–1355.

Gobbo-Neto, L. and Lopes, N.P. 2007. Medicinal plants: factors of influence on the content of secondary metabolites. *Quim. Nova*, **30:** 374–381.

Gould, K.S., McKelvie, J., Markham, K.R. 2002. Do anthocyanins function as antioxidants in leaves? Imaging of H_2O_2 in red and green leaves after mechanical injury. *Plant Cell Environ.*, **25:** 1261-1269.

Gouvea, D.R., Gobbo-Neto, L., and Lopes, N.P. 2012. The influence of biotic and abiotic factors on the production on secondary metabolites in medicinal plants. In, "*Plant Bioactives and Drug Discovery*", Ed. V. Cechinel-Filho, John Wiley *and* Sons, Inc, New York, NY, USA, pp. 519–452.

Grace, S.G. and Logan, B.A. 2000. Energy dissipation and radical scavenging by the plant phenylpropanoid pathway. *Phil. Trans. R. Soc. Lond.*, *B* **355:** 1499-1510.

Graham, T.L. 1998. Flavonoid and flavonol glycoside metabolism in *Arabidopsis. Plant Physiol.Biochem.,* **36:** 135-144.

Granda, V., Delatorre, C., Cuesta, C., Centeno, M.L., Fernández, B., Rodríguez, A. and Feito, I. 2014. Physiological and biochemical responses to severe drought stress of nine *Eucalyptus globulus* clones: a multivariate approach. *Tree Physiol.* **34:** 778-786.

Grassmann, J., Hippeli, S. and Elstner, E.F. 2002. Plant's defence and its benefits for animals and medicine: role of phenolics and terpenoids in avoiding oxygen stress. *Plant Physiol. Biochem.*, **40:** 471-478.

Greenway, H. and Munns, R. 1980. Mechanism of salt tolerance in nonhalophytes. *Annu. Rev. Plant Physiol*, **31:** 149-190.

Grieve, C.M. and Maas, E.V. 1984. Betaine accumulation in salt-stressed sorghum. *Physiol Plant*, **61***:* 167–171.

Griffith, M. and Yaish, M.W.F. 2004. Antifreeze proteins in overwintering plants: a tale of two activities. *Trends Plant Sci.* **9:** 399–405.

Groppa, M.D., Tomaro, M.L. and Benavides, M.P. 2001. Polyamines as protectors against cadmium or copper-induced oxidative damage in sunflower leaf discs. *Plant Sci.*, **161:** 481–488.

Gu, X.D., Sun, M.Y., Zhang, L., Fu, H.W., Cui, L., Chen, R.Z., Zhang, D.W. and Tian, J.K. 2010. UV-B induced changes in the secondary metabolites of *Morus alba* L. Leaves. *Molecules,* **15:** 2980-2993.

Gu, J., Weber, K., Klemp, E., Winters, G., Franssen, S.U., Wienpahl, I., Huylmans, A.K., Zecher, K., Reusch, T.B., Bornberg-Bauer, E. and Weber, A.P. 2012. Identifying core features of adaptive metabolic mechanisms for chronic heat stress attenuation contributing to systems robustness. *Integr. Biol.,* **4:** 480-493.

Güçlü-Üstündað, Ö., and Mazza., G. 2007. Saponins: properties, applications and processing. *Critical Reviews in Food Science and Nutrition*, **47(3):** 231-258.

Gumul, D., Korus, J. and Achremowicz, B. 2007. The influence of extrusion on the content of polyphenols and antioxidant/antiradical activity of rye grains (*Secale cereal* L.). *Acta Sci Pol.*, **6:** 103–111

Gutbrodt, B., Dorn, S., Unsicker, S.B. and Mody, K. 2012. *Species-specific responses of herbivores to within-plant and environmentally mediated between-plant variability in plant chemistry. Chemoecology,* **22:** 101–111.

Hada, H., Hidema, J., Maekawa, M. and Kumagai, T. 2003. Higher amounts of anthocyanins and UV-absorbing compounds effectively lowered CPD photorepair in purple rice (*Oryza sativa* L.). *Plant, Cell and Environment*, **26:** 1691–1701.

Hadacek, F., 2002. Secondary metabolites as plant traits: current assessment and future perspectives. *Crit. Rev. Plant Sci.*, **21:** 273-322.

Häder, D.-P., Kumar. H.D., Smith. R.C., orrest, R.C. 2007. Effects of solar UV radiation on aquatic ecosystems and interactions with climate change. *Photochem Photobiol Sci*, **6:** 267–285.

Hagemeyer, J. 1999. Ecophysiology of plant growth under heavy metal stress. In, "*Heavy Metal Stress in Plants*", Eds. M.N.V. Prasad and J. Hagemeyer, Berlin, Springer, pp. 222.

Haghighi,, Z., Karimi, N., Modarresi, M. and Mollayi, S. 2012. Enhancement of compatible solute and secondary metabolites production in *Plantago ovata* Forsk. by salinity stress. *Journal of Medicinal Plants Research*, **6(18):** 3495-3500.

Hahlbrock, K., Bednarek, P., Ciolkowski, I., Hamberger, B., Heise, A., Liedgens, H., Logemann, E., Nürnberger, T., Schmelzer, E. and Somssich, I.E. 2003. Non-self recognition, transcriptional reprogramming, and secondary metabolite accumulation during plant/ pathogen interactions. *Proc Natl Acad Sci U S A.*, **100(2):** 14569–14576.

Hall, J.L. 2002. Cellular mechanisms for heavy metal detoxification and tolerance. *Journal of Experimental Botany,* **53(366):** 1–11.

Hamerlynck, E.P., Huxman, T.E., Loik, M.E. and Smith, S.D. 2000. Effects of extreme high temperature, drought and elevated CO_2 on photosynthesis of the Mojave Desert evergreen shrub, *Larrea tridentata*. *Plant Ecol.*, **148**:183–193.

Hamilton, J.G., Zangerl, A.R., DeLucia, E.H. and Berenbaum, M.R. 2001. The carbon-nutrient balance hypothesis: Its rise and fall. *Ecol. Lett.,* **4:** 86–95.

Harada, E., Kim, J.-A., Meyer, A.J., Hell, R., Clemens, S. and Choi, Y.-E. 2010. Expression profiling of tobacco leaf trichomes identifies genes for biotic and abiotic stresses. *Plant and Cell Physiology,* **51(10):** 1627–1637.

Harborne, J.B. 1988. Introduction to ecological biochemistry, Third edition. Academic press, New York. Zarborne, J.B. 1989. Recent advances in chemical ecology, *Natural Products Reports*, **6:** 85-109.

Harvell, C.D. and Tollrian, R. 1999. Why inducible defenses? In, "*The ecology and evolution of inducible defenses",* Eds. R. Tollrian and C.D. Harvell, Princeton, New Jersey: Princeton University Press, pp: 3–9.

Hatata, M.M. and Abdel-Aal, E.A. 2008. Oxidative stress and antioxidant defense mechanisms in response to cadmium treatments. *American-Eurasian Journal of Agricultural and Environmental Sciences*, **4(6):** 655–669.

Havaux, M. 1998. Carotenoids as membrane stabilizers in chloroplasts. *Trends Plant Sci.*, **3:** 147–151.

Hauser, M.T. 2014. Molecular basis of natural variation and environmental control of trichome patterning. *Frontiers in Plant Science*, **5(320):** 1–7.

Heath, M.C., 2000. Hypersensitive response-related death, *Plant Mol. Biol.*, **44:** 321-334.

Hectors, K., Prinsen, E., Coen, W.D., Jansen, M.A.K. and Guisez, Y. 2007. *Arabidopsis thaliana* plants acclimated to low dose rates of ultraviolet B radiation show specific changes in morphology and gene expression in the absence of stress symptoms. *New Phytol.,* **175:** 255–270.

Heim, K.E., Tahliaferro, A.R. and Bobilya, D.J. 2002. Flavonoid antioxidants: chemistry, metabolism and structure-activity relationships. *J. Nutr. Biochem*., **13:** 572-584.

Heinonen, I.M., Meyer, A.S. and Frankel, E.N. 1998. Antioxidant activity of berry phenolics on human low- density lipoprotein and liposome oxidation. *J. Agric. Food Chem.*, **46:** 4107–4112.

Herms, D.A. and Mattson, W.J. 1992. The dilemma of plants: to grow or defend. *Quarterly Review of Biology*, **67(3):** 283–335.

Herrmann, K., and Nagel, C.W. 1989. Occurrence and content of hydroxycinnamic and hydroxybenzoic acid compounds in foods. *Critical Reviews in Food Science and Nutrition,* **28(4):** 315-347.

Hemm, M.R., Rider, S.D., Ogas, J., Murry, D.J. and Chapple, C. 2004. Light induces phenylpropanoids metabolism in *Arabidopsis* roots. *Plant J.*, **38:** 765–778.

Hernández, I., Alegre, L. and Munne-Bosch, S. 2006. Enhanced oxidation of flavan-3-ols and proanthocyanidin accumulation in water-stressed tea plants. *Phytochemistry*, **67:** 1120-1126.

Högy, P. and Fangmeier, A. 2009. Atmospheric CO_2 enrichment affects potatoes: 2. Tuber quality traits. *European Journal of Agronomy*, **30(2):** 85–94.

Hollósy, F. 2002. Effect of ultraviolet radiation on plant cells. *Micron*, **33:** 179–197.

Horborne, J.B. and Williams, C.A. 2000. Advances in flavonoid research since. *Phytochemistry,* **55:** 481–504.

Horton, P. 2002. Crop improvement through alteration in the photosynthetic membrane., ISB News Report. Virginia Tech, Blacksburg, VA. Horv´ath, G., Arellano, J.B., Droppa, M.

and Bar′on, M., 1998. Alterations in Photosystem II electron transport as revealed by thermoluminescence of Cu-poisoned chloroplasts. *Photosyn. Res.*, **57:** 175–181.

Hossain, M.A., Piyatida, P., da Silva, J.A.T. and Fujita, M. 2012. Molecular mechanism of heavy metal toxicity and tolerance in plants: central role of glutathione in detoxification of reactive oxygen species and methylglyoxal and in heavy metal chelation. *Journal of Botany*, **2012:** Article ID 872875, 37 pages.

Hummel, I.. El-Amrani, A., Gouesbet, G., Hennion, F. and Couee, I. 2004. Involvement of polyamines in the interacting effects of low temperature and mineral supply on *Pringlea antiscorbutica* (Kerguelen cabbage) seedlings. *J Exp Bot.*, **399:** 1125–1134.

Ibrahim, M.H., Jaafar, H.Z.E., Rahmat, A. and Z. A. Rahman, Z.A. 2011. The relationship between phenolics and flavonoids production with total non structural carbohydrate and photosynthetic rate in *Labisia pumila* Benth. under high CO_2 and nitrogen fertilization. *Molecules*, **16(1):** 162–174.

Ibrahim, M.H. and Jaafar, H.Z.E. 2011a. Involvement of carbohydrate, protein and phenylanine ammonia lyase in up-regulation of secondary metabolites in *Labisia pumila* under various CO_2 and N_2 levels. *Molecules*, **16(5):** 4172–4190.

Ibrahim, M.H. and Jaafar, H.Z.E 2011b. Increased carbon dioxide concentration improves the antioxidative properties of the malaysian herb Kacip Fatimah (*Labisia pimila* Blume). *Molecules*, **16(7):** 6068–6081.

Ibrahim, M.H. and Jaafar, H.Z.E 2011c. Enhancement of leaf gas exchange and primary metabolites under carbon dioxide enrichment up-regulates the production of secondary metabolites in *Labisia pumila* seedlings. *Molecules*, **16(5):** 3761–3777.

Ibrahim, M.H., Jaafar, H.Z.E., Rahmat, A. and Rahman, Z.A. 2012. Involvement of nitrogen on flavonoids, glutathione, anthocyanin, ascorbic acid and antioxidant activities of malaysian medicinal plant *Labisia pumila* Blume (Kacip Fatimah). *International Journal of Molecular Sciences*, **13(1):** 393–408.

Ibrahim, M.H., Jaafar, H.Z.E, Karimi, E. and Ghasemzadeh, A. 2014. Allocation of secondary metabolites, photosynthetic capacity, and antioxidant activity of Kacip Fatimah (*Labisia pumila* Benth) in response to CO_2 and Light Intensity. *The Scientific World Journal,***2014:** Article ID 360290, 13 pages.

Idso, S.B. 1988. Three phases of plant response to atmospheric CO_2 enrichment. *Plant Physiology* **87**: 5-7.

Idso, C.D. and Idso, K.E. 2000. Forecasting world food supplies: The impact of rising atmospheric CO_2 concentration. *Technology.* **7:** 33–55.

Idso, S.B., Kimball, B.A., Pettit III, G.R., Garner, L.C. and Backhaus, R.A. 2000. Effects of atmospheric CO_2 enrichment on the growth and development of *Hymenocallis littoralis* (Amaryllidaceae) and the concentrations of several antineoplastic and antiviral constituents of its bulbs. *The American Journal of Botany*, **87(6):** 769–773.

Inze, D. and van Montagu, M. 1995. Oxidative stress in plants. *Curr. Opin. Biotechnol.* **6:** 153–158.

Iqbal, M., Ashraf, M. Jamil, A. and Ur-Rehman, S. 2006. Does seed priming induce changes in the levels of some endogenous plant hormones in hexaploid wheat plants under salt stress? *Integrative Plant Biol*, **48:** 181-189.

Jaafar, H.Z.E., Ibrahim, M.H. and Karimi, E. 2008. Phenolics and flavonoids compounds, phenylanine ammonia Lyase and antioxidant activity responses to elevated CO_2 in *Labisia pumila)* (Myrisinaceae). *Molecules*, **17(6):** 6331–6347.

Jaafar, H.Z., Mohamed Haris, N.B. and Rahmat, A. 2008. Accumulation and partitioning of total phenols in two varieties of *Labisia pumila* Benth. under manipulation of greenhouse irradiance. *Acta Hortic.,* **797:** 387-392.

Jaafar, H.Z., Ibrahim, M.H. and Mohamad Fakri, N.F. 2012. Impact of soil field water capacity on secondary metabolites, phenylalanine ammonialyase (PAL), maliondialdehyde (MDA) and photosynthetic responses of Malaysian kacip fatimah (*Labisia pumila* Benth). *Molecules*, **17(6):** 7305–7322.

Jaakola, L., Maatta-Riihinen, K., Karenlampi, S. and Hohtola, A. 2004. Activation of flavonoid biosynthesis by solar radiation in bilberry (*Vaccinium myrtillus* L.) leaves. *Planta*, **218:** 721–728.

Jacobsen, S., Hauschild, M. and Rasmussen, U. 1992. Induction by chromium ion of chitinases and polyamines in barley (*Hordeum vulgare* L.) and rape (*Brassica napus* L ssp. *oleifera*). *Plant Sci.* **84:** 119–128.

Jaleel, C.A., Sankar, B., Murali, P.V., Gomathinayagam, M., Lakshmanan, G.M.A. and Panneerselvam, R. 2008. Water deficit stress effects on reactive oxygen metabolism in *Catharanthus roseus*; impacts on ajmalicine accumulation. *Colloid. Surfaces B*, 62: 105–111.

Janska, A., Marsik, P., Zelenkova, S. and Ovesna, J. 2010. Cold stress and acclimation – what is important for metabolic adjustment? *Plant Biol.*, **12:** 395–405.

Jensen, C.R., Mogensen, V.O., Mortensen, G., Fieldsend, J.K., Milford, G.F.J., Andersen, M.N. and Thage, J.H. 1996. Seed glucosinolate, oil and protein contents of field-grown rape (*Brassica napus* L.) affected by soil drying and evaporative demand. *Field Crops Res.*, **47:** 93–105.

Jansen, M.A.K. 2002. Ultraviolet-B radiation effects on plants: induction of morphogenic responses. *Physiologia Plantarum*, **116:** 423–429.

Jansen, M.A.K., Hectors, K., O'Brien, N.M., Guisez, Y. and Potters, G. 2008. Plant stress and human health: Do human consumers benefit from UV-B acclimated crops? *Plant Sci.*, **175:** 449–458.

Jin, C.W., You, G.Y., He, Y.F., Tang, C.X., Wu, P. and Zheng, S.J. 2007. Iron deficiency induced secretion of phenolics facilitates the reutilization of root apoplastic iron in red clover. *Plant Physiol.*, **144:** 278–285.

Jin, C.W., You, G.Y. and Zheng, S.J. 2008. The iron deficiency-induced phenolics secretion plays multiple important roles in plant iron acquisition underground. *Plant Signal. Behav.*, **3:** 60–61.

Jingjin, Y., Hongmei, D., Ming, X. and Bingru Huang, B. 2012. Metabolic responses to heat stress under elevated atmospheric CO_2 concentration in a cool-season grass species. *Journal of the American Society for Horticultural Science*, **137(4):** 221-228.

John, R., Ahmad, P., Gadgil, K. and Sharma, S. 2009. Heavy metal toxicity: effect on plant growth, biochemical parameters and metal accumulation by *Brassica juncea* L. *International Journal of Plant Production*, **3(3):** 65–76.

Johnson, R.H. and Lincoln, D.E. 1991. Sagebrush carbon allocation patterns and grasshopper nutrition: the influence of CO_2 enrichment and soil mineral limitation. *Oecologia*, **87(1):** 127–134.

Jochum, G.M., K.W. Mudge and R.B. Thomas, 2007. Elevated temperatures increase leaf senescence and root secondary metabolite concentrations in the understory herb *Panax quinquefolius* (Araliaceae). *Amer. J. Bot.*, **94:** 819–826

Joshee, N., Patrick, T.S., Mentreddy, R.S. and Yadav, A.K. 2002. Skullcap: Potential medicinal crop. In, "*Trends in New Crops and New Uses", Eds. J.* Janick, and A. Whipkey. ASHS Press, Alexandria, VA, USA, pp. 580-586.

Joshi, P.N., Ramaswamy, N.K., Iyer, R.K., Nair, J.S., Pradhan, M.K., Gartia, S., Biswal, B. and Biswal, U.C. 2007. Partial protection of photosynthetic apparatus from UV-Binduced damage by UV-A radiation. *Environ Exp Bot*, **59(2):** 166–172.

Jung, C.H., Maeder, V., Funk, F., Frey, B., Sticher, H. and Frosserd, E. 2003. Release of phenols from *Lupinus albus* L. roots exposed to Cu and their possible role in Cu detoxification. *Plant Soil*, **252:** 301.

Jwa, N.-S., Agrawal, G.K., Tomogami, S., Yonekura, M., Han, O., Iwahashi, H. and Rakwal, R. 2006. Role of defence/stress-related marker genes, proteins, and secondary metabolites in defining rice self defence mechanisms. *Plant Physiol Biochem* . **44:** 261–273. varieties diûering in chilling sensitivity. *Plant Sci*. **161:** 719–726

Kakani, V.G., Reddy, K.R., Zhao, D. and Sailaja, K. 2003. Field crop responses to ultraviolet-B radiation: a review. *Agricultural and Forest Meteorology*, **120:** 191–218.

Katerova, Z.., Todorova, D., Tasheva, K. and Sergiev, I. 2012. Influence of ultraviolet radiation on plant secondary metabolite production. *Genetics and Plant Physiology*, **2(304):** 113-144.

Khan, A.G., Kuek, T.M., Chaudhury, T.M., Khoo, C.S. and Hayes, W.J. 2000. Role of plants, mycorrhizae and phytochelators in heavy metal contaminated land remediation. *Chemosphere*, **41:** 197.

Khan, N.A., Syeed, S., Masood, A., Nazar, R. and Iqbal, N 2010. Application of salicylic acid increases contents of nutrients and antioxidative metabolism in mung bean and alleviates adverse effects of salinity stress. *Int. J. Plant Biol.*, **1:** e1.

Kinnersley, A.M. and Turano, F.J. 2000. Gamma aminobutyric acid (GABA) and plant responses to stress. *Cr Rev Plant Sci.*, **19:** 479–509.

Kitao, M., Yoneda, R., Tobita, H., Matsumoto, Y., Maruyama, Y., Arifin, A. Asan, A. and Muhamad, M. 2006. Susceptibility to photoinhibition in seedlings of six tropical fruit tree species native to Malaysia following transplantation to a degraded land. *Trees,* **20:** 601–610.

Kirkillis, C.G., Pasias, I.N., Miniadis-Meimaroglou, S., Thomaidis, N.S. and Zabetakis, I. 2012. Concentration levels of trace elements in carrots, onions and potatoes cultivated in Asopos Region, Central Greece. *Analytical Letters,* **45(5-6):** 551-562.

Kleinwächter, M. and Selmar, D. 2014. Influencing the product quality by applying drought stress during the cultivation of medicinal plants. In, "*Physiological mechanisms and adaptation strategies in plants under changing environment -volume 1*", Eds. P. Ahmad and M.R. Wani, Springer New York, pp 57-73.

Kleinwachter, M. and Selmar, D. 2015. New insights explain that drought stress enhances the quality of spice and medicinal plants: potential applications. *Agron. Sustain. Develop.,* **35(1):** 121-131.

Kolb, C.A., K¨aser, M.A., Kopeck´y, J., Zotz, G., Riederer, M. and Pf¨undel, E.E. 2001. Effects of natural intensities of visible and ultraviolet radiation on epidermal ultraviolet screening and photosynthesis in grape leaves. *Plant Physiol.* **127:** 863–875.

Kolb, C.A., Kopeck´y, J., Riederer, M. and Pf¨undel, E.E. 2003. UV screening by phenolics in berries of grapevine (*Vitis vinifera*). *Funct. Plant Biol.* **30:** 1177–1186.

Kogan S.B., Kaliya, M., and Froumin, N. 2006. Liquid phase isomerization of isoprenol into prenol in hydrogen environment. *Applied Catalysis A: General*, **297(2):** 231-236.

Kovács, E. and Keresztes, V. 2002. Effect of gamma and UV-B/C radiation on plant cells. *Micron*, **33:** 199–210.

Kováèik, J., Klejdus, B., Baèkor, M. and Repèák, M. 2007. Phenylalanine ammonia-lyase activity and phenolic compounds accumulation in nitrogen-deficient *Matricaria chamomilla* leaf rosettes. *Plant Sci.,* **172:** 393–399.

Kováèik, J., Klejdus, B. and Baèkor, M. 2010. Physiological responses of *Scenedesmus quadricauda* (Chlorophyceae) to UV-A and UV-C light. *Photochemistry and Photobiology*, **86:** 612–616.

Kovacik, J. and Backor, M. 2007. Phenylalanine ammonia-lyase and phenolic compounds in Chamomile tolerance to cadmium and copper excess. *Water Air Soil Pollut.*, **185:** 185–193.

Kovács, Z., Simon-Sarkadi, L., Szucs, A. and Kocsy, G. 2011. Differential effects of cold, osmotic stress and abscisic acid on polyamine accumulation in wheat. *Amino Acids.*, **38:** 623–631.

Kranner, I., Beckett, R.P., Wornik, S., Zorn, M. and Pfeifhofer, H.W. 2002. Revival of a resurrection plant correlates with its antioxidant status. *Plant J*, **31:** 13–24.

Kranner, I., Minibayeva, F.V., Backett, R.P. and Seal, C.E. 2010. What is stress? Concepts, definitions and applications in seed science. *New Phytol*, **188:** 655–673.

Kramer, D., Breitenstein, B., Kleinwächter, M. and Selmar, D. 2010. Stress metabolism in green coffee beans (*Coffea arabica* L.): expression of dehydrins and accumulation of GABA during drying. *Plant Cell Physiol.*, **51(4):** 546–553.

Krause, G.H., Schmude, C., Garden, H., Koroleva, O.Y. and Winter, K. 1999. Effects of solar ultraviolet radiation on the potential efficiency of photosystem II in leaves of tropical plants. *Plant Physiol.*, **121:** 1349–1358.

Kirkham, M.B. 2011. Elevated carbon dioxide: Impact on soil and plant water relations. CRC Press, Boca Raton, FL.,

Krishnamurthy, R. and Bhagwat, K.A.M. 1989. Polyamines as modulators of salt tolerance in rice cultivars. *Plant Physiol.* **91:** 500–504.

Krishnamurthy, R. and Bhagwat, K A M. 1990. Accumulation of choline and glycinebetaine in salt-stressed wheat seedlings. *Curr Sci.*, **59:** 111–112.

Krause, G.H., Koroleva, O.Y. and Dalling, W., 2001. Winter, K. Acclimation of tropical tree seedlings to excessive light in simulated tree-fall gaps. *Plant Cell Environ.*, **24:** 345–1352.

Krizek, D.T., Mirecki, R.M. and Britz, S.J. 1997. Inhibitory effects of ambient levels of solar UV-A and UV-B radiation on growth of cucumber. *Physiol. Plant* ,**100:** 886–893.

Krizek, D.T., Britz, S.J. and Mirecki, R.M. 1998. Inhibitory effects of ambient levels of solar UV-A and UV-B radiation on growth of cv. New Red Fire lettuce. *Physiol. Plant*, **103:** 1–7.

Krizek, D.T. 2004. Influence of PAR and UV-A in determining plant sensitivity and photomorphogenic responses to UV-B radiation. *Photochem. Photobiol.*, **79:** 307–315.

Krupa, Z., Baranowska, M. and Orzol, D. 1996. Can anthocyanins be considered as heavy metal stress indicator in higher plants? *Acta Physiol Plant.* **18:** 147–151.

Ksouri, R., Megdiche, W., Debez, A., Falleh, H., Grignon, C. and Abdelly, C. 2007. Salinity effects on polyphenol content and antioxidant activities in leaves of the halophyte *Cakile maritima. Plant Physiol Biochem.,* **45:** 244–249.

Kucera, B., Leubner-Metzger, G. and Wellmann, E. 2003. Distinct ultravioletsignaling pathways in bean leaves. DNA damage is associated with β-1,3-glucanase gene induction, but not with flavonoid formation. *Plant Physiology*, **133:** 1–8.

Kurata, H., Matsumura, S. and Furusaki, S. 1997. Light irradiation causes physiological and metabolic changes for purine alkaloid production by a *Coffea arabica* cell suspension culture. *Plant Sci.,* **123:** 197–203.

Laisk, A., Eichelmann, H., Oja, V., Rasulov, B., Padu, E., Bichele, I., Pettai, H. and Kull, O. 2005. Adjustment of leaf photosynthesis to shade in a natural canopy: Rate parameters. *Plant Cell Environ.*, **28:** 378–388.

Lambers, H. 1985. Respiration in intact plants and tissues: its regulation and dependence on environmental factors, metabolism and invaded organisms. In: Encyclopedia of Plant Physiology, New series, Vol. 18, Springer Verlag. Berlin. pp. 418-473.

Lambreva, M., Christov, K. and Tsonev, T. 2006. Short-term effect of elevated CO_2 concentration and high irradiance on the antioxidant enzymes in bean plants. *Biologia Plantarum,,* **50(4):** 617–623.

Larsson, S. Wiren, A., Lundgren, L. and Ericsson, T. 1986. "Effects of light and nutrient stress on leaf phenolic chemistry in *Salix dasyclados* and susceptibility to *Galerucella lineola* (Coleoptera). *Oikos*, **47(2):** 205–210.

Laisk, A., Eichelmann, H., Oja, V., Rasulov, B., Padu, E., Bichele, I., Pettai, H. and Kull, O. 2005. Adjustment of leaf photosynthesis to shade in a natural canopy: Rate parameters. *Plant Cell Environ.*, **28:** 378–388.

Lattanzio, V., di Venere, D., Linsalata, V., Bertolini, P., Ippolito, A. and Salerno, M. 2001. Low temperature metabolism of apple phenolics and quiescence of *Phlyctaena vagabunda. J. Agric. Food Chem.,* **49:** 5817–5821.

Lattanzio, V., Cardinali, A. and Linsalata, V. 2012. Plant phenolics: A biochemical and physiological perspective. In *Recent Advances in Polyphenols Research*, Cheynier, V., Sarni-Manchado, P., Quideau, S., Eds., Wiley-Blackwell Publishing: Oxford, UK, 2012, Volume 3, pp. 1–39.

Larson, R.A. 1988. The antioxidants of higher plants. *Phytochemistry*, **27:** 969–978.

Lavid, N., Schwartz, A., Yarden, O. and Tel-or, E. 2001. The involvement of polyphenols and peroxidase acitivities in heavy metal accumulation by epidermal glands of water lily (*Nymphaeceaea*). *Planta*, **212:** 323.

Lavola, A., Aphalo, P.J., Lahti, M. and JulkunenTiitto, R. 2003. Nutrient availability and the effect of increasing UV-B radiation on secondary plant compounds in Scots pine. *Environm Exp Botany*, **49:** 49–60.

Leakey, A.D.B., Uribelarrea, M., Ainsworth, E.A., Naidu, S.L., Rogers, A., Ort, D.R. and Long, S.P. 2006. Photosynthesis, productivity, and yield of maize are not affected by open-air elevation of CO_2 concentration in the absence of drought. *Plant Physiol.* **140**:779–790.

Le Rudulier, D. 2005. Osmoregulation in rhizobia: The key role of compatible solutes. *Grain Legume*, **42:** 18-19.

Lee, S., Moon, J.S., Domier, L.L. and Korban, S.S. 2002. Molecular characterization of phytochelatin synthase expression in transgenic *Arabidopsis. Plant Physiology and Biochemistry*, **40(9):** 727–733.

Lei, X.Y., Zhu, R.Y., Zhang, G.Y. and Dai, Y.R. 2004. Attenuation of cold induced apoptosis by exogenous melatonin in carrot suspension cells: the possible involvement of polyamines. *J Pineal Res.* **36:** 126–131.

Leopoldini, M., Russo, N., Chiodo, S. and Toscano, M. 2006. Iron chelation by the powerful antioxidant flavonoid quercetin. *J. Agric. Food Chem.*, **54:** 6343-6351.

Lesniewska, E., Adrian, M., Klinguer, A. and Pugin, A. 2004. Cell wall modification in grapevine cells in response to UV stress investigated by atomic force microscopy. *Ultramicroscopy*, **100(3– 4):** 171–178.

Levy, D.D.E. and Tang. P.C. 1995. The Chemistry of C-glycosides. Vol. 13. Elsevier.

Leyva, A., Jarillo, J.A., Salinas, J. and Martinez-Zapater, J.M.1995. Low temperature induces the accumulation of phenylalanine ammonia-lyase and chalcone synthase mRNAs of *Arabidopsis thaliana* in a light-dependent manner. *Plant Physiol.,* **108:** 39–46.

Liakopoulos, G. and Karabourniotis, G. 2005. Boron deficiency and concentrations and composition of phenolic compounds in *Olea europaea* leaves: A combined growth chamber and field study. *Tree Physiol.,* **25:** 307–315.

Liechti, R. and Farmer, E.E. 2002. The jasmonate pathway. *Science*, **296:** 1649–1650.

Lillo, C., Lea, U.S. and Ruoff, P. 2008. Nutrient depletion as a key factor for manipulating gene expression and product formation in different branches of the flavonoid pathway. *Plant Cell Environ.,* **31:** 587–601.

Lin, C.C. and Kao, C.H. 1999. Excess copper induces an accumulation of putrescine in rice leaves. *Bot Bull Acad Sin.*, **40:** 213–218.

Liu, C.Z., Guo, C., Wang, Y.C. and Ouyang, F. 2002. Effect of light irradiation on hairy root growth and artemisinin biosynthesis of Artemisia annua. *Process Biochem.*, **38:** 581-585.

Liu, W., Liu, C., Yang, C., Wang, L. and Li, S. 2010. Effect of grape genotype and tissue type on callus growth and production of resveratrols and their piceids after UV-C irradiation. *Food Chemistry*, **122:** 475–481.

Liu, H., Wang, X., Wang, D., Zou, Z. and Lianga, Z. 2011. Effect of drought stress on growth and accumulation of active constituents in *Salvia miltiorrhiza* Bunge. *Ind Crop Prod*, **33:** 84–88.

Lokko, Y., Okogbenin, E., Mba, C., Dixon, A., Raji, A. and Fregene, M. 2007. Cassava In, "*Pulses, sugar and tuber crop. Genome mapping and molecular breeding in plants, Volume 3*", Chittaranjan Kole, 2007, Springer.

Long, S.P., Humphries, S. and Falkowski, P.G. 1994. Photoinhibition of photosynthesis in nature. *Annu. Rev. Plant Physiol. Plant Mol. Biol.*, **45:** 633–662.

Loreto, F., Forster, A., Durr,M., Csiky, O. and Seufert, G. 1998. On the monoterpene emission under heat stress and on the increased thermotolerance of leaves of *Quercus ilex* L. fumigated with selected monoterpenes. *Plant Cell Environ*,.**21:** 101–107.

Loreto, F. and Velikova, V. 2001. Isoprene produced by leaves protects the photosynthetic apparatus against ozone damage, quenches ozone products, and reduces lipid peroxidation of cellular membranes. *Plant Physiol.* **127:** 1781-1787.

Lukas, B., Schmiderer, C., Franz, C. and Novak, J. 2009. *Composition of essential oil compounds from different Syrian populations ofOriganum syriacum L. (Lamiaceae). J. Agric. Food Chem.* **57:** 1362–1365.

Luo, Q., Yu, B. and Liu, Y. 2005. Differential sensitivity to chloride and sodium ions in seedlings of *Glycine max* and *G. soja* under NaCl stress. *J Plant Physiol*, **162:** 1003-1012.

Lütz, C. 2012. Cell physiology of plants growing in cold environments. *Protoplasma*, **244:** 53–73.

McKenzie, R.L., Aucamp, P.J., Bais, A.F., Björn, L.O. and Ilyas, M. 2007. Changes in biologically-active ultraviolet radiation reaching the Earth's surface. *Photochem Photobiol Sci*, **6:** 218–231.

Maestri, E., N. Klueva, C. Perrotta, M. Gulli, H.T. Nguyen and N. Marmiroli, 2002. Molecular genetics of heat tolerance and heat shock proteins in cereals. *Plant Mol. Biol.*, **48:** 667-681.

Maffei, M.E. 2010. Sites of synthesis, biochemistry and functional role of plant volatiles. *S. Afr. J. Bot.*, **76:** 612–631.

Maffei, M. E., Gertsch, J., and Appendino, G. 2011. Plant volatiles: production, function and pharmacology. *Natural Product Reports*, **28(8):** 1359-1380.

Mahajan, S. and Tuteja, N. 2005. Cold, salinity and drought stresses: An overview. *Arch Biochem Biophys*, **444:** 139–158.

Mahmood, S., Wahid, A., Rasheed, R., Hussain, I. and Basra, S.M.A. 2012. Possible antioxidative role of endogenous vitamins biosynthesis in heat stressed maize (*Zea mays*). *Int. J. Agric. Biol.*, **14:** 705-712.

Manara, A. 2012. Plant responses to heavy metal toxicity. In, "*Plants and Heavy Metals*", Ed. Furini, Springer Briefs in Molecular Science, pp. 27–53, Springer, Dordrecht, Netherlands.

Manach, C., Scalbert, A., Morand, C., Rémésy C. and Jiménez, L. 2004. Polyphenols: food sources and bioavailability. *The American Journal of Clinical Nutrition*, **79(5):** 727-747.

Mansfield, J.W. 2000. Antimicrobial compounds and resistance. The role of phytoalexins and phytoanticipins. In, "Mechanisms of resistance to plant diseases", Eds. A. Slusarenko, R. Fraser, Van Loon, L, Netherlands: Kluwer Academic Publishers. pp: 325–370.

Manukyan, A. 2011. Effect of growing factors on productivity and quality of lemon catmint, lemon balm and sage under soilless greenhouse production: I. Drought stress. *Med Aromat Plant Sci Biotechnol*, **5(2):** 119–125

Marschner, H. 1995. Mineral nutrition of higher plants, Academic Press, London, pp. 889.

Marchese, J.A. and Figueira, G.M. 2005. The use of pre and post-harvest technologies and good agricultural practices in the production of medicinal and aromatic plants. *Braz. J. Med. Plant*, **7:** 86–96.

Marron, N., Dreyer, E., Boudouresque, E., Delay, D., Petit, J.-M., Delmotte, F.M. and Brignolas, F. 2003. Impact of successive drought and re-watering cycles on growth and specific leaf area of two *Populus × canadensis* (Moench) clones, 'Dorskamp' and Luisa Avanzo. *Tree Physiol*, **23:** 1225–1235.

Martinez, C.A. and Maestri, M. 1995. Photoinhibition of photosynthesis in andean potato (*Solanum* spp.) species differing in frost resistance. In, "*Photosynthesis: From Light to Biosphere*", Ed. P. Mathis, Kluwer Academic Publishers: Dordrecht, The Netherlands, 1995, pp. 179–182.

Massacci, A.M., Nabiev, L., Pietrosanti, S.K., Nematov, T.N., Chernikova, K. and Leipner, J. 2008. Response of the photosynthetic apparatus of cotton (*Gossypium hirsutum*) to the onset of drought stress under field conditions studied by gas-exchange analysis and chlorophyll fluorescence imaging. *Plant Physiol Biochem*,.**46:** 189–195.

Matsuki, M. 1996. Regulation of plant phenolic synthesis: from biochemistry to ecology and evolution. *Australian Journal of Botany*, **44(6):** 613–634.

Mazza, C.A., Boccalandro, H.E., Giordano, C.V., Battista, D., Scopel, A.L. and Ballaré, C.L. 2000. Functional significance and induction by solar radiation of ultra-violet-absorbing sunscreens in field-grown soybean crops. *Plant Physiol.,* **122:** 117–125.

Memon, A.R., Aktoprakligil, D., Zdemur, A. and Vertii, A. 2001. Heavy metal accumulation and detoxification mechanisms in plants. *Turkish Journal of Botany*, **25(3):** 111–121.

Mercure, S.A., Daoust, B. and Samson, G. 2004. Causal relationship between growth inhibition, accumulation of phenolic metabolites, and changes of UV-induced fluorescences in nitrogen-deficient barley plants. *Can. J. Bot.,* **6:** 815–821.

Mes, J.J., van Doorn, A.A., Wijbrandi, J., Simons, G., Cornelissen, B.J.C. and Haring, M.A.2000. Expression of the Fusarium resistance gene I-2 colocalizes with the site of fungal containment. *The Plant Journal*, **23:** 183-193.

Michalak, A. 2006. Phenolic compounds and their antioxidant activity in plants growing under heavy metal stress. *Polish Journal of Environmental Studies*, **15(4):** 523–530. ,

Millic, B.L., Djilas, S.M. and Canadanovic-Brunet, J.M. 1998. Antioxidative activity of phenolic compounds on the metal-ion breakdown of lipid peroxidation system. *Food Chem.*, **61:** 443.

Miranda, M.A.F.M., Varela, R.M., Torres, A., Molinillo, J.M.G., Gualtieri, S.C.J. and Macias, F.A. 2015. *Phytotoxins from Tithonia diversifolia*. *J. Nat. Prod.* **78:** 1083"1092.

Mithöfer, A., Schulze, B. and Boland, W. 2004. Biotic and heavy metal stress response in plants: evidence for common signals. *FEBS Lett.*, **566:** 1.

Mittler, R. 2002. Oxidative stress, antioxidants and stress tolerance. *Trends Plant Sci.*, **7:** 405–410.

Mittler, R., Vanderauwera, S., Gollery, M. and van Breusegem, F. 2004. Reactive oxygen gene network of plants. *Trends Plant Sci.,* **9:** 490–498.

Mittova, V., Volokita, M., Guy, M. and Tal, M. 2000. Activities of SOD and the ascorbate-glutathione cycle enzymes in subcellular compartments in leaves and roots of the cultivated tomato and its wild salt-tolerant relative *Lycopersicon pennellii*. *Physiol. Plant.*, **110:** 45.

Mizukami, H., Konoshima, M. and Tabata, M. 1977. Effect of nutritional factors on shikonin derivative formation in Lithospermum callus cultures. *Phytochemistry*. **16:** 1183–1186.

Monclus, R., Villar, M., Barbaroux, C., Bastien, C., Fichot, R., Delmotte, F.M., Delay, D., Petit, J.-M., Bréchet, C., Dreyer, E. and Brignolas, F. 2009. Productivity, water-use efficiency and tolerance to moderate water deficit correlate in 33 poplar genotypes from a *Populus deltoides × Populus trichocarpa* F1 progeny. *Tree Physiol*, **29:** 1329–1339.

Morales, C.G., Pino, M.T. and del Pozo, A. 2013. Phenological and physiological responses to drought stress and subsequent rehydration cycles in two raspberry cultivars. *Sci Hortic* , **162:** 234– 241.

Morcuende, R., Bari, R., Gibon, Y., Zheng, W., Pant, B.D., Bläsing, O., Usadel, B., Czechowski, T., Udvardi, M.K., Stitt, M., and Scheible, W.R. 2007. Genome-wide reprogramming of metabolism and regulatory networks of *Arabidopsis* in response to phosphorus. *Plant Cell Environ.*,**30:** 85–112.

Mori, K., Sugaya, S. and Gemma, H. 2005. Decreased anthocyanin biosynthesis in grape berries grown under elevated night temperature condition. *Sci. Hortic.*, **105:** 319-330.

Mori, K., Goto-Yamamoto, N., Kitayama, M. and Hashizume, K. 2007. Loss of anthocyanins in red-wine grape under high temperatura. *Exp. Bot.*, **58:** 1935-1945.

Morison, J.I.L. and Lawlor, D.W. 1999. Interactions between increasing CO_2 concentration and temperature on plant growth. *Plant Cell Environ.* **22:** 659–682.

Morgan, J.F., Klucas, R.V., Grayer, R.J., Abian, J. and Becana, M. 1997. Complexes of iron with phenolic compounds from soybean nodules and other legume tissues: prooxidant and antioxidant properties. *Free Radic. Biol. Med.*, **22:** 861.

Mosaleeyanon, K., Zobayed, S.M.A., Afreen, F. and Kozai T. 2005. Relationships between net photosynthetic rate and secondary metabolite contents in St. John's wort. *Plant Sci.*, **169:** 523–531.

Mourato, M., Reis, R. and Martins, L.L. 2012. Characterization of plant antioxidative system in response to abiotic stresses: a focus on heavy metal toxicity. In, "*Advances in Selected Plant Physiology Aspects*", Eds. G. Montanaro and B. Dichio, InTech, Vienna, Austria, pp. 23–44.

Muhammad, Z. and Hussain, F. 2010. Vegetative growth performance of five medicinal plants under NaCl salt stress. *Pak. J. Bot.*, **42:** 303-316.

Munns, R. 2002. Comparative physiology of salt and Water stress. *Plant Cell Environ.*, **25:** 239–250.

Muthukumarasamy, M., Gupta, S.D. and Pannerselvam, R. 2000. Enhancement of peroxidase, polyphenol oxidase and superoxide dismutase activities by tridimefon in NaCl stressed *Raphanus sativus* L. *Biol Plant.* **43:** 317–320.

Mutlu, F. and Bozcuk, S. 2007. Salinity-induced changes of free and bound polyamine levels in sunflower (*Helianthus annuus* l.) roots differing in salt tolerance. *Pak J Bot.*, **39:** 097–1102.

Nacif de Abreu, I. and Mazzafera, P. 2005. Effect of water and temperature stress on the content of active constituents of *Hypericum brasiliense* Choisy. *Plant Physiol Biochem.*, **43:** 241–248.

Nadeau, P., Delaney, S. and Chouinard, L. 1987. Effects of cold hardening on the regulation of polyamine levels in wheat (*Triticum aestivum* L.) and Alfalfa (*Medicago sativa* L.). *Plant Physiol.*, **84:** 73–77.

Nadeau, F., Gaudreau, A., Angers, P. and Arul, J. 2012. Changes in the level of glucosinolates in broccoli florets (*Brassica oleracea* var. Italica) during storage following postharvest treatment with UV-C. *Acta Horticulturae*, **945:** 145–148.

Nagajyoti, P.C., Lee, K.D. and Sreekanth, T.V.M. 2010. Heavy metals, occurrence and toxicity for plants: a review. *Environmental Chemistry Letters*, **8(3):** 199–216.

Nakabayashi, R., Yonekura-Sakaibara, K., Urano, K., Suzuki, M., Yamada, Y., Nishizawa, T., Matsuda, F., Kojima, M., Sakakibara, H., Shinozaki, K.,Michael, A.J., Tohge, T., Yamazaki, M. and Saito, K. 2014. Enhancement of oxidative and drought tolerance in *Arabidopsis* by overaccumulation of antioxidant flavonoids. *Plant J.*, **77:** 367–379.

Narayan, M.S., Thimmaraju, R.and Bhagyalakshmi, N. 2005. Interplay of growth regulators during solid-state and liquid-state batch cultivation of anthocyanin producing cell line of *Daucus carota. Process Biochem.*, **40:** 351–358.

Nascimento, N.C. and Fett-Neto, F. 2010. Plant secondary metabolism and challenges in modifying its operation: an overview. *Methods Mol Biol*, **643:** 1–13.

Navarro, J.M., Flores. P., Garrido. C. and Martinez, V. 2006. Changes in the contents of antioxidant compounds in pepper fruits at ripening stages, as affected by salinity. *Food Chem.,* **96:** 66–73.

Nawaz, A., Farooq, M., Cheema, S.A. and Wahid, A. 2013. Differential response of wheat cultivars to terminal heat stress. *Int. J. Agric. Biol.,***15:** 1354-1358.

Negro, C., Tommasi, L. and Miceli, A. 2003. Phenolic compounds and antioxidant activity from red grape marc extracts. *Bioresour Technol*, **87:** 41–44.

Newman, R.A., Yang, P., Pawlus, A. D. and Block, K.I. 2008. Cardiac glycosides as novel cancer therapeutic agents. *Molecular Interventions*, **8(1):** 36-49.

Newshamm K,K,, Lewis, G.C., Greenslade, P.D. and McLeod, A.R. 1998. Neotyphodium lolii, a fungal leaf endophyte, reduces fertility of *Lolium perenne* exposed to elevated UV-B radiation. *Annals of Botany*, **81:** 397–403.

Nicolaou, K.C., Chen, J.S. and Corey, E.J. 2011. Classics in Total Synthesis. Further Targets, Strategies, Methods III. Weinheim: Wiley-VCH.

Niinemets, U. and Way, D. 2016. Uncovering the hidden facets of drought stress: secondary metabolites make the difference. *Tree Physiol.,* **36(2):** 129-132.

Noctor, G. and Foyer, C.H. 1998. Ascorbate and glutathione: keeping active oxygen under control. *Annu. Rev. Plant Physiol. Plant Mol. Biol.*, **49:** 249–279.

Nogués, S., Allen, D.J., Morison, J.I.L. and Baker, N.R. 1998. Ultraviolet-B radiation effects on water relations, leaf development, and photosynthesis in droughted pea plants. *Plant Physiol.*, **117:** 173–181.

Nowak, M., Manderscheid, R., Weigel, H.-J., Kleinwächter, M. and Selmar, D. 2010. Drought stress increases the accumulation of monoterpenes in sage (*Salvia officinalis*), an effect that is compensated by elevated carbon dioxide concentration. *J Appl Bot Food Qual.*, **83:** 133–136

Obrenovic, S. 1990. Effect of Cu (11) D-penicillanine on phytochrome mediated betacyanin formation in*Amaranthus caudatus* seedlings. *Plant Physiol Biochem.*, **28:** 639–646.

Ogaya, R. and Peñuelas, J. 2003. Comparative field study of *Quercus ilex* and *Phillyrea latifolia*: photosynthetic response to experimental drought conditions. *Environ Exp Bot.*, **50:** 137–148.

Ogaya, R. and Peñuelas, J. 2006. Contrasting foliar responses to drought in *Quercus ilex and Phillyrea latifolia. Biol Plant*, **50**:373–382.

Oh, M.-M., Trick, H.N. and Rajashekar, C.B. 2009. *Secondary metabolism and antioxidants are involved in environmental adaptation and stress tolerance in lettuce. J. Plant Physiol.,* **166:**180–191.

Ohlsson, A.B. and Berglund, T. 1989. Effect of high MnSO4 levels on cardenolide accumulation by *Digitalis lanata* tissue cultures in light and darkness. *J Plant Physiol.*, **135:** 505–507.

Okogbenin, E., Setter, T.L., Ferguson, M., Mutegi, R., Alves, A.C., Ceballos, H. and Fregene, M. 2010. Phenotyping cassava for adaptation to drought. In, "*Drought phenotyping in crops: From theory to practice*", Eds. P. Monneveux and J.M. Ribaut, CIMMYT / Gen, Challenge Prog. Mex. City, pp. 381-400.,

Osbourn, A.E., Qi, X., Townsend, B. and Qin, B. 2003. Dissecting plant secondary metabolism – constitutive chemical defences in cereals. *New Phytol.*, **159:**101-108.

Oszmañski, J. 1995. Polyphenols as antioxidants in food. *Przem Spoz.*, **3:**94–96.

Ozkur, O., F. Ozdemir, M. Bor and I. Turkan, 2009. Physiochemical and antioxidant responses of the perennial xerophyte *Capparis ovate* Desf to drought. *Environ. Exp. Bot.*, **66:** 487–492.

Paez, A., Gebre, G.M., Gonzalez, M.E. and Tschaplinski, T.J. 2000. Growth, soluble carbohydrates, and aloin concentration of Aloevera plants exposed to three irradiance levels. *Environ. Exp. Bot.*, **44:** 133–139.

Palevitch, D. 1987. Recent advances in the cultivation of medicinal plants. *Acta Hort.*, **208:** 29–35.

Perez-Ilzarbe, J., Hernandez, T., Estrella, I. and Vendrell, M. 1997. Cold storage of apples (cv. Granny Smith) and changes in phenolic compounds. *Z Lebensm Unters Forsch.*, **204:** 52 55.

Palo, R.T. 1984. Distribution of birch (*Betula* spp), willow (*Salix*spp), and poplar (*Populus*spp) secondary metabolites and their potential role as chemical defense against herbivores. *Journal of Chemical Ecology*, **10(3):** 499–520.

Pant, B.-D., Pant, P., Erban, A., Huhman, D., Kopka, J. and Scheible, W.-R. 2015. Identification of primary and secondary metabolites with phosphorus status-dependent abundance in *Arabidopsis*, and of the transcription factor PHR1 as a major regulator of metabolic changes during phosphorus limitation. *Plant Cell Environ.,* **38:** 172–187.

Parida, A/K/ and Das, A.B. 2005. Salt tolerance and salinity effects on plants: A review. Ecotoxicol. *Environ. Safety*, **60:** 324-349.

Park, E.S., Moon, W.S., Song, M.J., Kim, M.N., Chung, K.H., and Yoon, J.S. 2001. Antimicrobial activity of phenol and benzoic acid derivatives. *International biodeterioration and biodegradation*, **47(4):** 209-214.

Parvaiz, A,and Satyavati, S. 2008. Salt stress and phyto-biochemical responses of plants- a review. *Plant Soil Environ.*, **54:** 89–99.

Passioura, J. 2007. The drought environment: physical, biological and agricultural perspectives. *J Exp Bot.*, **58:** 113–117.

Patra, M., Bhowmik, N., Bandopadhyay, B. and Sharma, A. 2004. Comparison of mercury, lead and arsenic with respect to genotoxic effects on plant systems and the development of genetic tolerance. *Environmental and Experimental Botany*, **52(3):** 199–223.

Paul, N. 2001. Plant responses to UV-B: time to look beyond stratospheric ozone depletion? *New Phitol*, **150:** 5–8.

Pavarini, D.P., Pavarini, S.P., Niehues, M. and Lopes, N.P. 2012. *Exogenous influences on plant secondary metabolite levels*. *Anim. Feed Sci. Technol.*, **176:** 5–16.

Pedrazani, H., Racagni, G., Alemano, S., Miersch, O., Ramirez, I., Pena-Cortes, H., Taleisnik, E., Machado-Domenech, E. and Abdala, G. 2003. Salt tolerant tomato plants show increased levels of jasmonic acid. *Plant Growth Regul.*, **412:** 149–158.

Pedranzani, H., Sierra-de-Grado, R., Vigliocco, A., Miersch, O. and Abdala, G. 2003. Cold and water stresses produce changes in endogenous jasmonates in two populations of *Pinus pinaster* Ait. *Plant Growth Regul.*, **52:** 111–116.

Peng, Y. and Bo, Y. 2012. Extraction mechanism of total flavonoid on red kidney bean by microwave and light wave radiation. *Adv. Mater. Res.*, **361:** 707–711.

Pengelly, A. 2004. The constituents of medicinal plants: an introduction to the chemistry and therapeutics of herbal medicine, CABI Publishing.

Peñuelas, J. and Estiarte, M. 1998. Can elevated CO_2 affect secondary metabolism and ecosystem function? *Trends in Ecology and Evolution*, **13(1):** 20–24.

Petropoulos, S/A/, Daferera, D., Polissiou, M.G. and Passam, H.C. 2008. The effect of water deficit stress on the growth, yield and composition of essential oils of parsley. *Sci Hortic-Amsterdam*, **115:** 393–397.

Petrusa, L.M. and Winicov, I. 1997. Proline status in salt tolerant and salt sensitive alfalfa cell lines and plants in response to NaCl. *Plant Physiol Biochem.*, **35:** 303–310.

Pitta-Alvarez, S.I., Spollansky, T.C. and Giullietti, A.M. 2000. The influence of different biotic and abiotic elicitors on the production and profile of tropane alkaloids in hairy root cultures of *Brugmansia candida. Enzyme Microb Technol.*, **26:** 252–258.

Polle, A. 1997. Defense against photo oxidative damage in plants. In, "*Oxidative Stress and the Molecular Biology of Antioxidant Defenses*", Ed. J.G. Scandalios, Cold Spring Harbor Laboratory Press, Cold Spring Harbor, USA. pp. 623–666.

Polt. R.L. 1995. Method for making amino acid glycosides and glycopeptides, U.S. Patent No. 5,470,949. Washington, DC: U.S. Patent and Trademark Office.

Posmyk, M.M., Balabusta, M., Wieczorek, M., Sliwinska, E. and Janas, K.M. 2009. Melatonin applied to cucumber (*Cucumis sativus* L.) seeds improves germination during chilling stress. *J Pineal Res.*, **46:** 214–223.

Prasad, M.N.V. 2004. Phytoremediation of metals in the environment for sustainable development. *Proceedings of the Indian National Science Academy*, **70(1):** 71–98. ,

Prigent, S.V.E., Voragen, A.G.J., Visser, A.J.W.G., Van Koningsveld, G.A., Grruppen, H. 2007. Covalent interactions between proteins and oxidation products of caffeoylquinic acid (chlorogenic acid). *J. Sci. Food Agric.*, **87:** 2502-2510.

Prior, R.L., Cao, G., Martin, A., Sofic, E., McEwen, J., O'Brien, C., Lischner, N., Ehlenfeldt, M., Kalt, W., Krewer, G. and Mainland, C.M. 1998. Antioxidant capacity as influenced by total phenolic and anthocyanin content, maturity, and variety of *Vaccinium* species. *J. Agric. Food Chem.,* **46:** 2686–2693.

Procházková, D. and Wilhelmova, N. 2007. The capacity of antioxidant protection during modulated ageing of bean (*Phaseolus vulgaris* L.) cotyledon 1. The antioxidant enzyme activities. *Cell Biochem Funct*, **25:** 87–95.

Qaderi, M.M., Kurepin, L.V. and Reid, D.M. 2006. Growth and physiological responses of canola (*Brassica napus*) to three components of global climate change: Temperature, carbon dioxide and drought. *Physiol. Plant.* **128**:710–721.

Radwan, A., Hara, M., Kleinwächter, M. and Selmar, D. 2014. Dehydrin expression in seeds and maturation drying: a paradigm change. *Plant Biol.,* **16:** 853–855.

Rai, V., Vaypayee, P., Singh, S.N., Mehrotra, S. 2004. Effect of chromium accumulation on photosyn-thetic pigments, oxidative stress defence system, nitrate reduction, proline level and eugenol content of *Ocimum tenuiflorum* L . *Plant Sci..*, **167:** 1159.

Rajendra, L., Ravishankar, G.A., Venkataraman, L.V. and Prathiba, K.R. 1992. Anthocyanin production in callus cultures of Daucuscarota as influenced by nutrient stress and osmoticum. *Biotechnol Lett.*, **14:** 707–712.

Rahimmalek, M., Tabatabaei, B.E.S., Etemadi, N., Goli, S.A.H., Arzani, A. and Zeinali, H. 2009. *Essential oil variation among and within six Achillea species transferred from different ecological regions in Iran to the field conditions. Ind. Crops Prod.* **29:** 348–355.

Ralphs, M.H., Manners, G.D. and Gardner, D.R. 1998. Influence of light and photosynthesis on alkaloid concentration in larkspur. *J. Chem. Ecol.*, **24:** 167–182.

Rama Devi, S., Prasad, M.N.V. 1998. Copper toxicity in *Ceratophyllum demeresum* L. Coontail), a free floating macrophyte: Response of antioxidant enzymes and antioxidants. *Plant Sci.*, **138:** 157.

Ramakrishna, A. and Ravishankar, G.A. 2011. Influence of abiotic stress signals on secondary metabolites in plants. *Plant Signal Behav*, **6(11):** 1720–1731.

Rao, S.R. and Ravishankar, G.A. 2002. Plant cell cultures: chemical factories of secondary metabolites. *Biotechnol Adv.*, **20:** 101–153.

Rascio, N. and Navari-Izzo, F. 2011. Heavy metal hyperaccumulating plants: how and why do they do it? And what makes them so interesting? *Plant Science*, **180(2):** 169–181,

Rasmann, S. and Turlings, T.C.J. 2008. First insights into specificity of belowground tritrophic interactions. *Oikos*, **117:** 362–369.

Rastgoo, L., Alemzadeh, A. and Afsharifar, A. 2011. Isolation of two novel isoforms encoding zinc- and copper-transporting P1B-ATPase from Gouan (*Aeluropus littoralis*). *Plant Omics Journal*, **4(7):** 377–383.

Reddy, G.V.P. and Guerrero, A. 2004. Interactions of insect pheromones and plant semiochemicals. *Trends Plant Sci.*, **9:** 253–261.

Reddy, A.R., Rasineni, G.K. and Raghavendra, A.S. 2010. The impact of global elevated CO_2 concentration on photosynthesis and plant productivity. *Curr. Sci.* **99**:46–57.

Rellán-Álvarez, R., Ortega-Villasante, C., Álvarez-Fernández, A., Campo, F.F.D. and Hernández, L.E. 2006. Stress responses of Zea mays to cadmium and mercury. *Plant and Soil*, **279 (1-2):** 41–50.

Rhodes, D., Verslues, P.E. and Sharp, R.E. 1999. Role of amino acids in abiotic stress resistance. In, "*Plant amino acids: biochemistry and biotechnology*", Ed. B.K. Singh, Marcel Dekker Inc, New York, Basel, Hong Kong, pp 319-356

Rivero, R.M., Ruiz, J.M., Garcia, P.C., Lopez-Lefebre, L.R., Sanchez, E., and Romero, L. 2001. Resistance to cold and heat stress: accumulation of phenolic compounds in tomato and watermelon plants. *Plant Sci.*, **160:** 315–321.

Rice-Evans, C.A., Miller, N.J. and Paganga, G. 1997. Antioxidant properties of phenolic compounds. *Trends Plant Sci.* **2:** 152.

Rice-Evans, C.A. and Packer, L. 1998. *Flavonoids in Health and Disease*; Marcel Dekker Inc.: New York, NY, USA, 1998.

Rice-Evans, C.A., Miller, N.J. and Paganga, G. 2001. Structure-antioxidant activity relationships of flavonoids and phenolic acids. *Free Rad. Biol. Med.*, **335:** 166-180.

Rosenthal, G.A. 1991. The biochemical basis for the deleterious effects of L-canavanine. *Phytochemistry*, **30:** 1055-1058.

Rozema, J., Staaij, J.vd., Bjorn, L.O. and Caldwell, M. 1997. UV-B as an environmental factor in plant life: stress and regulation. *Trends Ecol. Evol.*, **12:** 22–28.

Rubio, V., Linhares, F., Solano, R., Martin, A.C., Iglesias, J., Leyva, A. and Paz-Ares, J.A 2001. conserved MYB transcription factor involved in phosphate starvation signaling both in vascular plants and in unicellular algae. *Genes Dev.*, **15:** 2122–2133.

Rudrappa, T., Neelwarne, B. and Aswathanarayana, R.G. 2004. *In situ* and *ex situ* adsorption and recovery of betalains from hairy root cultures of *Beta vulgaris. Biotechnol Prog.*, **20:** 777–785.

Saba, P.D., Iqbal, M. and Srivastava, P.S. 2000. Effect of ZnSO4 and CuSO4 on regeneration and lepidine content in *Lepidium sativum. Biol Plant.*, **43:** 253–256.

Sachray, L., Weiss, D., Reuveni, M., Nissim-Levi, A. and Shamir, M.O. 2002. Increased anthocyanin accumulation in aster flowers at elevated temperatures due to magnesium treatment. *Physiol. Plant*, **114:** 559–565.

Sahw, B.P., Sahu S.K. and Mishra, R.K.2004. Heavy metal induced oxidative damage in terrestrial plants. In, "*Heavy Metal Stress in Plants: From Biomolecules to Ecosystems, Second ed*", Ed. M.N.V. Presad, Springer, Berlin, pp. 84–126.

Sagoe, R. 2006. Climate changes and root and tuber production in Ghana. A report for the Environmental Protection Agency, Accra-Ghana.

Sakamoto, A., Murata, N. 2000. Genetic engineering of glycinebetaine synthesis in plants: current status and implications for enhancement of stress tolerance. *J Exp Bot*, **51:** 81–88.

Samuni-Blank, M., Izhaki, I., Dearing M.D., Gerchman, Y., Trabelcy, B., Lotan, A., Karasov, W. H., and Arad, Z. 2012. Intraspecific directed deterrence by the mustard oil bomb in a desert plant. *Current Biology*, **22(13):** 1218-1220.

Satya Narayan, V. and Nair, P.M. 1990. Metabolism, enzymology and possible roles of 4-aminobutyrate in higher plants. *Phytochemistry*, **29:** 367–375.

Saunders, J. and O'Neill, N.R. 2004. The characterization of defense responses to fungal infection in alfalfa. *Biocontrol*, **49:** 715–728.

Savithramma, N., Linga Rao, M., and Suhrulatha, D. 2011. Screening of medicinal plants secondary metabolites. *Middle-East J. Sci. Res.*, **8:** 579-584.

Schafer, H. and Wink, M. 2009. Medicinally important secondary metabolites in recombinant microorganisms or plants: progress in alkaloid biosynthesis. *Biotechnology Journal*, **4(12):** 1684-1703.

Schneider, S.H. 1993. Scenarios of global warming," in Biotic Interactions and Global Change, Sinauer, Sunderland, UK, pp. 9–23.

Scheible, W.-R., Morcuende, R., Czechowski, T., Fritz, C., Osuna, D., Palacios-Rojas, N., Schindelasch, D., Thimm, O., Udvardi, M.K. and Stitt, M. 2004. Genome-wide reprogramming of primary and secondary metabolism, protein synthesis, cellular growth processes, and the regulatory infrastructure of *Arabidopsis* in response to nitrogen. *Plant Physiol.*, **136:** 2483–2499.

Schmitz-Hoerner, R. and Weissenböck, G. 2003. Contribution of phenolic compounds to the UV-B screening capacity of developing barley primary leaves in relation to DNA damage and repair under elevated UV-B levels. *Phytochemistry*, **64:** 243–255.

Schreiner, M., Mewis, I., Huyskens-Keil, S., Jansen, M.A.K., Zrenner, R., Winkler, J.B., O'Brien, N. and Krumbein, A. 2012. UV-Binduced secondary plant metabolites - potential benefits for plant and human health. *Crit Rev Plant Sci*, **31:** 229– 240.

Schultz, J.C. 2002. Biochemical ecology: how plants fight dirty. *Nature*, **416:** 267.

Schützendübel, A. and Polle, A. 2002. Plant responses to abiotic stresses: heavy metal-induced oxidative stress and protection by mycorrhization. *The Journal of Experimental Botany,* **53(372):** 1351–1365.

Seigler, D.S. 1998. Plant Secondary Metabolism. Chapman and Hall, Kluwer Academic Publishers, Boston, MA, pp. 711.

Selmar, D. 2008. Potential of salt and drought stress to increase pharmaceutical significant secondary compounds in plants. *Landbauforsch Völk*, **58:**139–144.

Selmar, D. and Kleinwächter, M. 2013a. Influencing the product quality by deliberately applying drought stress during the cultivation of medicinal plants. *Ind Crop Prod.*, **42:** 558–566.

Selmar, D. and Kleinwächter, M. 2013b. Stress enhances the synthesis of secondary plant products: the impact of stress-related over-reduction on the accumulation. *Plant Cell Physiol*, **54(6):** 817–826.

Sfendla, R., Desilets, H., Laliberté, S. and Olivier, A. 2008. The effects of CO_2 enrichment, increased light irradiance and reduced sucrose concentration on acclimatization of micropropagated American ginseng plantlets. *Journal of Herbs, Spices and Medicinal Plants*, **13(3):** 97–106.

Singh, S. and Sinha, S. 2005. Accumulation of metals and its effects in *Brassica juncea* (L.) Czern. (cv. Rohini) grown on various amendments of tannery waste. *Ecotoxicol Environ Saf.*, **62:** 118–127.

Smeets, K., Cuypers, A., Lambrechts, A., Semane, B., Hoet, P., Van Laere, A., Van Gronsveld, J. 2005. Induction of oxidative stress and antioxidative mechanisms in *Phaseolus vulgaris* after Cdapplication. *Plant Physiol. Biochem.,* **43:** 437.

Shanker, A.K., Cervantes, C., Loza-Tavera, H. and Avudainayagam, S. 2005. Chromium toxicity in plants. *Environment International*, **31(5):** 739-753.

Sharkey, T.D. 2005. Effects of moderate heat stress on photosynthesis: importance of thylakoid reactions, rubisco deactivation, reactive oxygen species, and thermotolerance provided by isoprene. *Plant Cell Environ.*, **28:** 269–277.

Sharma, S., Sharma, K.P. and Uppal SK. 1997. Influence of salt stress on growth and quality on sugarcane. *Indian J Plant Physiol*, **2:** 179-180.

Sharma S.S. and Dietz, K.-J. 2006. The significance of amino acids and amino acid-derived molecules in plant responses and adaptation to heavy metal stress. *The Journal of Experimental Botany*, **57(4):** 711–726.

Sharma, R.K., Agrawal, M. and Marshall, F.M. 2009. Heavy metals in vegetables collected from production and market sites of a tropical urban area of India. *Food and Chemical Toxicology*, **47(3):**, 583- 591.

Sharma, P., Jha, A.B., Dubey, R.S. and M. Pessarakli, M. 2012. Reactive oxygen species, oxidative damage, and antioxidative defense mechanism in plants under stressful conditions. *Journal of Botany*, **2012:** Article ID 217037, 26 pages.

Shibata, M., Amano, M., Kawata, J. and Uda, M. 1988. Breeding process and characteristics of 'Summer Queen', a spray-type chrysanthemum. *Bull. Natl. Inst. Veg. Ornamental Plants Tea.* Ser. A **2:** 245–255.

Shohael, A.M., Ali, M.B., Yu, K.-W., Hahn, E.-J. and Paek, K.-Y. 2006. Effect of temperature on secondary metabolites production and antioxidant enzyme activities in *Eleutherococcus senticosus* somatic embryos. *Plant Cell, Tissue and Organ Culture*. **85:** 219–228.

Siemens, D.H., Garner, S.H., Mitchell-Olds, T. and Callaway, R.M. 2002. Cost of defense in the context of plant competition: *Brassica rapa* may grow and defend. *Ecology*, **83(2):** 505–517.

Snow, M.D., Bard, R.R., Olszyk, D.M., Minster, L.M., Hager, A.N. and Tingey, D. 2003. Monoterpene levels in needles of Douglas firexposed to elevated CO_2 and temperature. *Physiol Plant.*, **117:** 352–358.

Solanki, R. and Dhankhar, R. 2011. Biochemical changes and adaptive strategies of plants under heavy metal stress. *Biologia,* **66(2):** 195–204.

Solecka, D. 1997. Role of phenylpropanoid compounds in plant responses to different stress factors. *Acta Physiol Plant*, **19:** 257–268.

Solecka, D., Boudet, A.M. and Kacperska, A. 1999. Phenylpropanoid and anthocyanin changes in low-temperature treated oilseed rape leaves. *Plant Physiol. Biochem.,* **37:** 491–496.

Solecka, D. and Kacperska, A. 2003. Phenylpropanoid deficiency affects the course of plant acclimation to cold. *Physiologia Plantarum* **119**: 253-262.

Soliz-Guerrero, J.B., de Rodriguez, D.J., Rodriguez-Garcia, R., Angulo-Sanchez, J.L. and Mendez-Padilla, G. 2002. Quinoasaponins: concentration and composition analysis. In, "*Trends in new crops and new uses*", Eds. J. Janick and A. Whipkey, ASHS Press, Alexandria, pp. 110.

Solovchenko, A., and Schmitz-Eiberger, M. 2003. Significance of skin flavonoids for UV-B-protection in apple fruits. *Journal of Experimental Botany*, **54(389):** 1977–1984.

Somasegaran, P. and Hoben, H.J. 1994. Handbook for rhizobia: methods in legume-rhizobium technology. New York: Springer-Verlag.

Song, B., Lei, M., Chen, T., Zheng, Y., Xie, Y., Li, X. and Gao, D. 2009. Assessing the health risk of heavy metals in vegetables to the general population in Beijing, China, *Journal of Environmental Science*, **21(12):** 1702-1709.

Stamp, N. 2003. Out of the quagmire of plant defense hypotheses. *The Quarterly Review of Biology*, **78(1):** 23-55.

Stapleton, A.E. 1992. Ultraviolet radiation and plants: burning questions. *Plant Cell*, **4:** 1353–1358.

Stashenko, E.E., Martinez, J.R., Ruiz, C.A., Arias, G., Duràn, C., Salgar, W. and Cala, M. 2010. *Lippia origanoides chemotype differentiation based on essential oil GC-MS and principal component analysis*. *J. Sep. Sci.* **33:** 93–103.

Stewart, A.J., Chapman, W., Jenkins, G.I., Graham, I., Martin, T. and Crozier, A. 2001. The effect of nitrogen and phosphorous deficiency on flavonol accumulation in plant tissues. *Plant Cell Environ.*, **24:** 1189–1197.

Stutte, G.W., Eraso, I. and Rimando, A.M. 2008. Carbon dioxide enrichment enhances growth and flavoniod content of two Scutellaria species. *Journal of the American Society for Horticultural Science* **133**: 631-638.

Styger, G., Prior, B., and Bauer, F.F. 2011. Wine flavor and Aroma. *Journal of Industrial Microbiology and Biotechnology*, **38(9):** 1145-1159.

Sytar,O., Kumar, A., Latowski, D., Kuczynska, P., Strza³ka, K. and Prasad, M.N.V. 2013. Heavy metal-induced oxidative damage, defense reactions, and detoxification mechanisms in plants. *Acta Physiologiae Plantarum*, **35(4):** 985–999.

Szabo, B., Tyihak, E., Szabo, L.G. and Botz, L. 2003. Mycotoxin and drought stress induced change of alkaloid content of Papaver somniferum plantlets. *Acta Bot Hung.*, **45:** 409-417.

Szakiel, A., P¹czkowski, C. and Henry, M. 2010. *Influence of environmental abiotic factors on the content of saponins in plants. Phytochem. Rev.*, **10:** 471–491.

Taiz, L. and Zeiger, E. 2006. Plant Physiology. Sinauer Associates Inc. Publishers, Massachusetts.

Takeuchi Y, Inoue, T., Takemura, K., Hada, M., Takahashi, S., Ioki, M., Nakajima, N. and Kondo, N. 2007. Induction and inhibition of cyclobutane pyrimidine dimmer photolyase in etiolated cucumber (*Cucumis sativus*) cotyledons after ultraviolet irradiation depends on wavelength. *J Plant Res*, **120(3):** 365–374.

Tangahu, B.V., Abdullah, S.R.S., Basri, H., Idris, M., Anuar, N. and Mukhlisin, M. 2011. A review on heavy metals (As, Pb, and Hg) uptake by plants through phytoremediation. *Int. J. Chem. Engg.,* **2011:** Article ID 939161, 31 pages.

Tanveer, H., S. Ali and M.R. Asi, 2012. Appraisal of an important flavonoid, quercetin in callus cultures of *Citrullus colocynthis. Int. J. Agric. Biol.*, **14:** 528–532

Tari, I., Kiss, G., Deer, A.K., Csiszar, J., Erdei, L., Galle, A., Gemes, K., Horvath, F., Poor, P., Szepesi, A. and Simon, L.M. 2010. Salicylic acid increased aldose reductase activity and sorbitol accumulation in tomato plants under salt stress. *Biol Plant.*, **54:** 677–683.

Tausz, M., Šircelj, H. and Grill, D. 2004. The glutathione system a stress marker in plant ecophysiology: is a stress-response concept valid? *J Exp Bot.*, **55(404):** 1955–1962.

Telascrea, M., de Araüjo, C.C., Marques, M.O.M., Facanali, R., de Moraes, P.L.R. and Cavalheiro, A.J. 2007. Essential oil from leaves of *Cryptocarya mandioccana Meisner (Lauraceae):* Composition and intraspecific chemical variability. *Biochem. Syst. Ecol.* **35:** 222–232.

Tevini, M. and Teramura, A.H. 1989. UVB effects on terrestrial plants. *Photochem. Photobiol.*,**50:** 479–487.

Theis, N. and Lerdau, M. 2003. *The Evolution of Function in Plant Secondary Metabolites. Int. J. Plant Sci.* **164:** S93–S102.

Thimmaraju, R., Bhagyalakshmi, N., Narayan, M.S. and Ravishankar, G.A.. 2003. Kinetics of pigment release from hairy root cultures of Beta vulgaris under the influence of pH, sonication, temperature and oxygen stress. *Process Biochem.*, **38:** 1069–1076.

Thrane, U. 2001. Development in the taxonomy of *Fusarium* species based on secondary metabolites. In, "*Fusarium*", Ed. B.A. Summerel, Paul E. Nelson Memorial symposium, St.Paul, Minnesota: APS Press, pp. 29-49.

Tohge, T., Nishiyama, Y., Hirai, M.Y., Yano, M., Nakajima, J.-I., Awazuhara, M., Inoue, E., Takahashi, H., Goodenowe, D.B., Kitayama, M.,Noji, M., Yamazaki, M. and Saito, K. 2005. Functional genomics by integrated analysis of metabolome and transcriptome of *Arabidopsis*plants over-expressing an MYB transcription factor. *Plant J.,* **42:** 218–235.

Tomana, T. and Yamada, H. 1988. Relationship between temperature and fruit quality of apple cultivars grown at different locations. *J. Jpn. Soc. Hortic. Sci.*, **56:** 391–397.

Toncheva-Panova, T., Pouneva, I., Chernev, G. and Minkova, K. 2010. Incorporation of *Synechocystis salina* in hybrid matrices. Effect of UV-B radiation on the copper and cadmium biosorption. *Biotechnol and Biotechnol Eq*, **24(3):** 1946–1949.

Trejo-Tapia, G., Jimenez-Aparicio, A., Rodriguez-Monroy, M., De Jesus-Sanchez, A. and Gutierrez-Lopez, G. 2001. Influence of cobalt and other microelements on the production of betalains and the growth of suspension cultures of *Beta vulgaris. Plant Cell Tissue Organ Cult.*, **67:** 19–23.

Treutter, D. 2005. *Significance of flavonoids in plant resistance and enhancement of their biosynthesis. Plant Biol. (Stuttg).* **7:** 581–591.

Tuomi, J., Niemala, P., Chapin, F.S., Bryant, J.P. and Sirin, S. 1988. Defensive responses of trees in relation to their carbon/nutrient balance. In, "*Mechanisms for Woody Plant Defenses Against Insects Search for Pattern*", Springer, New York, pp. 57–72.

Tuteja, N. and Mahajan, S. 2007. Calcium signaling network in plants: an overview. *Plant Signal Behav.*, **2:** 79–85.

Uprety, D., Hejcman, M., Száková, J., Kunzová, E. and Tlustoš, P. 2009. Concentration of trace elements in arable soil after long-term application of organic and Inorganic fertilizers. *Nutrient Cycling in Agroechosystems*, **85(3):** 241-252.

Van Etten, H., Temporini, E. and Wasmann, C. 2001. Phytoalexin (and phytoanticipin) tolerance as a virulence trait: why is it not required by all pathogens? *Physiological and Molecular Plant Pathology*, **59:** 83-93.

Varshney, K.A. and Gangwar, L.P. 1988. Choline and betaine accumulation in *Trifolium alexandrinum* L. during salt stress. *Egypt J Bot.* **31:** 81–86.

Vasconsuelo, A. and Boland, R. 2007. Molecular aspects of the early stages of elicitation of secondary metabolites in plants. *Plant Sci.*, **172:** 861-875.

Vass, I., Szilárd, A. and Sicora, C. 2005. Adverse Effects of UV-B light on the structure and function of the photosynthetic apparatus. section xiii: photosynthesis under environmental stress conditions. In, "*Handbook of Photosynthesis*", Second Edition Ed. M. Pessarakli, CRC Press.

Velikova, V., Edreva, A. and Loreto, F. 2004. Endogenous isoprene protects *Phragmites australis* leaves against singlet oxygen. *Physiol. Plant.*, **122:** 219-225.

Velikova, V., Edreva, A., Tsonev, T. and Jones, H.G. 2007. A study on the singlet oxygen quenching ability of polyamines and their hydroxycinnamic conjugates. *Z. Naturforsch.C,* **62(11-12):** 833-838.

Velikova,V. and Loreto, F. 2005. On the relationship between isoprene emission and thermotolerance in *Phragmites australis* leaves exposed to high temperatures and during the recovery from a heat stress. *Plant Cell Environ.*, **28:** 318–327.

Velikova, V., Pinelli, P., and Loreto, F. 2005a. Consequences of inhibition of isoprene synthesis in *Phragmites australis* leaves exposed to elevated temperatures. *Agric. Ecosyst. Env.,* **106:** 209-217.

Velikova, V., Edreva, A. and Loreto, F. 2005b. Endogenous isoprene protects *Phragmites australis* leaves against singlet oxygen. *Plant Cell Environ.*, **28:** 318–327.

Verpoorte, R. 1998, Exploration of Nature's Chemodiversity: The Role of Secondary Metabolites as Leads in Drug Development. *Drug Discovery Today* ,**3(5):** 232-238.

Verstraeten, S.V., Keen, C.L., Schmitz, H.H., Fraga, C.G. and Oteiza, P.I. 2003. Flavan-3-ols and procyanidins protect liposomes against lipid oxidation and disruption of the bilayer structure. *Free Radic. Biol. Med.*, **34:** 84.

Vettakkorumakankav, N.N., Falk, D., Saxena, P. and Fletcher, R.A. 1999. A crucial role for gibberellins in stress protection of plants. *Plant Cell Physiol.*, **40:** 542–548.

Viehweger, K. 2014. How plants cope with heavy metals. *Botanical Studies*, **55(35):** 1–12.

Vigani, G., Zocchi, G., Bashir, K., Phillipar, K. and Briat, J.F. 2013. Cellular iron homeostasis and metabolism in plants. *Front. Plant Sci.,* **4:** 6–8.

Vilela, E.C., Durate, A.R., Naves, R.V., Santos, S.C., Seraphin, J.C. and Ferri, P.H. 2013. *Spatial chemometric analyses of essential oil variability in Eugenia dysenterica. J. Braz. Chem. Soc.* **24:** 873–879.

Wahid, A. and Ghazanfar, A., 2006. Possible involvement of some secondary metabolites in salt tolerance of sugarcane. *J. Plant Physiol.*, **163:** 723–730.

Wahid, A. 2007. Physiological implications of metabolites biosynthesis in net assimilation and heat stress tolerance of sugarcane sprouts. *J. Plant Res.* **120:** 219–228.

Wang, H., Cao, G. and Prior, R.L. 1996. Total antioxidant capacity of fruits. *J. Agric. Food Chem.*, **44:** 701–705.

Wang, W., B. Vinocur and A. Altman, 2003. Plant responses to drought, salinity and extreme temperature towards genetic engineering for stress tolerance. *Planta,* **218:** 1–14

Wang, D.H., Du, F., Liu, H.Y. and Liang, Z.S. 2010. Drought stress increases iridoid glycosides biosynthesis in the roots of *Scrophularia ningpoensis* seedlings. *J Med Plants Res,* **4:** 2691-99.

Wang Y., Gao, L., Wang, Z., Liu, Y., Sun, M., Yang, D., Wei, C. and Xia, T. 2012. Light induced expression of genes involved in phenylpropanoid biosynthesis pathway in callus of tea. *Sci. Hort.*, **133:** 72–83.

Waterman, P.G. and Mole, S. 1989. Extrinsic factors influencing production of secondary metabolites in plants. In, "*Insect-Plant Interactions*", Vol. 1, CRC Press, Boca Raton, Fla, USA, pp. 108–134.

Waterman, P.G. 1992. Roles for secondary metabolites in plants. In, "*Proceedings of the 171st Ciba Foundation Symposium on Secondary Metabolites: Their Function and Evolution*", 255-275.

Weidner, S., Karamaæ, M., Amarowicz, R., Szypulska, E. and Golgowska, A. 2007. Changes in composition of phenolic compounds and antioxidant properties of *Vitis amurensis* seeds germinated under osmotic stress. *Acta Physiol Plant*, **29:** 238–290.

Weidner, S., Karolak, M., Karamaæ, M., Kosiñska, A. and Amarowicz, R. 2009a. Phenolic compounds and properties of antioxidants in grapevine roots (*Vitis vinifera*) under drought stress followed by regeneration. *Acta Soc Bot Pol*, **78:** 97–103.

Weidner, S., Kordala, E., Brosowska-Arendt, W., Karamaæ, M., Kosiñska, A. and Amarowicz, R. 2009b. Phenolic compounds and properties of antioxidants in grapevine roots followed by recovery. *Acta Soc Bot Pol*, **78:** 279–286.

Weinberg, E.D. 1971. Secondary Metabolism: Raison d'etre. *Perspectives in Biology and Medicine*, **14(4):** 565-577.

Williams, R.S., Lincoln, D.E. and Thomas, R.B. 1994. Loblolly pine grown under elevated CO_2 affects early instar pine sawfly performance. *Oecologia*. **98:** 64–71.

Wink, M. 1988. Plant breeding: importance of plant secondary metabolites for protection against pathogens and herbivores. *Theor. Appl. Genet.,* **75:** 225–233.

Wink, M. 1999. Functions of plant secondary metabolites and their exploitation in biotechnology. *Annual plant reviews*, **3:** Boca Raton, Florida: CRC Press.

Wink, M. 2010. Introduction: biochemistry, physiology and ecological functions of secondary metabolites. In, "*Biochemistry of Plant Secondary Metabolism*", Ed. M. Wink, Wiley-Blackwell, pp 1-19

Winkel, S.B. 2002. Biosynthesis of flavonoids and effects of stress. *Curr. Opin. Plant Biol.*, **5:** 218–223.

Winkel-Shirley, B. 2001. Flavonoid biosynthesis, A colorful model for genetics, biochemistry, cell biology, and biotechnology. *Plant Physiol.* **126:** 485–93.

Weinstein, L.H., Kaur-Sawhney. R., Venkat Rajam. M., Wettlaufer. S.H. and Galston, A.W. 1986. Cadmium-induced accumulation of putrescine in oat and bean leaves. *Plant Physiol..*, **82:** 641–645.

Wojtaszek, P. 1997. Oxidative burst: an early plant response to pathogen infection. *Biochem. J.*, **322:** 681.

Wong, H.L., Sakamoto, T., Kawasaki, T., Umemura, K. and Shimamoto, K. 2004. Down-regulation of metallothionein, a reactive oxygen scavenger, by the small GTPase OsRac1 in rice. *Plant Physiology*, **135(3):** 1447–1456.

Wróbel, M., Karmaæ, M., Amarowicz, R., Fr'czek, E. and Weidner, S. 2005. Metabolism of phenolic compounds in *Vitis riparia* seeds during stratification and during germination under optimal and low temperature stress conditions. *Acta Physiol Plant*, **27(3A):** 313–320.

Wuana, R.A. and Okieimen, F.E. 2011. Heavy metals in contaminated soils: a review of sources, chemistry, risks and best available strategies for remediation. *ISRN Ecology*, **2011:** Article ID 402647, 20 pages.

Wuyts, N., Swennen, R. and De Waele, D. 2006. Effects of plant phenylpropanoid pathway products and selected terpenoids and alkaloids on the behaviour of the plant-parasitic nematodes *Radopholus similis*. *Pratylenchus penetrans* and *Meloidogyne incognita*. *Nematology*, **8:** 89-101.

Xia, L., Yang, W. and Xiufeng, Y. 2007. Effects of water stress on Berberine, Jatrorrhizine and Palmatine contents in *Amur corktree* seedlings. *Acta Ecol Sin.*, **27(1):** 58–64.

Xie, B.D. and Wang, H.T. 2006. Effects of light spectrum and photoperiod on contents of flavonoid and terpene in leaves of *Ginkgo biloba* L. *Nanjing For. Univ.*, **30:** 51-54.

Xu, L., Han, L.. and Huang, B. 2011. Antioxidant enzyme activities and gene expression patterns in leaves of Kentucky bluegrass in response to drought and post-drought recovery. *J. Amer. Soc. Hort. Sci.*, **136:** 247–255.

Xu, Y., Gao, S., Yang, Y., Huang, M., Cheng, L., Wei, Q., Fei, Z., Gao, J. and Hong, B. 2013. Transcriptome sequencing and whole genome expression profiling of chrysanthemum under dehydration stress. *BMC Genomics*, **14:** 662.

Yamane, T., Jeong, S.T., Goto-Yamamoto, N., Koshita, Y. and Kobayashi, S. 2006. Effects of temperature on anthocyanin biosynthesis in grape berry skins. *Am. J. Enol. Vitic.*, **57:** 54-59.

Yamasaki, H., Sakihama, Y. and Ikehara, N. 1997. Flavonoidperoxidase reaction as a detoxification mechanism of plant cell against H_2O_2. *Plant Physiol.*, **115:** 1405.

Yang, Y., Shah, J., and Klessig, D.F. 1997. Signal perception and transduction in plant defense responses. *Genes and Development*, **11:** 1621-1639.

Yin, X., Zhou, J., Jie, C., Xing, D. and Zhang, Y. 2004. Anticancer activity and mechanism of *Scutellaria barbata* extract on human lung cancer cell line A549. *Life Science* **75**: 2233-2244.

Yu, K., Niranjana Murthy, H., Hahn, E. and Paek, K. 2005. Ginsenoside production by hairy root cultures of *Panax ginseng*: influence of temperature and light quality. *Biochem Eng J.*, **23:** 53–56.

Yu, J., Chen, L., Xu, M. and Huang, B. 2012. Effects of elevated CO_2 on physiological response of tall fescue (*Festuca arundinacea*) to elevated temperature, drought stress, and the combined stresses. *Crop Sci.*, **52:** 1848-1858.

Zacchini, M, and de Agazio, M. 2004. Spread of oxidative damage and antioxidative response through cell layers of tobacco callus after UV-C treatment. *Plant Physiol Biochem*, **42:** 445–450.

Zavala, J.A. and Ravetta, D.A. 2001. Allocation of photoassimilates to biomass, resin and carbohydrates in *Grindelia chiloensis* as affected by light intensity. *Field Crops Res.* **69:** 143-149.

Zeid, I.M. 2009. Effect of arginine and urea on polyamines content and growth of bean under salinity stress. *Act. Physiol Plant.* **35:** 65–70.

Zengin F.K. and Munzuroglu, O. 2005. Effects of some heavy metals on content of chlorophyll, proline and some antioxidant chemicals in bean (*Phaseolus vulgaris* L.) seedlings. *Acta Biologica Cracoviensia Series Botanica*, **47(2):** 157–164.

Zenk, M.H. 1996. Heavy metal detoxification in higher plants–a review. *Gene*, **179:** 21.

Zhan, J.C., Wang, L.J. and Huang, W.D. 2002. Effects of low light environment on the growth and photosynthetic characteristics of grape leaves. *J. China Agric.*, **7:** 75–78.

Zhang, W., Seki, M. and Furusaki, S. 1997. Effect of temperature and its shift on growth and anthocyanin production in suspension cultures of strawberry cells. Plant Sci., **127:** 207-214.

Zhang, J.-Y., Cruz de Carvalho, M.H., Torres-Jerez, I. Kang, Y., Allen, S.N., Huhman, D.V., Tang, Y., Murray, J., Summer, L.W. and Udvardi, M.K. .2014. Global reprogramming of transcription and metabolism in *Medicago truncatula*during progressive drought and after rewatering. *Plant Cell Environ*, **37:** 2553–2576.

Zhao, Y., Qi, L.. Wei- Ming, Wang, W., Saxena, P.K. and Chun-Zhao Liu, C. 2011. Melatonin improves the survival of cryopreserved callus of *Rhodiola crenulata. J Pineal Res.*, **50:** 83–88.

Zhong, J.J. and Yoshida, T. 1993. Effects of temperature on cell growth and anthocyanin production by suspension cultures of *Perilla frutescens* cells. *J. Ferment. Bioeng.*, **76:** 530–531.

Zhou, B., Yao, W., Wang, S., Wang, X. and Jiang, T. 2014. The metallothionein gene, TaMT3, from *Tamarix androssowii* confers Cd^{2+} tolerance in tobacco. *International Journal of Molecular Sciences,* **15(6):** 10398–10409.

Zhu, J., Xiong, L., Yu, B. and Wu, J. 2005. Apoptosis induced by a new member of saponin family is mediated through caspase-8-dependent cleavage of Bcl-2. Mol. *Pharmacol.*, **68:** 1831-1838.

Zobayed, S.M.A., Afreen, F. and Kozai, T. 2005. Temperature stress can alter the photosynthetic efficiency and secondary metabolite concentrations in St. John's wort. *Plant Physiol. Bioch.,* **43:** 977–984.

Zornoza, P., Vázquez, S., Esteban, E., Fernández-Pascual, M. and Carpena, R. 2002. Cadmium-stress in nodulated white lupin: strategies to avoid toxicity. *Plant Physiol. Biochem.*, **40:** 1003.

Zu, Y.G., Wei, X.X., Yu, J.H., Li, D.W. and Pang, H.H. and Tong, L. 2011. Responses in the physiology and biochemistry of Korean pine (*Pinus koraiensis*) under supplementary UV-B radiation. *Photosynthetica*, **49(3):** 448–458.

3

Microbes: Support and Protect Plants Against Abiotic Stresses

Preeti[1] *and J.D.S Panwar*[2]

Abiotic Stress in Plants

World population is growing at an alarming rate and may reach about 6 billion by the end of the year 2050. Though, the agricultural productivity is also being increased world over, yet it is not keeping pace to meet the demand of food availability. The main reasons for limiting the agricultural productivity are water shortage, depleting soil fertility and mainly various abiotic stresses and climatic change. Minimizing these loses is of primary concern for all nations to cope with the increasing food requirement. The abiotic stress is defined as the negative impact of non living factors on the living organism in a specific environment and abiotic stress is the most harm full factor concerning the growth and productivity of crops worldwide. Researches have shown that abiotic stressors are at their most harm full when they occur together in combination of abiotic stress factors (Mittler, 2006; Bianco and Defez, 2011). Numerous stresses caused by complex environmental conditions *i.e.* drought, too high and too low temperatures, freezing, salinity, UV light, heavy metals and recently through smog (air pollution) leading to substantial crop losses worldwide (Mahajan and Tuteja, 2005). The abiotic stresses might increase in the near future even because of global climatic change.

Air pollution seems to have a direct and negative impact on grand production in India as a result of increasing smog with decreasing projected yield by 50%, in wheat in 2010. This decrease in yield has been reported up to 90% of the decrease in potential food production possibly linked to smog and 10% to global warming and precipitation level (Proceedings of National Academy of Science,

[1]R.K.P.G. College, Shamli, Distt. Muzaffarnagar, Uttar Pradesh – 247 776
[2]Ex Vice Chancellor, Mahatma Gandhi University, Meghalaya &
Vice Chancellor, Venkateshwara Open University, Nahar Lagun, Itanagar
Arunachal Pradesh - 791111

Report Hindustan Times, 5th November 2014). Among the abiotic factors that are shaping plant evolution, water availability is most important (Kijne, 2006). The plants, however, develops some tolerance mechanisms through involving biochemical and metabolic means which are in turn regulated by genes. Water stress is the predominant stress among all the abiotic factors, among all the abiotic stresses which causes enormous loss in production of crops because water stresses usually accompanied by other stresses like salinity, high temperature and nutrient deficiency. The droughts in India are being reported frequently based on the rainfall. In 2013-2014, it was reported 40% deficit rainfall that was recovered up to 10% deficient as a result of more floods and heavy rainfall in the later stage. The distribution of rainfall in a cropping duration should be considered for calculating the drought and not the total rainfall.

3.1. Role of Microbes in Mitigating Stresses

Various abiotic factors affect the plant growth, development and productivity of the plants adversely but the impact of stresses can be mitigated to some extent with use of different beneficial rhizobacteria, fungi, *Azotobacter, Acetobacter, Azospirillum*, phosphate mobilizer, phosphate solubilizers, zinc solubilizers. Microorganism's particularly beneficial bacteria and fungi can improve plant performance under stress environment and consequently enhance yield both directly and indirectly (Dimkpa *et al.* 2009a). Some plant growth promoting rhizobacteria may exert a direct stimulation on plant growth and development by providing plants with fixed nitrogen, phytohormones, iron that has been sequestered by bacterial siderophores and soluble phosphates (Hayat *et al.* 2010). Others do this indirectly by protecting the plant against soil born diseases which are caused by pathogenic fungi (Lugtenberg and Kamolova, 2009). Common adaptation mechanisms of plants exposed to environmental stresses, such as temperature extremes, high salinity, drought and nutrient deficiency or heavy metal toxicity includes changes in root morphology (Potters *et al.* 2010) and or through production of phytohormones (Spaepen and Vanderladen, 2010) or through the root associated bacteria by producing indole 3 acetic acid (Hayat *et al.* 2010).

In addition to the beneficial rhizobacteria, the fungi (AM fungi), *Azotobacter, Acetobacter, Azospirillum*, phosphate mobilizer, phosphate solubilizers, Zn solubilizers also help in mitigating the drought response. The application of FYM compost, vermicompost and green manuring has also found beneficial in improving the abiotic stress tolerance in crop plants.

3.2. Beneficial *Rhizobacteria*

The population of microorganism lives in close contact with the plant root zone called rhizosphere. The plant root acts as a major source of nutrients for microorganisms and plant supply organic carbon to their surroundings in the form of root exudates and rhizobacteria response this exudation by means of chemotaxis towards the exudates (Hardoim *et al.* 2008). Soil bacteria beneficial to plant growth are usually referred to as plant growth promoting rhizobacteria (PGPR), capable of promoting plant growth by colonizing the plant root (Panwar and Swarnalakshmi, 2005; Hayat *et al.* 2010). Bacteria of diverse genera such as *Arthrobacter, Azatobacter, Azospirillum, Bacillus, Entterobacter, Pseudomonas* and Serratia (Gray and Smith, 2005), as well as *Streptomyces* spp. (Dimkpa *et al.* 2008, 2009b; Tokala *et al.* 2002) were identified as PGPR.

PGPR can be divided into two groups.(1) which live inside the plant cells and are localized in the nodules, and (2) ePGPR which live outside the plant cells and do not produce organs like nodules, but still prompt plant growth (Gray and Smith, 2005). Although the exact mechanisms of plant growth stimulation remain largely speculative, possible explanation includes: (1) production of hormones like abscisic acid, gibberellic acid, cytokinins, and auxin, (IAA); (2) production of essential enzymes, 1-aminocyclopropane-1-carboxylate (ACC) deaminase to reduce the level of ethylene in the root of developing plants; (3) nitrogen fixation; (4) production of siderophores; (5) solubilization and mineralization of nutrients, particularly mineral phosphate; (6) improvement of abiotic stresses resistance (Hayat *et al.* 2010).

3.2.1. Rhizobium

It is a well known fact that successful production of legumes depends upon effective nodulation and nitrogen fixation. Inoculation of seed with *Rhizobium* at the time of sowing is imperative. Beneficial responses of legumes to inoculation with *Rhizobial* cultures have been reviewed by several workers. Kumar and Elamathi (2007) studied the effect of nitrogen levels and *Rhizobium* application methods on yield attributes, yield and economics of black gram. The maximum pod number, plant height, number of leaves, dry weight of plant, nodule number and pod number were obtained under the application of *Rhizobium.* Kalita *et al.* (2006) studied the evaluation of native *Rhizobium* from acid soils of Assam. *Rhizobium* isolate AR1 was the most resistant against 9 antibiotics. *Rhizobium* isolate AR1 was found to be more effective in terms of number of nodules (75 and 100), nodule dry weight (0.41g and 0.38g) and percentage of nitrogen content both in black gram and green gram. Performance of green gram with all the isolates was better with respect to the parameters than black gram. *Rhizobium* isolate AR1 also showed the highest nitrogenase activity in the root nodules of black gram and green gram (4.02 and 4.25 mole/g/h).

3.2.2. Azotobacter

Azotobacter belongs to family ***Azotobacteriaceae***, is a free living aerobic microorganisms, grow in the rhizosphere and fix atmospheric nitrogen non-symbiotically and make it available to particularly cereals. Inoculation with these biofertilizer have been found to increase the yield ranging from 2-50 percent over uninoculated control depend upon the strains used, crop type, soil and environmental conditions and crop management practices. *Azotobacter*s are present in neutral or alkaline soils and *A. chroococcum* is the most commonly occurring species in aerable soils. *A. vinelandii, A. beijerinckii, A. insignis* and *A. macrocytogenes* are other reported species. The number of *Azotobacter* rarely exceeds of 10^4 to 10^5 g^{-1} of soil due to lack of organic matter and presence of antagonistic microorganisms in soil. The bacterium produces anti-fungal antibiotics which inhibits the growth of several pathogenic fungi in the root region thereby preventing seedling mortality to a certain extent (Subba Rao, 2001). *Azotobacter* also to known to synthesize biologically active growth promoting substances such as vitamins of B- group, indole acetic acid (IAA) and gibberellins. Many strains of *Azotobacter* also exhibited fungi static properties against plant pathogens such as *Fusarium, Alternaria and Helminthosporium.* The population of *Azotobacter* is generally low in the rhizo-sphere of the crop plants and in uncultivated soils.The occurrence of this organism has been reported from the rhizosphere of a number of crop plants such as rice, maize, sugarcane, bajra, vegetables and plantation crops, (Arun, 2007).

3.2.3. Acetobacter inoculant

In the eighties Dobereiner and coworkers in Brazil discovered this association and named *Acetobacter diazotrophicus. Acetobacter diazotrophious* is one of the newly discovered endophytic diazotrophs isolated from leaves, stems and roots of sugarcane. These live in xylem vessels, intercellular space of root, shoot or leaf, ensuring proper supply of nutrients for nitrogen fixation. It was widely studied and used as a model system to assess the bacterial endophyte plant interaction. After its discovery, it was reported from a variety of crops like coffee (Jimnez-Salgado *et al.* 1997), ragi (Loganathan *et al.* 1999) and pineapple (Herandez *et al.* 2000). and a latest report throw light on *Gluconacetobacter* sp as a natural colonizer of the wild rice (*Porteresia cocarctata Tateoka, formerly Oryza coarctata Roxbi*) and a salt tolerant pokali rice variety (Loganathan and Nair, 2003). In India its nitrogen fixing ability and plant growth promoting rhizobacteria (PGPR) production ability have been described, tested and found promising (Saravanan *et al.* 2008).

Besides, the bacterium secretes plant growth promontory substances such as indole acetic acid (IAA) which favours the germination and root development, which in turn help in the absorption of plant nutrients effectively from soil. The application of *Acetobacter* biofertilizers can benefit the sugarcane growing areas in the country by increasing sugar contents in cane from 0.3 to 0.5%. *A. diazotrophicus* could enhance- rice plant growth and that growth promotion- might be related to the transfer of biologically fixed nitrogen although other factors such as auxin production could be involved reported by Sevilla and Kennedy, (2000). *Gluconoacetobacter diazotrophicus* colonizes the internal root tissue of sugarcane and fix nitrogen. Since it occupies vascular tissue, it has the obvious advantage of being first in line and thus solves the problem of competition by non diazotrophs. *G diazotrophicus* is also found in *Pennisetum purpureum, Ipomoea batatas* and *Coffea arabica*—non legume plants. The largest effect in this group was obtained with sugarcane, which can obtain up to 150 kg N/ha from BNF (Dobereiner, 1997). In Tamilnadu, 24 strains of *G. diazotrophicus* were isolated from sugarcane (root, stem and leaves) and screened for nitrogenase activity. The strain isolated from sugarcane stem (SoS2) showed maximum acetylene reduction assay, 410.92 n moles of C_2H_4/hr/mg cell protein followed by SoL3 from sugarcane leaf and hence the strain from sugarcane stem was recommended as an effective biofertilizer (Meenakshisundaram *et al.* 2010).

3.2.4. Azospirillum

Azospirillum, a member of *Spirillaceae,* is associative nitrogen fixing microorganism beneficial for non-leguminous plants. Associative diazotrophs are those in which there is some independence between the partners but they can grow satisfactory apart. They are found in association with cereals, grasses, vegetables, and oil seeds either freely in soil or indeed the root system. They not only fix nitrogen but also benefit plants by supplying growth hormones and vitamins. *Azospirillum* with farm yard manure (FYM), led to saving of 15.25 kg equivalent of nitrogen also the above ground portion of plant through associative symbiosis. The *Azospirillum* inoculation helps the plants in better vegetative growth saving nitrogenous fertilizers by 25-30% and fix nitrogen from 10 to 40 kg/ha. *Azospirillum* are micro-aerophilic in nature (Panwar and Sirohi, 1989).

Following three species of *Azospirillum have* been identified as (i) *A. lipoferum,* (ii) *A. brasilense, and* (iii) *A. amazonenes. Azospirillum brasilense* is abundant in Indian soil. *Azospirillum* is often used with Mycorrhizal fungi for optimum growth and yield. One of the characteristics of *Azsopirillum* is its ability to reduce and denitrify, therefore it is these characters. Inoculations with *Azospirillum* have registered increase in different vegetable crops.

It also secretes phyto-*hormones* in the plant root regions, which in turn enhances the root growth. The results of the various experiments conducted throughout India have clearly shown that *Azospirillum* can be used as a potential biofertilizer in both expensive and intensive agriculture. In developing countries like India, the use of *Azospirillum* as a biofertilizer would not only to the nitrogen supplementation to crops but also help in improving the fertility of soil in the long run. Its application is common practice in vegetables in Tamil Nadu. Increased yield in pear miller obtained by inoculation of *Azospirillum* was due to production of indole acetic acid (IAA), gibberellins, any cytokinins like substances by the bacterium and their subsequent effect on the plant (Kumar *et al.* 2010).

Researchers have demonstrated the feasibility of *Azospirillum* inoculation to mitigate negative effects of NaCl on plant growth parameters. This beneficial effect of *Azospirillum* inoculation was previously observed in wheat (*Triticum aestivum* cv. 'Buck Ombú') seeds, where a mitigating effect of salt stress was also evident. (Creus *et al.* 1997). *Azospirillum*- inoculated wheat (*T. aestivum*) seedlings subjected to osmotic stress developed significant higher coleoptiles, with higher fresh weight and better water status than non-inoculated seedlings. (Alvarez *et al.* 1996; Creus *et al.* 1998).

Azospirillum spp. is not considered to be a classic bio-control agent of soil-borne plant pathogens. However, there have been reports on moderate capabilities of *A. brasilense* in bio-control of crown gall-producing *Agrobacterium* (Bakanchikova *et al.* 1993), bacterial leaf blight of mulberry (Sudhakar *et al.* 2000) and bacterial leaf and/ or vascular tomato diseases. (Bashan and Bashan, 2002a, 2002b). In addition, *A. brasilense* can restrict the proliferation of other nonpathogenic rhizosphere bacteria (Holguin and Bashan, 1996). These anti- bacterial activities of *Azospirillum* could be related to its already known ability to produce bacteriocins (Oliveira and Drozdowicz, 1987) and siderophores (Tapia-Hernández *et al.* 1990; Shah *et al.* 1992). It was recently reported that *A. brasilense* can synthesize phenylacetic acid (PAA), an auxin like molecule with antimicrobial activity. Biofertilizers made from *Azospirillum* is suitable for C_4 crops such as sugarcane, maize, bajra, sorghum; and other cereals like rice, wheat, barley, ragi and various horticulture crops. In this context, practices and potentialities still have a wider gap (it is not as popular as *Rhizobium)* and a lot can be done in sustaining cereal production.

3.3. Phosphorus Mobilizers

Phosphorus mobilizers facilitate the mobilization of soluble phosphorus from distant places in soil, where plant roots can not reach and thus increase availability of P to plants. Mycorrhizas are prominent P mobilizers. Mycorrhizae is a symbiotic association between plant roots and a few fungus. The fungal partner is benefited by obtaining its carbon requirements from host's photosynthates and the plant in turn gains the much needed nutrients especially phosphorus, calcium, copper and zinc which would otherwise be inaccessible to the host. This uptake of nutrients is facilitated with the help of fine absorbing hyphae of the fungus. These fungi are associated with majority of agricultural crops. There are seven genera of these fungi that produce Arbuscular mycorrhizal symbiosis with plants, they are *Glomus, Gigaspora, Scutellospora, Acaulospora Entrophospora, Archaeospora* and *Paraglomus.* They account for 5–50% of the biomass of soil microbes (Olsson *et al.* 1999). Hyphal biomass of AM fungi may amount to 54–900 kg/ha (Zhu and Miller, 2003), and some products formed by them may account for another 3000 kg/ha (Lovelock, *et al.* 2004). Pools of organic carbon such as glomalin produced by AM fungi may even exceed soil microbial biomass by a factor of 10–20 (Rillig *et al.* 2001). Approximately 10–100 m mycorrhizal mycelium can be found per cm root (Mc Gonigle and Miller, 1999). The mechanism that is generally accepted is a wider physical exploration of the soil by mycorrhizal fungi (hyphae) rather than by roots. A speculative mechanism to explain P uptake by mycorrhizal fungi involves the production of glomalin (Lovelock *et al.* 2004) The micro-organisms always participate in cycling of phosphorus in the environment hence there is recycling and not exhaustion.

AM fungi play an important role in water economy of plants.Their association improves the hydraulic conductivity of the root at lower soil water potentials, and this improvement is one of the factors contributing towards better uptake of water by plants (Mahdi *et al.* 2010). A few proposed mechanisms by which AM fungi also help in activation of plant defense systems include changes in exudation patterns and concomitant changes in mycorrhiza sphere populations, increased lignifications of cell walls, and competition for space for colonization and infection sites (Kasiamdar *et al.* 2001)

3.3.1. Phosphate solubilizer

Phosphorus is second only to nitrogen in mineral nutrients most commonly limiting growth of crops, and an essential element for plant development and growth making up about 0.2% of plant dry weight. Plants acquire P from soil solution as phosphate anions, however, these are extremely reactive and may be immobilized through precipitation with cations such as Ca^{2+}, Mg^{2+}, Fe^{3+} and Al^{3+} depending on the particular properties of a soil. In these forms, P is highly insoluble and a large portion of soluble inorganic phosphate applied to the soil as chemical fertilizer is immobilized rapidly and becomes unavailable to plants (Goldstein, 1986). Hence, the amount available to plants is usually a small pro- portion of this total application.

Several bacteria, particularly as *Bacillus polymixa, Bacillus subtilis, and Pseudomonas striata,* and fungi *as Penicillium digitatum* and *Asprergillus awamori* have the ability to convert the insoluble inorganic phosphorus into the soluble form, which can be utilized by crop plants. The inoculants of these microorganisms are called PSB (Phosphate solubilizing bacteria) or PSM (Phosphate solubilizing microorganisms) inoculants. They also produce siderophores (iron chelating substances, *e.g.* pseudobactin), which chelate with iron and make it unavailable to harmful fungi as *Erwinia* in rhizosphere, and produce plant growth hormones like lndole acetic acid and gibberellic acid etc. These cultures can be used in all crops of legumes, cereals and vegetables. Normally these bacteria can solubilize about 15 kg P /ha/season. Their inoculation was found to increase the yield of crops by 10 %. Phosphate rock minerals are often too insoluble to provide sufficient P for crop uptake. Use of phosphate solubilizing microorganisms increase crop yields up to 70 per cent (Verma, 1993). Combined inoculation of AM and PSB enhanced uptake of both native P from soil and P coming from the phosphatic rock (Goenadi *et al.* 2000; Cabello *et al.* 2005). Microorganisms with phosphate solubilizing potential increase the availability of soluble phosphate and enhance the plant growth by improving biological nitrogen fixation (Kucey *et al.* 1989; Ponmurugan and Gopi, 2006).

3.4. Zinc Solubilizers

The nitrogen fixers like *Rhizobium, Azospirillum, Azotobacter, BGA* and Phosphate solubilizing bacteria like *B. magaterium, Pseudomonas striata,* and phosphate mobilizing Mycorrhizae have been widely accepted as bio-fertilizers (Subba Roa, 2001a). However, these supply only major nutrients but a host of microorganism that can transform micronutrients are there in soil that can be used as bio-fertilizers to supply micronutrients like zinc, iron, copper etc., zinc being utmost important is found in the earth's crust to the tune of 0.008 per cent but more than 50 per cent of Indian soils exhibit deficiency of zinc with content must below the critical level of 1.5 ppm of available zinc.

Theplant constraints in absorbing zinc from the soil are overcome by external application of soluble zinc sulphate (Zn SO_4). But the fate of applied zinc in the submerged soil conditions is pathetic and only 1-4% of total available zinc is utilized by the crop and 75% of applied zinc is transformed into different mineral fractions (Zn-fixation) which are not available for plant absorption (crystalline iron oxide bound and residual zinc). There appears to be two main mechanisms of zinc- fixation, one operates in acidic soils and is closely related with cation exchange and other operates in alkaline conditions where fixation takes by means of chemisorptions, (chemisorptions of zinc on calcium carbonate formed a solid-solution of $ZnCaCO_3$), and by complexation by organic ligands (Alloway, 2008). The zinc can be solubilized by microorganisms viz., *B. subtilis, Thiobacillus thioxidans and Saccharomyces* sp. These microorganisms can be used as bio-fertilizers for solubilization of fixed micronutrients like zinc. The results have shown that a *Bacillus* sp. (Zn solubilizing bacteria) can be used as bio-fertilizer for zinc or in soils where native zinc is higher or in conjunction with insoluble cheaper zinc compounds like zinc oxide (ZnO), zinc carbonate ($ZnCO_3$) and zinc sulphide (ZnS) instead of costly zinc sulphate (Mahdi *et al.* 2010).

3.5. Plant Growth Promoting Rhizobacteria (PGPR)

The relationship between IAA and ethylene precursor ACC (Dimkpa *et al.* 2009a), the positive effects of IAA on root growth can be either direct or indirect through the reduction of ethylene levels (Lugtenberg and Kamilova, 2009). Indeed, under stress conditions, including drought and salinity, the plant hormone ethylene endogenously regulates plant homeostasis and results in

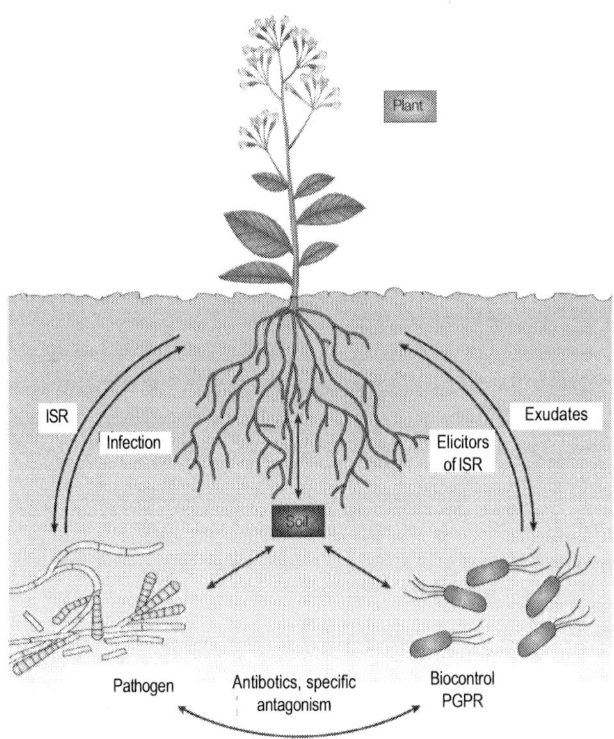

reduced root and shoot growth. It has been shown that plants produce ethylene at two different phases in response to stressful stimuli. In the first phase, the small amount of ethylene produced promotes the activity of stress-related genes. In the second phase (1–3 days after stimulus application) the larger amount of ethylene produced lead to inhibition of growth and harmful effects on plants including senescence, chlorosis, and abscission (Glick *et al.* 2007).

Degradation of the ethylene precursor ACC into 2-oxobutanoate and ammonia by bacterial ACC-deaminase lowers the ethylene concentration in plant roots, relieves the ethylene repression of auxin response factors synthesis, and indirectly increases plant growth (Glick *et al.* 2007; Kang *et al.* 2010). It has been proposed that ACC might be exuded from plant roots and that soil bacteria containing ACC-deaminase could convert this for their growth. As result, the hydrolyzed ACC products would enhance bacterial growth. Taken together, the ACC-deaminase function seems to be mutually beneficial between plants and PGPR, because ethylene in plants can be reduced by continuous ACC secretion and degradation by bacteria, and bacteria can use metabolized ACC (Glick *et al.* 1998). Mayak *et al.* (2004) reported that *Achromobacter piechaudii* having ACC deaminase activity significantly increased the fresh and dry weights of tomato seedlings grown in the presence of NaCl salt (up to 172 mM). *Pseudomonas fluorescens* strain TDK1 containing ACC deaminase activity enhanced the saline resistance in groundnut plants and increased yield as compared to plants inoculated with *Pseudomonas* strains lacking ACC deaminase activity (Saravanakumar and Samiyappan, 2007). *Pseudomonas putida* UW4, which produces IAA and ACC deaminase, protected canola seedling from growth inhibition by high levels of salt. Siddikee *et al.* (2010) have also confirmed that inoculation with 14 halotolerant bacterial strains ameliorate salt stress in canola plants through the reduction of ethylene production *via* ACC deaminase activity. Inoculation of maize plants with *Pseudomonas fluorescens* containing ACC deaminase boosted root elongation and fresh weight significantly under saline conditions (Kausar and Shahzad, 2006). Inoculation with *Pseudomonas* spp. containing ACC-deaminase partially eliminates the effects of drought stress on growth, yield, and ripening of pea (*Pisum sativum* L.) (Ashad *et al.* 2008), Nadeem *et al.* (2010) reported that rhizobacteria capable of producing ACC deaminase mitigate salt stress in wheat.

Medicago plants nodulated by the IAA-overproducing RD64 strain (*Mt*-RD64) showed a phytohormones re-modulation, with a higher IAA content in nodules and roots and a reduced accumulation of IAA in the shoot strain as compared to plants nodulated by the wild-type strain 1021 (*Mt*-1021).

Transcriptional analysis of the main ethylene signaling genes showed that, when compared to *Mt*-1021 plants, *Mt*-RD64 plants did not showed and induction of this pathway when 150 mm NaCl was applied, which means less plants stress damages (Bianco and Defez, 2009).

3.6. Improving Mechanism of Stress Response by *Rhizobacteria*

Dehydration, salinity, low as well as high-temperature stresses and other abiotic stresses lead to metabolic toxicity, membrane disorganization, generation of ROS, inhibition of photosynthesis, reduced nutrient acquisition and altered hormones levels. Accumulation of osmoprotectants, production of superoxide radical scavenging mechanisms, exclusion or compartmentation of ions by efficient transporter and symporter systems, production of specific enzymes involved in the regulation of plant hormones are some of the mechanisms that plants have evolved for adaptation to abiotic stresses (Des Marais and Juenger, 2010, Mahajan and Tuteja, 2005; Parida and Das, 2005; Santner *et al* 2009, Shao *et al.* 2009).

3.6.1. Phytohormones synthesis and modulation

Plant faces a wide variety of environmental constrains during their growth and development period. This is due to the continuously active shoot and root meristems and their capability to generate new organs after embryogenesis (Wolter and Jurgens, 2009).They have developed an extensive array of defensive responses that includes changes in the root morphology. The root architecture and by the pattern of root branching and by the rate and direction of growth of individual roots (Malamy, 2005), constitutes an important model to study how developmental plasticity is translated into growth responses under several environmental stresses. Morphogenesis is tightly linked to hormonal homeostasis, with several hormones controlling cell elongation, cell division and re-orientation of growth.

The physiologically most active auxin in plants is indole-3-acetic acid (IAA), and the fact that no fully auxin-deficient mutant plants have been identified so far reflects the importance of auxin in plant development. There is a high capacity for auxin biosynthesis not only in young aerial tissues, but also in roots, particularly in the meristematic primary root tip (Teale *et al.* 2006). Auxin, and its fine concentration gradients have powerful effects on plant development and in particular on lateral root formation and branching, two key components of the response phenotype induced in plants under stress conditions (Potters *et al.* 2007,2009). Alteration in the pattern of lateral root formation and emergence in response to P availability is mediated by changes in auxin sensitivity in *Arabidopsis thaliana* roots. These changes alter the expression of auxin-

Table 1: Microbes mediated plant tolerance to abiotic stress

Stress type	Microbial inoculate	Plant Species	Reference
Drought stresss	*Pseudomonas* spp, *A. brasilense, Glomus mosseae*	Maize (*Zea mays* L. cv. Kaveri)	Sandhya *et al.*, 2010; Casanovas *et al.*, 2002; Abdelmoneim *et al.*, 2014
	Pseudomonas spp.	Asparagus(*Aspaagus officinalis* L.)	Liddycoat *et al.*, 2009
	Pseudomonas mendocina, Bacillus	Lettuce (*Lactuca sativa* L.)	Kohler *et al.*, 2008; Arkhipova *et al.*, 2007
	Rhizobium tropici,Paenibacillus polymyxa,Ensifer meliloti bv.*mediterranense, A. brasilense*	Common bean (*Phaseolus vulgaris* L.)	Figueiredo *et al.*,2008;Mnasri *et al.*, 2007; German *et al.*, 2000
	Bradyrhizobium elkanii	Flat crown (*Albizia adianthifolia*)	Swaine *et al.*, 2007
	Achromobacter piechaudii	Tomato (*L. esculentum*),pepper (*Capsicum annuum*)	Mayak *et al.*, 2004
	Azospirillum	Wheat (*T. aestivum*)	Creus *et al.*, 2004
Salt stress	*Pseudomonas pseudoalcaligenes,*	Rice (*Oryza sativa*) *Bacillus pumilus*	Jha *et al.*, 2010
	Azospirillum brasilense	Barley(*Hordeum vulgare*)	Omar *et al.*, 2009
	Pseudomonas mendocina, Azospirillum	Lettuce (*L. sativa* L. cv. Tafalla)	Kohler *et al.*, 2009; Barassi *et al.*, 2006
	Azospirillum sp.	Pea (*Phaseolus vulgaris*)	Dardanelli *et al.*, 2008
	Bacillus subtilis	*Arabidopsis thaliana*	Zhang *et al.*, 2008
	Pseudomonas syringae, Pseudomonas fluorescens,Enterobacter aerogenes, Azospirillum, Bacillus megaterium	Maize (*Zea maize*)	Nadeem *et al.*, 2007; Hamdia *et al.*, 2004; Marulanda *et al.*, 2010
	P. fluorescens	Groundnut (*Arachis hypogaea*)	Saravanakumar and Samiyappan, 2007
	Achromobacter piechaudii	Tomato (*Lycopersicon esculentum*)	Mayak *et al.*, 2004
	Aeromonas hydrophila/caviae, Bacillus insolitus, Bacillus sp. *Glomus fasciculatum*	Wheat (*Triticum aestivum*)	Ashraf *et al.*, 2004 and Bheemareddy *et al.*, 2010

Contd.

Stress type	Microbial inoculate	Plant Species	Reference
	A. brasilense	Chickpeas (*Cicer arietinum*), faba beans (*Vicia faba* L.)	Hamaoui *et al.*, 2001
Temperature stress	*Burkholderia phytofirmans*	Grapevine (*Vitis vinifera*)	Barka *et al.*, 2006
	Pseudomonas fluorescens, Pantoea agglomerans, Mycobacterium sp.	Wheat (*Triticum aestivum*)	Egamberdiveya and Hoflich, 2003
	B. phytofirmans	Potato (*Solanum tuberosum*)	Bensalim *et al.*, 1998
	Aeromonas hydrophila, Serratia liquefaciens, Serratia proteamaculans	Soy bean (*Glycine max*)	Zhang *et al.*, 1997
Osmotic stress	*Bacillus subtilis*	Arabidopsis	Zhang *et al.*, 2010
	A. brasilense	Rice (*Oryza sativa* L.)	Cassan *et al.*, 2009
	Arthrobacter sp., *Bacillus sp.*	Pepper (*C. annuum*)	Sziderics *et al.*, 2007
	Azospirillum	Wheat (*T. aestivum*)	Pereyra *et al.*, 2006
Nutrient deficiency	*Azospirillum* sp.,*Azotobacter chroococcum, Mesorhizobium ciceri, Pseudomonas fluorescens*	Chickpea (*Cicer arietinum* L.)	Rokhzadi and Toashih, 2011
	Azotobacter coroocoocum, Azospirillum brasilens, Pseudomonas putida, Bacillus lentus Bacillus sp., *Burkholderia* sp., *Streptomyces platensis Bacillus* sp., *Bacillus polymyxa, Mycobacterium phlei, Pseudomonas alc aligenes*	*Zea maize* L.	Yazdani *et al.*, 2009;Oliveira *et al.*, 2009;Adesemoye *et al.*, 2008; Egamberdiveya 2007

Contd.

Stress type	Microbial inoculate	Plant Species	Reference
	Pseudomonas fluorescrns, *Burkholderia capacia,* *Glomus etunicatum*	Wheat(*Triticum aestivum*)	Minaxi *et al.*,2013
Heavy metals toxicity	*Sanguibacter* sp.,*Pseudomonas* sp.	*Nicotina tabacum*	Mastretta *et al.*, 2009
	Bacillus subtilis,Pantoea agglomerans	Oat (*Avena sativa*)	Pishchik *et al.*, 2009
	Pseudomonas fluorescens, *Microbacterium* sp.	Rape (*Brassica napus*)	Sheng *et al.*, 2008
	Methylobacterium oryzae, *Burkholderia* sp.	Tomato (*Lycopersiconesculentum* L.)	Madhaiyan *et al.*, 2007
	Bacillus subtilis, *Bacillus megaterium,* *Bacillus* sp.	Rice (*O. sativa*)	Asch and Padham 2005; Terré *et al.*, 2007
Flooding stress	*Enterobacter cloacae,* *Pseudomonas putida*	Tomato (*L. esculentum*)	Grichko and Glick 2001

responsive genes and stimulate pericycle cells to proliferate (Perez-Torres *et al.* 2008).

Auxin application results in formation of branched root and, similarly, mutants that accumulate high levels of auxin, or mutants with an altered auxin distribution, produce excess of lateral roots. A broad range of abiotic stresses induce lateral root formation, therefore, auxin may be an intermediate between the action of a stressor and the realization of response phenotype. Several mechanisms have been proposed to explain stress-induced changes in auxin metabolism and/or receptiveness; however, evidences for stress-induced changes in auxin transport and catabolism are predominantly found in literature. For example, water and osmotic stresses impact on auxin transport by altering the expression of PIN genes and/or by inhibition of polar auxin transport (Potters *et al.* 2009). Moreover, auxin conjugates and the respective hydrolases were shown to be involved in the reaction of plant to stress (Muller, 2011). Interestingly, overexpression of an auxin-amidohydrolase in *Arabidopsis* is associated with a reduced inhibition of root elongation and increased resistance to salt stress. This effect was probably due to the increase in the content of free auxin sufficiently to provide a protective effect against salt stress (Junghans *et al.* 2006).

An increased number of IAA-producing PGPR strains are detected inside the plant tissue (Spaepen *et al.* 2007). Various plant species inoculated with such bacteria showed increased root growth and/or formation of lateral roots and roots hairs (Dimkpa *et al.* 2009a). For example, the stimulatory effect of *Azospirillum* strains on the development of roots is well documented. Morphological plant root changes have been observed repeatedly upon *Azospirillum* inoculation and have been attributed to the production of plant-growth promoting substances: auxins, cytokinins and gibberellins, with auxin production being quantitatively the most important (Spaepen *et al.* 2008). Specific evidences for the involvement of auxins produced by *Azospirillum* in roots proliferation were obtained in many cases. Addition of filter-sterilized culture supernatants of *A. brasiliense* to rice roots grown in hydroponic tanks increased root elongation, root surface area, root dry matter, and development of lateral roots and root hairs, compared with untreated roots (El-Khawas and Adachi, 1999). Similarly, a cell-free supernatant of *A. brasiliense* Cd applied to soybean plants induced many roots and increased root length (Molla *et al.* 2001a). Exogenous application of IAA to bean roots resembled responses of these plants to inoculation with *Azospirillum* (Remans *et al.* 2008a). More direct evidence for the importance of IAA was provided when several IAA-attenuate mutants were compared with their parental wild types for their effect on plant growth. A mutant of *A. brasiliense* with low production of phytohormones, but high N_2-fixing activity, did not enhance root growth over uninoculated controls (Kundu *et al.* 1997).

3.7. Release of Protective Compounds and Control of Plant Diseases

Several studies correlated accumulation of nitrogen-containing compounds (NCC) with drought and salt tolerance in plants (Parida and Das,2005). The most frequently accumulating NCC includes amino acids, amides, imino acids, proteins, quaternary ammonium compounds and polyamines.

Very high accumulation of cellular proline (up to 80% of the amino acids pool under stress and 5% under normal conditions) due to increased synthesis and decreased degradation under a variety of stress conditions such as salt and drought has been documented in many plant species (Szabados and Savoure, 2009). Several comprehensive studies using transgenic plants or mutants demonstrate that proline metabolism has a complex effect on development and stress responses. Proline has been proposed to act as a compatible osmolyte and to be a way to store carbon and nitrogen. Saline and drought are known to induce oxidative stress. Several studies showed that proline may have an antioxidant activity acting as a ROS scavenger. Proline may also function as molecular chaperones able to stabilize the structures of proteins and enhance the activity of different enzymes, and its accumulation play a role in maintenance of cytosolic pH and regulation of intracellular redox potential (Hare and Cress,1997; Kavi Kishor *et al.* 2005;Verbruggen and Hermans, 2008).

Under abiotic stress conditions, increased proline biosynthesis was observed for various plant species inoculated with different PGPR(Barka *et al.* 2006; Jha *et al.* 2010; Kohler *et al.* 2009; Sandhya *et al.* 2010; Vardharajula *et al.* 2011). The synthesis of proline as well as other compatible solutes require an energy cost (41 moles of ATP) and occur at the expense of plant growth, but may allow the plant to survive and recover from the presence of high external salt concentration (Munns and Tester, 2008).We found a significant correlation between reduced symptoms of senescence, such as chlorosis, necrosis and drying, and 2-fold increased proline content in the shoot of *Mt*-RD64 as compared to *Mt*-1021 plants, after exposure to 150 mM NaCl (Bianco and Defez, 2009).

Several microorganisms are exploited directly to control fungi, bacteria and viruses whereas for indirect control several plant products, farm yard manure and compost biofertilizers are used which augment the process of plant disease control. By utilizing these, there is reduction in the pathogenic propagules and also their competitive saprophytic ability. In these several bioproducts such as green manure, farm yard manure, green and dry leaves, non-edible cakes, Neem cakes and saw dust can be incorporated at the time ploughing the fields and then irrigate the several organic acids which are toxic to the plant pathogens and thus a significant control is achieved. Organic matter enhances the saprophytic microflora growth and activities which releases antibiotic substances which are toxic to pathogenic fungi. The saprophytic microflora also compete for food with the pathogenic fungi. The diseases which are controlled significantly include root rot, collar rot and wits and other soil borne pathogen. Some of the microbial antagonists found effective in controlling of diseases are listed below.

Table 2: Plant Growth Promoting Rhizobacteria (PGPR) in control of root diseases of crop plants.

Crop	PGPR strain used	Disease
Rice	*Pseudomonas fluorescens, Pseudomonas putida, Bacillus polymyxa, Bacillus pumulus, Bacillus coagulans, Enterobacter agglomerans, Bacillus cereus*	Suppression of plant pathogens, blast, sheath blight, sheath rot, stem rot, bacterial blight, tungro disease, downey mildew
Sorghum	*Pseudomonas fluorescens*	Head mould
Pigeonpea	*Bacillus brevis, Bacillus subtilis, AF1, Pseudomonas fluorescens*	Wilt
Chickpea	*Bacillus subtilis, Pseudomonas aeruginosa, Pseudomonas fluorescens*	Wilt
Pea	*Azotobacter chrococcum, Azospirillum lipoferum, Pseudomonas fluorescens*	Root rot
Soybean	*Bacillus megaterium*	Damping off
Cowpea	*Bravibacterium linens, Pseudomonas fluorescens*	Stem and root rot
Bean	*Pseudomonas capacia*	Root rot
Groundnut	*Bacillus subtilis AF1*	Crown rot
Lentil	*Pseudomonas fluorescens*	Wilt
Urdbean	*Pseudomonas aeruginosa, Pseudomonas fluorescens*	Root rot
Mustard	*Trichoderma viride, Bacillus amylolique faciens*	White rust, Bud rot
Coconut	*Pseudomonas fluorescens*	Leaf rot

3.8. Production of Antioxidative Enzymes

In plants ROS such as superoxide ($^{\bullet}O_2^-$), hydrogen peroxide (H_2O_2), hydroxyl radical (OH^-), and singlet oxygen ($^1O_2$) are continuously produced as byproducts of various metabolic pathways localized in different cellular compartments (Apel and Hirt, 2004). A common feature of these species is their capacity to cause oxidative damage to proteins, DNA, and lipids. Since internal O_2 concentrations are high during photosynthesis, chloroplasts are especially prone to generate activated oxygen species (Gill and Tuteja, 2010). Under physiological steady-state conditions, these molecules are scavenged by different anti-oxidative defence components that are often confined to particular compartments (Apel and Hirt, 2004). Under normal growth conditions, the production of ROS in cells is low, whereas, during stress their rate of production is enhanced. ROS accumulation during stress results from the imbalance between production and scavenging of ROS. Major ROS-scavenging mechanisms of plants include SOD, APX and CAT enzymes. Antioxidants such as ascorbic acid and glutathione, which are found in high concentration in chloroplasts and other cellular compartments, are also crucial for plant defense against oxidative stress (Miller *et al.* 2010; Tyagi and Sairam, 2004). For the detoxification of excess ROS in plant, the overall balance between different antioxidants is crucial for determining the steady-state level of superoxide radicals and hydrogen peroxide, and has to be tightly controlled (Mittler, 2002).

Induction of antioxidant enzymes (catalase and total peroxidase) is involved in the alleviation of salinity stress in lettuce plants inoculated with PGPR strains (Kohler *et al.* 2010). Under non-saline conditions, inoculation with *Pseudomonas mendocina* and fertilization led to similar increases in plant growth (about 30% greater than the control plants). Salinity decreased the dry weight of the shoots and roots for all lettuce plants. However, the plants inoculated with *P. mendocina* had significantly greater shoot biomass than the control plants at both medium and high salinity levels. It was reported that salt-stressed *Mt*-RD64 plants showed much less oxidative damage (reduced chlorosis, necrosis, and drying) compared with salt-stressed *Mt*-1021 plants. These effects were connected to the enhanced activity of the antioxidant enzymes SOD, APX, GR and POX (Bianco and Defez, 2009).

The potential of PGPR strains in alleviating drought stress effects in maize. Maize plants inoculated with five drought tolerant plant growth promoting *Pseudomonas* spp. strains namely *P. entomophila*, *P. stutzeri*, *P. putida*, *P. syringae*, and *P. montelli* were subjected to drought stress and the effects of inoculation on growth, osmoregulation and antioxidant status was investigated. Inoculated plants showed significantly lower activity of antioxidant enzymes plants as compared to uninoculated plants (Sandhya *et al.* 2010). Reduction in the activity of antioxidant enzymes was also observed in barley plants. Omar *et al.* (2009) reported that, without inoculation, salinity led to a significant increase of catalase and peroxidase activities in salt-stressed leaves of two barley cultivars differing in salinity tolerance. Inoculation of the two cultivars with *Azospirillum brasilense* lowered the magnitude of increase and significantly ameliorated the deleterious effects of salinity improving crop productivity. The inoculated plants felt less stress as compared to uninoculated plants in wheat (Panwar *et al.* 2000).

3.9. Nutrient Supply

Survival and productivity of crop plants exposed to environmental stresses are dependent on their ability to develop adaptive mechanisms to avoid or tolerate stress (Munns and Tester, 2008). Accumulating evidence suggests that mineral nutritional status of plants greatly affects their ability to adapt to adverse environmental conditions and in particular to abiotic stress factors. Impairment of the mineral nutrition status of plants exacerbates the adverse effects of abiotic stresses and the exogenous addition of high levels of macronutrients can alleviate the adverse effects of stress on plant growth (Baligar *et al.* 2001; Endris and Mohammed, 2007; Grattan and Grieve, 1999; Heidari and Jamshid, 2010; Kaya *et al.* 2001; Kaya *et al.* 2002; Khoshgoftarmanesh *et al.* 2010).

After nitrogen, phosphorous is the second major nutrient for plant growth as it is an integral part of different biochemicals like nucleic acids, nucleotides, phospholipids and phosphoproteins. In most cases salinity decreased P accumulation in plant, which developed P-deficiency symptoms (Martinez and Lauchli, 1994; Navarro *et al.* 2001; Parida and Das, 2004; Rogers *et al.* 2003). The reduction in P availability in saline soils was suggested to be a result of ionic strength effects that reduce the activity of phosphate and the tight control of P concentrations by sorption processes and by low solubility of Ca-P minerals. The concentration of soluble P in soil is usually very low (1 ppm or less) (Hinsinger, 2001). The cell might take up several P forms but the major part is adsorbed in the forms of HPO_4^{-2} or $H_2HPO_4^{-1}$. Phosphorus exists in two forms in soil, as organic and inorganic phosphate, and like other nutrient elements such as potassium, iron, zinc and copper, possesses limited mobility in the soil (Hayat *et al.* 2010; Rodriguez and Fraga, 1999). The conversion of insoluble phosphate compounds (both organic and inorganic) in a form accessible to the plant is an important trait of PGPR strains. PGPR strains belonging to various genera have the ability to solubilize insoluble inorganic phosphate compounds such as tricalcium phosphate, dicalcium phosphate, hydroxyapatite, and rock phosphate (Richardson *et al.* 2009; Khan *et al.* 2009; Rodriguez and Fraga, 1999; Preeti *et al.* 2011). It is generally accepted that the major mechanism of mineral phosphate solubilization is the action of organic acids synthesized by soil microorganisms (Rodriguez and Fraga, 1999). Production of organic acids results in acidification of the microbial cell and its surroundings. Consequently, P may be released from a mineral phosphate by proton substitution for Ca^{2+}. The production of organic acids has been well documented for different PGPR genera such as *Pseudomonas, Erwinia*, *Rhizobium* and *Bacillus* (Rodriguez and Fraga, 1999).

Soil contains a wide range of organic substrates, which can be a source of P for plant growth. To make this form of P available for plant nutrition, it must be hydrolyzed to inorganic P. Mineralization of most organic phosphorous compounds is carried out by means of phosphatase enzymes (phosphohydrolases). Considering that the pH of most soils ranger from acid to neutral values, acid phosphatases should play the major role in this process (Rodriguez and Fraga, 1999).

Under P-limiting conditions, the IAA-overproducing RD64 strain showed high P-mobilizing activity that is connected to the synthesis of high levels of acid phosphatase enzymes and the secretion into the growth medium of malic, succinic, and fumaric acids in large quantities. As compared to *Mt*-1021 plants, *Mt*-RD64 plants released large amount of another P-solubilizing organic acid, 2-hydroxyglutaric acid and showed significant increase in both shoot and root fresh weights, when grown under P-deficient conditions (Bianco and Defez, 2010a).

3.10. Role of AM Fungi

Unlike *Arabidopsis*, more than 80% of higher plants associate with mycorrhizal fungi, which elicit profound changes in the root morphology of host plants (Hetrick, 1991). In particular, ectomycorrhizae suppress root elongation and induce dichotomous branching of short lateral roots, culminating in the formation of coralloid structures resulting from higher – order dichotomous branching. All of these anatomical structures are variable depending on the plant and fungal species.

Once the fungus is established, root branching is suppressed, which makes the plant more dependent on the nutrients provided by the fungus (Hetrick, 1991; Price *et al.* 1989). Whether this modification of root system architecture (RSA) is a direct consequence of symbiosis or an indirect effect of improved nutrient status of the plant is not clear. However, it appears that symbionts can triggers RSA changes by promoting lateral root initiation very early in the interaction (Harrison, 2005). Moreover, the maize mutant *lrt-1* normally lacks lateral roots, but displays extensive lateral root development following inoculation with the mycorrhizae *Glomus mosseae* (Paszkowski *et al.* 2002). Notably, many micro-organisms that interact with plants can produce plant hormone analogs. Thus, symbiotic association might employ hormone signaling pathways to regulate RSA.

3.10.1. VA-mycorrhizae and plant growth

As VAM are ubiquitous in nature: they are commonly associated with most plants important in agriculture, horticulture and forestry. In agriculture, potential hosts occur in the economically important families; *viz, Gramineae* and *Leguminoseae*. It is also found in many tropical plantation crops, trees, walnuts and all temperate fruit trees. Although VAM being omnipresent, the establishment of successful entry points on host trees, establishment, development and spreading of internal and external hyphae and its response to various kinds of hosts is dependent on prevailing fungi, plant and environmental factors interaction.

The main agronomic significance of VAM lies in their ability to supply the plants with extra nutrients, particularly phosphorus. Crush (1974) reported that endophyte host relationship varies between parasitism and mutualism depending on the available phosphorus levels. Maximal mycorrhizal effectiveness occurs when nutritional conditions for the host are sub-optimal but these become severely deficient and the fungus becomes ineffective.

Mycorrhizal colonization frequently alters root-to-shoot ratio and Busse and Ellis (1985) reported that root-to-shoot ratio could conceivably affect foliar water relations. It is likely that potted plants with different under ground biomass, will respond differently under conditions of drought, as proportionately different amounts of water would be available to transpiring plant tissues. Additionally, type, size and branching patterns of root system can affect exploitation of a given soil mass by plant through changes in hyraulic conductivity and water flow-rates (Nye and Tinker, 1977; Preeti *et al.* 2013).

VAM might alter the hormonal balance of the host (Brown *et al.* 1981). Various workers have offered this possibility as an explanation for changed stomatal behaviour and water flow rates in mycorrhizal plants (Allen, 1982; Levy and Krikun, 1980). Cooper (1984) stated that mycorrhizal infection may possibly alter membrane permeability through increased production of photosterios, thereby influencing symplastic water transport.

VAM can tolerate a wide spectrum of soil moisture, as evidenced by their presence in arid deserts and aquatic habitats (Cooper, 1984). Many workers reported that P-uptake in non-mycorrhizal plants may be diminished in soils of low water content (Nelson and Safir, 1982). The diffusion coefficient of phosphate in soil is linearly related to soil moisture content and therefore, mycorrhizal P-supplies are likely to be much more beneficial to hosts growing in dry than in moist soils (Fitter, 1985).

Colonization by *G. fasciculatum* could decrease resistance to water transport in *Bouteloua* up to 90% under drought conditions (Allen *et al.* 1981).

Leaf water potentials of mycorrhizal plants dropped more quickly than those of non-mycorrhizal plants as moisture stress increased and stomatal resistance remained lower in colonized plants. Cooper, (1984) suggested that this influence of *G. fasciculatum* might have resulted from improved water uptake, increased photosynthesis or elevated cytokinin levels.

The colonization of roots by AM fungi in various plant species induces proline accumulation when water is limiting (Goicoechea *et al.* 1998; Yooyongwech *et al.* 2013).The enhanced accumulation of proline in these studies was linked to AM-induced drought resistance with proline acting as osmoproctectant. Conversely, in several studies, while proline content increased in response to water deficit, a lower accumulation of proline has been observed in mycorrhizal plants relative to non mycorrhizal counterparts (Wu and Xia,2006; Aroca *et al.* 2008; Ruiz-Sanchez *et al.* 2010; Abbaspour *et al.* 2012; Fan and Liu, 2011; Asrar *et al.* 2012; Doubkova *et al.* 2013), suggesting that AM symbiosis enhanced host plant resistance to drought.

3.11. Co-inoculation of Different Microbes for Stress Tolerance

Co-inoculation is based on mixed inoculants, combination of microorganisms that interact synergistically, or when microorganisms such as *Azospirillum* are functioning as "helper" bacteria to enhance the performance of other beneficial microorganisms. In the rhizosphere the synergism between various bacterial genera such as *Bacillus*, *Pseudomonas* and *Rhizobium* has been demonstrated to promote plant growth and development. Compared to single inoculation, co-inoculation improved the absorption of nitrogen, phosphorus and mineral nutrients by plants (Figueiredo *et al.* 2010; Yadegari *et al.* 2010). A significant increase in root and shoot biomass was observed in chickpea plants when co-inoculated with *Mesorhizobium* and *Pseudomonas* (Sindhu *et al.* 2002a,b). Increased nodule weight, root and shoot biomass and total nitrogen of chickpea plants was also reported due to co-inoculation of *Rhizobium, Pseudomonas* and *Bacillus* (Parmar and Dadarweal, 1999). Co-inoculation with *Bradyrhizobium japonicum and Pseudomonas fluorescens* increased colonization *B japonicum* on soybean roots, nodule number and acetylene reduction assay (Tchebotar *et al.* 1998). Combined inoculation of *Rhizobium* with *Pseudomonas striata* or *Bacillus megaterium* led to increased dry matter, grain yield and phosphorus uptake significantly over the uninoculated control in legumes (Elkoca *et al.* 2008).Verma *et al.* (2010) have reported the application of *Rhizobium* spp. and plant growth promoting rhizobacteria on nodulation, plant biomass and yields of chickpea plants. In field studies, the grain and straw yield were significantly increased in co-inoculation of *Rhizobium* with *P. fluorescens* followed by *B. megaterium* and *Azotobacter chroococcum* over uninoculated control.

Co-inoculation of *Pseudomonas* spp. with *Rhizobium* improves growth and symbiotic performance of fodder galega (Egamberdiveya *et al.* 2010). The greenhouse experiment showed that co-inoculation of fodder galega with *R. galegae* and *P. trivialis* or with *R. galegae* and *P. extremorien-talis* improved plant growth, nodulation and nitrogen content compared to plants inoculated with *R. galegae* alone in potting soil containing low levels of nitrogen. Co-inoculation of plants with *P. trivialis* and *R. galegae* showed the highest stimulatory effect.

The mechanisms behind these effects are only partially understood. One of the mechanisms used by these PGPR strains is the production of phytohormones such as auxins, gibberellins and cytokinins, which steadily contribute to the plant auxin "pool" in a way that the effect of PGPR inoculation can be mimicked by exogenous auxin application. The endogenous IAA level in plant regulates growth of the shoots and roots, and in the case of legumes, nodules formation (Teale *et al.* 2006). It has been observed that low concentrations of exogenously given pure IAA stimulated shoot and root growth of wheat in non-saline and saline conditions, and similar effects were induced by IAA-producing PGPR strains (Egamberdiveya, 2009).

Bacteria of the genus *Azospirillum* are free-living, surface colonizing and, sometimes, endophytic diazotroph and plant growth promoting rhizobacteria. *Azospirillum* strains had no preference for crop plants or weeds, or for annual or perennial plants, and can be successfully applied to plants that have no previous history of *Azospirillum* in their roots. Although reports about isolating *Azospirillum* from graminaceous plants are common, other reports showed that the bacterium is a natural inhabitant of many non-graminaceous plants. It appears that *Azospirillum* is a general root colonizer and is not a plants-specific bacterium (Bashan and Holguin, 1997). *Azospirillum* strains are capable of increasing yield of important crops growing in various soils and climatic regions. It has been reported that root elongation rate, mineral N, P and K and microelements uptake are consequently improved after *Azospirillum* inoculation (Bashan *et al.* 2004), even under stressful environmental conditions (Askary *et al.* 2009). Dual inoculation of legumes with *Rhizobium* and *Azospirillum* signifycantly increase several plant-growth variables when compared with single inoculations (Hamaoui *et al.* 2001; Itzigsohn *et al.* 2000; Remans *et al.* 2007; Remans *et al.* 2008b; Tchebotar *et al.* 1998). *Azospirillum* is considered a *Rhizobium* helper by stimulating nodulation, nodule function, and possibly plant metabolism (Molla *et al.* 2001; Verma *et al.* 2010). Phytohormones produced by *Azospirillum* promote epidermal-cell differentiation in root hairs that increased the number of potential sites for rhizobial infection leading to the formation of more nodules. Morphological and physiological changes in root system are also stimulated (Bashan and Levanony,1990; Pacovsky, 1990; Sarig *et al.* 1992; Volpin and Kapulnik,1994). An increase in the number of lateral roots and root hairs cause addition of root surface available for nutrients and water uptake. Higher water and nutrient uptake by inoculated roots cause an improved water status of plant, which in turn could be the main factor enhancing plant growth (Boddey *et al.*1986; Dalla Santa *et al.* 2004; Fallik and Okon, 1996; Mostajeran *et al.* 2002). Positive effects of co-inoculation were also observed on symbiotic performance of common bean, which is usually considered a poor nitrogen-fixing legume. Poor nodulation and variable response to inoculation is mainly attributed to intrinsic characteristics of the host plant, particularly the great sensitivity to nodulation-limiting factors, such as high rate of nitrogen fertilizer used in intensive agriculture, high temperature and soil dryness (Bais *et al.* 2006; Egamberdiveya, 2007). Indeed, Yadegari and Rahmani, (2010) showed that co-inoculation of three *Phaseouls vulgaris* cultivars with two *Rhizobium* strains, *Pseudomonas fluorescens* and *A. lipoferum* resulted in increased seed yield, number of pods per plant, weight of seeds, seeds protein yield and number of seeds per pod. Preeti *et al.* (2012) reported the synergistic effect of *Rhizobium* and VAM on productivity in *Vigna mungo* under rainfed conditions.

Inoculation of common bean or alfalfa (*Medicago sativa*) with *Azospirillum brasilense* in the absence of *Rhizobium* resulted in a more persistent exudation of flavonoids by legumes roots. *Azospirillum-Rhizobium* co-inoculation positively affected the expression of *nod*-genes and production of nodulation factor patterns in *Rhizobium tropici* and *Rhizobium etli* in the presence or absence of NaCl at 50 mM. A significant increase of total and upper nodule numbers was observed at different concentrations of *Rhizobium* inoculum (Dardanelli *et al.* 2008). Several greenhouse and field experiments demonstrated the potential of co-inoculation to increase grain yield of various legumes (Bashan and Holguin, 1997; Galal *et al.* 2002; Itzigsohn *et al.* 2000; Sarig *et al.* 1986; Yahalom *et al.* 1989).

3.12. Organic Ammendments and Abiotic Stress

3.12.1. Green manure

Bio farming or organic farming uses bio inoculants such as *Rhizobia, Azotobacter, Mycorrhizae etc.* which have several functions. Besides there are other microbes such as *Trichoderma, Glichoderma, Glicoladium, Penicillium, Talaromyces* and *Pseudomonas* which are responsible for the control of plant diseases. However, their presence and action encouraged by the presence of organic matter, compost, farm yard manure, and other farm wastes.

By the use of green manure crops the soil is supplied with most essential by the plants. Plants contain carbon and nitrogen as main constituents. Both the nutrients are available from the atmosphere. They are available from the green manure crops on their turning. By the use of green manure crops the organic content of the soil is increased which is considered to be food of bio fertilizing agents. Green manure crops contain nodules in their roots in which *Rhizobium* bacteria are found which fixes atmospheric nitrogen in the root and also called the leguminous plants. By planting such legumes the nitrogen content of soil increases. In the presence of organic matter and nitrogen in the soil the vaporization of the water is drastically reduced; the presence of organic matter (OM) facilitates air circulation in the soil. The decomposition of the green manure crops releases carbon in the soil which in the presence of carbon dioxide forms various carbonic acids. These carbonic acids in the presence of phosphorous, calcium, potassium, magnesium, iron, manganese and zinc nutrients from various organic salts and also solubilises these salts from the various complex forms and which thus made available to the plants. These are food of microbes and thus the soil biota increased to the significant extent. In salt affected alkaline or acidic soil the pH is regularized to the desirable extent; in acid soil pH is increased whereas in the alkaline soil the pH is decreased, and the soil becomes normal. By the use of OM and green manure crops the adverse effects of chemical fertilizers is nullified. Thus both the soils, acid and alkaline, are corrected. Since the water holding and storage increased in the presence of OM and drainage also regularized. For green manure both the crops such as legumes as well as grasses or non legumes are used, (Table 3)

3.12.2. Compost

Compost is prepared through the action of microbes on wastes *viz.*, leaves, roots and stubbles, crop residues, straw, hedge, clippings, weeds, water hyacinth, kitchen wastes and human habitation wastes by a biochemical process. The microbes are aerobic (which require air or oxygen) and anaerobic (which function in the of air or free oxygen) types decompose OM into a final product known as compost. The materials undergo intensive decomposition under medium high temperatures in heaps or pits with adequate moisture. In about 3-6 months, an amorphous, brown to black humified material, compost is obtained. It is more stable in form, valuable source of plant nutrients, helps in maintenance of soil OM and improving soil physical conditions and biological activities. Composts are grouped as rural or urban. Important parameters which are required to be maintained at optimum level for efficient composting are C:N ratio, particle size, blending or proportioning of raw materials, moisture, aeration, temperature, destruction of pathogens and parasite and use of microbial activities.

Table 3: Some important green manure crops, seed rate, stage of turning, quantity of biomass and nutrient added per hectare

Crop	Seed rate and Turning phase (Days after sowing)	Biomass added/ha	Quantity of element added/ha
Sunhemp (*Crotolaria juncea*)	90-100kg seed/ha; after 6-8 weeks in soil	170 q/ha	75kg N/ha
Daincha (*Sesbania aculenta*)	65-100kg seed/ha; after 8 weeks	160 q/ha	70kg N/ha
Guwar (*Cyamopsis tetragonolobe*)	30kg seed/ha; after 50 days	160 q/ha	56kg N/ha
Lobia (*Dalichos lablab*)	20-25kg seed/ha; after 40 days	120 q/ha	50kg N/ha
Urd and Mung (*Phaseolus mungo* and *V. radiata*)	25-35kg seed/ha; after 60 days after plucking beans	64 q/ha	35kg N/ha
Cowpea (*Vigna sinensis*)	30-40kg seed/ha	150 q/ha	74kg N/ha
Senji (*Melilotus alba*)	4-5kg seed/ha	280 q/ha	163kg N/ha
Berseem (*Trifolium alexandrium*)	4-5kg seed/ha	155 q/ha	67kg N/ha

3.12.3. Farm yard manure

By the use of cow dung manure a substantial quality of chemical fertilizers can be reduced. Besides the various micro nutrients , the application improve structure of the soil and water holding capacity, water conserving capacity, aeration and biomass of soil is increased. The application of cow dung also improves soil salinity and alkalinity. The presence of farm yard manure (FYM) in fields helps in better nitrogen fixation through *Rhizobia*. The quality of FYM is influenced by several other factors for example: age of the cattle, its health, type of feed, and lastly the method of preparation. Desert areas cattle such as camel, sheep, and goat excreta are concentrated organic manure and rich in elements as compared to cow and buffalo manure. Those cattle which are fed into balanced nutritive diet excrete rich manure than those fed with unbalanced and incomplete diet. The heafers and old cattle manure is low in nutrients. In natural cycle healthy and productive soil and healthy cattle are complementary and depended on each other.

3.12.4. Vermiwash and vermicompost

Vermicomposting is the biological degradation and stabilization of organic waste by earthworms and microorganisms to form vermicompost. This is an essential part in organic farming today. It can be easily prepared, has excellent properties, and is harmless to plants. The earthworms fragment the organic waste substrates, stimulate microbial activity greatly and increase rates of mineralization. These rapidly convert the waste into humus-like substances with finer structure than thermophilic composts but possessing a greater and more diverse microbial activity. Vermicompost being a stable fine granular organic matter, when added to clay soil loosens the soil and improves the passage for the entry of air. The mucus associated with the cast being hydroscopic absorbs water and prevents water logging and improves water-holding capacity. The organic carbon in vermicompost releases the nutrients slowly and steadily into the system and enables the plant to absorb these nutrients. The soil enriched with vermicompost provides additional substances that are not found in chemical fertilizers (Kale, 1998). Vermicomposting offers a solution to tonnes of organic agro-wastes that are being burned by farmers and to recycle and reuse these refuse to promote our agricultural development in more efficient, economical and environmentally friendly manner. The role of earthworms in organic solid waste management has been well established since first highlighted by Darwin, (1881) and the technology has been improvised to process the waste to produce an efficient bio-product vermicompost (Kale *et al.* 1982; Ismail, 1993; Ismail, 2005). Epigeic earthworms like *Perionyx excavatus*, *Eisenia fetida, Lumbricus rubellus* and *Eudrilus eugeniae* are used for vermicomposting but the local species like *Perionyx excavatus* has proved efficient composting earthworms in tropical or subtropical conditions (Ismail, 1993; Kale, 1998). The method of vermicomposting involving a combination of local epigeic and anecic species of earthworms (*Perionyx excavatus* and *Lampito mauritii*) is called Vermitech (Ismail, 1993; Ismail, 2005). The compost prepared through the application of earthworms is called vermicompost and the technology of using local species of earthworms for culture or composting has been called Vermitech (Ismail, 1993). Vermicompost is usually a finely divided peat-like material with excellent structure, porosity, aeration, drainage and moisture holding capacity. The nutrient content of vermicompost greatly depends on the input material. It usually contains higher levels of most of the mineral elements, which are in available forms than the parent material. Vermicompost improves the physical, chemical and biological properties of soil (Kale, 1998). There is a good evidence that vermin compost promotes growth of plants (Lalitha *et al.* 2000) and it has been found to have a favorable influence on all yield parameters of crops like wheat, paddy and sugarcane (Ismail, 2005). By the use of vermicompost we can lower the cost of food production, reduced usage of chemical pesticides and cut cost, reduced usage of water for irrigation and cut cost, improvement of growth and higher yield.

3.13. Futuristic Approach

Global climatic change is the matter of concern in affecting the crop and human health adversely. The sudden changes in environmental and edaphic factors that causes the abiotic stresses is the direct cause of concern in affecting the crop productivity world over. The role of microbes is quite important and efforts should be made to increase the efficiency of microbes by using the biotechnological tools and through genetic engineering of plant spp and microbes to incorporate new traits and screening of efficient beneficial microorganisms will be of immense help to the plants facing the abiotic stresses. The minimal use of inorganic fertilizers and insecticides and pesticides should be advocated by replacing it with organic farming practices. The use of green manuring, compost, FYM and vermicompost should be encouraged among the farming community and modern concept in improving the traditional farming should be encouraged.

References

Abbaspour, H., Saeid-Sar, S., Afshari, H. and Abdel-Wahhab, M.A. 2012. Tolerance of mycorrhiza infected Pistachio (*Pistacia vera* L.) seedlings to drought stress under glasshouse conditions. *Journal of Plant Physiology,* **169**: 704–709.

Abdelmoneim, T.S., Tarek, A.A.M., Almaghrabi, O.A., Hassan, S.A. and Abdelbagi, I. 2014. Increasing plant tolerance to drought stress by inoculation with arbuscular mycorrhizal fungi. *Life Science Journal,* **11(1):** 10-17.

Adesemoye, A.O., Torbert, H.A. and Kloepper, J.W. 2008. Enhanced plant nutrient use efficiency with PGPR and AMF in an integrated nutrient management system. *Can. J. Microbiol.,* **54**:876-888.

Allen, M.F., Smith, W.K., Moore, T.S. and Christensen, M. (1981): Comparative water relations and photosynthesis of mycorrhizal and non-mycorrhizal *Bouteloue gracilis* (H.B.K.) Lag ex Strend. *The New Phytologist.* **88:** 683-693.

Allen, M.F. 1982. Influence of vesicular-arbuscular mycorrhiza on water movement through *Bouteloua gracilis* (H.B.K.) Lag ex strend. *New Phytol,* **91**: 191-196.

Alloway, B.J. 2008. Zinc in soils and crop nutrition, *2nd edition. IZA and IFA, Paris, France.*Alvarez, M.I., Sueldo, R.J. and Barassi, C.A. (1996): 'Effect of *Azospirillum* on coleoptiles growth in wheat seedlings under water stress'. *Cereal Research Communications.* **24:** 101–107.

Apel, K. and Hirt, H. 2004. Reactive oxygen species: metabolism, oxidative stress and signal transduction. *Annual Review of Plant Biology,* **55**:373-399.

Arkipova, T.N., Prinsen, E., Vaselov, S.U., Martinenko, E.V., Melentiev, A.I. and Kudoyaroya, G.R. 2007. Cytokin in producing bacteria enhance plant growth in drying soil. *Planta and Soil,* **292**: 305-315.

Aroca, R., Vernieri, P. and Ruiz-Lozano, J.M. 2008. Mycorrhizal and nonmycorrhizal *Lactuca sativa* plants exhibit contrasting responses to exogenous ABA during drought stress and recovery. *Journal of Exp. Botany,* **59**: 2029–2041.

Arun, K.S. 2007. Bio-fertilizers for sustainable agriculture. Mechanism of P-solubilization. 6th *Edition,* Agribios Publishers, Jodhpur, India, pp. 196-197.

Asch, F. and Padam, J.L. 2005. Root associated bacteria suppress symptom of iron toxicity in lawland rice. In "The Global Food and Product Chain-Dynamics,Innovations, Conflicts, Strategies" Eds. E. Tielkes, Hulsebusch, I. Hauser, A. Deininger, and K. Becker, 276 MDD GmbH, Stuttgart, Germany.

Ashad, M., Shaharoona, B. and Mahmood, T. 2008. Inoculation with *Pseudomonas* spp. containing ACC-deaminase partially eliminates the effect of drought stress on growth, yield and ripening of pea (*Pisum sativum* L.). *Pedosphere,* **18**: 611-620.

Ashraf, M., Hasnain, S., Berge, O. and Mahmood, T. 2004. Inoculating wheat seeds with exopolysaccharide-producing bacteria restricts sodium uptake and stimulates plant growth under salt stress. *Biology and Fertility of Soils,* **40**: 157-162.

Askary, M., Mostajeran, A., Amooaghaei, R. and Mostajeran, M. 2009. Influence of the co-inoculation *Azospirillum brasilense* and *Rhizobium meliloti* plus 2,4–D on grain yield and N,P,K content of *Triticum aestivum (*Cv.Baccros and Mahdavi). *American-Eurasian Journal of Agricultural and Environmental,* **5**: 296-307.

Asrar, A.A., Abdel-Fattah, G.M. and Elhindi, K.M. 2012. Improving growth, flower yield, and water relations of snapdragon (*Antirhinum majus* L.) plants grown under well-watered and waterstress conditions using arbuscular mycorrhizal fungi. *Photosynthetica* **50**: 305–316.

Bais, H.T., Perry, L.G., Simon, G. and Vivanco, J.M. 2006. The role of root exudates in rhizosphere interactions with plants and other organisms. *Plant Biology,* **57**: 233-266.

Bakanchikova, T.I. Lobanok, E.V., Pavlova-Ivanova, L.K., Redkina, T.V., Nagapetyan, Z.A. and Majsuryan, A.N. 1993. Inhibition of tumor formation process in dicotyledonous plants by *Azospirillum brasilense* strains'. *Mikrobiologiya.* **62:** 515–523.

Baligar, V.C., Fageria, N.K. and He, Z.L. 2001. Nutrient use efficiency in plants. *Communication in Soil Science and Plant Analysis,* **32**: 921-950.

Barassi, C.A., Ayrault, C.M., Creus, C.M., Sueldo, R.J. and Sobrero, M.T. 2006. Seed inoculation with *Azospirillum* mitigate NaCl effects on lettuce. *Scientia Horticulturae,* **109**: 109-814.

Barka, E.A., Nowak, J. and Clement, C. 2006. Enhancement of chilling resistance of inoculated grapevine plantlets with a plant growth-promoting rhizobacterium. *Burkholderia phytofirmas* strain Ps. *JN Applied and Environmental Microbiology,* **72**: 7246-7252.

Bashan, Y. and de-Bashan, L.E. 2002a.: 'Reduction of bacterial speck *(Pseudomonas syringae* pv. tomato) of tomato by combined treatments of plant growth-promoting bacterium *Azospirillum brasilense*, streptomycin sulfate, and chemo- thermal seed treatment'. *European Journal of Plant Pathology.* **108**: 821–829.

Bashan, Y. and de-Bashan, L.E. 2002b. Protection of tomato seedlings against infection by *Pseudomonas syringae* pv. tomato by using the plant growth-promoting bacterium *Azospirillum brasilense'*. *Applied and Environmental Micro- biology.* **68**: 2637–2643.

Bashan, Y. and Holguin, G. 1997. *Azospirillum*-plant relationships: environmental and physiological advances 1990-1996. *Canadian Journal of Microbiology,* **43**: 103-121.

Bashan, Y. and Levanony, H. 1990. Current status of *Azospirillum* as a challenge for agriculture. *Canadian Journal of Microbiology,* **36**: 591-608.

Bashan, Y., Holguin, G. and de-Bashan, L.E. 2004. *Azospirillum*-plant relationships: physiological, molecular, agricultural, environmental and advances (1997-2003). *Canadian Journal of Microbiology,* **50**: 521-577.

Bensalim, S., Nowak, J. and Asiedu, S.K. 1998. A plant growth promoting rhizobacterium and temperatures effects on performance of 18 cyclones of potato. *American Journal of Potato Research,* **75**: 145-152.

Bianco, C. and Defez, R. 2009. Medicago truncatula improves salt tolerance when nodulated by an indole-3-acetic acid-overproducing *Sinorhizobium meliloti* strain. *Journal of Experimental Botany,***60(11):** 3097-3107.

Bianco, C. and Defez, R. 2010a. Improvement of phosphate solubilization and Medicago plant yield by an indole-3-acetic acid-overproducing strain of *Sinorhizobium meliloti*. *Applied and Environmental Microbiology,* **76**: 4626-4632.

Bianco, C. and Defez, R. 2011. Soil bacteria support and protect plants against abiotic stresses. In "Agricultural and Biological science - Abiotic Stress in Plants-echanisms and Adaptations", Ed. A. Shankar and B. Venkateswarlu, ISBN 978-953-307-394-1.

Bheemareddy, V.S., Byatanal, M.B. and Lakshman, H.C. 2010. Effect of salt and acid stress on *Triticum aestivum* L. Var. inoculated with *Glomus fasciculatum. Libyan Agriculture Research Centre Journal International,* **1(5):** 325-331.

Boddey, R.M., Baldani, V.L.D., Baldani, J.I. and Dobereiner, J. 1986. Effect of inoculation of *Azospirillum spp.* on nitrogen accumulation by field grown wheat. *Plant and Soil,* **95**:109-121.

Brown, R.W., Brown, C.L. and Kormanil, P.P. 1981. Effects of VA-mycorrhizae and soil phosphorus on cytokinins and nutrient levels in *Plantanus occidentalis.* In "Absts. of 5th North America. Conf. on Mycorrhizae", *Quebec. 12.*

Busse, M.D. and Ellis, J.R. 1985. Vesicular arbuscular mycorrhizal (*Glomus fasciculatum*) influence on soya bean drought tolerance in high phosphorus soil. *Can. J. Bot.* **63**: 2290-2294.

Cabello, M., Irrazabal, G., Bucsinszky A.M., Saparrat, M. and Schalamuck, S. 2005. Effect of an arbuscular mycorrhizal fungus, *G. mosseae and* a rock-phosphate-solubilizing fungus, *P. thomii* in *Mentha piperita* growth in a soil less medium'. *Journal of Basic Microbiology.* **45**: 182–189.

Casanovas, E.M. Barassi, C.A. and Sueldo, R.J. 2002. *Azospirillum* inoculation mitigates water stress effects in maize seedlings. *Cereal Research Communication,* **30**: 343-350.

Cassan, F., Maiale, S. Masciarelli, O., Vidal, A., Luna, V. and Ruiz, O. 2009. Cadaverine production by *Azospirillum brasilense* and its possible role in plant growth promotion and osmotic stress mitigation. *European Journal of Soil Biology,* **45**: 12-19.

Cooper, K.M. 1984. Physiology of VA-mycorrhizal association. 155-203. In "VA mycorrhiza". Ed. C.L. Powell, and D.J., Bagyaraj; Boca Roton, Fla: CRC Press,London, pp. 113-130.

Creus, C.M., Sueldo, R.J. and Barassi, C.A. (1997): 'Shoot growth and water status in *Azospirillum*-inoculated wheat seedlings grown under osmotic and salt stresses'. *Plant Physiology and Biochemistry.* **35**: 939–944.

Creus, C.M., Sueldo, R.J. and Barassi, C.A. 1998. 'Water relations in *Azospirillum* inoculated wheat seedlings under osmotic stress'. *Canadian Journal of Botany.* **76**: 238–244,

Creus, C.M., Sueldo, R.J. and Barassi, C.A. 2004. Water relations and yield in *Azospirillum*-inoculated wheat exposed to drought in the field. *Canadian Journal of Botany,* **82**: 273-281.

Crush, J.R. 1974. Plant growth responses to vesicular-arbuscular mycorrhiza VII. Growth and nodulation of some herbage legumes. *New Phytol.* **73**: 743-749.

Dalla Santa, O.R., Hernandez, R.F., Alvarez, G.L.M., Ronzelli, P. and Soccol, C.R. 2004. *Azospirillum sp*. Inoculation in wheat, barley and oats seeds greenhouse experiments. *Brazilian Archives of Biology and Technology,* **47:** 843-850.

Dardanelli, M.S., Fernandez de, C.F.J., Espuny, M.R., Caryaial, M.A.R., Diaz, M.E.S., Serrano, A.M.G., Okon, Y. and Megias, M. 2008. Effect of *Azospirillum brasilense* coinoculated with *Rhizobium* on *Phaseolus vulgaris* flavonoids and Nod factor production under salt stress. *Soil Biology and Biochemistry,* **40**: 2713-2721.

Darwin, C. 1881. The formation of vegetable mould through the action of worms, with observations on their habitats. *Murray, London.* 326.

Dimkpa, C., Weinand, T. and Ash, F. 2009a. Plant-rhizobacteria interactions alleviate abiotic stress conditions. *Plant, Cell and Environment,* **32**: 1682-1694.

Dimkpa, C.O., Merten, D., Svatos, A., Buchel, G. and Kothe, E. 2009b. Metal induced oxidative stress impacting plant growth in contaminated soil is alleviated by microbial siderophores. *Soil Biology and Biochemistry,* **41**:154-162.

Dimkpa, C.O., Svatos, A., Merten, D., Buchel, G. and Kothe, E. 2008. Hydroxamate siderophores produced by *Streptomyces acidiscabies* E_{13} bind nickel and promotes growth in cowpea (*Vigna Unguiculata* L.) under nickel stress. *Canadian Journal of Microbiology,* **54**:163-172.

Dobereiner, J. 1997. Biological Nitrogen fixation in tropics: Social and economic contributions. *Soil Biology and Biochemistry*. **29(5-6):** 771-774.

Doubkova, P., Vlasakova, E. and Sudova, R. 2013. Arbuscular mycorrhizal symbiosis alleviates drought stress imposed on *Knautia arvensis* plants in serpentine oil. *Plant & Soil,*. **345(1-2):** 325-338.

Edwards, C. A. and Bohlen, P. J. 1996. Biology and ecology of earthworm. (*3rd edn.), Chapmanand Hall,* London. 426.

Egamberdiveya, D. 2007. The effect of plant growth promoting bacteria on growth and nutrient uptake of maize in two different soils. *Applied Soil Ecology,* **36**:184-189.

Egamberdiveya, D. 2009. Alleviation of salt stress by plant growth regulators and IAA producing bacteria in wheat. *Acta Physiologiae Plantarum,* **31**:861-884.

Egamberdiveya, D. and Hoflich, G. 2003. Influence of growth-promoting bacteria on the growth of wheat in different soil temperatures. *Soil Biology and Biochemistry,* **35**: 973-998.

Egamberdiveya, D., Berg, G., Lindstrom, K. and Rasanen, L.A. 2010. Inoculation of *Pseudomonas spp.* With *Rhizobium* improves growth and symbiotic performance of fodder galega (*Galega orientalis* Lam.). *European Journal of Soil Biology,* **46**:269-272.

El-Khawas, H. and Adachi, K. 1999. Identification and quantification of auxins in culture media of *Azospirillum* and *Klebsiella* and their effect on rice roots. *Biology and Fertility of Soils,* **28**: 377-381.

Elkoca, E., Kantar, F. and Sahin, F. 2008. Influence of nitrogen fixing and phosphate solubilizing bacteria on nodulation, plant growth and yield of chickpea. *Journal of Plant Nutrition,* **33**: 157-171.

Endris, S. and Mohammed, M.J. 2007. Nutrient acquisition and yield response of barley exposed to salt stress under different levels of potassium nutrition. *International Journal of Environmental Science and Technology,* **4**: 323-330.

Fallik, E. and Okon, Y. 1996. Inoculants of *Azospirillum brasilense*:biomass production, survival and growth promotion of *Setaria italica* and *Zeamays*. *Soil Biology and Biochemistry,* **28**: 123-126.

Fan, Q.J. and Liu, J.H. 2011. Colonization with arbuscular mycorrhizal fungus affects growth, drought tolerance and expression of stress-responsive genes in *Poncirus trifoliata*. *Acta Physiol Plant,* **33**: 1533–1542.

Figueiredo, M.V.B., Burity, H.A., Martinez, C.R. and Chanway, C.P. 2008. Alleviation of drought stress in common bean (*Phaseolus vulgaris* L.) by co-inoculation with *Paenibacillus polymyxa* and *Rhizobium tropici*. *Applied Soil Ecology,* **40**: 182-188.

Figueiredo, M.V.B., Seldin, L., de Arauio, F.F. and Mariano, R.L.R. 2010. Plant growth promoting rhizobacteria fundamentals and applications. In "Plant Growth and Health Promoting Bacteria, Ed. D.K. Aheshwari, **11**: 21-43.

Fitter, A.H. 1985. Functioning of vesicular - arbucular mycorrhizas under field conditions *New Phytol,* **99**: 257-265.

Galal, Y., El-Ghandour, I. and Al-Akel, E. 2002. Stimulation of wheat growth and nitrogen fixation through *Azospirillum* and *Rhizobium* inoculation: a field trial with ^{15}N techniques. *Development in Plant and Soil Science,* **92**:.*666-667.*

German, M.A., Burdman, S., Okon, Y. and Kigel, J. 2000. Effects of *Azospirillum brasilense* of root morphology of common bean (*Phaseolus vulgaris* L.) under different water regimes. *Biology and Fertility of Soils,* **32**: 294-264.

Gill, S.S. and Tuteja, N. 2010. Reactive oxygen species and antioxidant machinery in abiotic stress tolerance in crop plants. *Plant Physiology and Biochemistry,* **48**: 909-930.

Glick, B.R., Cheng, Z., Czarny, J. and Duan, J. 2007. Promotion of plant growth by ACC deaminase-producing soil bacteria. *European Journal of Plant Pathology,* **119**: 329-399.

Glick, B.R., Penrose, D.M. and Li, J. 1998. A model for the lowering of plant ethylene concentrations by plant growth-promoting bacteria. *Journal of Theoritical Biology,* **190**: 63-68.

Goenadi, D.H., Siswanto and Sugiarto, Y. 2000. Bioactivation of poorly soluble phosphate rocks with a phosphate solubilizing fungus'. *Soil Science Society of America Journal.* **64**: 927-932.

Goldstein, A.H. 1986. Bacterial solubilization of mineral phosphates: historical perspectives and future prospects'. *Amer ican Journal of Alternate Agriculture.* **1**: 57–65.

Grattan, S.R. and Grieve, C.M. 1999. Mineral nutrient acquisition and response of plants grown in saline environments. In: "*Handbook of Plant and Crop Stress*". Ed. M. Pessarakli, MarcelDekker Press Inc., New York, pp. 203-229.

Gray, E.J. and Smith, D.L. 2005. Intracellular and extracellular PGPR:commonalities and distinctions in the plant-bacterium signalling pr ocess. *Soil Biology and Biochemistry,* **37**: 395-412.

Goicoechea, N., Szalai, G., Antolín, M.C., Sánchez-Díaz, M. and Paldi, E. 1998. Influence of arbuscular mycorrhizae and *Rhizobium* on free polyamines and proline levels in water-stressed alfalfa. *Journal of Plant Physiology,* **153**: 706–711.

Grichko, E.J. and Glick, B.R. 2001. Amelioration of flooding stress by ACC deaminase-containing plant growth-promoting bacteria. *Plant Physiology and Biochemistry,* **39**: 11-17.

Hamaoui, B., Abbadi, J.M., Burdman, S., Rashid, A., Sarig, S. and Okon, Y. 2001. Effects of inoculation with *Azospirillum brasilense* on chickpeas (*Cicer arietinum*) and faba beans (*Vicia faba*) under different growth conditions. *Agronomic,* **21**:553-560.

Hamdia, A.B.E., Shaddad, M.A.K. and Doaa, M.M. 2004. Mechanism of salt tolerance and interactive effects of *Azospirillum brasilense* inoculation on maize cultivars grown under salt stress conditions. *Plant Growth Regulation,* **44**:165-174.

Hardoim, P.R., Van Overbeek, L.S. and Van Elsas, J.D. 2008. Properties of bacterial endophytes and their proposed role in plant growth. *Trends in Microbiology,* **16**: 463-471.

Hare, P.D. and Cress, W.A. 1997. Metabolic implications of stress-induced proline accumulation in plants. *Plant Growth Regulation,* **21**: 79-102.

Harrison, M.J. 2005. Signaling in the arbuscular mycorrhizal symbiosis. *Ann. Rev. Microbiol.* **59**:19 42.

Hayat, R., Ali, S., Amara, U., Khalid, R. and Ahmed, I. 2010. Soil beneficial bacteria and their role in plant growth promotion: *a review. Annals of Microbiology,* **60**:579-598.

Heidari, M. and Jamshid, P. 2010. Interaction between salinity and potassium on grain yield, carbohydrate content and nutrient uptake in pearl millet. *Journal of Agricultural and Biological Science,* **5**: 39-46.

Herandez, A.T., Bustilos-cristales, M.R., Jimnez-salgado, T., Caballero-Mellado, J. and Fuentez-Ramirez, L.E. 2000. Natural endophytic occurrence of *Acetobacter* diazotrophicus in pineapple plants'. *Microbial Ecology.* **39**: 49–55.

Hetrick, B.A.D. 1991. Mycorrhizas and root architecture.*Experientia,***47**: 355-362. Hinsinger, P. 2001. Bioavailability of soil inorganic P in the rhizosphere as affected by root-induced chemical changes *a review. Plant and Soil,* **237**: 173-195.

Holguin, G. and Bashan, Y. (1996): 'Nitrogen-fixation by *Azospirillum brasilense* Cd is promoted when co-cultured with a mangrove rhizosphere bacterium (*Staphylococcus* sp)'. *Soil Biology and Biochemistry.* **28**: 1651–1660.

Ismail, S. A. 1993. Keynote Papers and Extended Abstracts. *Congress on traditional sciencesand technologies of India, I.I.T., Mumbai.* **10**: 27-30.

Ismail, S.A. 2005. The Earthworm Book. *Other India Press, Mapusa, Goa.* 101.

Itzigsohn, R., Burdman, S., Okon, Y., Zaddy, E., Yonatan, R. and Pereyolotsky, A. 2000. Plant-growth promotion in natural pastures by inoculation with *Azospirillum brasilense* under suboptimal growth conditions. *Arid Soil Research and Management,* **14**:151-158.

Jha, Y., Subramanian, R.B. and Patel, S. 2010. Combination of endophytic and rhizospheric plant growth promoting rhizobacteria in *Oryza sativa* shows higher accumulation of osmoprotectant against saline stress. *Acta Physiologiae Plantarum,* **33**: 797-802.

Jimnez-Salgado, T., Fuentes-Ramirez, L.E., Tapia- Herandez, A., Mascarua, M.A., Martinez-Romero, E. and Caballero-Mellado, J. 1997. Coffea Arabica L.; a new host plant for Acetobacter diazotrophicus and isolation of other nitrogen fixing cyanobacteria'. *Applied and Environmental Microbiology.* **63**: 3676–3683.

Junghans, U., Polee, A., Duchting, P., Weiler, E., Kuhlman, B., Grubber, F. and Teichmann, T. 2006. Adaptation to high salinity in poplar involves changes in xylem anatomy and auxin physiology. *Plant, Cell and environment,* **29**:1519-1531.

Kale, R. D. 1998. Earthworm Cinderella of Organic Farming. *Prism Book Pvt Ltd, Bangalore, India.* 88.

Kale, R. D., Bano, K. and Krishnamoorthy, R. V. 1982. Potential of *Perionyx excavatus* for utilising organic wastes. *Pedobiologia.* **23**: 419-425.

Kang, B.G., Kim, W.T., Yun, H.S. and Chang, S.C. 2010. Use of plant growth promoting rhizobacteria to control stress responses of plant roots. *Plant Biotechnology Reports,* **4**: 179-183.

Kasiamdar, R.S., Smith, S.E., Smith, F.A. and Scott, E.S. 2001. Influence of the mycorrhizal fungus, *Glomus coronatum*, and soil phosphorus on infection and disease caused by binucleate Rhizoctonia and *Rhizoctonia solani* on mung bean (*Vigna radiata*)'. *Plant and Soil,* **238**: 235–244.

Kausar, R. and Shahzad, S.M. 2006. Effect of ACC-deaminase containing rhizobacteria on growth promotion of maize under salinity stress. *Journal of Agriculture and Social Science,* **2**: 216-218.

Kavi Kishor, P.B., Sangam, S., Amrutha, R.N., Sri Laxmi, P., Naidu, K.R., Rao, K.R.S.S., Rao, S., Reddy, K.J., Theriappan, P. and Sreenivasulu, N. 2005. Regulation of proline biosynthesis, degradation, uptake and transport in higher plants: its implications in plant growth and abiotic stress tolerance. *Current Science,* **88**: 424-438.

Kaya, C., Kirvak, H. and Giggs, D. 2001. Enhancement of growth and normal growth parameters by foliar application of potassium and phosphorus on tomato cultivars grown at high (NaCl) salinity. *Journal of Plant Nutrition,* **24**: 357-367.

Kaya, C., Kirvak, H., Giggs, D. and Saltali, K. 2002. Supplementry calcium enhances plant growth and fruit yield in strawberry cultivars grown at high (NaCl) salinity. *Scientia Horticulturae,* **93**: 65-74.

Khan, A.A., Jilani, G., Akhtar, M.S., Naqyi, S.M.S. and Rasheed, M. 2009. Phosphorous solubilizing bacteria:occurrence, mechanism and their role in crop production. *Research Journal of Agriculture and Biological Sciences*, **1**: 48-58.

Khoshgoftarmanesh, A.H., Schulin, R., Claney, R.L., Daneshbakhsh, B. and Afyuni, M. 2010. Micronutrient-efficient genotypes for crop yield and nutritional quality in sustainable agriculture. *A review.Agronomy For Sustainable Development*, **30**: 83-107.

Kijne, J.W. 2006. Abiotic stress and water scarcity: identifying and resolving conflicts from plant level to global level. *Field Crop Research,* **97**: 3-18.

Kohler, J., Hernandez, J.A., Caravaca, F. and Roldan, A. 2008. Plant-growth-promoting rhizobacteria and arbuscular mycorrhizal fungi modify alleviation biochemical mechanisms in water-stressed plants. *Functional Plant Biology,* **35**: 141-151.

Kohler, J., Hernandez, J.A., Caravaca, F. and Roldan, A. 2009. Induction of antioxidant enzymes is involved in the greater effectiveness of a PGPR versus AM fungi with respect to increasing the tolerance of lettuce to serve salt stress. *Environmantal and Experimental Botany,* **65**: *245-252*.

Kohler, J., Caravaca, F. and Roldan, A. 2010. An AM fungus and a PGPR intensify the adverse effects of salinity on the stability of rhizosphere soil aggregates of *Lactuca sativa. Soil Biology and Biochemistry,* **42**: 429-434.

Kucey, R.M.N., Janzen, H.H. and Legget, M.E. (1989): Microbial mediated increase in plant available phosphorus'. *Advances in Agronomy.* **42**:199–228.

Kumar, V., Chandra, A., and Singh, G. 2010. Efficacy of fly-ash based biofertilizers vs perfected chemical fertilizers in wheat (*Triticum aestivum*). *Int. J. Sci. Technol.,* **2**: 31-35.

Kundu, B.S., Sangwan, P., Sharma, P.K. and Nandwal, A.S. 1997. Response of pearlmillet to phytohormones produced by *Azospirillum brasilense. Indian Journal of Plant Physiology,* **2**:101-104.

Lalitha, R., Fathima, K. and Ismail, S. A. 2000. Impact of biopesticides and microbial fertilizers on productivity and growth of *Abelmoschus esculentus. Vasundhara: The Earth*, **1-2:** 4-9.

Levy, I. and Krikun, J. 1980. Effect of vesicular-arbuscular mycorrhiza in *Citrus jambhiri* water relations. *New Phytol,* **85**: 25-32.

Liddycoat, S.M., Greenberg, B.M. and Wolyn, D.J. 2009. The effect of plant growth-promoting rhizobacteria an asparagus seedling and germinating seeds subjected to water stress under greenhouse conditions. *Canadian Journal of Microbiology,* **55**: 388-394.

Loganathan, P. and Nair, S. 2003. Crop-specific endophytic colonization by a novel, salt tolerant, nitrogen fixing and phosphate solubilizing *Gluconacetobacter* sp. from wild rice'. *Biotechnology letter.* **25**: 497–501.

Loganathan, P., Sunitha, R., Parida, A.K. and Nair, S. 1999. 'Isolation and characterization of two genetically distant groups of *Acetobacter diazotrophicus* from a new host plant *Eleusine coracana L.'Applied Microbiology.* **87**: 167–172.

Lovelock, C.E., Wright, S.F., Clark, D.A. and Ruess, R.W. 2004. Soil stocks of glomalin produce by arbuscular mycorrhizal fungi across a tropical rain forest landscape'. *Journal of Ecology.* **92**: 278–287.

Lutgtenberg, B. and Kamilova, F. 2009. Plant growth promoting rhizobacteria. *Annual Review of Microbiology,* **63**:541-556.

Madhaiyan, M., Poonguzhali, S. and Sa, T. 2007. Metal tolerating methylotropic bacteria reduces nickel and cadmium toxicity and promotes plant growth of tomato (*Lycopersicon esculentum* L.).*Chemosphere,* **69**: 220-228.

Mahajan, S. and Tutcja, N. 2005. Cold, salinity and drought stresses: an overview. *Archives of Biochemistry and Biophysics,***444**: 139-158.

Mahdi, S.S., Hassan, G.I., Samoon S.A., Rather, H.A., Dar, Showkat A. and Zehra, B. 2010: 'Bio-Fertilizers In Organic Agriculture'. *Journal of Phytology,* **2(10):** 42–54.

Malamy, J. 2005. Intrinsic and environmental response pathways that regulate root system architecture. *Plant, Cell and Environment,* **28**: 67-77.

Martinez, V. and Lauchli, A. 1994. Salt-induced inhibition of phosphate-uptake in plants of cotton (*Gossipium hirsutum* L.). *New Phytologist,* **126**: 609-614.

Marulanda, A., Azcon, R., Chaumont, F., Ruiz-Lozano, J.M. and Aroca, R. 2010. Regulation of plasma membrane aquaporins by inoculation with *Bacillus megaterium* strain in maize (*Zea mays* L.) plants under unstressed and salt stressed conditions. *Planta,* **232**: 533-543.

Mastretta, C., Taghavi, S., Vander Lelie, D., Mengoni, A., Galardi, F., Gonelli, C., Barac, T., Boulet, J., Weyens, N. and Vangronsveld, J. 2009. Endophytic bacteria from seeds of *Nicotiana tobacum* can reduce cadmium phytotoxicity. *International Journal of Phytoremediation,* **11**: 251-267.

Mayak, S., Tirosh, T. and Glick, B.R. 2004. Plant growth-promoting bacteria confer resistance in tomato plants to salt stress. *Plant Physiology and Biochemistry,* **42**: 565-572.

McGonigle, T.P. and Miller, M.H. 1999.'Winter survival of extra radical hyphae and spores of *Arbuscular* mycorrhizal fungi in the field'. *Applied Soil Ecology.* **12**: 41–50.

Meenakshisundaram, M. and Karrupagnaniar, Santhaguru. 2010. Isolation and nitrogen fixing efficiency of a novel endophytic diazotroph *Gluconacetobacter diazotrophicus* associated with *Saccharum officinarum* from southern districts of Tamilnadu'. *International Journal of Biology and Medical Research.* **1(4):** 298–300.

Minaxi, Saxena, J., Chandra, S. and Nain, L. 2013. Synergistic effect of phosphate solubilizing rhizobacteria and arbuscular mycorrhiza on growth and yield of wheat plants. *Journal of Soil Science and Plant Nutrition,* **13(2):** 511-525.

Miller, G., Susuki, N., Ciftci-Yilmaz, S. and Mittler, R. 2010. Reactive oxygen species homeostasis and signaling during drought and salinity stresses. *Plant, Cell and Environment,* **33**: 453-467.

Mittler, R. 2002. Oxidative stress, antioxidants and stress tolerance. *Trends in Plant Science,* **7**: 405-410.

Mittler, R. 2006. Abiotic stress, the field environment and stress combination. *Trends in Plant Science,* **11(1):** 15-19.

Mnasri,B., Aouani, M.E. and Mhamdi, R. 2007. Nodulation and growth of common bean (*Phaseolus vulgaris*) under water deficiency. *Soil Biology and Biochemistry,***39**: 1744-1750.

Molla, A.H., Shamsuddin, Z.H. and Saud ,H.M. 2001b. Mechanism of root growth and promotion of nodulation in vegetable soybean by *Azospirillum brasilense. Communication in Soil Science and Plant Analysis*, **32**:2177-2187.

Molla, A.H., Shamsuddin, Z.H., Halimi, M.S., Morziah, M. and Puteh, A.B. 2001a. Potential for enhancement of root growth and nodulation of soybean co-inoculated with *Azospirillum* and *Bradyrhizobium* in laboratory systems. *Soil Biology and Biochemistry*, **33**:457-463.

Mostajeran, A., Amooaghaie, R. and Emtiazi, G. 2002. Root hair density and deformation of inoculated roots of wheat cultivars by *Azospirillum brasilense* and role of IAA in this phenomenon. *Iranian Journal of Biology,* **13**: 18-28.

Muller, J.L. 2011. Auxin conjugates: their role for plant development and in the evolution of land plants. *Journal of Experimental Botany,* **62**: 1757-1773.

Munns, R. and Tester, M. 2008. Mechanism of salinity tolerance. *Annual Review of Plant Biology,* **59**: 651-681.

Nadeem, S.M., Zahir, Z.A., Nayeed, M. and Arshad, M. 2007. Preliminary investigations on inducing salt tolerance in maize through inoculation with rhizobacteria containing ACC deaminase activity. *Canadian Journal of Microbiology,* **53**: 1141-1149.

Nadeem, S.M., Zahir, Z.A., Nayeed, M., Asghar, H.N. and Arshad, M. 2010. Rhizobacteria capable of producing ACC-deaminase activity may mitigate salt stress in wheat. *Soil Science of American Journal,* **74**: 533-542.

Navarro, J.M., Botella, M.A., Cerda, A. and Martinez, V. 2001. Phosphorus uptake and translocation in salt-stressed melon plants. *Journal of Plant Physiology,* **158**: 375-381.

Nelson, C.E. and Safir, G.R. 1982. Increased drought tolerance of mycorrhizal onion plants caused by improved phosphorous nutrition. *Planta.* **59**: 401-413.

Nye, P.H. and Tinker, P.B. 1977. In "Solute movement in the soil-root systems". *Blackwell Scientific Publications, London.*

Oliveira, R.G.B. and Drozdowicz, A. 1987. Inhibition of bacteriocin producing strains of *Azospirillum lipoferum* by their own bacteriocin'. *Zentralblatt für Mikrobiologie.* **142**: 387–390.

Oliveira, C.A., Alyes, V.M.C., Marriel, I.E., Gomes, E.A., Scotti, M.R., Carneiro, N.P., Guimaraes, C.T.,

Schaffert, R.E. and Scotti, M.R. 2009. Phosphate solubilizing microorganisms isolated from rhizosphere of maize cultivated in an oxisol of the Brazilian Cerrado Biome. *Soil Biology and Biochemistry,* **41**: 1782-1787.

Olsson, P.A., Thingstrup, I., Jakobsen, I. and Baath, E, 1999. Estimation of the biomass of *Arbuscular* mycorrhizal fungi in a linseed field'. *Soil Biology and Biochem*istry, **31:** 1879–1887.

Omar, M.N.A., Osman, M.E.H., Kasim, W.A. and Abd-Daim El,I.A. 2009. Improvement of salt tolerance mechanisms of barley cultivated under salt stress using *Azospirillum brasilense. Tasks for Vegetation Science,* **44**:133-147.

Pacovsky, R.S. 1990. Development and growth effects in the Sorghum-*Azospirillum* association. *Journal of Applied Bacteriology,* **68**: 555-563.

Panwar, J. D. S., Ompal, S., and Singh, K. 2000. Responce of *Azospirillum* and *Bacillus* on growth and yield of wheat under field conditions. *Indian Journal of Plant Physiology,* **5**: 108-110.

Panwar, J.D.S. and Sirohi, G.S. 1989. *Azospirillum* : An important biofertilizers for improving crop productivity. *Farmers and Parliament,* **24**: 23-24.

Panwar, J.D.S. and Swarnalakshmi, K. 2005. PGPR: A New Group of Biofertilizers for Sustainable Crop Production. In "*Developments in Physiology, Biochemistry and Molecular Biology of Plants*," Eds. Bandana Bose and A. Hemantrajan, **1**: 159-180, New India Publishing Agency, New Delhi.

Panwar, J.D.S. and Swarnalakshmi, K. 2005. Biological Nitrogen Fixation in Cereals and Pulses, In "*Developments in Physiology, Biochemistry and Molecular Biology of Plants*". Eds. Bandana Bose and A. Hemantrajan, **1**:125-158, New India Publishing Agency, New Delhi.

Parida, A.K. and Das, A.B. 2004. Effect of NaCl stress on nitrogen and phosphorus metabolism in a true mangrove *Bruguiera paryiflora* grown under hydroponic culture. *Journal of Plant Physiology,* **161**: 921-928.

Parida, A.K. and Das, A.B. 2005. Salt tolerance and salinity effects on plants. *A review. Ecotoxicology and Environmental Safety* , **60**: 324-349..

Parmar, N. and Dadarwal, K.R. 1999. Stimulation of nitrogen fixation and induction of flavonoids like compounds by rhizobacteria. *Journal of Applied Microbiology,* **86**:36-44.

Paszkowski, U., Kroken, S., Roux, C. and Briggs, S.P. 2002. Rice phosphate transporters include an evolutionarily divergent gene specifically activated in *Arbuscular mycorrhizal* symbiosis. *Ptoc. Natl.Acad.Sci.U.S.A.* **99:**13324-13329.

Pereyra, M.A., Zlazar, C.A. and Barassi, C.A. 2006. Root phospholipids in *Azospirillum*-inoculated wheat seedlings exposed to water stress. *Plant physiology and Biochemistry,* **44**: 873-879.

Perez-Torres, C.A., Lopez-Bucio, J., Cruz-Ramirez, A., Ibarra-Laclette, E., Dharmasiri, S., Estelle, M. and Herrera-Estrella, L. 2008. Phosphate availability alters lateral root development in *Arabidopsis* by modulating auxin sensitivity via a mechanism involving the TIR1 auxin receptor. *The Plant Cell,* **20(12):** 258-3272.

Pishchik, V.N., Provorov, N.A., Vorobyoy, N.I., Chizevskaya, E.P., Safronova, V.I., Tuev, A.N. and Kozhemvakoy, A.P. 2009. Interactions between plants and associated bacteria in soils contaminated with heavy metals. *Microbiology,* **78**: 785-793.

Ponmurugan, P. and Gopi, C. 2006. Distribution pattern and screening of phosphate solubilizing bacteria isolated from different food and forage crops'. *Journal of Agronomy.* **5:** 600–604.

Potters, G., Pasternak, T.P., Guisez, Y. and Jansen, M.A.K. 2009. Different stresses, similar morphogenic responses: integrating a plethora of pathways. *Plant, Cell and Environment,* **32:** 158-169.

Potters, G., Pasternak, T.P., Guisez, Y., Palmae, K.J. and Jansen, M.A.K. 2010. Stress-induced morphogenic responses, growing out of trouble. *Trends in Plant Science,* **12**: 98-105.

Price, N.S., Roncadori, R.W. and Hussey, R.S. 1989. Cotton root growth as influenced by Phosphorus –nutrition and vesicular arbuscular mycorrhizas.*New Phytol,* **111:** 61-66.

Preeti, Kumar, S. and Panwar, J.D.S. 2011. Effect of *Rhizobium* and AM interaction on growth and yield of Urd bean cultivars under rainfed condition. *Journal of Plant Development Sciences,* **3(3-4):**283-286.

Preeti, Kumar, S. and Panwar, J.D.S. 2012. Synergistic effect of *Rhizobium* and AM fungi interaction on photosynthesis, RPA and grain quality in urd bean (*Vigna mungo* (L.) Hepper) under rainfed conditions. *Journal of Plant Development Sciences,* **4(2):** 325-328.

Preeti, Kumar, S. and Panwar, J.D.S. 2013. Mycorrhiza: its potential use for augmenting soil fertility and crop productivity. *Advances in Plant Physiology,* **14**: 111-172.

Remans, R., Beebe, S., Blair, M., Manrique, G., Tovar, E., Rao, I.,Croonenborghs, A.,Torres-Gutierrez, R., El-Howeity, M., Michiels, J. and Vanderleyden, J. 2008a. Physiological and genetic analysis of root responsiveness to auxin-producing plant growth-promoting bacteria in common bean (*Phaseolus vulgaris* L.). *Plant and Soil,* **302:** 149-161.

Remans, R., Croonenborghs, G.R.T., Michiels, J. and Vanderleyden, J. 2007. Effect of plant growth-promoting rhizobacteria on nodulation of *Phaseolus vulgaris* L. are dependent on plant P nutrition. *European Journal of Plant Pathology,* **119**: 341-351.

Remans, R., Ramaekers, L., Shelkens, S., Hernandez, G., Garcia, A., Reyes, G.L., Mendez, N., Toscano, V., Mullin, M., Galvez, L. and Vanderleyden, J. 2008b. Effect of *Rhizobium-Azospirillum* co-inoculation on nitrogen fixation and yield of two contrasting *Phaseolus vulgaris* L. genotypes cultivated across different environments in Cuba. *Plant and Soil,* **312**: 25-37.

Richardson, A.E., Barea, M.J., McNeil, A.M. and Prigent-Cobaret, C. 2009. Acquisition of phosphorus and nitrogen in the rhizosphere and plant growth promotion by microorganisms. *Plant and Soil,* **321**: 305-339.

Rillig, M.C., Wright, S.F., Nichols, K.A., Schmidt, W.F. and Torn, M.S. 2001. Large contribution of *Arbuscular mycorrhizal* fungi to soil carbon pools in tropical forest soils'. *Plant and Soil,* **233**: 167–177.

Rodriguez, H. and Fraga, R. 1999. Phosphate solubilizing bacteria and their role in plant growth promotion. *Biotechnologies Advances,* **17**: 319-339.

Rogers, M.C., Grieve, C. and Shannon, M. 2003. Plant growth and ion relations in lucerne (*Medicago sativa* L.) in response to the combined effects of NaCl and P. *Plant and Soil,* **253**: 187-194.

Rokhzadi, A. and Toashih, V. 2011. Nutrient uptake and yield of chickpea (*Cicer arietinum* L.) inoculated with plant growth promoting rhizobacteria. *Australian Journal of Crop Science,* **5**: 44-48.

Ruiz-Sánchez, M., Aroca, R., Muñoz, Y., Armada, E., Polón, R. and Ruiz-Lozano, J.M. 2010. The arbuscular mycorrhizal symbiosis enhances the photosynthetic efficiency and the antioxidative response of rice plants subjected to drought stress. *Journal of Plant Physiology,* **167**: 862–869.

Sandhya, V., Ali, Sk.Z., Grover, M., Reddy,G. and Venkateswarlu, B. 2010. Effect of plant growth promoting *Pseudomonas spp*. on compatible solutes, antioxidant status and plant growth of maize under drought stress. *Plant Growth Regulation,* **62**: 21-30.

Saravanakumar, D. and Samiyappan, R. 2007. ACC deaminase from *Pseudomonas fluorescens* mediated saline resistance in groundnut (*Arachis hypogea*) plants. *Journal of Applied Microbiology,* **102**: 1283-1292.

Saravanan, V.S., Madhaiyan, M., Osborne, Jabez., Thangaraju, M. and Sa, T.M. 2008. Ecological Occurrence of *Gluconacetobacter diazotrophicus* and Nitrogen-fixing Acetobacteraceae Members: Their Possible Role in Plant Growth Promotion'. *Microboial Ecology.* **55**: 130–140.

Sarig, S., Kapulnik, Y. and Okon, Y. 1986. Effect of *Azospirillum* inoculation on nitrogen fixation and growth of several winter legumes. *Plant and Soil,* **90**: 335-342.

Sarig, S., Okon, Y. and Blum, A. 1992. Effect of *Azospirillum brasilense* inoculation on growth dynamics and hydraulic conductivity of *Sorghum bicolor* roots. *Journal of Plant Nutrition,* **15**: 805-819.

Sevilla, M. and Kennedy, C. 2000. Colonization of Rice and other cereals by *Acetobactor diazotrophicus*, an endophyte of sugarcane. In "The Quest for Nitrogen Fixation in Rice" *Eds. J.K. Ladha and P.M. Reddy, International Rice Research Institute,Manila, Philippines,* **15**: 1-165.

Shah, S., Karkhanis, V. and Desai, A. 1992. 'Isolation and characterization of siderophore, with antimicrobial activity, from *Azospirillum lipoferum* M'. *Current Microbiology,* **25**: 34–35.

Sheng, X.F., Xia, J.J., Jiang, C.Y., He, L.Y. and Qian, M. 2008. Characterization of heavy metal-resistant endophytic bacteria from rape (*Brassica napus*) roots and their potential in promoting the growth and lead accumulation of rape. *Environmental Pollution,* **15**: 1164-1170.

Siddikee, M.A., Chauhan, P.S., Anandham, R., Han, G.H. and Sa, T. 2010. Isolation, characterization and use for plant growth promotion under salt stress, of ACC deaminase-producing halotolerant bacteria derived from costal soil. *Journal of Microbiology and Biotechnology,* **20**: 1577-1584.

Sindhu, S.S., Gupta, S.K., Suneja, S. and Dadarwal, K.R. 2002a: Enhancement of green gram nodulation and growth by Bacillus species. *Biologia Plantarum,* **45**: 117-120.

Spaepen, S., and Vanderleyden, J. 2010. Auxin and plant-microbe interactions. *Cold Spring Harbor Perspectives in Biology, doi: 10.1101/cshperspect.aoo1438.*

Spaepen, S., Boddelaere, S., Croonenborghs, A. and Vanderleyden, J. 2008. Effect of *Azospirillum brasilense* indole-3-acetic acid production on inoculated wheat plants. *Plant and Soil,* **312**: 15-23.

Spaepen, S., Vanderleyden, J. and Remans, R. 2007. Indole-3-acetic acid in microbial and microorganisms-plant signaling. *FEMS Microbiology Reviews,* **31**: 425-448.

Subba Roa, N.S. 2001. An appraisal of biofertilizers in India. In "The Biotechnology of Biofertilizers" Ed. *S.Kannaiyan,* Narosa Pub. House, New Delhi.

Sudhakar, P., Gangwar, S.K., Satpathy, B., Sahu, P.K., Ghosh, J.K. and Saratchandra, B. 2000. 'Evaluation of some nitrogen fixing bacteria for control of foliar diseases of mulberry' (*Morus alba*). *Indian Journal of Sericulture.* **39**: 9–11.

Swaine, E.K., Swaine, M.D. and Kilham, K. 2007. Effects of drought on isolates of *Bradirhizobium elkanii* cultured from *Albizia adianthifolia* seedlings on different provenances. *Agroforestry Systems,* **69**: 135-145.

Szabados, L. and Savoure, A. 2009. Proline: a multifunctional amino acid. *Trends in Plant Science,* **15**: 89-97.

Sziderics, A.H., Rasche, F., Trognitz, F., Sessitsch, A. and Wilhelm, E. 2007. Bacterial endophytes contribute to abiotic stress adaptation in pepper plants (*Capsicum annum* L.). *Canadian Journal of Microbiology,* **53**: 1195-1202.

Tapia-Hernández, A., Mascarúa-Esparzá, M.A. and Caballero- Mellado, J. 1990. 'Production of bacteriocins and siderophore- like activity in *Azospirillum brasilense'. Microbios.* **64**: 73–83.

Tchebotar, V.K., Kang, U.G., Jr Asis, C.A. and Akao, S. 1998. The use of GUS-reporter gene to study the effect of *Azospirillum-Rhizobium* co-inoculation on nodulation of white clover. *Biology and Fertility of Soils,* **27**: 349-352.

Teale, W.D., Paponoy, I.A. and Palmae. K. 2006. Auxin in action: signalling, transport and control of plant growth and development. *Nature Reviews Molecular Cell Biology,* **7**: 847-859.

Terre, S., Asch, F., Padham, J., Sikora, R.A. and Becker, M. 2007. Influence of root zone bacteria on root iron plaque formation in rice subjected to iron toxicity. *In:* "Utilization of Diversity in Land Use Systems: Sustainable and Organic Approaches to Meet Human Needs", Ed. E. Tielkes, Tropentag, Witzenhausen, Germany, pp. 446

Tokala, R.K., Strap, J.L., Jung, C.M., Crawford, D.L., Salove, H., Deobald, L.A., Bailey, F.J.and Morra, M.J. 2002. Novel plant-microbe rhizosphere interaction involving *S.lydicus* $WYEC_{108}$ and the pea plant (*Pisum sativum*). *Applied and Environmental Microbiology,* **68**: 2161-2171.

Vardharajula, S., Ali, S.Z., Grover, M., Reddy, G. and Bandi,V. 2011. Drought-tolerant plant growth promoting *Bacillus spp.*: effect on growth, osmolytes and antioxidant status of maize under drought stress. *Journal of Plant Interactions,* **6**: 1-14.

Verbruggen, N. and Hermans, C. 2008. Proline accumulation in plants: *a review. Amino Acids,* **35**: 753-759.

Verma, L.N. 1993. Biofertiliser in agriculture'. In "Organics in soil health and crop production" Ed. P.K. Thampan, Peekay Tree Crops Development Foundation, Cochin, India, 152–183.

Verma, J., Yadav, J. and Tiwari, K.N. 2010. Impact of plant growth promoting rhizobacteria on crop production. *International Journal of Agricultural Research,* **11**: 954-983.

Volpin, H. and Kapulnik, Y. 1994. Interaction of *Azospirillum* with beneficial soil microorganisms, 111-118. In "*Azospirillum*/ Plant associations", Ed. Y. Okon, Boca Raton, Fla.

Wolter, H. and Jurgens, G. 2009. Survival of the flexible: hormonal growth control and adaptation in plant development. *Nature Reviews Genetics,* **10**: 305-317.

Yadegari, M. and Rahmani, A. 2010. Evaluation of bean (*Phaseolus vulgaris*) seeds inoculation with *Rhizobium phaseoli* and plant growth promoting rhizobacteria (PGPR) on yield and yield components. *African Journal of Agricultural Research.* **5**: 792-799.

Wu, Q.S., Xia, R.X. and Zou, Y.N. 2006a. Reactive oxygen metabolism in mycorrhizal and non mycorrhizal citrus (*Poncirus trifoliata*) seedlings subjected to water stress. *Journal of Plant Physiology,* **163**: 1101–1110.

Wu, Q.S., Zou, Y.N. and Xia, R.X. 2006b. Effects of water stress and arbuscular mycorrhizal fungi on reactive oxygen metabolism and antioxidant production by citrus (*Citrus tangerine*) roots. *European Journal of Soil Biology,* **42**: 166–172.

Yadegari, M., Rahmani, A., Noormohammadi, G. and Ayneband, A. 2010. Plant growth promoting rhizobacteria increase growth, yield and nitrogen fixation in *Phaseolus vulgaris*. *Journal of Plant Nutrition,* **33**: 1733-1743.

Yahlom, E., Okon, Y. and Dovrat, A. 1989. Possible mode of action of *Azospirillum brasilense* strain Cd on the root morphology and nodule formation in burr medic (*Medicago polymorpha*). *Canadian Journal of Microbiology,* **36**: 10-14.

Yazdani, M., Bahmanyar, M.A., Pirdashti, H. and Esmaili, M.A. 2009. Effect of phosphate solubilization microorganisms (PSM) and plant growth promoting rhizobacteria (PGPR) on yield and yield components of corn (*Zea mays* L.). *World Academy of Science, Engineering and Technology,* **49**: 90-92.

Yooyongwech, S., Phaukinsang, N., Cha-Um, S. and Supaibulwatana, K. 2013. *Arbuscular mycorrhiza* improved growth performance in *Macadamia tetraphylla* L. grown under water deficit stress involves soluble sugar and proline accumulation. *Plant Growth Regululations,* **69**: 285-293.

Zhang, F., Dashti, N., Hynes, R.K. and Smith, D.L. 1997. Plant growth-promoting rhizobacteria and soybean [*Glycine max* (L.) Merr] growth and physiology at suboptimal root zones temperatures. *Annals of Botany,* **79**: 243-249.

Zhang, H., Kim, M.S., Sun, Y., Dowd, S.E., Shi, H. and Pare, W. 2008. Soil bacteria confer plant salt tolerance by tissue-specific regulation of the sodium transporter HKT1. *Molecular Plant-Microbe Interactions,* **21**: 737-744.

Zhang, H., Murzello, C., Sun, Y., Kim-Seong, Mi., Xie, X., Jeter, R.M., Zak, J.C., Dowd, S.E. and Pare, P.W. 2008. Choline and osmotic-stress tolerance induced in Arabidopsis by the soil microbe *Bacillus subtilis* (GBo_3). *Molecular Plant-Microbe Interactions,* **23**: 1097-1104.

Zhu, Y.G. and Miller, R.M. 2003. 'Carbon cyclingby arbuscular mycorrhizal fungi in soil plant systems. *Trends Plant Sciences.* **8:** 407–409.

4

Abiotic Stress and Physiological Traits in Plants

A. Bhattacharya

Plants that are affected by various abiotic stresses react differently as a result such as the reduction in organ size or a change in the antioxidant enzyme system, and while drought and heat are often studied separately, in nature they often occur together, meaning it is essential to carry out joint studies. Furthermore, and related to the above when we consider high temperature stress, one must always consider the use of water and nutrients by the plant and other physiological functions. The effects of heat stress can also be analyzed by the technique that measures lipid peroxidation in polyunsaturated acids of the plasma membrane. The accumulation of metals in plants, affects the normal processes of plant metabolism. The study shows that there is a relationship between high metal contents in plants and their modified morphology: strong reduction of leaf thickness, modified parenchyma structure, and decreased mitochondria organization were ascertained, although toxic symptoms were apparently absent. Water-deficit stress has a significant effect on plants growth and development, with primary affects on plant structure, leaf morphology and cell ultrastructure. Physiological processes such as stomatal conductance, photosynthesis and respiration are consequently impaired with further implications on the metabolic functions such as carbohydrate and energy production as well as carbohydrate translocation and utilization. Even though plant possesses mechanisms to anticipate the negative effects of water-deficit stresses their protective capacity depend not only on the extent of the stress, but also on the timing of the stress as well as on the way the stress occurs (sudden or gradual). Yield reductions and its quality compromises are inescapable when water-deficit stress conditions override the plant's protective mechanisms. However, advances are being made at the physiological level entailing identification of exogenous or endogenous substances that can ameliorate the negative effects of drought and at the molecular level identification of genes involved with increased drought tolerance.

Abiotic stresses are often interrelated, either individually or in combination, they cause morphological, physiological, biochemical, and molecular changes that adversely affect plant growth and productivity, and ultimately yield. Heat, drought, cold, and salinity are the major abiotic stresses that induce severe cellular damage in plant species, including crop plants. Fluctuations in temperature occur naturally during plant growth and reproduction. However, extreme variations during hot summers can damage the intermolecular interactions needed for proper growth, thus impairing plant development and fruit set. The increasing threat of climate change is already having a substantial impact on agricultural production worldwide as heat waves cause significantly yield losses with great risks for future global food security (Christensen and Christensen, 2007). Climatological extremes including very high temperatures are predicted to have a general negative effect on plant growth and development, leading to catastrophic loss of crop productivity and resulting in wide spread famine. Future agricultural production and thus global food security will encounter additional challenges from human population growth (Bita and Gerats, 2013).

Populations from developing countries are likely to be the most seriously affected as nearly 50% rely entirely on agriculture. In addition, 75% of the world's poor live in rural areas. Thus, as population expands, crop production will have to be tailored to sustain food security and it has been suggested that world food production will have to increase by 70% to meet the demand of an expected population of 9 billion in 2050. Despite the predicted increase in global food production, at present there is a major global food deficit and the relative rates of yield increase for the major cereal crops are declining (Fischer and Edmeades, 2010). In many crop species, the effects of high temperature stress are more prominent on reproductive development than on vegetative growth and the sudden decline in yield with temperature is mainly associated with pollen infertility (Young *et al.* 2004; Zinn *et al.* 2010). Adding in the surging demand for food from the emerging economies such as China and India reveals an even more extreme challenge for plant breeders and farmers to such an extent that by 2050, the expected decline in calorie availability will aggravate malnutrition in children by 20% (Chhetri and Chaudhary, 2011).

4.1. Physiological Traits

Invasive plants often grow more vigorously and attain higher abundances in their introduced range compared to conspecifics in their native range (Bossdorf *et al.* 2005). To identify factors contributing to their invasive success, a number of studies have investigated morphological and physiological traits of invasive plants by comparing them with the native species they displace or non-invasive congeners (Reichard and Hamilton, 1997). In these studies and others, greater relative growth rates (RGR) of introduced species are generally associated with lower root : shoot ratios (RSR), higher specific leaf areas (SLA; leaf area per unit leaf mass), leaf area ratios (LAR; total leaf area per unit plant mass) and higher net CO_2 assimilation (A) as well as lower respiration costs (RD) (Wilsey and Polley, 2006). These functional traits, common to invasive plants, may reflect their inherent properties due to previous evolution in the native range before its introduction or new adaptation as a result of an evolutionary response to escape from natural enemies in the introduced range (Erfmeier and Bruelheide, 2005; Güsewell *et al.* 2006). Some invasive plants may be innately better competitors because they evolved in a more competitive environment (Callaway and Aschehoug, 2000; Davis *et al.* 2000). Once established in the introduced range, they may gain a systematic advantage over competitively inferior native plants. On the other hand, since allocation to defence may be as costly as herbivore damage (Baldwin *et al.* 1990), plants that escape their enemies in an introduced range could gain a selective benefit from decreasing their defensive investment. As a consequence, they may evolve to be fast-growing and low herbivore-defence plants (EICA hypothesis, Mooney and Cleland, 2001). The distinction between environmentally induced phenotypic differences and a genetic change could be revealed by common garden experiments in which both native and invasive individuals are grown together in the same environment (Erfmeier and Bruelheide, 2005; Güsewell *et al.* 2006).

While the majority of studies have compared invasive plants with natives or non-invasive congeners, only a few common garden studies have concentrated on variation in morphological and physiological traits between native and invasive populations of exotic plants (Bastlová and Kvêt, 2002; DeWalt *et al.* 2004; Erfmeier and Bruelheide 2004, 2005; Buschmann *et al.* 2005; Güsewell *et al.* 2006). A greenhouse study revealed that introduced Hawaiian and native Costa Rican populations of the tropical shrub *Clidemia hirta* displayed no significant differences in RGR, Amax or SLA (DeWalt *et al.* 2004). Common garden and greenhouse experiments showed significant but not always consistent differences in growth and reproductive characteristics between native and introduced ranges of invasive Brassicaceae species except *Bunias orientails* (Buschmann *et al.*2005). Güsewell *et al.* (2006) found that invasive European plants produced more shoots than native American plants of *Solidago gigantea*, but they did not differ in shoot size, leaf traits and litter decomposition. A four month greenhouse experiment indicated that total leaf area (TLA) and SLA were significantly greater for plants from invasive populations than from native populations of *Lythrum salicaria*, but no significant differences in LAR or A were found between them (Bastlová and Kvêt, 2002). In a study on *Rhododendron ponticum* provided evidence for a genetic shift in invasive populations towards an increased investment in growth relative to native populations (Erfmeier and Bruelheide, 2005). Chinese tallow tree (*Sapium sebiferum* L. Roxb., Euphorbiaceae, synonyms include *Triadica sebifera, Sapium' henceforth*) is native to China (Zhang and Lin, 1994), and has recently become a severe invader that aggressively displaces native plants and forms mono-specific stands in the south-eastern USA (Bruce *et al.* 1997). Results of studies on *S. sebiferum* generally support the EICA hypothesis (Rogers and Siemann, 2004, 2005; Zou *et al.* 2006), suggesting that *Sapium* has evolved to be a faster-growing, less herbivore-resistant plant in response to low herbivore loads in its introduced range. In a 14-year common garden study, if the EICA hypothesis holds for *Sapium*, however, native and invasive populations may also differ in some functional traits that typically characterize invasive plant species: biomass production, total number of leaves (TLN) and leaf area (TLA), relative stem height growth rate (RHR), relative growth rate (RGR), below- and above-ground biomass allocation (RSR), non-photosynthetic and photosynthetic tissue allocation (SLA and LAR), net CO_2 assimilation (A) and dark respiration rate (RD).

Tolerance to abiotic stresses is complex at the whole plant and cellular level (Ashraf and Harris, 2004, Munns and Teaster, 2008, Grewal, 2010). This is in part due to the complexity of interactions between stress factor and various molecular, biochemical and physiological phenomena affecting plant growth development (Zhu, 2002).

4.2. High Temperature Stress

The constantly rising global air temperature @ ~0.2 °C per decade is raising apprehension regarding crop productivity and food security. This is expected up to ~4.0 °C higher than the current level by 2100 (Rana *et al.* 2011; Rao *et al.* 2011; Srivastava *et al.* 2011). While temperature and other abiotic stresses are clearly limiting factors for crops cultivated on marginal lands, crop productivity everywhere is often at the mercy of random environmental fluctuations. Current speculation about global climate change is that most agricultural regions will experience more extreme environmental fluctuations (Solomon *et al.* 2007). Effect of high temperature stress has been extensively studied and reviewed (Chinnusamy *et al.* 2007; Wahid *et al.* 2007). With industrialization, natural environment deterioration and climate change, heat stress has become an increasingly important factor affecting crop growth, and it is thought that crop production may be severely affected by an increase in mean global temperature (Grover *et al.* 2009; Han *et al.* 2009). For example, models indicate that a mean temperature rise of 1°C could reduce wheat yields by 10% in some regions (1a) which could be highly relevant in the context of estimated projections for global mean surface temperature rise of up to 4.8°C published by the

Intergovernmental Panel on Climate Change (1b), due to multiple factors that appear to include increases in the concentration of greenhouse gases such as carbon dioxide, methane, nitrous oxide and chlorofluorocarbons.

Various physiological injuries have been observed under elevated temperatures, such as scorching of leaves and stems, leaf abscission and senescence, shoot and root growth inhibition or fruit damage, which consequently lead to decreased plant productivity (Vollenweider and Günthardt-Goerg, 2005). In many cases, plant architecture changes and hypocotyls and petioles elongate resembling the morphological responses of shade avoidance (Tian *et al.* 2009). However, high temperatures reduce plant growth by affecting the shoot net assimilation rates and thus the total dry weight of the plant (Wahid *et al.* 2007). The impact of temperature stress is a complex function of intensity, duration, and rate of temperature change (Wahid *et al.* 2007; Thakur *et al.* 2010). Both hot and cold stresses can alter multiple aspects of cellular physiology. For example, temperature can dramatically change membrane fluidity, nucleic acid and protein structures, as well as metabolite and osmolyte concentrations (Wang *et al.* 2003; Chinnusamy *et al.* 2007). Both hot and cold temperature stresses induce the production of ROS (reactive oxygen species), which at elevated concentrations will result in oxidative damage and, potentially, cell death (Apel and Hirt, 2004).

High temperature stress causes various physiological changes in plants such as scorching of leaves and stems, leaf abscission and senescence, shoot and root growth inhibition or reduction in number of flowers, pollen tube growth and pollen infertility, fruit damage, leading to catastrophic loss of crop yield (Bita and Gerats, 2013; Teixeria *et al.* 2013; Song *et al.* 2013; Hemantaranjan *et al.* 2014). Heat stress affects the photosynthesis, respiration, water relations and membrane stability, and modulates levels of hormones, and primary and secondary metabolites (Hemantaranjan *et al.* 2014). While comparing the photosynthetic rate and photosynthesic efficiency in chickpea under normal and late sowing condition Bhattacharya and Singh (1999) reported that under late sown condition, photosynthetic efficiency decline drastically during both pre and post flowring stages. Under late sown condition of chickpea genotypes it was reported that amongst functional traits, leaf dry matter allocation at different growth stages expressed maximum association with seed yield (Ganguly *et al.* 1999). Heat stress impairs the stability of proteins, membrane integrity, RNA and activity of enzymes in chloroplast and mitochondria, resulting in an imbalance in the metabolic homeostasis (Mittler *et al.* 2012; Hemantaranjan *et al.* 2014). Low yield in chickpea genotypes under late sown condition has been attributed to sub optimal concentrations of total nitrogen, soluble sugars and free amino acids in developing grains during seed filling stages (Bhattacharya *et al.* 1999). The disturbance in metabolic homeostasis leads to the accumulation of toxic by-products, such as reactive oxygen species (ROS) (Mittler *et al.* 2012). To survive or maintain steady-state balance of metabolic processes under heat stress, plants reprogram their transcriptome, proteome, metabolome and lipidome, thereby adjusting their composition of certain transcripts, proteins, metabolites and lipids (Mittler *et al.* 2012). Many heat shock proteins are accumulated in plants to mitigate the adverse effect of heat stress on plant metabolism (Lavania *et al.* 2015a, b). Exposure of plants to growth-limiting temperatures induces the depolymerization of microtubules and microfilaments (Hardham and Gunning, 1978; Müller *et al.* 2007). These two structures are intimately involved in cell morphogenesis (Fowler and Quatrano, 1997) and its rearrangement may explain variations of the leaf shape in plants growth under extreme temperatures (Campitelli and Stinchcombe, 2013). Probably, the cellular component most sensitive to temperature fluctuations is the photosynthetic apparatus. The primary targets of thermal stress on the photosynthetic apparatus in plants are the photosystem II (PSII) and the carbon fixation by Rubisco (Salvucci and Crafts-Brandner, 2004). An early effect of temperature in the

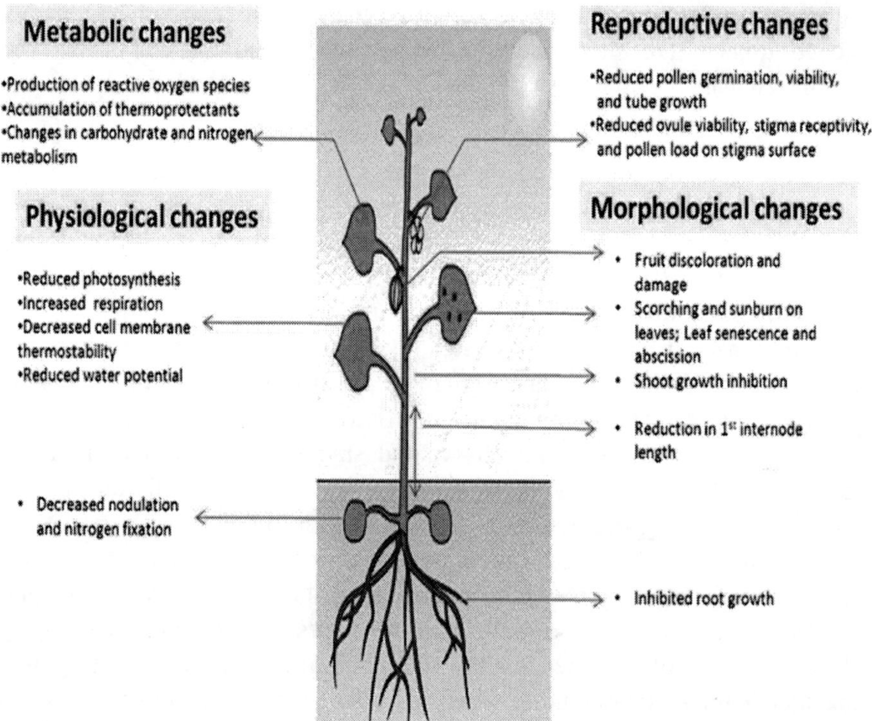

Fig. 1: Morphological, physiological, metabolic, and reproductive changes in plants under heat stress

photosynthetic apparatus is the inhibition of the activity of the PSII. Consideration of abiotic stresses in crop species is of vital importance due to the widespread presence of such stresses on agricultural land, the probable increase in their severity and incidence due to global climactic change and other anthropogenic activities, and the frequent deleterious effects such stresses have on crop productivity. These effects are the result of processes that can be observed at different levels of plant behavior, *i.e.* morphological, physiological and biochemical/molecular changes (Wang *et al.* 2003). At the morphological level, abiotic stress can cause altered shoot, root and leaf growth, as well as developmental changes that result in altered life cycle duration and fewer and/ or smaller organs (Schwarz *et al.* 2010; Wang *et al.* 2003). Physiological processes are also affected, such as photosynthetic rate, respiration and the partitioning of assimilates to different organs within the plant (Schwarz *et al.* 2010). At the cellular level, cell membranes can be damaged, thylakoid structures disorganized, cell size reduced, stomatal guard cell function altered, degree of cellular hydration modified and programmed cell death promoted. And finally, at the biochemical / molecular level, effects include enzyme inactivation, the production of reactive oxygen species, osmotic damage, changes in primary and secondary metabolite profiles, changed water and ion uptake or movement and altered hormone concentrations. These effects are in general potentially deleterious for plant performance and plants possess an array of strategies for combating them aimed at maintaining growth and productivity (Schwarz *et al.* 2010; Wang *et al.* 2003), within the limits of ecology, timing, severity and crop stage (Han *et al.* 2009).These strategies include an antioxidative defense system, the activation of protein synthesis for specific tasks involved in ameliorating the effects, such as heat shock proteins, ethylene production, detoxification, osmoprotection, stabilization of enzymes and membranes, avoidance of stresses that occur at fixed times during the year and the ability to acclimatize to stresses. The antioxidative defense system protects against reactive oxygen species (Norman, 1998). Ascorbate, glutathione and α-tocopherol act as antioxidants in aerobic cells, and carotenoids have important antioxidant effects in photosynthetic systems (Halliwell, 1987; Larson, 1988; Norman, 1998). In addition, an antioxidant enzyme system for scavenging the toxic oxygen species acts in various plant cell compartments, which includes catalase (CAT), superoxide dismutase (SOD). Halliwell-Asada pathway enzymes, ascorbate peroxidase (APx), peroxidise (POD), glutathione reductase (GR) and dehydroascorbate reductase (DHAR) (Norman, 1998). Both natural and artificial stress provoke increased production of toxic oxygen species, and, in response, the capacity of the antioxidative defence system is increased (Kairola *et al.* 2006; Norman, 1998).

Emission of green-house gases such as carbon dioxide, methane and nitrous oxide from agricultural systems is one of the major concerns contributing to this global increase of temperature (Smith and Olesen, 2010). Under high temperature conditions, plants accumulate different metabolites such as antioxidants, osmoprotectants, heat-shock proteins and metabolites from different pathways (Bokszczanin and Fragkostefanakis, 2013). Reactive oxygen species may damage cellular components and act as signaling molecules, leading to the expression of antioxidant enzymes, Heat shock protein and a rebalancing of osmolyte concentrations that perturb cell-water balance (Bohnert *et al.* 2006). Heat shock proteins play a role in stress signal transduction, protecting and repairing damaged proteins and membranes, protecting photosynthesis as well as regulating cellular redox state. Expression of various heat shock proteins is known to be an adaptive strategy in heat tolerance.

The major sites of thermal damage are the oxygen-evolving complex (OEC) along with associated cofactors in photosystem II, carbon fixation by Rubisco and the ATP-generating system. High temperature stress also reduces the efficiency of electron transport and consequently leading to increased production of reactive oxygen species in plant cells. Plants under high temperature stress usually accumulate more reactive oxygen species in both chloroplasts and mitochondria, which can severely damage DNA and cause cell membrane lipid peroxidation (LPO). Thus, plant protection against high temperature is closely correlated to increased capacity of scavenging and detoxifying the reactive oxygen species. Induction of thermotolerance may be ascribed to the maintenance of better membrane thermostability, and low level of reactive oxygen species accumulation (Hameed *et al.* 2012) due to improved antioxidant capacity (Chakraborty and Pradhan, 2011). To overcome stress, plants are equipped with different protective mechanisms including the maintenance of cell membrane stability, capturing the reactive oxygen species, synthesis of antioxidants, accumulation and osmoregulation of osmoticum and upregulation of heat shock protein synthesis.

Role of membranes in heat tolerance

Modification in the membrane function under high temperature stress is mainly due to the alteration of membrane fluidity. Three commonly used parameters are related to membrane-based processes which include plasmalemma (cell membrane stability assay), photosynthetic membranes (chlorophyll fluorescence assay) and mitochondrial membranes (cell viability assay based on 2,3,5-triphenyl tetrazolium chloride, TTC reduction test). Membrane lipid saturation is considered an important element in high temperature tolerance. High temperature causes an increase in fluidity of membranes which can lead to disintegration of the lipid bilayer. Membrane damage is sometimes taken as a stress parameter to determine the level of lipid destruction. It has been recognized that lipid peroxidation products are formed from polyunsaturated precursors that include small hydrocarbon fragments such as ketones, malondialdehyde (MDA) and compounds related to them (Garg and Manchanda, 2009). Lipid peroxidation, in both cellular and organelle membranes, takes place when above-threshold reactive oxygen species levels are reached, thereby affecting normal cellular functioning (Montillet *et al.* 2005).

Heat stress-induced decrease of the duration of developmental phases leading to fewer and smaller organs, lower light perception due to a reduced life cycle and altered carbon assimilation is of major importance for cereal yields losses (Han *et al.* 2009). *Vigna radiata* grown and maturing at higher temperature (summer season) showed different biochemical constituents and slower rate of germination (Vijaylaxmi, 2013) as compared with those maturing in lower temperature (rainy season). It has been reported that high temperature may slow down or prevent germination, depending on plant species and stress intensity, and, at later stages, may adversely affect photosynthesis, respiration, water relations and membrane stability, as well as modulate levels of hormones and primary and secondary metabolites. Furthermore, throughout plant ontogeny, enhanced expression of a variety of heat shock and other stress-related proteins, and enhanced reduction of reactive oxygen species constitute major plant responses to heat stress (Wahid *et al.* 2007). High temperatures reduce photosynthesis by changing the structural organization of thylakoids (Wahid *et al.* 2007). In general, it is evident that high temperature considerably affects anatomical structures not only at the tissue and cellular levels, but also at the sub-cellular level. The cumulative effects of all these changes may result in poor plant growth and productivity (Wahid *et al.* 2007). Plants can adapt to changing environmental conditions by a series of strategies aimed at the maintaining of cellular metabolism, molecular activities, growth and development, through a variety of molecular networks that allow a rapid and efficient sensing of the stress, resulting in the triggering of a response activation (Rampino *et al.* 2006).

Earlier studies have revealed the nature of heat stress effects on mature, well- developed green leaves (Dash and Mohanty, 2001). However, the impact of heat stress on the development of the photosynthetic system during seedling establishment and leaf growth has not been studied extensively. Exposure to high temperature adversely affects germination and seedling emergence in wheat (Dash and Mohanty, 2001) and retards shoot and leaf growth in *Lolium* (Dash and Mohanty, 2001) and sorghum (Dash and Mohanty, 2001). The impact of heat stress on seedling growth and leaf development has also been established based upon pigmentation sensitivity (Kariola *et al.* 2006; Mathews *et al.* 2008) and Photosystem II (PS II) function in wheat (Dash and Mohanty, 2001; Nieto-Sotelo *et al.* 1999). Genotypes of wheat exhibited considerable variation in their sensitivity to heat stress (Dash and Mohanty, 2001). Differences in the sensitivity of chloroplast photoreactions to heat stress, however, could not be established detected between wild and cultivated wheat species (Dash and Mohanty, 2001) or between temperate and tropical cereals cultivars, including wheat (Dash and Mohanty, 2001). Studies intended to establish cultivar response to temperature suggest that it is not possible to generalize the relationship between temperature and sensitivity of all developmental phases and in all wheat varieties (Dash and Mohanty, 2001; Porter and Gawith, 1999) prior to assessment of wheat cultivars for thermo-tolerance. Moreover it has been demonstrated that the higher the temperature the plant tolerates, the more protective machineries are involved (Han *et al.* 2009). Heat stress has a complex impact on cell function, suggesting that many processes are involved in thermotolerance. Some processes may be specific to basal thermotolerance, others may be induced during acquired thermotolerance, and many may be involved in both. High temperatures are known to affect membrane-linked processes due to alterations in membrane fluidity and permeability (Larkindale *et al.* 2005; Sangwan *et al.* 2002).

Some studies have demonstrated that the responsiveness of cultivars to elevated temperature during greening of wheat seedlings is differential, as judged from the analysis of lipid peroxidation, pigmentation, light-induced changes in chlorophyll a fluorescence and photosynthetic electron transport (Dash and Mohanty, 2001). Heat-induced alterations in enzyme activity can lead to imbalance in metabolic pathways and can cause complete enzyme inactivation due to protein denaturation (Krastanov, 2010). Membrane and protein damage lead to the production of active oxygen species that cause heat induced oxidative stress (Larkindale *et al.* 2005). Heat stress causes, like other abiotic stresses, a series of complex morphological, physiological, biochemical and molecular changes that adversely affect plant growth and productivity (Schwarz *et al.* 2010). Among the many deleterious effects described are growth reduction, decreased photosynthetic rate, increased respiration, assimilate partitioning towards the fruits, osmotic and oxidative damage, reduced water and ion uptake/movement, and cellular dehydration. On the other hand, plants activate stress-responsive mechanisms, such as shifts in the aforementioned protein synthesis detoxification, osmoprotection, and stabilization of enzymes and membranes (Schwarz *et al.* 2010).

Moreover, supraoptimal root zone temperatures severely reduce root elongation and increase average root diameter (Qin *et al.* 2007; Schwarz *et al.* 2010). Increased ethylene production also seems to play a role as well in this stress response (Schwarz *et al.* 2010). It is known that heat stress may provoke multiple mineral deficiencies (P and Fe) in roots and shoots, which can both increase ethylene production (Schwarz *et al.* 2010; Ward *et al.* 2008). Heat can also promote programmed cell death (Larkindale *et al.* 2005; Vacca *et al.* 2004). In plants, these different types of damage translate into reduced photosynthesis, impaired translocation of assimilates and reduced carbon gain, leading to altered growth and reproduction (Larkindale *et al.* 2005).

Hot and cold temperature stresses have several major effects on reproductive tissues that contribute to poor seed set yield: (i) early or delayed flowering, (ii) asynchrony of male and female reproductive development, (iii) defects in parental tissue, and (iv) defects to male and female gametes. Temperature stress can trigger either early or delayed flowering, depending on the species and other environmental conditions. One important modifier is the photoperiod, which provides seasonal information in which a stress can be interpreted in the appropriate context (Putterill *et al.* 2004; Craufurd and Wheeler, 2009). Nevertheless, moderate heat stress will often accelerate flowering, which may cause reproduction to occur before plants accumulate adequate resources (*i.e.* biomass) for allocation to developing seeds. Mechanisms of heat-stimulated bolting and flowering in *Arabidopsis* (*Arabidopsis thaliana*) are being uncovered (Balasubramanian *et al.* 2006; Tonsor *et al.* 2008). Cold temperatures will typically delay flowering, which may cause seeds to develop later in the growing season under suboptimal temperatures.

Temperature stress can sometimes have different effects on male and female structures, thereby creating asynchrony between male and female reproductive development (Herrero, 2003; Hedhly *et al.* 2008). In maize (*Zea mays*), floral asynchrony is a significant problem under conditions of combined stress from heat and water deficit (Barnabãs *et al.* 2008). In addition, high temperature stress can shorten the period of time in which the stigmas in the flowers are receptive to pollen, and thereby decrease the chances for a successful fertilization. For example, the stigmas in peach (*Prunus persica* L.) at 30^0 C lost their ability to support pollen germination after 3 days, whereas at 20^0 C they were viable for 8 days (Hedhly *et al.* 2005). A third category of temperature stress effects includes defects in the structure and function of parental tissues (*i.e.* corollas, carpels, and stamens). Heat stress can reduce the number, decrease the size, and cause deformity of floral organs (Morrison and Stewart, 2002). For example, a high frequency of flower abortion in response to low temperatures is well documented for chickpea (Croser *et al.* 2003).

There are relatively few examples in which the effects of temperature stress on female reproductive organs have been investigated. For heat stress, ovary abnormalities were observed in wheat (*Triticum aestivum*, variety 'Gabo') (Saini *et al.* 1983). In *Arabidopsis*, heat stress reduced the total number of ovules and increased ovule abortion (Whittle *et al.* 2009). For cold stress, a study on chickpeas showed reduced ovule size, reduced ovule viability, missing embryo sacs, and impaired pistil function in temperature-sensitive cultivars (Sirinivasan *et al.* 1999). In rice (*Oryza sativa*) and maize, evidence suggests that ovule/female fertility is fairly tolerant of a moderate cold stress (Dupuis and Dumas, 1990).

More is known of the effects of temperature stress on male reproductive structures (Barnabãs *et al.* 2008; Thakur *et al.* 2010). For example, in wheat, heat stress during the period of microspore meiosis can induce tapetum degradation (Sakata *et al.* 2000). This degradation of the nutritive tissues of the tapetum leads to pollen sterility. High temperatures cause poor anther dehiscence characterized by tight closure of the locules, which was shown to reduce pollen dispersal in rice and tomato (*Solanum lycopersicum*) (Matsui and Omasa, 2002; Sato *et al.* 2002). In the final category of temperature-induced reproductive effects, temperature stress may directly affect the development of male and female gametes. The effects of temperature stress on male gametes are well documented for numerous plant species. Pollen maturation, viability, germination ability, and pollen tube growth can be negatively affected by heat (Aloni *et al.* 2001; Young *et al.* 2004).

Rate of plant growth and development is dependent upon the temperature surrounding the plant and each species has a specific temperature range represented by a minimum, maximum, and optimum. These values were summarized by Hatfield *et al.* (2008, 2011) for a number of different species typical of grain and fruit production. The expected changes in temperature over the next 30-50 years are predicted to be in the range of 2-3^0 C IPCC (2007). Heat waves or extreme temperature events are projected to become more intense, more frequent, and last longer than what is being currently been observed (Meehl *et al.* 2007). Extreme temperature events may have short-term durations of a few days with temperature increases of over 5^0 C above the normal temperatures. Extreme events occurring during the summer period would have the most dramatic impact on plant productivity; however, there has been little research conducted to document these effects as found by (Kumudini *et al.* 2014). Review by Barlow *et al.* (2015) on the effect of temperature extremes, frost and heat, in wheat (*Triticum aestivum* L.) revealed that frost caused sterility and abortion of formed grains while excessive heat caused reduction in grain number and reduced duration of the grain-filling period. Analysis by Meehl *et al.* (2007) revealed that daily minimum temperatures will increase more rapidly than daily maximum temperatures leading to the increase in the daily mean temperatures and a greater likelihood of extreme events and these changes could have detrimental effects on grain yield. If these changes in temperature are expected to occur over the next 30 years then understanding the potential impacts on plant growth and development will help develop adaptation strategies to offset these impacts.

Responses to temperature differ among crop species throughout their life cycle and are primarily the phenological responses, *i.e.*, stages of plant development. For each species, a defined range of maximum and minimum temperatures form the boundaries of observable growth. Vegetative development (node and leaf appearance rate) increases as temperatures rise to the species optimum level. For most plant species, vegetative development usually has a higher optimum temperature than for reproductive development. Cardinal temperature values for selected annual (non-perennial) crops are given by Hatfield *et al.* (2008, 2011) for different species. The range of temperatures in the Figure 2 affecting plant response is species dependent. For example, an extreme event for maize (*Zea mays* L.) is warmer than for a cool season vegetable (broccoli, *Brassica oleracea* L.) where the maximum temperature for growth is 25^0 C compared to 38^0 C.

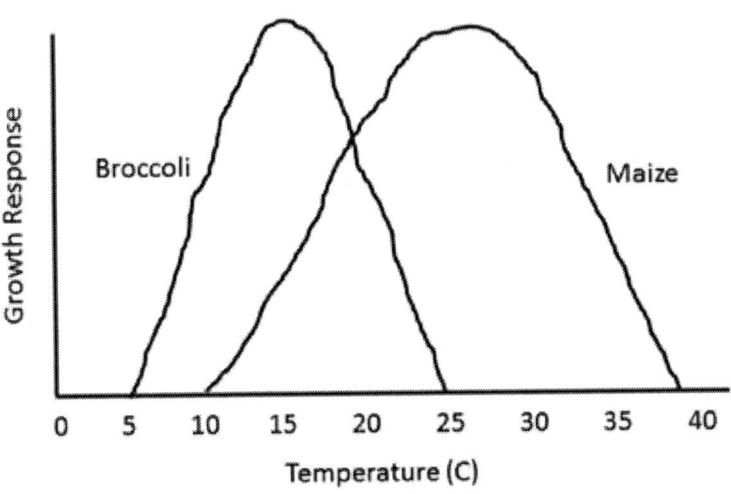

Fig. 2: Temperature response for maize and broccoli plants showing the lower, upper and optimum temperature limits for the vegetative growth phase

Faster development of non-perennial crops results in a shorter life cycle resulting in smaller plants, shorter reproductive duration, and lower yield potential. Temperatures which would be considered extreme and fall below or above specific thresholds at critical times during development can significantly impact productivity. Photoperiod sensitive crops, *e.g.*, soybean, would also interact with temperature causing a disruption in phenological development. In general, extreme high temperatures during the reproductive stage will affect pollen viability, fertilization, and grain or fruit formation Hatfield *et al.* (2008, 2011). Chronic exposures to extreme temperatures during the pollination stage of initial grain or fruit set will reduce yield potential. However, acute exposure to extreme events may be most detrimental during the reproductive stages of development. The impacts of climate change are most evident in crop productivity because this parameter represents the component of greatest concern to producers, as well as consumers. Changes in the length of the growth cycle are of little consequence as long as the crop yield remains relatively consistent. Yield responses to temperature vary among species based on the crop's cardinal temperature requirements. Warming temperatures associated with climate change will affect plant growth and development along with crop yield.

One of the more susceptible phenological stages to high temperatures is the pollination stage. Maize pollen viability decreases with exposure to temperatures above 35^0 C (Dupuis and Dumas, 1990). The effect of temperature is enhanced under high vapor pressure deficits because pollen viability (prior to silk reception) is a function of pollen moisture content which is strongly dependent on vapor pressure deficit (Fonseca and Westgate, 2005). Exposure to temperatures above 30^0 C damaged cell division and amyloplast replication in maize kernels which reduced the size of the grain sink and ultimately yield (Commuri and Jones, 2001). Rice (*Orzya sativa* L.) shows a similar temperature response to maize because pollen viability and production declines as daytime maximum temperature (T_{max}) exceeds 33° C and ceases when T_{max} exceeds 40^0 C (Kim *et al.* 1996). Cultivars of rice, flower near mid-day which makes T_{max} a good indicator of heat-stress on spikelet sterility. These exposure times occur quickly after anthesis and exposure to temperatures above 33° C within 1-3 hours after anthesis (dehiscence of the anther, shedding of pollen, germination of pollen grains on stigma, and elongation of pollen tubes) cause negative impacts on reproduction (Satake and Yoshida, 1978). Close observations in rice reveal that anthesis occurs between about 9 to 11 am (Prasad *et al.* 2006b) and exposure to high temperatures may already be occurring and will increase in the future. There is emerging evidence that differences exist among rice cultivars for flowering times during the day (Sheehy *et al.* 2005). Given the negative impacts of high temperatures on pollen viability, observations from Shah *et al.* (2011) suggest flowering at cooler times of the day would be beneficial to rice grown in warmer environments. They proposed that variation in flowering times during the day would be a valuable phenotypic marker for high-temperature tolerance. As daytime temperatures increased from 30 to 35⁰ C, seed set on male-sterile, female fertile soybean [*Glycine max* (L.) Merr] plants decreased (Wiebbecke *et al.* 2012). This confirms earlier observations on partially male-sterile soybean in which complete sterility was observed when the daytime temperatures exceeded 35⁰ C regardless of the night temperatures and concluded that daytime temperatures were the primary factor affecting pod set (Caviness and Fagala, 1973). Crop sensitivity to temperature extremes depends upon the length of anthesis. Maize, for example, has a highly compressed phase of anthesis for 3-5 days, while rice, sorghum [*Sorghum bicolor* (L.) Moench] and other small grains may extend anthesis over a period of a week or more. In soybean, peanut (*Arachis hypogaea* L.), and cotton (*Gossypium hirsutum* L.) anthesis occurs over several weeks and avoid a single occurrence of an extreme event affecting all of the pollening flowers. For peanut (and potentially other legumes) the sensitivity to elevated temperature for a given flower, extends from 6 days prior to opening (pollen cell division and formation) up through the day of anthesis (Prasad *et al.* 2001). Therefore, several days of elevated temperature may affect fertility of flowers in their formative 6-day phase or anthesis. Singh *et al.* (2015) reported differences in the threshold temperature for grain sorghum among genotypes and differences in the percentage of seed set in response to high temperatures. Pollination processes in other cereals, maize and sorghum, may have a similar sensitivity to elevated daytime temperature as rice. Rice and sorghum have exhibited similar sensitivities of grain yield, seed harvest index, pollen viability, and success in grain formation in which pollen viability and percent fertility is first reduced at instantaneous hourly air temperature above 33 °C and reaches zero at 40 °C (Kim *et al.* 1996; Prasad *et al.* 2006a *and* 2006b). Diurnal max/min day/night temperatures of 40/30 °C (35 °C mean) cause zero yield for those two species with the same expected response for maize.

Projected air temperature increases throughout the remainder of the 21st century suggests that grain yields will continue to decrease for the major crops because of the increase temperature stress on all major grain crops (Hatfield *et al.* 2011). Beyond a certain point, higher air temperatures adversely affect plant growth, pollination, and reproductive processes (Klein *et al.* 2007; Sack and Kucharik, 2011). However, as air temperatures rise beyond the optimum, instead of falling at a rate commensurate with the temperature increase, crop yield losses accelerate. For example, an analysis by Schleker and Roberts, (2009) indicated yield growth for corn, soybean, and cotton would gradually increase with temperatures up to 29°C to 32°C and then sharply decrease with temperature increases beyond this threshold.

Increases of temperature may cause yield declines between 2.5% and 10% across a number of agronomic species throughout the 21st century (Hatfield *et al.* 2011). Other evaluations of temperature on crop yield have produced varying outcomes. Lobel *et al.* (2011) showed estimates of yield decline between 3.8% and 5%; and Schleker and Roberts, (2009) used a statistical approach to estimate wheat, corn, and cotton yield declines of 36% to 40% under a low CO_2 emissions scenario, and between 63% to 70% for high CO_2 emission scenarios. These estimates of yield loss did not consider the positive effects of rising atmospheric CO_2 on crop growth, variation among crop genetics, impact of biotic stresses on crop growth and yield, or the use of adaptive management strategies, *e.g.,* fertilizers, rotations, tillage, or irrigation. These analyses assumed that air temperature increased without regard to the potential negative effects of temperature extremes. The current evaluations of the impact of changing temperature have focused on the effect of average air temperature changes; however, increases in minimum air temperature may be more significant in their effect on growth and phenology (Hatfield *et al.* 2011). Minimum air temperatures are more likely to increase under climate change (Knowles *et al.* 2006). While maximum temperatures are affected by local conditions, especially soil water content and evaporative heat loss as soil water evaporates (Alfaro *et al.* 2006), minimum air temperatures are affected by mesoscale changes in atmospheric water vapor content. Hence, in areas where changing climate is expected to cause increased rainfall or where irrigation is predominant, large increases of maximum temperatures are less likely to occur than in regions prone to drought. Minimum air temperatures affect nighttime plant respiration rates and can potentially reduce biomass accumulation and crop yield (Hatfield *et al.* 2011). Welch *et al.* (2010) found higher minimum temperatures reduced grain yield in rice, while higher maximum temperature raised yields; because the maximum temperature seldom reached the critical optimum temperature for rice. However, under the scenario of future temperatures increases, they found maximum temperatures could decrease yields if they are near the upper threshold limit.

Similar responses have been found in annual crops in which temperature is the major environmental factor affecting production with specific stresses, such as periods of hot days, overall growing season climate, minimum and maximum daily temperatures, and timing of stress in relationship to developmental stages having the greatest effect (Sonsteby and Heide, 2008; Dufault *et al.* 2009). When plants are subjected to mild heat stress (1 °C to 4 °C above optimal growth temperature), there was moderately reduced yield (Wagstaffe and Battey, 2006; Tesfaendrias *et al.* 2010). In these plants, there was an increased sensitivity heat stress 7 to 15 days before anthesis, coincident with pollen development. Subjecting plants to a more intense heat stress (generally greater than 4 °C above optimum) resulted in severe yield loss extending to complete crop failure (Kadir *et al.* 2006; Gote and Padgham, 2009; Tesfaendrias *et al.* 2010). Tomatoes under heat stress fail to produce viable pollen while their leaves remain active. The non-viable pollen does not pollinate flowers causing failure in fruit set (Sato *et al.* 2000). If the same stressed plants are cooled to normal temperatures for 10 days before pollination, and then returned to high heat, they are able to develop fruit. There are some heat tolerant tomatoes which perform better than others related to their ability to successful pollinate even under adverse conditions (Peet *et al.* 2003; Sato, 2006).

Perennial plants are also susceptible to exposure to increasing temperatures similar to annual plants. These responses and the magnitude of the effects are dependent upon individual species. Exposure to high temperatures, >22⁰ C, for apples during reproduction increases the fruit size and soluble solids but decreases firmness as a quality parameter (Warrington *et al.* 1999). In cherries, increasing the temperature 3° C above the 15° C optimum mean temperature decreases fruit set (Beppu *et al.* 2001). Optimum temperature range in citrus (*Citrus sinensis* L. Osbeck) is 22-27° C and temperatures greater than 30° C increased fruit drop (Cole and McCloud, 1985). During fruit development when the temperatures exceed the optimum range of 13-27° C with temperatures over 33⁰ C there is a reduction in Brix (sugar content), acid content, and fruit size in citrus (Hutton and Landsberg, 2000). Temperature stresses on annual and perennial crops have an impact on all phases of plant growth and development.

Perennial crops have a more complex relationship to temperature than annual crops. Many perennial crops have a chilling requirement in which plants must be exposed to a number of hours below some threshold temperature before flowering can occur. For example, chilling hours for apple (*Malus domestica* Borkh.) range from 400 to 2900 h (5-7⁰ C base, Hauaqqe, 2010) while cherry trees (*Prunus avium)* require 900 to 1500 h with the same base temperature (Seif and Gruppe, 1985). Grapes (*Vitis vinifera* L.) have a lower chilling threshold that other perennial plants with some varieties being as low at 90 h (Reginato *et al.* 2010). Increasing winter temperatures may prevent chilling hours from being obtained and projections of warmer winters in California revealed that by mid-21st century, plants requiring more than 800 h may not be exposed to sufficient cooling except in very small areas of the central Valley (Luedeling *et al.* 2009). Climate change will impact the chilling requirements for fruits and nut trees. Hatfield *et al.* (2014) showed that under a warming climate, adequate chilling hours for perennial crops for fruit development may not be met. Innovative adaptation strategies will be required to overcome this effect because of the long time requirements for genetic selection and fruit production once perennial crops are established.

Since warmer temperatures are typically associated with a reduction in leaf area index and green area duration, the main targets for improving leaf area under warm conditions include canopy establishment and architecture and the maintenance of green area.

Rapid Ground Cover and Canopy Structure

Rapid ground cover or early vigor characterizes the capacity of genotypes to develop leaf area or aboveground biomass. Genotypic variability in rapid green cover has been associated with differences in seedling emergence rate and/or specific leaf area, grain and embryo size, and tillering capacity (Richards and Lukacs, 2002). These characteristics are relatively heritable, making them easy breeding targets (Rebetzke *et al.* 2008). Wheat shows enormous diversity in canopy architecture, and it has long been proposed that optimized light distribution could improve radiation use efficiency as well as light interception (Murchie *et al.* 2009). Further modification in canopy architecture may not be easy to address genetically, since the canopy is a highly complex structure; furthermore, light extinction interacts with sun angle and light intensity, making it a constantly moving target.

Stay-Green

Leaf senescence is characterized initially by structural changes in the chloroplast, followed by a controlled vacuolar collapse, and a final loss of integrity of plasma membrane and disruption of cellular homeostasis (Lim *et al.* 2007). Delay in the expression of senescence-related genes permits Stay-Green genotypes to maintain photosynthesis (Lim *et al.* 2007). While stay-green is recognized as an adaptive physiological traits for stress conditions, the optimal pattern of senescence/pigment loss in terms of improving grain yield under heat stress has not been identified. This is partly because chlorosis is an integral part of programmed senescence, where there are unavoidable tradeoffs between maintaining photosynthetic area and remobilization of nitrogen to the maturing grain (Vijaylakshmi *et al.* 2010). Since chlorosis is expressed heterogeneously on all aboveground organs, even within a single organ such as a leaf, it is not straightforward to parameterize. However, an easy and integrated approach to estimate stay-green using the spectral reflectance Normalized Difference Vegetation Index showed significant association with yield under heat stress in two large mapping populations, making it a reliable tool for large-scale screening and gene discovery work (Lopes and Reynolds, 2012). Although the Normalized Difference Vegetation Index, which is a reliable indicator of greenness integrating all chlorophyll, is associated with heat tolerance, studies in other species suggest that chlorophyll *a* degrades sooner than chlorophyll *b* (Keskitalo *et al.* 2005).

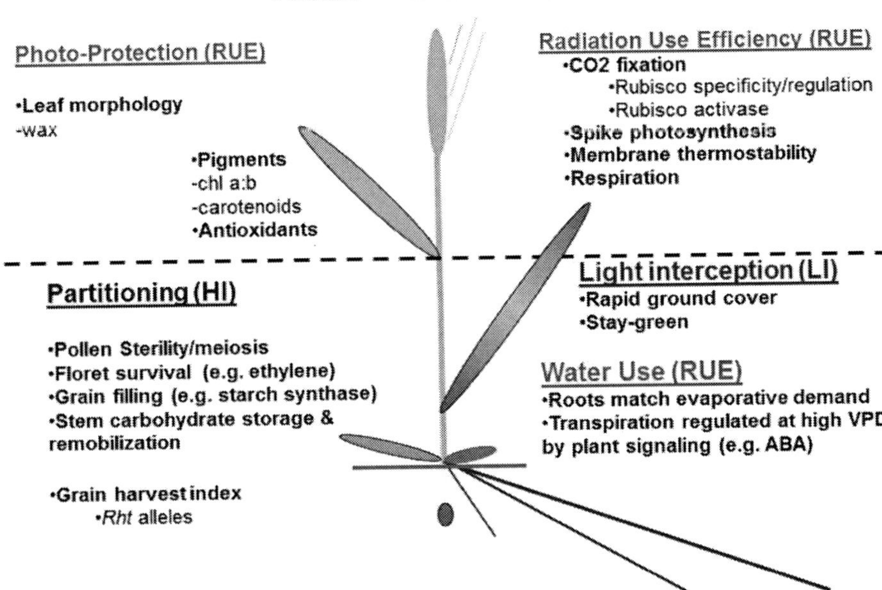

Fig. 3: Conceptual model of heat-adaptive traits grouped by three main drivers of yield (i.e. yield = LI × RUE × HI) in the absence of water limitations (adapted from Reynolds *et al.*, 2007). ABA, Abscisic acid

Therefore, more specific spectral indices related to functional stay-green would be worth developing as a screening tool for heat tolerance.

Improving photosynthetic capacity and efficiency are key targets for increasing the yield potential of crops under heat stress. Interestingly, increased affinity of Rubisco for CO_2 versus oxygen and a better catalytic rate (Parry *et al.* 2011), are likely to be particularly beneficial at warmer temperatures. It is well known that Rubisco's affinity for CO_2 decreases with temperatures, therefore, increasing affinity would simultaneously improve adaptation to warmer conditions, the proof of concept coming from C_4 species, in which it is achieved by concentrating CO_2 (Sage, 2002). Spike photosynthesis has a number of characteristics that make it a worthwhile target for increasing heat adaptation: (1) spikes intercept up to 40% of incident radiation in wheat canopies and operate at higher temperatures than leaves; (2) their transpiration efficiency is higher than that of leaves; and (3) genetic variation exists in spike senescence. The main difficulties in analyzing spike photosynthesis include the interpretation of results that are potentially confounded by CO_2 assimilation/recycling within the spike, the standardization of the units of carbon fixation given the complex geometry of spikes, and the development of heat stress tolerant chlorophyll fluorescence represents a good candidate (Parry *et al.* 2011).

High temperature stress and flowering

The timing of flowering, a critical stage of development in the life cycle of most plants when seed number is determined, is important for adaptation both to the abiotic stresses of temperature and water deficit, and to biotic (pest and disease) constraints (Curtis, 1968) within the growing season. For example, in many annual crops, brief episodes of hot temperatures (>32–36 °C) can greatly reduce seed set, and hence crop yield, if they coincide with a brief critical period of only 1–3 day around the time of flowering (Vara Prasad *et al.* 2000; Jagdish *et al.* 2008). Therefore, the moderation of crop development will be critical to the impacts of climate change on yield in two ways: through determining the season length, and hence the availability of radiation, water, and nutrient resources for growth; and by affecting the exposure of the crop to climate extremes. Adaptation to moderate changes in climate that influence temperature, season length, and planting dates, as well as the occurrence of abiotic stress, can be achieved by selecting varieties with appropriate flowering times and crop durations (Richards, 2006; Ludlow and Muchow, 1990). Farmers (landraces) and plant breeders (cultivars) have very successfully selected/ manipulated life cycle duration and phenology to maximize the range of environments in which crops grow as well as their yield (Evans, 1993; Roberts *et al.* 1996) at least for current climates. A major challenge for crop improvement is how to plan for future climate change. Relationship between grain yield and time to flowering of a range of sorghum landrace genotypes grown at four locations varying in rainfall in northern Nigeria. At the highest rainfall location, Kaffin Maiyaki, yield is proportional to duration to flowering. As rainfall decreases and season length is reduced, the yield of later flowering genotypes is reduced and the optimum flowering time ranges from 90 days to ~70 days (Flower, 1996).

Heat stress affects flowering by reducing flower number and size, and causing deformed floral organs (Morrison and Stewart, 2002) leading to loss of flowers and young pods and hence reduced yield as observed in mungbean and chickpea (Tickoo *et al.* 1996). common bean Suzuki *et al.* 2001), and rapeseed (Angadi *et al.* 2000). Hall (1992) reported cowpea to be susceptible to high night temperatures during early flowering and pod set which suppressed pod set due to anther indehiscence and low pollen viability (Warrag and Hall, 1984). However, in pea, day temperature has a greater effect than night temperature on dry matter production (Stanfield *et al.* 1966). Exposure to high temperature (32/27° C) for 10 days before and during anthesis reduced pod set in common beans (Gross and Kigel, 1994), which is in agreement with similar studies on cowpea (Hall, 1992), peanut (Prasad *et al.* 1999), apricot (Rodrigo and Herrero, 2002), and sweet cherry (Hedhly *et al.* 2007). Brown and Zeiher (1998) reported several flower abnormalities such as smaller flowers, asynchronous development of male and female reproductive structures, failure of anthers to release pollen and the presence of elongated stigmas in cotton plants under heat stress (above 32°C). Heat stress during the reproductive phase in legumes is generally linked with reduced or no pollination, and abscission of flower buds, flowers, and pods with substantial yield loss (Nakano *et al.* 1998).

Generally, pollen grain development is more sensitive to heat stress, at all stages, compared to the female gametophyte as reported in maize (Herrero and Johnson, 1980), cowpea (Ahmed *et al.* 1992), chickpea (Devasirvatham *et al.* 2012; 2013; Kaushal *et al.* 2013), rice (Wassmann *et al.* 2009) and tomato (Giorno *et al.* 2013). The number and morphology of pollen grains, anther dehiscence, pollen viability, metabolism, and composition are affected by heat stress, as observed in common beans (Gross and Kigel, 1994) groundnut (Prasad *et al.* 1999), soybean (Djanaguiraman *et al.* 2013), and chickpea (Devasirvatham *et al.* 2013). Endo *et al.* (2009) found that although high-temperature-treated pollen grains had a normal round shape, some tapetal functions, and pollen adhesion to the stigma and its subsequent germination were negatively affected. In chickpea, reduced pollen germination and tube growth in the style was observed in the heat-sensitive genotype ICC 5912 at 35/20^0 C due to sterile pollen (Devasirvatham *et al.* 2010; Kaushal *et al.* 2013). Djanaguiraman *et al.* (2013) observed significant reductions in *in vitro* pollen germination in soybean plants when exposed to 38/28^0 C for 14 days at flowering. In cowpea, anthers were rendered indehiscent under heat stress (33/30^0 C) condition which was attributed to degeneration of the tapetal layer (Ahmed *et al.* 1992) Heat stress decreased the concentration of soluble sugars in the anther walls of developing and mature pollen grains in cowpea (Ismail and Hall, 1999) and the findings were further confirmed by Suzuki *et al.* (2001) in common bean.

Female gametophytic tissue is less sensitive to heat stress than male gametophytic tissue (Devasirvatham *et al.* 2012; 2013; Kaushal *et al.* 2013). Saini and Aspinal, (1983) reported that heat stress (30^0 C) during meiosis in wheat can reduce yield due to abnormal ovary development resulting in reduced pollen growth and seed set. Heat stress can potentially harm style length and induce abnormalities in ovary development, as reported in mango (Sukhvibul *et al.* 1999) chickpea (Srinivasan *et al.* 1999) and apricot (Rodrigo and Herrero, 2002) Stigma receptivity significantly reduced at 40/30 and 45/35^0 C in chickpea (Kumar *et al.* 2013). Unfavorable high temperature (30^0 C) reduced ovule number and viability in common beans (Suzuki *et al.* 2001) and *Arabidopsis thaliana* (Whittle *et al.* 2009) Reduced length of stigmatic receptivity under temperature stress has been reported in sweet cherry (Hedhly *et al.* 2003). High temperature results in an exerted style which in turn inhibits pollen germination and thus the reproductive process (Wahid *et al.* 2007). Adverse effects of high temperature on the development of reproductive organs has been observed in 'Hakuho' peach trees. and rapeseed (Polowick and Sawhney, 1988). Young *et al.* (2004) studied the response of both microgametophytes and megagametophytes to heat stress (28/23° C) in rapeseed. When male gametes were subjected to heat stress (35/18° C) and female gametes were unstressed, seed set decreased by 88% whereas, in case of unstressed male gametes and heat-stressed female gametes, seed set decreased by 37% but when both male and female gametes were subjected to high temperature, seed set loss was the highest, *i.e.* it decreased by 97%. (Young *et al.* 2004)

4.3. Low Temperature Stress

Among various environmental stresses, low temperature is one of the most important factors limiting the productivity and distribution of plants. Low temperatures, defined as low but not freezing temperatures (0–15° C), are common in nature and can damage many plant species. In order to cope with such conditions, several plant species have the ability to increase their degree of freezing tolerance in response to low, non-freezing temperatures, a phenomenon known as cold acclimation. It is well established that some of the molecular and physiological changes that occur during cold acclimation are important for plant cold tolerance (Zhu *et al.* 2007). Accordingly, it has been concluded that cold tolerance that develops in initially insensitive plants is not entirely constitutive and at least some of it is developed during exposure to low temperatures.

Plants respond and survive under stress conditions by bringing changes at the molecular and cellular levels as well as at the biochemical and physiological levels (Xin and Browse, 2000). Low temperature stress inhibits seedling establishment effecting early growth stages of rice and resulting in poor crop maturation. In order to gain stable rice production cold tolerance at the seedling stage is an important character. One of the most effective ways to avoid the low-temperature damage is to develop cold tolerant genotype (Lou *et al.* 2007). Mineral nutrition acquisition and assimilation are strongly influenced by both high and low temperature stress in plants (Rivero *et al.* 2006). Some essential nutrients such as nitrogen (N), sulfur (S) phosphorus (P), magnesium (Mg), calcium (Ca) are structurally important for the proteins, nucleic acids, chlorophylls, certain secondary metabolites and defense related micro and macromolecules, while others have both structural and functional roles (Epstein and Bloom, 2005; Taiz and Zeiger, 2006).

As a result of exposure to low temperatures, many physiological and biochemical cell functions have been correlated with visible symptoms (wilting, chlorosis, or necrosis) (Ruelland and Zachowski, 2010). Often, these adverse effects are accompanied by changes in cell membrane structure and lipid composition (Matteucci *et al.* 2011), cellular leakage of electrolytes and amino acids, a diversion of electron flow to alternate pathways (Seo *et al.* 2010), alterations in protoplasmic streaming and redistribution of intracellular calcium ions (Knight *et al.* 1998). It also involve changes in protein content and enzyme activities (Ruelland and Zachowski, 2010) as well as ultrastructural changes in a wide range of cell components, including plastids, thylakoid membranes and the phosphorylation of thylakoid proteins, and mitochondria (Zhang *et al.* 2011). Brief exposures to low temperatures may only cause transitory changes, and plants generally survive. However, prolonged exposure to stress causes plant necrosis or death. To overcome stresses generated by exposure to low non-freezing temperatures, plants can trigger a cascade of events that cause changes in gene expression and thus induce biochemical and physiological modifications that enhance their tolerance (Zhu *et al.* 2007). This phenomenon is known as chilling or cold acclimation.

Establishment of seedling is a critical process to plant growth, especially under adverse environmental conditions (Bohnert *et al.* 1995). Seedlings adapt to stress environment by different mechanisms, including changes in morphological and developmental pattern as well as physiological and biochemical processes. Adaptation is associated with maintaining osmotic homeostasis by metabolic adjustments that lead to the accumulation of metabolically compatible compounds such as soluble sugar, malondialdehyde (MDA) and proline. It also includes modification of related enzyme activity and cell membrane stability (Liang *et al.* 2009; Chinnusamy *et al.* 2007). Moreover, it is well known that plant structural modifications and growth pattern adjustments are useful indices of the consequences of stress environment (Rauf, 2008; Ushio *et al.* 2008). Therefore, plants adapt to low temperature and water stress by mediations of these substances. the effects of water or low temperature stress on the change of physiological indexes are documented (Yang and Lin, 2008; Cao *et al.* 2009)

Mechanisms of acclimation to low non-freezing temperatures

The primary mechanisms involved in cold acclimation are related to a number of processes discussed below. These include molecular and physiological modifications occurring in plant membranes, the accumulation of cytosolic Ca^{2+}, increased levels of ROS and the activation of ROS scavenger systems, changes in the expression of coldrelated genes and transcription factors, alterations in protein and sugar synthesis, proline accumulation, and biochemical changes that affect photosynthesis.

Modifications to plant cell membranes

Membranes are a primary site of cold-induced injury. Several studies have demonstrated that membrane rigidification, coupled with cytoskeletal rearrangements, calcium influxes, and the activation of MAPK cascades, triggers low temperature responses (Sangwan *et al.* 2002). The lipid composition of the plasma membrane and chloroplast envelopes in acclimated plants changes such that the threshold temperature for membrane damage is lowered relative to that for non-acclimated plants (Uemura and Steponkus, 1999). This is achieved by increasing the coldadapted membranes' unsaturated fatty acid content, which makes them more fluid (Vogg *et al.* 1998). The process of cold acclimation promotes the stabilization of membranes, which prevents damage leading to cell death. The acclimation process also activates mechanisms that protect membrane fluidity by ensuring the optimal activity of associated enzymes (Matteucci *et al.* 2011). At the physiological level, photosynthesis is strongly affected by exposure to cold. The cessation of growth resulting from cold stress reduces the capacity for energy utilization, causing feedback inhibition of photosynthesis (Ruelland and Zachowski, 2010). In cold-acclimated winter annuals, photosynthetic activity is maintained by increases in the abundance and activity of several Calvin cycle enzymes (Goulas *et al.* 2006). This recovery is associated with elevated levels of thylakoid plastoquinone A and a concomitant rise in the apparent size of the intersystem electron donor pool to PSI (Baena-Gonzalez *et al.* 2001). Consequently, non-photochemical quenching increases in cold-stressed leaves in parallel with increased zeaxanthin levels to compensate for the reduced electron consumption by photosynthesis. Ruelland and Zachowski (2010) reported that energy dissipation via nonphotochemical quenching and electron transport was not only enhanced following cold acclimation but also contributed to protection from oxidative damage.

Xanthophylls

Although they are not considered photosynthetic pigments *per se*, the xanthophylls (notably, violaxanthin, antheraxanthin, and zeaxanthin) help in protecting the photosystems and their abundance increases at low temperatures (Ivanov *et al.* 2006). Xanthophylls have structural roles and act as natural antioxidants, quenching triplet chlorophyll and singlet oxygen, which are potentially harmful to the chloroplast (Passarini *et al.* 2009; Han *et al.* 2010). It has also been postulated that unbound zeaxanthin and other carotenoids may also stabilize thylakoid membranes against putative peroxidative damage and heat stress (Laugier *et al.* 2010).

Flavonoids

These accumulate in leaves and stems in response to low temperatures. They are synthesized via the phenylpropanoid pathway, which is controlled by key enzymes, including phenylalanine ammonia-lyase and chalcone synthase (Sharma *et al.* 2007). It has been reported that cold stress induces transcriptomic modifications that increase flavonoid biosynthesis, including reactions involved in anthocyanin biosynthesis and the metabolic pathways that supply it (Crifo *et al.* 2011).

Role of reactive oxygen species in acclimation to low temperatures

The role of ROS in abiotic stress management has become a subject of considerable research interest, particularly since ROS have been reported to be involved in processes leading to plant stress acclimation (Suzuki *et al.* 2011). This finding indicates that ROS are not simply toxic by-products of metabolism, but act as signalling molecules that modulate the expression of various genes, including those encoding antioxidant enzymes and modulators of H_2O_2 production (Suzuki *et al.* 2011). In addition, low temperature stress has been reported to cause significant increases in the levels of the soluble non-enzymatic antioxidants ascorbate and glutathione, as well as the activity of the main NADPH-generating dehydrogenases (Airaki *et al.* 2011).

Chilling or cold stress leads to reduced growth and leaf expansion (Sowinski *et al.* 2005; Ryman *et al.* 2007), wilting (Bagnall *et al.* 1983) and chlorosis (Yoshida *et al.* 1996) in plants and may even leads to necrosis (tissue death) of plants (Yadav, 2010a). Low temperature both directly by metabolic inhibition (decreased metabolism, freezing the cell membrane and loss of membrane (Jewell *et al.* 2010), and indirectly, through the reaction caused by cold osmotic (dehydration and freezing due to cold) prevent the full expression of genetic potential of plants. The destruction of plasma membrane of plant cells and pigment especially in thylakoid membranes (Wang, 1990) reported as the major side effects of cold stress in plants (Steponkus *et al.* 1993). It has generally been noticed that plants native to warm habitats such as soybean, corn and cotton exhibit symptoms of injury upon exposure to low non-freezing temperatures or hypothermia (Jenabiyan *et al.* 2014). The degree of damage, however, depends on the sensitivity of plants to cold stress and may vary from one plant to another plant. Angadi *et al.* (2000) observed that temperatures below 10^0 C negatively influenced the growth and stem elongation in *B. napus*, *B. rapa* and *Raphanus sativus* plants. Also Rivero *et al.* (2001) reported that low temperature stress decreased the plant biomass and antioxidant enzymes activity in tomato and watermelon plants. Genotype response to different environments has a great deal of importance in selecting process of these lines against abiotic stress tolerance (Hazer Jaribi *et al.* 2013). Soybean is sensitive to temperatures below 15° C which is reflected in the metabolism, growth, development and yield changes (Balestrasse *et al.* 2010). According to Hume and Jackson (1981) findings, a single night of dark chilling, with minimum temperature of 8^0 C can reduce plant growth and prevent pod formation in soybean. The damage will be more sever if cold stress occurs with high light intensity. Furthermore, the combination of high light intensities and low temperatures, such as those experienced on cold but sunny mornings in spring, can cause irreversible damage to young soybean seedlings (Balestrasse *et al.* 2010). Moosavi (2007) on forage sorghum and Mohammadi Nikpoor (1995) on safflower reported an increase in plant height with the decrease of light intensity.

Genes expressed under stress conditions may increase tolerance to growth retardation, cold, high salt concentration during normal conditions in rice transgenic and *Arabidopsis* (Ito *et al.* 2006). It has been determined that in comparison with other cereals like wheat and barley, rice is more sensitive to cold stress (Wen *et al.* 2002). Low temperature stress induces many genes, acting either as protectants towards stressinduced damage or regulates expression of other genes and transduce signal (Shinozaki *et al.* 2003). Davletova *et al.* (2011) investigated that in contrast to many signaling and regulatory genes that are stress specific, Zat12 which is a zinc-finger protein gives response to various abiotic and biotic stresses and plays an important role in abiotic stress signaling in *Arabidopsis*. Similarly, MYB transcription factors play a central role in plant growth and respond to stress conditions (Yang *et al.* 2012). OsMYB2 encodes a MYB transcription factor (stress responsive element), which is involved in regulation of various pathways leading to tolerance of rice to dehydration stress, salt and cold. Low temperature exposure often induces a variety of biochemical, physiological and enzymatic change in plant, which can result in an acclimation response (Hughes and Dunn, 1996). Rice is sensitive to chilling stress and its persistence leads to poor germination, stunted seedling, yellowing or withering and decreased tillering (Mukhopadhyay *et al.* 2004).

Cold temperatures can induce pollen sterility, which may be due to disruption of sugar metabolism in the tapetum, ultimately abolishing starch accumulation (*i.e.* energy reserves) in the pollen grains (Oliver *et al.* 2005). Evidence suggests that cold-induced disruption of anther sugar transport and corresponding pollen sterility is signalled by abscisic acid, in part by down-regulating expression of cell wall invertase and monosaccharide transporters (Oliver *et al.* 2007). Heat stress reduces carbohydrate deposition in pollen grains (Jain *et al.* 2007). The pre-existing carbohydrate reserves within the pollen grains fuel tube growth, but pollen later switch to using carbohydrates provided by the transmitting tract of the style (Herrero and Arbeloa, 1989). Cotton (*Gossypium hirsutum*) flowers exposed to moderately high temperatures showed reduced carbohydrate reserves (particularly sucrose) and ATP production in the pistils, which correlated with a heat stress-induced decline in net photosynthesis (Snider *et al.* 2009). Cold stress can disrupt mitosis I and II, thus preventing rice microspores from maturing into tricellular pollen grains (Sataka and Hayase, 1970). Cold temperatures inhibited pollen germination and shortened pollen tubes in *Trifolium repens* and chickpea (Srinivasan *et al.* 1999).

Under low temperature, plasma membrane is thought to be the primary site of injury because of its central role in regulation of various cellular processes (Takahashi *et al.* 2013). Low-temperature stress will initiate peroxidation of the membrane lipids and cause the accumulation of malondialdehyde, which will bind with some proteins and destructs the membrane system (Chen. 1991) and therefore has been considered as an important indicator of membrane system injuries (Fan *et al.* 2012). Some studies showed that more malondialdehyde will be generated in plants with less tolerance to stress (Zhai *et al.* 2013). Plants under abiotic stress have evolved a defense system against oxidative stress by increasing the activities of ROS-scavenging enzymes, such as superoxide dismutase, catalase, and peroxidase (Miller *et al.* 2010). SOD plays a crucial role in antioxidant defense because it catalyzes the dismutation of $O_{2''}$ into H_2O_2 and catalase and peroxidase can reduce the H_2O_2 to H_2O and protect super oxide dismutase from oxidation (Tian *et al.* 2012). Therefore, the activities of these enzymes may be connected with plant tolerance to stress.

The accumulation of osmolytes is also related to the low-temperature tolerance. The free proline, soluble proteins, and soluble sugar are among the major osmolytes. Accumulation of proline was reported in many plant species under diverse abiotic stress conditions and its role in plant stress tolerance was widely investigated and discussed (Delauney and Verma, 1993). The amount of soluble protein increased under low-temperature conditions (Bravo *et al.* 1999) and their abundance is correlated to the degree of freezing tolerance (Karimzadeh *et al.* 2006).

4.4. Low Soil Moisture (Drought) Stress

Since the Green Revolution, the yields of cereals and other crops have increased considerably in many regions of the world as a result of genetic improvement and better agronomic practices. The yield potential, *i.e.*, the yield achieved when the best available technology is used, has also increased almost linearly since the sixties, particularly in more favorable environments where soil water availability is not limited (Zhou *et al.* 2007; Fischer and Edmeades, 2010; Matus *et al.* 2012; del Pozo *et al.* 2014). Yield under water-limiting conditions, such those of the rainfed environments, has also increased during the past decades (Sànchez-Gracia *et al.* 2013). Notwithstanding the possible need for phenological adjustment (earliness) a higher yield potential may also translate into a higher performance under water stress (Nouri *et al.* 2011; Hawkeford *et al.* 2013). However, the potential yield and water-limited yield needs to continue increasing in order to cope with future demand for food, which is a consequence of the growing population and changes in social habits (Hawkeford *et al.* 2013), and also to reduce the negative impacts on crop productivity of global climate change (Lobell *et al.* 2008; Lobell and Gourdji, 2012).

Under the climatic changing context, drought has been, and is becoming an acute problem most constraining plant growth, terrestrial ecosystem productivity, in many regions all over the world, particularly in arid and semi-arid area (Knapp *et al.* 2001; Chaves *et al.* 2003; Fischlin *et al.* 2007). Based on the fourth assessment report by IPCC, global surface average temperature will have a 1.1–6.4^0 C range increase by the end of this century (Fischlin *et al.* 2007). It is indicated that a warming above 3^0 C would eliminate thoroughly fixed carbon function of global terrestrial vegetation, shift a net carbon source. With global warming, it is expected that water deficit would be escalated by increasing evapotranspiration, increasing the frequency and intensity of drought with an increase from 1% to 30% in extreme drought land area by 2100 (Fischlin *et al.* 2007), which would offset the beneficial effect from the elevated CO_2 concentration, further limiting the structure and function of the terrestrial ecosystem. The global climate models may forecast the precipitation regimes including its distribution and amount, but the complicated responses of terrestrial ecosystem to climate change may adversely affect the predict accuracy (Knapp *et al.* 2001; Swemmer *et al.* 2007). Water is an increasingly scarce resource given current and future human population and societal needs, putting an emphasis on sustainable water use (Rosegrant and Cline, 2003). Thus, an understanding of drought stress and water use in relation to plant growth is of importance for sustainable agriculture.

Plant would response to water stress by dramatically complex mechanisms from genetic molecular express, biochemical metabolism through individual plant physiological processes to ecosystem levels (Chaves *et al.* 2003; Izanloo *et al.* 2008; Xu *et al.* 2009) which may mainly includes six aspects: (1) drought escape via completing plant life cycle before severe water deficit. For example, earlier flowering in annuals species before the onset of severe drough (Greber and Dawson, 1990) (2) drought avoidance via enhancing capacity of getting water. Forexample, developing root systems or conserving it such as reduction of stomata and leaf area/canopy cover (Jackson *et al.* 2000) (3) drought tolerance mainly via improving osmotic adjustment ability and increasing cell wall elasticity to maintain tissue turgidity (Morgan, 1984) (4) drought resistance via altering metabolic path for life survives under severe stress (*e.g.*, increased antioxidant metabolism) (Bartoli *et al.* 1999; Peñuelas *et al.* 2004) (5) drought abandon by removing a part of individual, *e.g.,* shedding elder leaves under water stress (Chaves *et al.* 2003) (6) drought-prone biochemical-physiological traits for plant evolution under long-term drought condition via genetic mutation and genetic modification (Hoffmann and Merilä, 1999; Sherrard *et al.* 2009; Maherali *et al.* 2010). The processes may be involved in multi-aspects simultaneously in responses of plants to drought stress and thereafter rewatering. Leaf rolling, induced by loss of turgor and poor osmotic adjustment represents an important drought-avoidance mechanism (Richards, 1996). Under drought condition, leaf rolling decreased stomatal closure (O'Toole *et al.* 1979). The erectophile leaf canopy has been also proposed as a trait that could increase crop yield potential by improving radiation use efficiency in high radiation environments (Reynolds *et al.* 1999). Peduncle length has been also suggested as useful indicator of yield capacity in dry environments. Kaya *et al.* (2002) have been found a strong positive correlation between peduncle length and grain yield. In other cases, such relationship has been found inverse (Briggs and Aytenfisu, 1980) or no relationship (Villegas *et al.* 2006) depending on the environment.

In the field context, there is always interval occurrence in drought and/or rewetting events, particular under climatic change conditions predicting more frequent drought and flooding events. The water cycle change may greatly impact plant growth, photosynthesis and many key metabolic functions, thereby ecosystem productivity and agricultural achievement (Izanloo *et al.* 2008; Xu *et al.* 2009; Chen *et al.* 2009). Actually, sporadic precipitation would become a critical issue for maintaining ecosystem structural stability and even it's surviving in arid and semi-arid area. For example, a small rainfall pulse can induce a rapid response in a desert ecosystem, which quickly triggers plant growth so that the plants can survive (Reynolds *et al.* 2004). In crops, water stress has been associated with reduced yields and possible crop failure. The effects of water stress however vary between plant species. As the plant undergoes water stress, the water pressure inside the leaves decreases and the plant wilts. The main consequence of moisture stress is decreased growth and development caused by reduced photosynthesis, a process in which plants combine water, carbon dioxide and light to make carbohydrates for energy (Sibomana *et al.* 2013). Chemical limitations due to reductions in critical photosynthetic components such as water can negatively impact plant growth.

Water deficit is the major abiotic factor limiting plant growth and crop productivity around the world. Approximately one third of the cultivated area of the world suffers from chronically inadequate supplies of water (Massacci *et al.* 2008). In all agricultural regions, yields of rain-fed crops are periodically reduced by drought, and the severity of the problem may increase due to changing world climatic trends (Le Houerou, 1996). Advances in irrigation technology have helped reduce the gap between potential and actual yield, but irrigation costs and limited water supplies constrain irrigation throughout the world. To counteract soil moisture deficit effects, plants have evolved a range of physiological and biochemical responses, including lowering the rates of cellular growth and net photosynthesis, stomatal closure, and the accumulation of organic solutes such as sugar alcohols, or osmolytes such as proline and/or quaternary ammonium compounds (Yamada *et al.* 2005). Proline is largely responsible for changes in the osmotic potential of plant cells during drought stress. Proline also buffers cells against the effects of water-deficit stress (Yamada *et al.* 2005). Some plant species are thought to accumulate proline under conditions of water stress in order to maintain growth under conditions of low soil water potential. Plants synthesise proline from glutamine using the enzyme, pyrroline-5-carboxylate synthase, in their leaves (Mahajan and Tujeta, 2005). Studies on tobacco have shown that over-expression of the gene for pyrroline-5-carboxylate synthase leads to an increase in proline concentration and improved growth under drought conditions (Parvaiz and Satyawati, 2008). Glycinebetaine (GB)

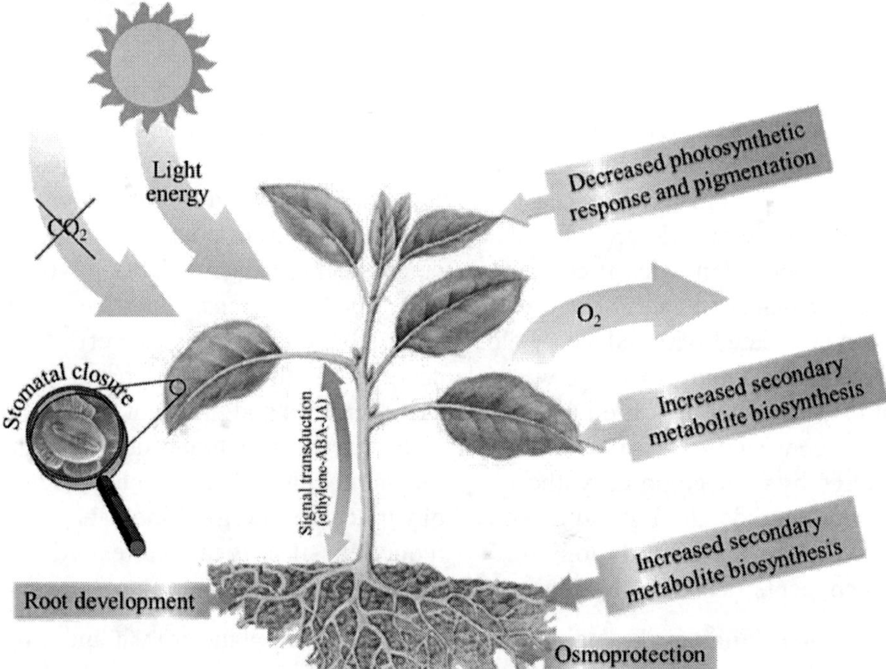

Fig. 4: Effect of soil moisture deficit on physiological processes of plant

is a quaternary ammonium compound that also accumulates in some plant species in response to drought stress (Ashraf and Iram, 2005). Studies have shown that the accumulation of GB in plants depends on genotype, the stage of growth, and the level of drought stress (Ashraf and Foolad, 2007). The level of accumulation of GB also varied considerably among plant species. Some, such as *Oryza sativa* and *Brassica* spp. did not accumulate the compound even when subjected to drought stress. Chen and Murata (2011) demonstrated that plants which accumulated GB normally contained lower endogenous levels of the compound and only accumulated GB when subjected to stress.

Water availability affect the growth and physiological processes of all plants since water is the primary component of actively growing plants ranging from 70-90% of plant fresh mass (Gardner *et al.* 1984). Due to its predominant role in plant nutrient transport, chemical and enzymatic reactions, cell expansion, and transpiration, water stresses result in anatomical and morphological alterations as well as changes in physiological and biochemical processes affecting functions of the plants (Kramer, 1980). Plant water deficits depend both on the supply of water to the soil and the evaporative demand of the atmosphere. In general, plant water stress is defined as the condition where a plant's water potential and turgor are decreased enough to inhibit normal plant function (Hsiao *et al.* 1973). The effects of water stress depend on the severity and duration of the stress, the growth stage at which stress is imposed, and the genotype of the plant (Kramer, 1983). In chickea it was reported that under low soil moisture condition yield is highly influenced by rate of dry matter pertitioning alongwith seed filling period, plant relative growth rate and net assimilation rate (Bhattacharya and Singh, 1997).

Water-deficit stress adversely affects plant performance and yield development throughout the world. Water-deficit stress reduces cell and leaf expansion, stem elongation, and leaf area index (Ball *et al.* 1994; Gerik *et al.* 1996). While comparing the efficiency of days to flowering and seed filling period for seed yield in lentil genotypes under irrigatedand non irrigated conditions Bhattacharya (1999) reported that under non irrigated condition both have lower per cent association as compared to irrigated condition. Leaf, stem and root growth rate are very sensitive to water stress because they are dependent on cell expansion (Hsiao, 1976; Hearn, 1994). Krieg and Sung (1986) reported that water stress caused a reduction in the whole plant leaf area by decreasing the initiation of new leaves, with no significant changes in leaf size of leaf abscission. Both the main stem and sympodial branches developed significantly less leaves; however, the effect was less severe on the main-stem leaves. Pettigrew (2004) reported that water-deficit stress resulted in a decrease in leaf size, but noted that this decrease was accompanied by an increase in the specific leaf weight (SLW), a phenomenon also observed by Wilson *et al.* (1987). Significantly fewer nodes and lower dry weights of stems and leaves of water-stressed plants compared to those of the control were reported by Pace *et al.* (1999), while McMichael and Quisenberry (1991) observed decreased shoot-to-root ratios of plants grown under conditions of severe water stress. Malik *et al.* (1979) reported that root growth appears to be less affected by drought than shoot growth. Several researchers (McMichael and Quisenberry, 1991; Ball *et al.* 1994; Pace *et al.* 1999) observed that seedlings of water stressed cotton showed increased root elongation, accompanied by a reduction in root diameter. The ability of roots to penetrate strong soil has been reviewed by Clark *et al.* (2003), the response of roots to soil physical stresses by Bengough *et al.* (2006), the management of root systems by Hoad *et al.* (2001), and means to sense the root environment by Clark *et al.* (2005). Biophysical processes in the rhizosphere have been comprehensively reviewed (Gregory, 2006; Hinsinger *et al.* 2009) and the process of root–shoot hormonal signalling by Dodd (2005) and Davies *et al.* (2000, 2002). While comparing the association of physiolological and phonological traits with chickpea seed yield Bhattacharya and Pandey (1999) reported that maximum and minimum temperature has definite relationship with normal seeding, but under late seeding no significant relationship was noticed. It was concluded that under late seeding condition, chickpea yield is not a function of temperature during different crop growth duration. Yielding ability of lentil genotypes under non irrigated condition dependes mainly on duration of seed filling period as increase in other physiological traits leads to higher vegetative and less reproductive growth (Bhattacharya and Chandra, 1997). Droght tolerance index (DTI) in lentil genotypes was found to be largely controlled by leaf area and its duration, leaf dry matter and leaf relative growth rate (Bhattacharya and Desh Raj, 1997).

A correlation between leaf abscission and low plant water potentials has been reported (Bruce *et al.* 1965), and McMichael *et al.* (1972) identified a linear relationship between the rates of leaf abscission and the levels of the imposed water deficit stress; however, leaf abscission occurred after the stress was relieved and not during the period of stress. In addition, McMichael *et al.* (1973) observed that younger leaves were not as prone to abscission as older ones. The effects of water deficit on different plant physiological processes are complex and interrelated. Cellular water content largely controls stomatal aperture, and stomatal conductance directly affects CO_2 diffusion and photosynthetic carbon fixation, which in turn affects metabolic functions such as respiration.

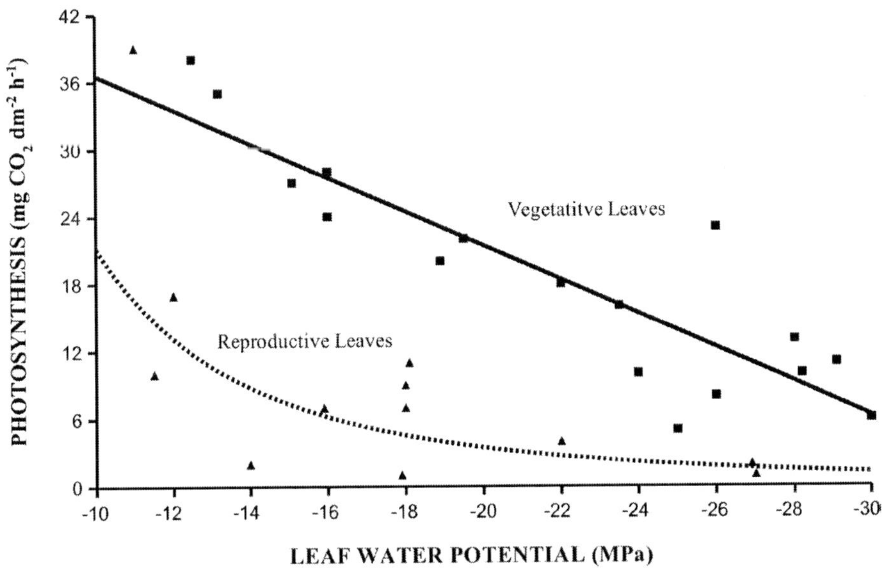

Fig. 5: Relationship between photosynthesis and leaf water potential of vegetative and reproductive cotton leaves. (Redrawn from Ackerson *et al.*, 1977a)

Photosynthesis plays a major role in determining crop productivity in all species and is directly affected by water stress. Photosynthetic rates of the leaves decrease as the relative water content and leaf water potential decrease (Lawlor and Cornic, 2002) (Fig. 5). The effects of water stress on photosynthesis are complex, and may include a combination of stomatal closure and the inhibition of metabolic processes, including ribulose bisphosphate synthesis and adenosine triphosphate synthesis. several reports have indicated that water stress causes a reduction in photosynthesis rates due to a combination of stomatal and non-stomatal limitations (Ephrath *et al.* 1990; Faver *et al.* 1996, Lacape *et al.* 1998; Leidi *et al.* 1999). However, there has been some controversy concerning the relative importance of these two processes responsible for photosynthetic impairment under water deficit (Flexas and Medrano, 2002; Lawlor and Cornic, 2002; Chaves *et al.* 2002, Lawlor, 2002). Chaves and Oliveira (2004) and Flexas *et al.* (2004) reported that decreased CO_2 diffusion from outside the plant to the site of carboxylation is the main cause for reduced photosynthetic rates under most water-stress conditions. Reduced CO_2 diffusion has been attributed to stomatal closure, reduced mesophyll conductance, or a combination of these factors (Flexas *et al.* 2002; Warren *et al.* 2004). Additionally, other factors, such as time of day, ambient CO_2 concentrations, nutrient levels, leaf type, growth stage, genotypic differences and abscisic acid (ABA) concentrations may affect photosynthetic rate in drought-stressed plants.

Respiration is the process by which a plant obtains energy by reacting oxygen with sugars (glucose) to produce water, carbon dioxide and adenosine 5-triphosphate (ATP). Dark respiration (in contrast to photorespiration and photosynthesis) occurs during the day and night, and its rates during the day vary between 25 and 100% of the respiratory activity during the night (Krômer, 1995). Of the CO_2 fixed each day by net photosynthesis, about 30-70% is released back to the atmosphere through dark respiration (Atkin *et al.* 1996) with 50-70% of whole plant respiration occurring in leaves (Atkin *et al.* 2007). However, Flexas *et al.* (2006) pointed out that the percentage of daily fixed carbon that is respired is expected to be higher in water stressed plants, mainly because of the inhibitory effect of water deficit has on photosynthesis. According to Atkin *et al.* (2009) the responses of respiration rates to water deficit vary by plant genotype, the type and the age of tissue (mature or still actively growing), the duration and severity of stress, changes in activity of respiratory enzymes, substrate availability, and ATP demand. De Vries *et al.* (1979) conducted studies in maize (*Zea mays* L.) and wheat (*Triticum aestivum* L.) and observed that while respiration rates remained unaffected at low or moderate water stress, they decreased at severe water stress. A similar pattern was also observed by McCree *et al.* (1984) in sorghum (*Sorghum bicolor* L.), and Boyer (1970) and Ribas-Carbo *et al.* (2005) in soybean (*Glycine max* L.). However, Boyer (1970, 1971) in studies with sunflower (*Helianthus annuus* L.), found a decrease in respiration rates when drought stress was imposed, while Ghashgaie *et al.* (2001) noticed an increase, and Lawlor and Fock (1977) reported no change.

Parida *et al.* (2007) found that total leaf soluble carbohydrate and leaf hexose concentrations were increased, while leaf starch contents decreased in both drought-tolerant and drought-sensitive cultivars. Increase in hexose and depletion of leaf starch concentration have also been reported in soybean (*Glycine max* L. Merr.) (Liu *et al.* 2004) and pigeonpea (*Cajanus cajan* L.) (Keller and Ludlow, 1993). It was observed that higher quantities of starch were accumulated in water-stressed cotton leaves compared to those of the control (Ackerson, 1981), additionally, acclimated young cotton leaves had the ability to export sucrose, whereas non-acclimated plants did not at the same low leaf water potential. It was speculated that translocation of photosynthates was greatly inhibited under conditions of water stress. In support of this observation, Timpa *et al.* (1986) reported that drought stress caused no change in leaf sucrose concentrations of non-flowering cotton strains, while glucose levels were significantly higher in the drought stressed leaves compared to the control, indicating that the source sink relationships are affected by drought.

Impairment of the photoassimilate translocation mechanism under conditions of water-deficit stress has been reported for crops, such as maize (*Zea mays* L.)

(Boyer and McPherson, 1975), and wheat (*Triticum aestivum*, L.) (Johnson and Moss, 1976). Liu *et al.* (2004) made a similar observation for soybean source-sink relationships and reported that sucrose and leaf starch concentrations decreased significantly under water stress resulting in a decrease in the rate of sucrose export from the leaves. Heitholt *et al.* (1994) reported that carbohydrate concentrations of the receptacle and ovary had no relationship with subsequent retention of 5-day-old floral buds or two-day-old bolls in cotton, however, the plants were not subjected to water stress. Similarly, Liu *et al.* (2004) failed to correlate pod abortion of water-stressed soybeans with pod carbohydrate concentrations. Zinselmeier *et al.* (1995, 1999) observed that accumulation of sucrose in young water-stressed maize ovaries paralleled the cessation of ovary growth and an additional decrease in hexose concentration. They speculated that the ratio of hexose to sucrose could play an important role in ovary development. An inhibition of invertase activity due to drought stress could also result in an increase in ovary sucrose content (Liu *et al.* 2004). This was also noted by Weber *et al.* (1998) for legume seed development.

Plants experiencing stressful conditions, such as drought, tend to actively accumulate highly soluble organic compounds of low molecular weight, called compatible solutes, as well as inorganic ions, *i.e.* K, in order to prevent water loss, maintain water potential gradients and reestablish cell turgor (Hsiao, 1973). This process is called osmotic adjustment and according to Boyer (1982) enables plants to: (1) continue normal leaf elongation but at a reduced rate, (2) adjust their stomatal and photosynthetic functions, (3) maintain the development of their roots and subsequently continue soil moisture extraction, (4) postpone leaf senescence, and (5) achieve better dry matter accumulation and yield production under adverse conditions. Osmotic adjustment has been reported in the leaves of a number of crops such as wheat (*Triticum aestivum* L.) (Morgan, 1987), maize (*Zea mays* L.) (Acevedo *et al.* 1979), sorghum (*Sorghum bicolor* L.) (Jones *et al.* 1978; Turner *et al.* 1978), rice (*Oryza sativa* L.) (Cutler *et al.* 1980), barley (*Hordeum vulgare* L.) (Matsuda *et al.* 1981), pearl millet (*Pennisetum americanum* L.) (Henson *et al.* 1982), sunflower (*Helianthus annuus* L.) (Turner *et al.* 1978), as well as cotton (*Gossypium hirsutum* L.) (Acevedo *et al.* 1979, Oosterhuis and Wullschleger, 1987). Interestingly, cotton appears to have a greater ability to osmotically adjust to water stress compared to other major crops (Ackerson *et al.* 1977; Oosterhuis and Wullschleger, 1988). Additionally, Oosterhuis *et al.* (1987) observed that primitive landraces and wild types of cotton exhibited higher osmotic adjustment compared to commercial cultivars. They investigated the osmotic adjustment of cotton roots under water deficit and demonstrated that cotton roots show a considerably larger percentage adjustment than the leaves, reinforcing the ability of the plant to maintain a positive turgor and hence continue normal growth under water stress. A similar pattern was observed in cotton flowers (Trolinder *et al.* 1993) and bolls (Van Iersel and Oosterhuis, 1996) wherein both flowers and fruits were found to be less affected by the water stress imposed than the subtending leaves. These authors concluded that cotton flowers and bolls are largely independent on the xylem connections for their water supply and that the phloem is the most important factor in water transport to the flowers and developing bolls. Ackerson and Hebert (1980) observed that cotton plants that had been subjected to consecutive water-stress cycles exhibited increased osmoregulation compared to plants that had not been subjected to stress previously. They reported that photosynthetic rates were higher due to higher stomatal conductance at low water potentials, but the opposite was observed under high water potentials.

Drought increases senescence, by accelerating chlorophyll degradation, leading to a decrease in leaf area and canopy photosynthesis. There is evidence that stay-green phenotypes with delayed leaf senescence can improve their performance under drought conditions (Rivero *et al.* 2007; Lopes and Reynolds, 2012). $\Delta^{13}C$ can be used as a selection criterion for high water use efficiency (Condon *et al.* 2004; Richards, 2006), but also can provide an indirect determination of the effective water used by the crop (Araus *et al.* 2002, 2008; Blum, 2009). In fact, kernel $\Delta^{13}C$ can be positively or negatively correlated with grain yield depending on soil water availability. Indeed, under moderate stress to well-watered Mediterranean conditions $\Delta^{13}C$ has been reported to be positively correlated with grain yield in wheat (Araus *et al.* 2003, 2008 for wheat) and barley (del Pozo *et al.* 2012), whereas the opposite trend has been reported under severe drought conditions (Araus *et al.* 1998).

Plant responses to water deficit stress are confounded by several factors such as time, intensity, duration and frequency of stress as well as by plant, soil and climate interactions (Reynolds and Tuberosa, 2008). In addition, the difficulty to establish well-defined and repeatable water stress conditions makes screening of drought tolerant genotypes more complex (Ramirez and Kelly, 1998). Therefore different indicators should be used for the phenotyping of drought tolerance (Tuberosa, 2012). Presently, a number of selection indicators, such as stress tolerance index (STI), water-use efficiency (WUE), drought susceptibility index (DSI), relative vigor index (RVI) and leaf wilting index (LWI), are widely used. However plants respond and adapt to drought stress by the induction of various morphological and physiological responses (Wang and Huang, 2004). Many physiological factors could be involved in the drought stress injury (Jiang and Huang, 2001) which may promise for characterizing drought resistance in screening studies. For example, water stress can be caused increase of proline content, stomata close and photosynthesis inhibit. Also, water stress induced a significant decrease and increase in chlorophyll contents and accumulation of proline in *Brassica* crops, respectively (Gibon *et al.* 2000). All these abnormalities as a result overall ultimately are decreased production of crop. Maliwal *et al.* (1998) have reported reduced yield in *Brassicas* in response to water stress. Kumar *et al.* (1984) and Singh *et al.* (1985) have reported close associations between osmotic adjustment and both stomatal conductance and canopy temperature in many *Brassica* species. Keles and Oncel (2004) suggested that the high relative water content is closely related to drought resistance. Result of a study shows that with increasing drought stress, amount of relative water content is reduced (Sepehri and Golparvar, 2011). Din *et al.* (2011) reported significant differences among the various canola genotypes for leaf chlorophyll a, chlorophyll b and proline accumulation. Previous studies reported that STI, MP, GMP and MSTI are useful indices for screening drought tolerant rapeseed genotypes under stress condition (Malekshahi *et al.* 2009; Shirani-rad and Abbasian, 2011; Yarnia *et al.* 2011; Khalili *et al.* 2012), however the association of these indices with physiological and morphological traits was not assessed.

4.5. Excess Soil Moisture (Water logging) Stress

Plants require water for growth but excess water that occurs during submergence or waterlogging is harmful or even lethal. A submerged plant is defined as "a plant standing in water with at least part of the terminal above the water or completely covered with water" (Figure 6). Submergence subjects plants to the stresses of low light, limited gas diffusion, effusion of soil nutrients, mechanical damage, and increased susceptibility to pests and diseases (Ram *et al.* 1999). Basically, flooding (*i.e.*, submergence) can be classified into "flash flooding" and "deepwater flooding" in accordance with the duration of flooding and the water depth (Bailey-Serres *et al.* 2010). Flash flooding, which generally lasts less than a few weeks, is caused by heavy rain but the depth is not very deep. On the other hand, deepwater flooding, which lasts for several months, occurs during the rainy season, and the water depth reaches several meters (Hattori *et al.* 2011). Waterlogging is defined as a condition of the soil in which excess water limits gas diffusion (Setter and Waters, 2003). Oxygen diffusivity in water is approximately 10,000 times slower than in air, and the flux of O_2 into soils is approximately 320,000 times less when the soil pores are filled with water than when they are filled with gas (Colmer and Flowers, 2008). The principal cause of damage to plants grown in waterlogged soil is inadequate supply of oxygen to the submerged tissues as a result of slow diffusion of gases in water and rapid consumption of O_2 by soil microorganisms. Oxygen deficiency in waterlogged soil occurs within a few hours under some conditions. In addition to the O_2 deficiency, production of toxic substances such as Fe^{2+}, Mn^{2+}, and H_2S by reduction of redox potential causes severe damage to plants under waterlogged conditions (Setter *et al.* 2009). Thus, growth and development of most plants, except for rice (*Oryza sativa* L.) and other wetland species, are impeded under waterlogged conditions.

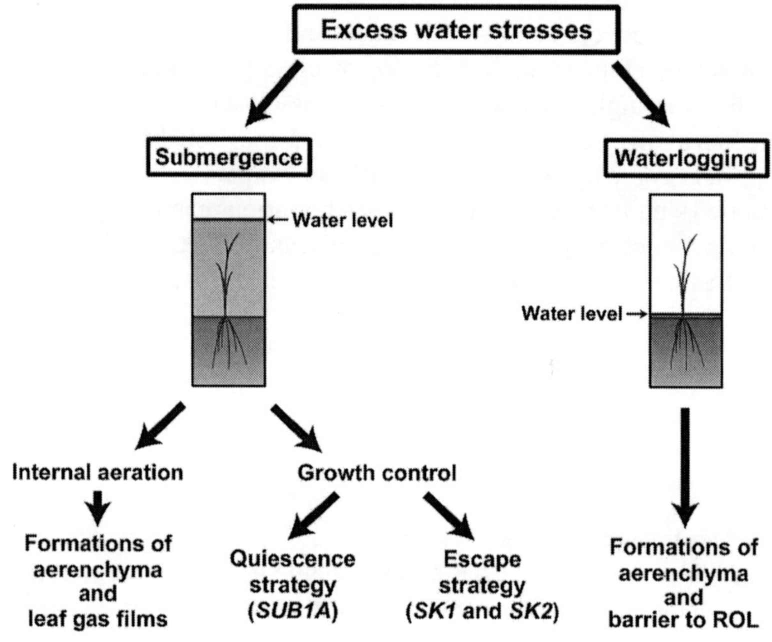

Fig. 6: Strategies of adaptation to excess water stresses in the form of submergence or waterlogging in rice plants

In some wetland plant species root lysigenous aerenchyma is constitutively formed under drained soil conditions, and its formation can be further enhanced during soil waterlogging (Figure 6) (Colmer *et al.* 2006; Shiono *et al.* 2011). In rice, aerenchyma formation is initiated at the apical parts of the roots and gradually expands to the basal parts of the roots (Figure 6) (Ranathunge *et al.* 2003). Fully developed aerenchyma, which is observed on the basal parts of roots, separates the inner stele from the outer cell layers (*i.e.*, sclerenchyma, hypodermis/ exodermis, and epidermis) of the roots (Figure 6) (Kozela and Regan, 2003; Ranathunge *et al.* 2003). Strands of remaining cells and cell walls separate gas spaces in the cortex, forming radial bridges, which are important for the structural integrity of the root and for both apoplastic and symplastic transport of nutrients (Figure 7). During aerenchyma formation in rice root, cell death begins at the cells in the mid-cortex and then spreads out radially to the surrounding cortical cells (Kawai *et al.* 1998). The epidermis, hypodermis/ exodermis, endodermis, and stele are unaffected, indicating that lysigenous aerenchyma formation occurs by closely controlled mechanisms (Yamauchi *et al.* 2011).

Rice can adapt to submergence by internal aeration and growth control. For internal aeration, rice develops longitudinally forming aerenchyma and leaf gas films. On the other hand, some rice cultivars can survive under submergence by using special strategies of growth control: a quiescence strategy or an escape strategy. The Submergence-1A (SUB1A) gene is responsible for the quiescence strategy, which is important for survival under flash-flood conditions. The SNORKEL1 (SK1) and SNORKEL2 (SK2) genes are responsible for the escape strategy, which is important for survival under deepwater-flood conditions. Rice can adapt to soil waterlogging by forming aerenchyma and a barrier to radial O_2 loss (ROL) in the roots ((Nishiuchi et al. 2012).

Excess moisture stress, waterlogging and flooding are the major problems worldwide causing severe damage to the crop plants (Perata *et al.* 2011). Morphological, anatomical and physiological adaptations in plants ensure their survival under stress conditions. The extent of damage depends upon the crop, growth stageand environmental conditionsat the time of stress (Shah *et al.* 2012). In maize, submergence of root for more than one day restrict the optimum production of the crop (Singh and Ghildyal, 1980). When maize plants are exposed to waterlogging stress at different crop growth stag, plant growth and yield were reduced at all growth stages. Maize is vey susceptible to waterlogging at an early vegetative stage (Mukhter *et al.* 1990; Lonc and Warsi, 2009). In maize, a positive correlation was found between yield and node-bearing adventitious roots (Lone and Warsi, 2009), and between flooding tolerance and the amount of adventitious roots (Zaidi *et al.* 2007; Lone and Warsi, 2009). Waterlogging induces aerenchyma formation in maize and that facilitate gaseous transport between roots and shoots and confers wterlogging resistance in the crop (Mano *et al.* 2006; Yadav, 2010). Flooding causes greater yield losses in maize (Lizaso and Ritchie, 1997; Ferreira *et al.* 2007) and wheat (Ghobadi and Ghobadi, 2010) when it occur early in the season. Reduction in maize yirld under waterlogging stress is mainly due to decrease kernel rows cob^{-1}, cob length, cob diameter, test weight (Tripathi, 2000) and grain number cob^{-1} (Schild *et al.* 1999).

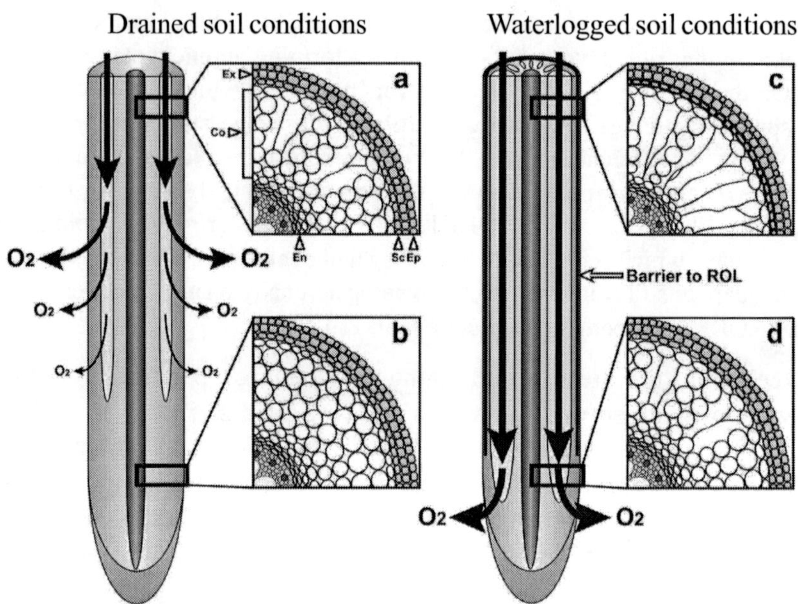

Fig. 7: Differences in lysigenous aerenchyma formation and patterns of radial O_2 loss (ROL) in rice roots under drained soil conditions and waterlogged soil conditions. Under drained soil conditions, lysigenous aerenchyma is constitutively formed, but a barrier to ROL is not formed; thus ROL at the basal part of the root decreases O_2 diffusion to the apical part. By contrast, under waterlogged soil conditions lysigenous aerenchyma formation is enhanced and formation of the barrier to ROL is induced, resulting in the promotion of longitudinal O_2 diffusion to the root apex. Under drained soil conditions, lysigenous aerenchyma is constitutively formed at the basal part of the roots (a), but it is not usually formed at the apical part of the roots (b). Under waterlogged soil conditions, lysigenous aerenchyma is induced at the basal part (c) and the apical part (d) of the roots. Lysigenous aerenchyma is more highly developed at the basal part of the roots (a, c) than at the apical part (b, d). Arrow thickness reflects the aMount of O_2 available. Ep, epidermis; Ex, exodermis; Sc, sclerenchyma; Co, cortex; En, endodermis. (Nishiuchi *et al.* 2012).

Permanent flooding is a severe stress to trees that grow in wetland areas and affects normal plant functioning (Jackson and Colmer, 2005). Plant responses to waterlogging are mediated by different morphological and physiological mechanisms (Bailey-Serres and Voesenek, 2008; Parent *et al.* 2008). Flooding suppresses leaf formation and expansion of leaves, lead to premature leaf abscission, chlorosis and senescence, and reduces leaf size, area and number of leaves (Kozlowski, 1997). Waterlogging negatively affects shoot growth, produces death and decay of roots (Kozlowski, 2002), alters the proportions allocated to below- and above-ground biomass compartments (Gonzalez *et al.*

2010) and is often associated with massive decreases in overall plant biomass increment and productivity (Kozlowski, 1997, 2002), water and nutrient uptake and reduced metabolism (Dat *et al.* 2004). Other physiological adaptations include reduction in stomata conductance and photosynthesis as well as root hydraulic conductivity (Parent *et al.* 2008), and decrease in chlorophyll and protein levels (Kozlowski, 2002). Partial and total submersion, that are extreme cases of flooding, may also induce in flood tolerant species the formation of adventitious roots, hypertrophied lenticels and aerenchyma development (Kozlowski, 1997, 2002; Dat *et al.* 2004; Glenz *et al.* 2006). In fact, the height of flood waters during plant submersion is an important issue for plant development. Flooding depths may affect the expression of the stress adaptability by means of change in growth, morphology and physiological processes in plants, as noted by Iwanaga and Yamamoto (2008) for *Alnus japonica.* The variety of plant strategies to cope with waterlogging conditions, however, may vary enormously among closely related species (Kozlowski, 1997; Voesenek *et al.* 2006). For example, *Populus nigra* a riparian tree that dominates the forest of many river courses in Southwestern Europe, is generally considered as a high flood-tolerant species, while one of its natural competitors, *Populus alba,* might be better classified as an intermediate flood-tolerant (Glenz *et al.* 2006). In the case of the genus *Alnus* previous studies have examined the tolerance to permanent flooding of *A. glutinosa* (Siebel and Blom, 1998), *A. incana* (Hughes *et al.* 1997; Kaelke and Dawson, 2003; Francis *et al.* 2006), *A. japonica* (Iwanaga and Yamamoto, 2008), *A. maritima* (Schrader *et al.* 2005), *A. rubra* Bong and *A. viridis* Sinuate (Batzli, 1997). The overall result of those studies is that *Alnus* spp. are high tolerant to permanent flooding but some species such as *A. glutinosa* may be relatively more than others in the genus. However, to our knowledge, there is neither study concerning the flood-tolerance of *A. subcordata* nor any information about the effects of flooding depth on growth, morphology and physiology of this species in the existing literature.

Water logging causes a condition of hypoxia (low oxygen concentration) in soils. Water logging can also cause the accumulation of ethylene and products of root and bacterial anaerobic metabolism (Barrett-Lennard, 2003). Plants tolerant to water logging stress exhibit certain adaptations, such as, formation of aerenchyma and adventitious roots. Furthermore, due to the interaction of plant hormones, auxin and ethylene, the formation of adventitious roots takes place (Ashraf, 2012). Rice and some associated species, such as, barnyard grass (*Echinochloa* sp.) will germinate quickly under anoxia once the seeds imbibed (Crawford, 2003). Some species are very sensitive to water logging at germination stage. A decrease in germination ability is due to oxygen deficiency in water logged soils. Respiration, electron transport and ATP formation are inhibited during germination when oxygen is short (Hsu *et al.* 2000). Plants under water logging condition exhibit growth reduction, low SLA (specific leaf area), photosynthesis reduction, stomatal closure, decrease in respiration and biomass production and protein degradation (Juan *et al.* 2009). Plants could get only two ATP molecules in fermentation, whereas 36 ATP molecules are produced per glucose molecule in aerobic respiration. Flood tolerant plants could retain their energy status by using fermentation. In waterlogged plants, initial decline in cytosolic pH attributed to the production of lactic acid fermentation. The initial decline in pH helps the plants to change from lactate to ethanol formation by establishment of alcohol dehydrogenase and inhibition of lactate dehydrogenase (Ashraf, 2012). The formation of aerenchyma and adventitious roots is an indicator of the presence of adoptive mechanisms in many flood-tolerant plants. The interaction of auxin and ethylene is important for the stimulation of adventitious root formation (Liao, 2000). Moreover, aerenchyma which Is thought to contribute to water logging tolerance, is develops in cortex of nez existing roots of plant species (Akhtar and Nazir, 2013). In rice crop the aerenchyma well developed and adopt flooded condition (Tepwadee *et al.* 2008). Many plants such as *Paspalum dilatatum* responded to flooding by increasing root and leaf sheath aerenchyma. Water logging under saline condition inhibit the ability to screen out salt at the root surface, therefore, there are large increase in salt uptake and in salt concentration in the shoots (Alam *et al.* 2010). Reduction of root respiration is one of the earliest responses of plants under the absence of oxygen, in spite of whether the plants are flooding-tolerant or intolerant (Wample and Davis, 1983). Oxygen deficiency inhibits the root respiration of plants, which results in considerable reduction in energy kind of root cells.

Excess soil moisture causes major changes in physical and chemical properties in rhizosphere (Zaidi *et al.* 2003). Gaseous diffusion rates in flooded soil are about 100 times lower than air (Kenedy *et al.* 1992) and respiration of plant roots, soil micro-flora and fauna leads rapid exhaustion of soil oxygen, and thereby causes hypoxia/anoxia condition. Therefore plant root suffer with extreme oxygen stress, hypoxia followed by anoxia (Zaidi *et al.* 2003). Research on plant response to anoxia has investigated a wide spectrum of changes at the molecular, biochemical, physiological, anatomical and morphological levels and reviewed (Kenedy *et al.* 1992; Perata and Alpi, 1993; Ricard *et al.* 1994; Drew, 1997). Sachs *et al.* (1980) reported that within the first hour of anaerobiosis, the synthesis of normal aerobic proteins was suppressed and a new set of protein, known as anaerobic proteins, were synthesized in maize. At physiological level, anoxia induces changes in the levels of all plant hormones including ethylene (Voesenek *et al.* 1992), abscissic acid (Jackson, 1990), cytokinin (Bradford, 1983), auxin and gibberellins (Reid and Bradford, 1984). Changes in plant anatomy and morphology in response to anaerobiosis have been well documented (Jackson, 1990). The formation of aerenchyma in root cortex, stem hypertrophy and adventitious root formation are the examples of adaptive acclimation responses to flooding stress. In maize, excess moisture stress suppressed plant growth and development, reduced dry matter accumulation, leaf area development, transpiration, affected anthesis and silking and poor yield (Zaidi and Singh, 2001; Zaidi *et al.* 2003).

The extent of damage due to excess moisture stress varies significantly with developmental stage (Zaidi *et al.* 2004). In some genotypes of maize, excessive soil moisture stress at later growth stages adventitious roots and morphological adaptation like aerenchyma formation in cortical region of adventitious roots have been reported (Zaidi *et al.* 2003). Considerable genetic variation have been observed in maize for tolerance to excess moisture (Rathore *et al.* 1998; Zaidi and Singh, 2001; Zaidi *et al.* 2003).

4.6. Soil Salinity Stress

Plants are exposed to many types of environmental stresses. Among these, salinity and drought are two of the most significant problems that limit plant growth (Raveh and Levy, 2005). Salinity is a problem of great concern, at which, about one third of the world's irrigated land is not in use (Giri and Mukerji, 2004) and about ten million hectares of irrigated agricultural land are abandoned annually. Salinity as an abiotic stress widely limit the crop production severely (Shannon, 1998). Salinity is the buildup of soluble salts by which saline soils are formed (Levy and Syvertsen, 2004). Salinity may be due to many reasons but some of the adverse effects of salinity have been attributed to increase in chloride and sodium ions (Zekri, 2001; 2004) in different plant organs hence these ions create the critical conditions for plants survival by intercepting different plant mechanisms (Raveh, 2005; Grieve *et al.* 2007). Sodium and chloride are the major ions, which cause many physiological disorders and limit plant growth and productivity (Kamal *et al.* 2003; Ashraf and Foolad, 2005). Excess of these salts also enhances the osmotic potential of soil matrix as a result of which water intake by plants is restricted (Garcia-Sanchez *et al.* 2002).

A saline soil is usually the reservoir of a number of soluble salts such as Ca^{2+}, Mg^{2+}, Na^{+} and anions SO_4^{2-}, Cl^{-}, HCO_3^{-} with exceptional amounts of K^{+}, CO_3^{2-}, and NO_3^{-}. A soil can be termed as saline if its EC is 4 dS/m or more (USDA-ARS, 2008), (equivalent to approximately 40 mM NaCl) with an osmotic pressure of approximately 0.2 MPa. Salinity is the condition when the EC is sufficient to cause yield reduction of most crops. The pH of saline soils generally ranges from 7-8.5 (Mengel *et al.* 2001). However, the pH in saturated soil can vary provoking severe crop damage. The arid and semi-arid zones, characterized by low precipitation and high evaporation are the most affected due to minimum lixiviation of salt from the soil profile resulting in increased salt accumulation.

Salinity prone areas found in the arid and semiarid zones are usually accounted to the accumulation of salts over ages. Moreover, weathering of the parental rocks has accelerated the process a lot (Rengasamy, 2002). Salinity is a well off natural phenomenon occurring near sea shores due to sea water flooding.

Table: Crop response to salinity influenced by electrical conductivity of saturated soils extract

Electrical conductivity (ds m^{-1} at 25^0 C)	Crop response
0-2	Salinity effect is practically zero
4-Feb	Reduction in yield of very sensitive crops
8-Apr	Reduction in yield of most crops
16-Aug	Only tolerant crops produce satisfactory yield
>16	Few highly tolerant crops produce satisfactory

Adapted from Menzel *et al.* 2001

Salinity has been a potential threat affecting almost 900 million ha of land which nearly accounts for 20% of the globally cultivated area and also half of the total irrigated land of the world (Munns, 2002; FAO, 2007). Globally salt affected area accounts to about 1 billion ha of land (Fageria *et al.* 2012). In India the scenario accounts for about 8.4 million ha land affected by salinity (Tyagi and Mitra, 1998). In India, irrigated land accounts for only 15% of total cultivated land, nevertheless, it has at least twice the productivity of rain-fed land and yields about one third of the world's consumption.

Plants, have evolved many different physiological and biochemical strategies for coping with water stress caused by NaCl or drought conditions. Salt stress is one of the most important abiotic stresses in arid and semiarid regions (Ashraf and Harris, 2004; Hebbara *et al.* 2003). Plants under stress have less dry matter, leaf area (Amirjani, 2011) and yield. Also it has been reported that salinity changes plant morphological characteristics (Zadeh and Naeini, 2007). According to FAO (2013) increased soil salinity, aquifer depletion and land degradation reduce achievable yields, thus putting at risk farmers' ability to bridge production gaps and improve food security. The impact of salt stress has been correlated with morphological and physiological traits like reduction in fresh and dry weight (Chartzoulakis and Klapaki, 2000). The deleterious effects of salinity on plant growth are associated with low osmotic potential, specific ion effects, and oxidative stress, causing nutritional imbalances due nutrient or element uptake (Ashraf and Harris, 2004). Studies of plant tolerance to salt stress cover many aspects on the influences of salinity on plant response, including alterations at the morphological, physiological and molecular levels.

Succulent plants with crassulacean acid metabolism (CAM) have adapted to arid conditions by conserving water. The most important benefit of CAM on the plant is the ability to maintain closed most of its leaf stomata during the day. This characteristic helps CAM plants be more resistant to drought stress (Rahimi-Dehgolan *et al.* 2012). Some experiments have been carried out in different geographical locations for the assessment of growth, biomass, soluble carbohydrates, cations, gel, and others variables or traits under saline stress (Tawfik *et al.* 2001; Sun *et al.* 2003; Silva *et al.* 2010; Kiran Kumari *et al.* 2012; Moghbeli *et al.* 2012; Rahimi-Dehgolan *et al.* 2012; Rahi *et al.* 2013). However, the majority of these studies have used high concentrations of NaCl (up to 200 mM), high electrical conductivities, some of them up to 12 dS m^{-1}, and in some cases, those studies have used 60 % of sea water as a salinity treatment, with the consequence of negative effects and damages in growth, yield, physiology, morphology, and other traits and characteristics of plants. Most research related to saline stress focus on glycophytes under salt stress in the laboratory, greenhouse or field. In general terms, when salinity is focused for selection of plants with superior tolerance to salinity, mild or moderate salinity treatments have been to use because of studies have been demonstrated that using them, the effects of salinity can be more easily evaluated because of if high salinity levels are used, the majority of the plants species died and the evaluation of yield, growth and other characteristics or variables cannot be measured. Moreover, when the studies of salt-tolerant plants use mild or moderate salinity levels, the antioxidative capacity of fruits increased (D'Amico *et al.* 2003), with some levels of enzymes increasing such as superoxide dismutase (Keutgen and Pawelzik, 2007), causing the quality of the fruit to improve (Awang *et al.* 1993a,b; Keutgen and Pawelzik, 2008). Other effects measurable at mild salt stress are Na or Cl exclusion (Orsini *et al.* 2012), increase of biomass production (Adolf *et al.* 2013). Hence, moderate salinities are probably the more suitable for measuring response to growth, yield and survival (Allen *et al.* 1994; Munns, 2002). Also, when mild or moderate salinity levels are used, this offers the possibility of genetic improvement by selection of suitable plant material to reclaim moderately salinized lands. To improve the agricultural management of *Aloe vera* under the arid and semi-arid conditions of northwest of Mexico, it is important to investigate how crop growth and yield might be affected by the physiological and morphometric responses to saline stress, because of many farmlands around the coastal zone of this region are affected by sea water intrusion. Therefore, to explore the response of morphometric and physiological traits of *Aloe* to salt stress, plants of one cultivar of *Aloe* were subjected to four NaCl concentrations (0, 30, 60, 90, and 120 mM) to assess the salt tolerance of the species by analyzing the plant growth and other variables associated with plant physiology and to provide scientific basis for safe and productive cultivation of *Aloe* under mild saline water irrigation.

Plant roots provide an ideal experimental system to investigate the effects of salinity on growth and other parameters given that, a) growth is restricted to a definite region, the millimetres immediately above the apical meristem, followed by a non-growing zone consisting of mature elongated cells located just some centimetres above the tip (Ishikawa and Evans, 1995); b) root cells can be directly exposed to different NaCl concentrations by changing the root medium (Hilal *et al.* 1998). Anatomical changes could compromise plant ability to conduct water and nutrients in high salinity. Ortega *et al.* (2006) observed a diminution of leaf protoxylem diameter in Rhodes grass leaves. Applying the law of Hagen-Poiseulle which relates water flux to the forth power of the xylem conduits radius, it can be deduced that a small variation in this value would imply a big increase in hydraulic resistance (Lewis and Bosse, 1995). A reduction in root hydraulic conductivity was observed by Peyrano *et al.* (1997) in tomato under salinity. Salt stress affects other aspects of plant metabolism, such as reduced growth (Choudhary and Srivastava, 2015). Photosynthesis is reduced because it is affected by leaf area and leaf duration, as well as by respiration per unit leaf area. Mineral uptake by roots is affected as a result of imbalance in the availability of different ions. Furthermore, a decrease in nitrate reductase activity and an inhibition of photosystem-II and chlorophyll breakdown are associated with increased Na^+ accumulation in plant tissues (Krishnamurthy and Bhagwat, 1995).

Salinity reduces root length and diameter (Neumann, 1995). Anatomically, it affects cell division and expansion processes (Zidan *et al.* 1990), reduces the size of apical meristems, cortex and vascular cylinder. Additionally, stimulates exodermis and endodermis suberization (Reinhardt and Rost, 1995; Sanderson *et al.* 1997; Ramos *et al.* 2004) or the occurrence of atypical structures such as rhizodermis with phi-thickenings. (Degenhardt and Gimmler, 2000). The most common anatomical response to salinity is related to cell wall modifications. In cotton, an accelerated deposition of suberin in cells of the Casparian strip was observed (Reinhardt and Rost, 1995). Zhong and Lauchli (1993) found a rise in uronic acid along with lower cellulose content per unit of dry matter in cell walls of primary roots of cotton as effect of elevated salts levels. These alterations could be the cause of the change observed in the relative proportions of root tissues (Ramos *et al.* 2004), which modify the shape of the organ and its function (Hauser *et al.* 1995). At a molecular level it is known that several genes are expressed upon salt exposure and a number of proteins involved in salt-tolerance have been identified (Bohnert and Jensen, 1996).

One of the known mechanisms of *Chloris gayana* (Kunth.) to cope with salinity is the presence of bi-cellular salt glands in its leaves which build up and excretes the sodium excess (Amarasinghe and Watson, 1989). Due to the biological and agronomical importance of *C. gayana,* the study of its mechanisms of tolerance is of special importance (Kobayashi *et. al.,* 2007). In this case, it exist a broad variability in function of cultivars. Significant reductions in the tetraploid cultivars productivities have been observed (Taleisnik *et al.* 1997; de Luca *et al.* 2001). Reduced yielding in these cultivars under salinity is manifested as a higher number of senescent leaves, a lower leaf area expansion and a minor number of stolons per plant (de Luca *et al.* 2001).

4.7. Heavy Metal Stress

Heavy metals belong to group of nonbiodegradable, persistent inorganic chemical constituents with the atomic mass over 20 and the density higher than 5 $g \cdot cm^{-3}$ that have cytotoxic, genotoxic, and mutagenic effects on humans or animals and plants through influencing and tainting food chains, soil, irrigation or potable water, aquifers, and surrounding atmosphere (Flora *et al.* 2008; Rascio and Navari-Izzo, 2011; Wuana and Okieimen, 2011). There are two kinds of metals found in soils, which are referred to as essential micronutrients for normal plant growth (Fe, Mn, Zn, Cu, Mg, Mo, and Ni) and nonessential elements with unknown biological and physiological function (Cd, Sb, Cr, Pb, As, Co, Ag, Se, and Hg) (Rascio and Navari-Izzo, 2011; Schützendübel, 2002; Tangahu *et al.* 2011; Zhou *et al.* 2014). Both underground and aboveground surfaces of plants are able to receive heavy metals (Patra *et al.* 2004). Essential elements play a pivotal role in the structure of enzymes and proteins. Plants require them in tiny quantities for their growth, metabolism, and development; however, the concentration of both essential and nonessential metals is one single important factor in the growing process of plants so that their presence in excess can lead to the reduction and inhibition of growth in plants. Heavy metals at toxic levels hamper normal plant functioning and act as an impediment to metabolic processes in a variety of ways, including disturbance or displacement of building blocks of protein structure, which arises from the formation of bonds between heavy metals and sulfhydryl groups (Hall, 2002), hindering functional groups of important cellular molecules (Hossaim *et al.* 2012), superseding or disrupting functionality of essential metals in biomolecules such as pigments or enzymes (Ali *et al.* 2013) and adversely affecting the integrity of the cytoplasmic membrane (Farid *et al.* 2013), resulting in the repression of vital events in plants such as photosynthesis, respiration, and enzymatic activities (Hossaim *et al.* 2012). Elevated levels of heavy metals are associated with the increased generation of reactive oxygen species (ROS) as well as cytotoxic compounds which can cause oxidative stress via disturbing the equilibrium between prooxidant and antioxidant homeostasis within the plant cells (Zengin and Munzuroglu, 2005; Hossaim *et al.* 2012; Sytar *et al.* 2013). This condition implicates the causation of multiple deteriorative disorders such as, oxidation of protein and lipids, ion leakage, oxidative DNA attack, redox imbalance, and denature of cell structure and membrane, ultimately resulting in the activation of programmed cell death (PCD) pathways.

Heavy metals present a risk for primary and secondary consumers and ultimately humans (Zeller and Feller, 1999). Among toxic metals, Lead (Pb) and Cadmium (Cd) appear to be the most dangerous to the environment (Malkowski *et al.* 2005). Pb^{2+} and Cd^{2+}, are growth inhibition, ion uptake and transport disturbances, enzyme activation or inhibition photosynthesis (Geebelen *et al.* 2002). Mercury (Hg) poisoning has become a problem of current interest as a result of environmental pollution on a global scale. The availability of soil Hg to plants is low and there is a tendency for Hg accumulation in the roots. Indicating that roots serve as a barrier to Hg uptake (Tripathi and Tripathi, 1999). Cereals in this case maize are known to be good accumulators of contaminants (Malgorzata and Andzej, 2005). The mechanisms of heavy metal toxicity on photosynthesis is still a matter of speculations, this may be partly due to the differences in experimental design, but some evidence points to the involvement of electron transport in lights reactions (Giardi *et al.* 1997). The mechanism of accumulation of heavy metals is still not completely understood (Malkowski *et al.* 2005). Lead and cadminum are the most widespread no nutrient heavy metals (Mihailovic, 2010).

Plants employ various inherent and extrinsic defense strategies for tolerance or detoxification whenever confronted with the stressful condition caused by the high concentrations of heavy metals. As a first step towards dealing with metal intoxication, plants adopt avoidance strategy to preclude the onset of stress via restricting metal uptake from soil or excluding it, preventing metal entry into plant root (Viehweger, 2014). This can be achieved by some mechanisms such as immobilization of metals by mycorrhizal association, metal sequestration, or complexation by exuding organic compounds from root (Patra *et al.* 2004; Dalvi and Bhalerao, 2013). At next stage, if these strategies fail and heavy metals manage to enter inside plant tissues, tolerance mechanisms for detoxification are activated which include metal sequestration and compartmentalization in various intracellular compartments (*e.g.*, vacuole) (Patra *et al.* 2004),, metal ions trafficking, metal binding to cell wall, biosynthesis or accumulation of osmolytes andosmoprotectants, for example, proline, intracellular complexation or chelation of metal ions by releasing several substances, for example, organic acids, polysaccharides, phytochelatins, and metallothioneins (Dalvi and Bhalerao, 2013; Prasard, 2004, John *et al.* 2009), and eventually if all these measures prove futile and plants become overwhelmed with toxicity of heavy metal, activation of antioxidant defense mechanisms is pursued (Manara, 2012).

One of the mechanisms adopted by plants to detoxify heavy metals is the production of short-chain thiol-rich repetitions of peptides of low-molecular weight synthesized from sulfur-rich glutathione by the enzyme phytochelatin synthase with the general structure of (γ-glutamyl-cysteinyl) *n*-glycine ($n = 2$ to 11) that have a high affinity to bind to heavy metals when they are at toxic levels (Wang *et al.* 2009; Shukla *et al.* 2013; Gupta *et al.* 2013). Phytochelatins, as a pathway for metal homeostasis and detoxification, have been identified in a wide range of living organisms from yeast and fungi to many different species of animals (Bundy *et al.* 2014). In plants, phytochelatins are found to be part of the defensive act not only against metal-related stresses but also in response to other stressors such as excess heat, salt, UV-B, and herbicide (Zagorchev *et al.* 2013). Cytosol is the place where phytochelatins are manufactured and actively shipped from there in the form of metal-phytochelatin complexes of high molecular weight to vacuole as their final destination (Song *et al.* 2014).

Metallothioneins are another family of small cysteine-rich, low-molecular-weight cytoplasmic metal-binding proteins or polypeptides that are found in a wide variety of eukaryotic organisms including fungi, as well as some prokaryotes (Du *et al.* 2012; Macovei *et al.* 2010). Contrary to phytochelatins that are the product of enzymatically synthesized peptides, metallothioneins are synthesized as a result of mRNA translation (Verkhlei *et al.* 2003). Phytochelatins in plants are mainly deal with Cd detoxification, metallothioneins appear to be capable of showing affinity with a greater range of metals such as Cu, Zn, Cd, and As (Yang and Chu, 2011). In addition to their role in heavy metals detoxification, metallothioneins are known to be active agents in a number of cellular-related events including ROS scavenger (Wong *et al.* 2004), maintenance of the redox level (Macovei *et al.* 2010), repair of plasma membrane (Mishra and Dubey 2006), cell proliferation, and its growth and repair of damaged DNA (Grennan 2011).

The assessment of soil contamination by metals has been extensively carried out through plant analysis (Blaylock *et al.* 2003; Brooks, 1998); both wild and cultivated plant species have been frequently used as (passive accumulative) bio indicators for large scale and local soil contamination (Zupan *et al.* 1995, 2003). Attention has been deserved to plants as tools to clean up metal-contaminated soils by the low cost and environmental friendly technique of phytoremediation (Adriano *et al.* 1995; Baker *et al.* 2000). This technology is focused on the ability of plants to accumulate high heavy metal concentrations (up to 100 times the normal concentration) in their aerial parts. The plant ability to uptake metals was firstly applied in phytomining projects (Brooks and Robinson, 1998; Helios-Rybicka, 1996; Mc Grath, 1998), and only successively when environmental contamination became a global concern, it was recognized as an useful tool for remediation projects (Bini, 2005, 2010; Bini *et al.* 2000b Mc Grath, 1998; Salt *et al.* 1995). Up to now, more than 400 plants that accumulate metals are reported, *Brassicaceae* being the family with the largest number of accumulator species (Bini, 2010; Marchiol *et al.* 2004). Heavy metal accumulation is known to produce significant physiological and biochemical responses in vascular plants (Mangabeira *et al.* 2001). Preeti and Tripathi, (2011) reported a direct relationship between chemical characteristics of soil, heavy metals' concentration and morphological and biochemical responses of plants. Yet, metabolic and physiological responses of plants to heavy metal concentration can be viewed as potentially adaptive changes of the plants during stress. Plants growing on abandoned mine sites and naturally metal enriched soils (*e.g.* serpentine soils) are of particular interest in this perspective, since they are genetically tolerant to high metal concentrations, as reported by several authors (Bini, 2005; Giuliani *et al.* 2008; Brooks, 1998; Maleci *et al.* 1999), who studied endemic serpentine flora (*Alyssum bertoloni, A. murale, Silene paradoxa, Stachys serpentini, Thymus ophioliticus*) at various sites in the world. All these authors agree that morphological, physiological and phytochemical characters of serpentine plants are strongly dependent on the substrate composition, and that they are likely metal accumulator or tolerant ecotypes.

Understanding the mechanisms of metal bioaccumulation by plants species and of metal bioreduction by microorganisms is a clue to the efficiency of phytoremediation techniques. The localization and the chemical form of metals in cells are key information for this purpose (Kidd *et al.* 2009; Sarret *et al.* 2001). After their assimilation by plants, heavy metals could interfere with metabolic processes and are potentially toxic (Lopareva-Pohu *et al.* 2011) phytotoxicity results in chlorosis, weak plant growth, yield depression, and may be accompanied by disorders in plant metabolism such as reduction of the meristematic zone (Maleci *et al.* 2001), plasmolysis and reduced chlorophyll and carotenoids production (Corradi *et al.* 1993). Mangabeira *et al.* (2001) studied the ultrastructure of different organs of tomato plants (root, stem, leaf) which showed visible symptoms of Cr toxicity, and argued that Cr (VI) induces changes in the ultrastructure of these organs. Similar findings were reported by Vasquez *et al.* (1991) for Cd in vacuoles and nuclei of bean roots. Since both these metals are known to be inessential to plant nutrition, it is suggested that they are likely confined in roots by a barrier-effect as defense strategy during stress. Conversely, essential metals such as Zn and Cu are easily translocated to the aerial parts, as reported by Fontana *et al.* (2010). Among wild plants, the common dandelion (*Taraxacum officinale* Web) has received attention (Bini *et al.* 2000a; Królak, 2003; Zupan *et al.* 2003) as bioindicator plant, and has been also suggested in remediation projects (Turuga *et al.* 2008), given its ability to uptake and store heavy metals in the aerial tissues. *T. officinale* is a very common species, widely diffused in Central and Southern Europe, easy to identify and greatly adaptable to every substrate (Keane *et al.* 2001; Malawska and Wilkomirski, 2001). Moreover, this species is commonly collected to be used in cooking as fresh salad or boiled vegetable, and is used also in ethnobotany and traditional pharmacopoeia (Rosselli *et al.* 2006). Therefore, when grown on heavily contaminated soils, it may be potentially harmful if introduced in dietary food, as it occurs in many countries.

Studies of (Bini *et al.* 2000a; Fontana *et al.* 2010) investigated the heavy metal concentration of soils developed from mine waste material, and the wild plants (*Plantago major, Silene dioica, Stachys alopecuros, Stellaria nemorum, T. officinale, Vaccinium myrtillus, Gymnocarpium dryopteris, Gymnocarpium robertianum, Salix caprea, Salix eleagnos, Salix purpurea*) growing on those contaminated soils, in order to determine the extent of heavy metal dispersion, and the uptake by both known and unreported metal-tolerant plant species. The results showed that *T. officinale* is a species tolerant to high metal concentrations, and suggested to use it as a bioindicator plant. Metals accumulated preferentially in roots, but also leaves proved accumulator organs, being able to store up to 200 mg kg^{-1} Pb and 160 mg kg^{-1} Zn, showing only little damages (*e.g.* reduced foliar surface, reduced plant development).

4.8. Elevated CO_2 Concentration Stress

Carbon dioxide is not only a major greenhouse gas, but essential to plant growth (Kimball, 2011). Global change factors, particularly increases in atmospheric CO_2 concentration and temperature, changes in the mean and variance of regional precipitation, and land-use changes, are predicted to have profound effects on ecosystem functioning in the future. There is evidence that some global change factors are already affecting current ecosystems. For example, there is strong evidence that plants have already responded to the 25% increase in atmospheric CO_2 concentration that has occurred since the onset of the Industrial Revolution (Dippery *et al.* 1995, Duquesnay *et al.* 1998). Furthermore, atmospheric CO_2 concentrations are projected to double from the current concentration of 350 ppm to 700 ppm within the next 80 years, which will further stimulate ecosystem responses. In contrast to non uniform water availability and temperature among ecosystems, CO_2 concentrations are similar among ecosystems as a result of more thorough atmospheric mixing (Schlesinger, 1997). In addition, similar increases in CO_2 are expected to occur in all ecosystems, making this change unique among global change factors. Because the predicted increase in atmospheric CO_2 concentration may affect biological processes at many levels of organization (Mooney *et al.* 1999), it is important to continue studying the direct effects of increasing CO_2 at levels ranging from the molecular to the global. Physiological and ecological controls on carbon sequestration in ecosystems were reviewed more than a decade ago by Strain (1985).

Table : Molecular, physiological and ecological controls on carbon sequestering in ecosystems. Modified from Strain (1985).

I. Molecular and physiological controls on carbon sequestering

A.Molecular responses
   1. Gene transcription

B.Primary physiological responses
   1. Photosynthesis,
   2. Photorespiration,
   3. Dark respiration
   4. Stomatal regulation

C. Secondary physiological responses
   1. Photosynthate concentration,
   2. Photosynthate translocation
   3. Plant water status
      a. Transpiration,
      b. Tissue water potential,
      c. Water-use efficiency
      d. Leaf temperature

D. Tertiary whole plant responses
   1. Growth rate
      a. Mass,
      b. Height,
      c. Leaf area,
      d. Node formation
   2. Growth form
      a. Height,
      b. Branch number,
      c. Leaf area and number
      d. Root architecture,
      e. Root versus shoot mass,
      f. Leaf specific mass
   3. Reproduction
      a. Flower number and size,
      b. Fruit number and size,
      c. Nectar production
      d. Seed size and number,
      e. Seed germination
   4. Phenology (development rate)
      a. Time to germinate,

Contd.

b. Time to reproduction,
c. Time to leaf senescence
d. Time to whole-plant senescence

II. Ecological controls on carbon sequestering

A. Primary organism interaction
1. Plant—plant
a. Competition
2. Plant—animal
a. Herbivory,
b. Pollination,
c. Shelter
3. Plant—microbes
a. Disease, b. Decomposition, c. Symbiosis

B. Secondary organism interaction
1. Evolutionary responses and genetic differentiation

C. Tertiary ecosystem responses
1. Integration of all effects through time

1.8.1. Molecular and physiological controls on carbon sequestration

Primary molecular and physiological responses

Effects of elevated CO_2 on C3 photosynthetic rates have been the subject of many CO_2 enrichment studies and have been reported in hundreds of papers. Most of these studies show that photosynthetic rate is increased following initial exposure to elevated CO_2 (hours to days). Increases in photosynthetic rate are brought about by increased availability of CO_2 at the chloroplasts and reductions in photorespiration resulting from an increased ratio of CO_2 to O_2 (Pearcy *et al.* 1987). However, many studies report that high photosynthetic rates are not maintained over long time periods and substantial reductions in photosynthesis (down-regulation) may occur within days to weeks after initial exposure to elevated CO_2 (Long *et al.* 1993, Sims *et al.* 1998). Therefore, short-term measurements of photosynthetic rate may overestimate the potential for carbon assimilation of plants subjected to long-term exposure to elevated CO_2 (Oechel and Strain, 1985).

At the physiological level, down-regulation of photosynthesis is most often related to reduced sink strength (Stitt, 1991) and low nutrient availability. For example, Myers *et al.* (1999) showed that reduced sink strength (experimentally induced by excising developing needles and by girdling of branches) resulted in reductions in leaf photosynthetic rate and carbohydrate storage in *Pinus taeda* L. within several days. Photosynthetic downregulation in response to elevated CO_2 is also common in habitats characterized by nutrient-poor soils, and it is well known that nutrient (particularly nitrogen) availability in soil can affect production of photosynthetic enzymes. Li *et al.* (1999) reported that Florida scrub-oak species in a nutrient-limited system showed initial increase in photosynthetic rate in response to elevated CO_2, and later exhibited photosynthetic down-regulation that varied among species and it was attributed to the interspecific differences in downregulation to possible variation in nutrient acquisition by roots of the different species. Photosynthetic downregulation does not always reduce photosynthesis to current rates. For example, Tissue *et al.* (1999) observed strong responses of photosynthetic down-regulation in *Pinus ponderosa* Dougl. *ex* Laws., but photosynthetic rates were still 53% higher in plants grown at elevated CO_2 for six years relative to plants grown at current ambient CO_2. Photosynthetic down-regulation may also be maintained over long time scales by selection, as indicated by reduced photosynthetic capacity of *Nardus stricta* L. growing near a CO_2 spring (that has emitted CO_2 for hundreds to thousands of years) relative to plants of the same species growing at a distance from the spring (Cook *et al.* 1998).

Over the last decade, progress has been made in determining the biochemical and molecular mechanisms by which photosynthesis is down-regulated in response to elevated CO_2. Photosynthetic down-regulation is characterized at the biochemical and leaf levels by reduced chlorophyll content, reduced Rubisco (ribulose-1,5-bisphosphate carboxylase-oxygenase) content and activity, limitations in RuBP and *P*i regeneration, higher leaf mass/leaf area ratios and decreased leaf nitrogen concentration on a leaf mass basis (Sage 1994, Tissue *et al.* 1995). These down-regulation responses are often associated with increased accumulation of carbohydrate in leaves that may result in feedback inhibition of photosynthesis at the molecular level. Strain and Thomas (1995) present a conceptual model of controls on CO_2 acquisition. On a molecular basis, glucose and other sugars are known to suppress the transcription of photosynthetic genes (Sheen, 1990), and genes encoding D1 and D2 of photosystem II, cyt *f*, Rubisco small subunit and Rubisco activase, and carbonic anhydrase appear to be most affected (van Oosten and Besford 1995, Griffin and Seemann 1996).

Studies of the direct effects of elevated CO_2 on plant respiration have become increasingly important as attempts are made to scale the physiological effects of elevated CO_2 from the biochemical to whole-plant level. Gonzàlez-Meler and Siedow (1999) point out that a doubling of atmospheric CO_2 concentration results in an average 15—20% reduction in mitochondrial respiration that varies both within and among species (some crops may show as much as a 20% increase in respiration). The direct effects of CO_2 on respiration also vary among leaves of different ages (Thomas and Griffin, 1994) and between developmental stages (pre versus post-reproductive, Griffin *et al.* 1999). The exact mechanisms that account for inhibition of respiration by elevated CO_2 have not been fully elucidated, in part because responses have been highly variable. Possible mechanisms include inhibition of enzymes involved in mitochondrial electron transport (Gonzàlez-Meler *et al.* 1996) and reduced activity of other enzymes in response to dissolved inorganic carbon (Amthor, 1991). Furthermore, Gonzàlez-Meler and Siedow (1999) provide evidence that other mechanisms besides inhibition of mitochondrial enzymes may be associated with reductions in respiration in response to elevated CO_2. Because many studies have demonstrated direct effects of CO_2 on respiration, it is important that high concentrations of CO_2 are maintained during both daytime and nighttime hours in elevated CO_2 studies in order to predict accurately the effects of elevated atmospheric CO_2 concentration on the carbon balance of plants (Griffin *et al.* 1999).

C3 vs C4 species

C4 species have evolved in a high CO_2 environment. This increases both their nitrogen and water use efficiency compared to C3 species. C4 plants have greater rates of CO_2 assimilation than C3 species for a given leaf nitrogen when both parameters are expressed either on a mass or an area basis (Ghannoum *et al.* 2011). Although the range in leaf nitrogen content per unit areas is less in C4 compared to C3 plants, the range in leaf nitrogen concentration per unit dry mass is similar for both C4 and C3 species. Even though leaf nitrogen is invested into photosynthetic components into the same fraction in both C3 and C4 species, C4 plants allocate less nitrogen to Rubisco protein and more to other soluble protein and thylakoids components. In C3 plants, the photosynthetic enzyme Rubisco accounts for up to 30% of the leaf nitrogen content, but accounts for only 4–21% of leaf nitrogen in C4 species (Evans and von Caemmerer, 2000). The lower nitrogen requirement of C4 plants results from their CO_2 concentrating mechanism, which raises the bundle sheath CO_2 concentration, saturating Rubisco in normal air and almost eliminating photorespiration. Without this mechanism, Rubisco in the C3 photosynthetic pathway operates at only 25% of its capacity (Lara and Andreo, 2011), and loses approximately 25% of fixed carbon to photorespiration. To attain comparable photosynthetic rates to those in C4 plants, C3 leaves must therefore invest more heavily in Rubisco and have a greater nitrogen requirement. Because the Rubisco specificity for CO_2 decreases with increasing temperature (Long, 1991), this difference between the C3 and C4 photosynthetic nitrogen-use efficiency is greatest at high temperatures (Long, 1999). The high photosynthetic nitrogen-use efficiency of C4 plants is partially offset by the nitrogen-requirement for CO_2-concentrating mechanism enzymes, but the high maximum catalytic rate of PEP-carboxylase means that these account for only approximately 5% of leaf nitrogen (Lomg, 1999). Improved leaf and plant water use efficiency in

C4 plants is due to both higher photosynthetic rates per unit leaf area and lower stomatal conductance, with the greater CO_2 assimilation contributing to a major extent (Ghannoum *et al.* 2011).

The advantages of greater nitrogen use efficiency and water use efficiency of C4 relative to C3 photosynthesis are fully realized at high light and temperature, where oxygenase reaction of Rubisco is greatly increased. It is worth noting, although in C4 plants energy loss due to photorespiration is eliminated, and additional energy is required to operate the C4 cycle (2 ATPs per CO_2 assimilated). In dim light, when photosynthesis is linearly dependent on the radiative flux, the rate of CO_2 assimilation depends entirely on the energy requirements of carbon assimilation (Long, 1999). However, when the temperature of a C_3 leaf exceeds approximately 25° C, the amount of light energy diverted into photorespiratory metabolism in C3 photosynthesis exceeds the additional energy required for CO_2 assimilation in C4 photosynthesis (Long, 1999). This is the reason why at temperatures below approximately 25–28° C, C4 photosynthesis is less efficient than C4 photosynthesis under light-limiting conditions. It is interesting to note, that while global distribution of C4 grasses is positively correlated with growing season temperature, the geographic distribution of the different C4 subtypes is strongly correlated with rainfall (Ghannoum *et al.* 2011). On the contrary, C4 plants are rare to absent in cold environments. Although there are examples of plants with C4 metabolisms that show cold adaptation, they still require warm periods during the day in order to exist in cold habitats (Sage *et al.* 2011). In consequence, C4 species are poorly competitive against C3 plants in cold climates (Sage and McKown, 2006).

Secondary physiological responses

Many studies have shown that regulation of stomata by guard cells can be directly affected by CO_2 concentration, but the exact mechanism underlying this response remains controversial (Zhu *et al.* 1998). With few exceptions, studies have shown that stomatal conductance of C3 plants is initially reduced in response to elevated CO_2 (higher *C*a), resulting in reduced transpiration and increased conservation of water (higher carbon assimilation per water lost); however, these responses are short-lived in some species (Bunce, 1992). Elevated CO_2 has also been shown to ameliorate the negative effects of drought stress (Tolley and Strain, 1985). Although most drought studies have been conducted in small pots where water was withheld and dry down of soil was rapid, these studies are useful for understanding short-term physiological responses to severe droughts that do not allow for long-term physiological adjustments. However, open-top chamber sites, free air carbon dioxide enrichment (FACE) sites, and CO_2 springs now provide excellent systems for studying long-term effects of drought under elevated CO_2 conditions with non-limiting soil volume. For example, Tognetti *et al.* (1999*a*) showed that, in *Quercus pubescens* Willd., water flux (relative to cross-sectional area) was reduced in trees that occurred near a CO_2 spring relative to trees that occurred far from the spring. However, they found that transpiration responses to elevated CO_2 showed strong seasonal effects, and the beneficial effects of elevated CO_2 were lowest during the most severe periods of drought (periods of highest vapor pressure deficit). Furthermore, Tognetti *et al.* (1999*b*) reported that there were interspecific differences in the hydraulic conductivity of Mediterranean tree species, but trees of the same species did not differ in their responses at a CO_2 spring site compared to a non-spring site, indicating that long-term exposure to elevated CO_2 may not result in differentiation of xylem hydraulic properties. Furthermore, Pataki *et al.* (1998) found that water flux per unit sap wood did not vary between current ambient and elevated CO_2 concentrations in *Pinus taeda*, but elevated CO_2 increased absolute water loss by increasing leaf and sap wood areas. In addition, Ellsworth *et al.* (1995) found that *P. taeda* generally did not undergo adjustments in stomatal conductance after 80 days of exposure to elevated CO_2 at a FACE site in the southeastern United States, although plants in elevated CO_2 did exhibit transient adjustments in stomatal conductance and reductions in water loss during cloudy conditions.

Tertiary whole-plant responses

It is well known that elevated CO_2 stimulates biomass production of C3 plants, and plants with indeterminate growth show higher growth enhancements in response to elevated CO_2 than plants with determinate growth, presumably because of differences in sink strength (Oechel and Strain 1985). Furthermore, plants often show higher growth responses to elevated CO_2 when other resources such as nutrients and water are not limiting (Curtis and Wang 1998). With respect to light, however, higher relative growth enhancements are sometimes observed under low light conditions than under high light conditions because elevated CO_2 increases the quantum yield (photosynthetic carbon gain per photons absorbed) of C3 species (Long and Drake 1991; Ehleringer *et al.* 1997). However, Lewis *et al.* (1999) point out that increases in irradiance may increase, decrease, or have no effect on the growth of plants under elevated CO_2 conditions. Initial stimulations in growth in response to elevated CO_2 may diminish over time, possibly because of down-regulation of photosynthesis or modifications in biomass allocation and phenology. Jach and Ceulemans (1999) observed that three-year-old *Pinus sylvestris* L. seedlings showed enhanced relative growth rates during the first season of exposure to elevated CO_2, but showed similar growth rates to control plants during the second season. Similarly, Tissue *et al.* (1997*b*) reported that, after four years of exposure to elevated

CO_2 in open-top chambers, *Pinus taeda* plants exhibited 90% more biomass than control plants grown at the current ambient CO_2 concentration. However, the greater final biomass production was attributed to the large increases in growth and leaf area that occurred only during the first season that compounded growth responses over time. This study indicates that long-term measurements of growth in response to elevated CO_2 are necessary for predicting the potential for carbon sequestration by terrestrial forest ecosystems. The rate of height growth and branching increase in some tree species exposed to elevated CO_2 (Curtis and Wang 1998). Increased height growth by some genotypes and species may give certain individuals a competitive advantage over others if light is a limiting resource. Furthermore, increased branching has been correlated with increases in leaf number (Oechel and Strain 1985). As leaf number increases, leaf area index (leaf area/land area) may also increase, resulting in higher carbon assimilation on an ecosystem level. Jach and Ceulemans (1999) found evidence for these responses in *Pinus sylvestris* seedlings grown at elevated CO_2 and they predicted that the increase in leaf area index would result in more rapid canopy closure. These results indicate that changes in growth form in response to elevated CO_2 may have a substantial effect on light interception.

Plants are generally predicted to allocate biomass to structures that are involved in the uptake of limiting resources. Therefore, relative limitations in nitrogen and other soil nutrients at elevated CO_2 were initially predicted to increase the allocation of biomass to roots. In the early CO_2 studies, conclusions about allocation of biomass were often based on measurements of root-to-shoot ratio (root biomass/shoot biomass); however, this measurement only relates information on growth form at one point in time and does not include information on allocation of biomass over time (Samson and Werk 1986). Plants grown at different CO_2 concentrations and even plants within the same CO_2 treatment are likely to differ in size and this may result in differences in root-to-shoot ratio that are independent of the effects of CO_2 treatment, and are only related to shifts in allometry during development. To remove the effects of plant size when testing for effects of CO_2 on biomass allocation, allometric statistical techniques (primarily analysis of covariance) can be used. The majority of studies based on the allometric technique have shown that elevated CO_2 rarely alters the allocation of biomass between roots and shoots when size effects are removed (Gebauer *et al.* 1996, Tissue *et al.* 1997*b*). However, more studies are needed to determine if differences in biomass allocation occur between fine and course root fractions in response to elevated CO_2 (King *et al.* 1996).

Increasing atmospheric CO_2 significantly increased the final plant biomass, aboveground biomass, leaf area and belowground biomass (Obrist and Arnone, 2003). Increased root growth contributes to root biomass and root dry weight under elevated atmospheric CO_2 regardless of species or study conditions. Roots often exhibit the greatest relative dry weight gain among plant organs (Norby *et al.* 1992), even more than aboveground biomass or leaf area production (Pritchard and Rogers, 2000; Bernacchi *et al.* 2000). However, increased allocation of carbon to plant-root systems in response to CO_2 levels may offset losses from increased activity of soil microorganisms (Suter *et al.* 2002).

Wheat (McMaster *et al.* 1999), sorghum (Chaudhuri *et al.* 1986), black gram (Vanaja *et al.* 2007), and soybean (Rogers *et al.* 1992), have shown increases in root dry weight under CO_2 enrichment (Rogers *et al.* 1994). Laboratory studies on cotton (*Gossypium hirsutum* L.) plants in a free air carbon dioxide enrichment project revealed dry weights, lengths, and volumes of taproots, lateral roots, and fine roots were higher for CO_2–enriched cotton plants with a short-term exposure to increased CO_2 for 6 wk (Prior *et al.* 1994a, 1994b). Observations from a spring wheat free air carbon dioxide enrichment study revealed a 37% increase in total root dry mass in the CO_2–enriched plot during early vegetative growth (stem elongation) and sustained root growth rates until anthesis (Wechsung *et al.* 1999). The combined effect of temperature and elevated CO_2 concentration affect growth and development of roots, Pilumwong *et al.* (2007) observed in groundnut (*Arachis hypogaea* L.) that fibrous root dry weight increased with increased CO_2 when plants were grown at 25 / 15^0 C, but decreased with increasing CO_2 at 35 / 25^0 C. At 35 / 25^0 C, there was greater root dry weight by 34% at 400, 14% at 600, and 7% at 800 ìmol mol^{-1} compared with 25 / 15^0 C.

Root growth of crop plants is often stimulated to a greater extent than other plant parts to increased CO_2 concentration (Heinemann *et al.* 2006; Vanaja *et al.* 2007). Pilumwong *et al.* (2007) reported increased root growth of groundnut due to increasing CO_2 concentration. With no limitations to water and nutrients, elevated CO_2 increased root and shoot growth of most plant species including tree species (Obrist and Arnone, 2003) but the response varied among species (Hanley *et al.* 2004). Salsman *et al.* (1999) found increased CO_2 affected root and shoot growth of *Phaseolus acutifolius* but not *P. vulgaris* due to increased concentration of starch in roots of *P. acutifolius* by 10-fold, while root concentrations of abscisic acid (ABA) doubled and caused more carbon to be allocated to root growth. Growth and morphology of four native chalk grassland herbs roots responded differently to increasing CO_2 (Ferris and Taylor, 1993). Root growth and biomass was stimulated by elevated CO_2 for *Sanguisorba minor* and *Lotus corniculatus* with no significant effect on growth and root biomass for *A. vulneraria* or *P. media.* They suggested these growth and root responses could potentially lead to changes in the structure of this plant community with continued increases in CO_2.

Most elevated CO_2 studies on reproductive output have been conducted on herbaceous species. However, the mechanisms underlying these responses are probably similar in tree species, although they may occur over a longer life cycle. Reproductive output is associated with fitness (more so than any of the previously discussed measurements) and may influence the effects of elevated CO_2 on long-term evolutionary processes. Changes in reproduction in response to elevated CO_2 are important for quantifying effects on crop yields, for determining changes in the fitness of genotypes in natural systems, and for predicting changes in species composition that may be manifested through differences in reproduction. Most studies have shown that elevated CO_2 increases reproductive output (flower number, fruit number, and seed production) of herbaceous species (Curtis *et al.* 1996, Ward and Strain 1997).

Elevated CO_2 can affect seed germination, seed quality, and seedling viability. Studies of the effects of CO_2 enrichment on germination have been inconsistent, with some showing significant effects of CO_2 enrichment (Andalo *et al.* 1996) Concentrations of CO_2 within soil and on the surface of soil may be several times that of the atmosphere because of the high rates of microbial respiration. Consequently, seeds are often exposed to very high CO_2 concentrations under field conditions, and may not show changes in germination in response to the relatively small increases in CO_2 concentration predicted for the next 50—100 years (Andalo *et al.* 1998). Interestingly, however, Andalo *et al.* (1996) found that the germination percentage of field-collected *Arabidopsis thaliana* (L.) Heynh. seeds was unaffected by high CO_2 concentrations unless the seeds were borne by mother plants that had developed at high CO_2, in which case, germination was reduced and was slower. The authors later reported that this response may be a result of increases in the C:N ratio of seed tissue induced by a maternal effect at elevated CO_2 that reduced growth of the radical (Andalo *et al.* 1998). If these effects occur in tree species and vary among species, the impact on reproductive success may result in progressive changes over successive generations in the species composition of forest communities. Relatively few studies have measured the effects of elevated CO_2 on seedling survival. Increase in atmospheric CO_2 concentration may have direct effects on the carbon balance of seedlings and may affect early seedling survival, because respiration often exceeds photosynthesis during the early stages of growth (Larcher, 1995). Polley *et al.* (1996) showed that, under drought-stress conditions and low plant density, the survival of *Prosopis glandulosa* Torr. seedlings was increased from 0 to 40% as a result of increases in CO_2 concentration above the current ambient concentration, suggesting that increased CO_2 reduced the negative effects of drought stress on seedling survival. However, when *Abutilon theophrasti* Medic. was grown at high density, elevated CO_2 reduced survival because it increased the intensity of competition

(Bazzaz *et al.* 1992). Therefore, differences in microhabitats may alter the effects of high CO_2 on seedling survival. Variations in the effects of elevated CO_2 on survival have also been documented between closely related species. Rochefort and Bazzaz (1992) found that, of four birch species, only *Betula alleghaniensis* Britt. showed increased survival when the concentration of CO_2 was increased from 350 to 700 ppm. This result suggests that shifts in community composition may result from the effects of increasing CO_2 on seedling survival. Elevated CO_2 can alter plant phenology (development rate) and time to senescence at both the leaf and whole-plant levels. For example, Jach and Ceulemans (1999) observed that, in *Pinus sylvestris*, needle fall occurred earlier at elevated CO_2 than at current ambient CO_2, which the authors attributed to possible changes in transpiration rate or earlier translocation of nutrients away from leaves. In addition, *Nardus stricta* growing near a CO_2 spring exhibited earlier leaf senescence compared to plants growing far from the spring, indicating that long-term exposure to elevated CO_2 may decrease the life span of leaves (Cook *et al.* 1998). In contrast, delayed leaf senescence in response to elevated CO_2 occurred in *Glycine max* (L.) Merrill (Hardy and Havelka 1975), and leaves of *Eriophorum vaginatum* L. maintained high photosynthetic rates later in the season when grown at elevated CO_2 compared to current ambient CO_2 (Tissue and Oechel 1987). These studies demonstrate that effects of elevated CO_2 on leaf phenology may alter leaf-level carbon assimilation throughout the growing season in ways that may be difficult to predict.

Elevated CO_2 can also affect whole-plant development rate (most often measured as time to reproduction). Species may exhibit slower (Carter and Peterson 1983), faster (St. Omer and Horvath 1983, Garbutt and Bazzaz 1984), or similar (Garbutt and Bazzaz 1984) rates of development in response to elevated CO_2. In some species, changes in time to reach reproduction were caused by faster growth rates that resulted in earlier attainment of the minimum size required for reproduction. In other species, however, elevated CO_2 altered the size at which plants initiated reproduction (Reekie and Bazzaz 1991). Changes in the timing of reproduction may have implications at the community level by disrupting the timing of plant—pollinator interactions (Garbutt and Bazzaz 1984). Alterations in flowering time may also affect the reproductive output of both herbaceous species and tree species, particularly in regions where frost or drought events limit seed production at the end of the growing season.

Much emphasis has been placed on predicting the degree to which terrestrial ecosystems will sequester or release carbon in future scenarios of increasing atmospheric CO_2 concentration. (Mooney *et al.* 1999). However, predictions about the effects of elevated CO_2 on the carbon uptake of C3 plants have generally assumed that plant responses will remain stable over long time scales, and relatively little attention has focused on the consequences of increasing CO_2 on plant evolution. Recent studies have demonstrated that C3 plants exhibit genetic variation in response to elevated CO_2 for photosynthesis (Curtis *et al.* 1996), stomatal characters (Case *et al.* 1998), growth (Norton *et al.* 1995, Zhang and Lechowicz 1995, Schmid *et al.* 1996), and reproduction (Curtis *et al.* 1994, Bazzaz *et al.* 1995). The existence of genetic variation in the responses to elevated CO_2 strongly suggests that plants may undergo rapid directional selection in response to increasing CO_2 concentration (Strain, 1991). The effects of selection may result in phenotypic changes in development rate and biomass production that may alter the capacity for carbon sequestration by terrestrial systems. More specifically, selection for high seed number at elevated CO_2 over multiple generations resulted in plants that initiated reproduction at a younger age and had lower final biomass production as a result of earlier senescence relative to control plants. If such responses occur in tree species, evolutionary responses may not further increase the capacity for trees to utilize increased CO_2 concentrations as has been suggested in the literature (Curtis *et al.* 1994). The results of future studies will improve our understanding of the effects of elevated CO_2 on the evolution of tree species.

It is well known that response to high atmospheric CO_2 concentration differs from one species to another due to above and below ground environmental conditions. The interactive effects of increasing CO_2 concentration with growth influencing factors may alter the plant structure and productivity (Poorter *et al.* 1997) and have a greater impact on agriculture, affecting both growth and development of crops and ultimately impacting yield and food production (Cox *et al.* 2000; Hansen *et al.* 2000). The effect of CO_2 concentration enrichment on crops varies under different soil moisture regimes (Ewert *et al.* 2002). Most studies were carried out under favorable water conditions (Amthor, 2001). However, data on the interactive effects of CO_2 concentration and soil moisture on plants are scarce and often contradictory (Amthor, 2001). Many controlled environment studied the effects and interactions of these climate change factors on plants (Pickering *et al.* 1994), which showed alterations in physiology, growth and development in many crops (Allen and Boote, 2000; Reddy *et al.* 2004). Therefore, knowledge of vegetative responses should aid in ascertaining how environmental factors affect the phenology, growth and development of soybean plants along with genotypic response to cope with global climate change (Madhu and Hatfield, 2015).

It is important that elevated CO_2 studies consider ecosystem responses over long time scales (Oechel *et al.* 1994). Changes in net primary production, carbon storage in soils, and nutrient cycling may occur during long-term exposure to elevated CO_2 that may alter the degree to which vegetation and soils sequester carbon. Hungate *et al.* (1997) found that, following three years of exposure to elevated CO_2 in open-top chambers, serpentine and sandstone grasslands exhibited increased carbon uptake by vegetation. However, carbon was partitioned to soil pools with rapid turnover that reduced the potential for long-term carbon storage in these systems (Ward and Strain, 1999). Following long-term exposure to elevated CO_2, the model predicted relatively low carbon uptake by vegetation in response to elevated CO_2 because of reduced nitrogen availability and lower production of leaf biomass and fine root biomass in older stands. The model predicted that carbon storage would be increased by only 4% in trees and 9% in soils at the 10th year of exposure to elevated CO_2. Despite these low values for carbon storage in vegetation and soils of old forests, long-term accumulation of carbon in soils was predicted to increase by 20% throughout forest development at elevated CO_2 relative to current ambient CO_2.

4.9. Low Light Intensity Stress

Light intensity varies throughout the year, causing a significant effect on the growth and productivity of crops (Castilla *et al.* 2010). A reduction of around 20-50% of the solar radiation has been reported in rice-growing areas at certain periods of the year in Colombia (Garcés *et al.* 2005). The yield of field-grown rice mainly depends on the solar radiation throughout the growth period, especially during the reproductive and/or grain filling stages (Fageria, 2007). Low irradiance during the reproductive and/or ripening stages has an adverse effect on potential yield because the photosynthetic activity in the leaves of rice cultivars decreases (Srivastava, 2011). In addition, plants in low irradiance environments have shown physiological responses such as: changes in chlorophyll and rubisco content (Hidema *et al.* 1991).

Agronomical, morphological and physiological responses of rice under low irradiance conditions have been widely reported on by several rsearchers (Lakshmi-Prada *et al.* 2004; Fageria, 2007; Moula, 2009). Restrepo and Garcés (2013) studied the influence of two irradiance levels (0 and 50% shading) during initiation of the panicle primordium, flowering or grain-filling stages on yield components and chlorophyll fluorescence, stomatal conductance, and photosynthesis in two rice cultivars under Colombian conditions. It was reported that the leaf chlorophyll content (SPAD readings) was higher in leaves under low irradiance. The chlorophyll content from the shade treatment apparently remained constant until the grain filling phase, then decreased slightly in the "F50" rice plants (susceptible to low irradiance) and remarkably in the other cultivar. Stomatal conductance was negatively affected by shading, with the effect being more adverse in other cultivar. At the flowering and grain filling phases, grain yield was reduced by the low light treatments by around <"20% in the "F50" plants. While, in other cultivar, they were only affected by shading at the grain filling stage, causing a decrease of around 25%. It was concluded that other cultivar may have a better capacity for partitioning dry matter than "F50" in spite of the fact that the gas exchange characteristics were conditioned by low irradiance conditions at the reproductive and ripening phases.

Low light conditions result in significantly increased leaf length, leaf width, leaf area and growth duration, and the increases are enhanced with a reduction of light intensity (Ren *et al*, 2002; Ding *et al*, 2004). Leaf area only increases by 5.76% under 50% of natural light, however, it increases by 29.83% under 20% of natural light. Conversely, mesophyll thickness and the number of cells per square millimeter in leaves decrease by 14.61% and 15.86%, respectively, when rice plants are grown under 20% of natural light (Liu *et al.* 2014). Chlorophyll a and b are important pigments involved in the absorption and transmission of solar energy, with part of chlorophyll a involved in converting solar energy into electrochemical energy (Wang, 2011). Differences exist in the chlorophyll content produced in response to low light among varieties (Zhu *et al*, 2008; Liu *et al*, 2009). When subjected to low light for 15 days (when treatment had commenced at the initial heading stage), varieties that are tolerant to low light exhibit higher chlorophyll b and lower chlorophyll a/b content in their leaves when compared with those perform poorly in low light (Zhu *et al*, 2008). Similarly, leaf chlorophyll a and b content during the grain-filling stage is markedly enhanced in low light tolerant varieties after being treated by low light from the transplanting to the booting stages, whereas the opposite is found in varieties that perform poorly in low light (Liu *et al*, 2009). These results suggest that tolerant varieties capture as much solar energy as possible under low light conditions through increased leaf area and higher chlorophyll b content, demonstrating the morphological and physiological responses of rice plants when they experience low light stress (Ren *et al*, 2002). Low light negatively affects stomatal conductance (fewer stomata are produced per square millimeter) while it results in enhanced concentrations of intercellular CO_2 in rice leaves (Yang *et al*, 2011). Stomatal conductance decreases by 24.31% and 29.23% when light intensity decreases to 45% and 15% of natural light, respectively. A similar reduction is found in the number of stomata per square millimeter (10.29% and 12.52%), while the intercellular CO_2 concentration increases by 11.11% and 16.67%, respectively, under the same conditions. The net photosynthetic and respiration rates decline by 79.84% and 34.33% under low light, respectively, as compared to those under natural light. The respiration rate decreases more than the net photosynthetic rate, thus resulting in a higher ratio of respiration to net photosynthetic rates under low light than under natural light (Sato and Kim, 1980). Photosynthesis is a complex physiological procedure in plants, including light absorption, energy conversion, electron transfer, adenosine triphosphate synthesis, key regulating enzyme activities, *etc*. Shi *et al* (2006) noted that the ribulose bisphosphate carboxylase activity in chloroplasts declines dramatically under low light conditions. As an indispensable enzyme that regulates the biochemical process of photosynthesis, Rubisco plays an essential role in determining the photosynthetic rate of leaves (Okada and Katoh, 1998). Jiao and Li (2001) showed that light intensity alters the rates of non-photochemical quenching, electron transfer and quantum yield of PS II.

Low light results in a reduction in the amount of dry matter produced in shoots and roots, as well as a lower overall dry matter weight for rice plants (Yamamoto *et al*, 1995). The rate of dry matter weight in shoots to the total (shoots + roots) in rice plants increases under low light conditions, suggesting that dry matter weight in roots decreases more than that of shoots. Similarly, an increased rate of dry matter weight in culm to the total shoot (culm + tillers) indicates a lower reduction in the dry matter weight of the culm when compared with that of tillers (Yamamoto *et al*. 1995). When low light intensity decreases, culm-sheath dry matter exports, and exportation and translocation rates all decrease, which leads to an increase in the rate of culm-sheath dry matter weight compared with the total dry matter weight of shoots as well as a reduction in the dry matter weight of panicles (Sun *et al*. 2012). Additionally, low light results in an increase in the rate of leaf and culm-sheath dry matter weight produced compared with the total dry matter weight of aboveground (leaves + culm + sheaths + panicles). Moreover, low light also decreases the dry matter weight in the panicles, showing that most of the dry matter produced is used to sustain the growth of leaves, culm and sheaths rather than being allocated to panicles (Ren *et al*. 2003a). Cao *et al*. (2001) noted that the post-anthesis photosynthetic production of leaves and pre-anthesis dry matter stored in the culm and sheaths are the main sources of nutrients for developing rice panicles, accounting for about 60% and 40% of the total dry matter in the panicles, respectively. Consequently, the reduction of dry matter weight in panicles at maturity is believed to be the primarily cause of the decrease in photosynthetic assimilates and dry matter translocation under low light conditions (Zhu *et al*. 2008).

Low light in rice also decreases the amount of nitrogen transported from culm and sheaths to panicles, which triggers an increase in the percentage of nitrogen in leaves and culm-sheaths as well as a decrease in panicles compared with the total amount of nitrogen in aboveground (leaves + culm + sheaths + panicles). The results show that the amount of nitrogen allocated to panicles under low light conditions is less than that under natural light, and the amount of nitrogen used for the development of leaves and culm-sheath increases under low light conditions (Ren *et al.* 2003b). When grown under low light conditions from the tillering to booting stages, soluble sugar and soluble protein contents in rice leaves as well as the number of fertile panicles decline significantly. However, soluble sugar and protein content recover to a normal level when the low light treatment is terminated, indicating that the ratio of source to sink capacity is enhanced after the shading treatment. When encountering low light from the heading to maturity stages, the photosynthetic ability of rice plants is inhibited, which is verified by the decreased chlorophyll content and impaired photosynthetic rate in leaves, additionally, the seed-setting rate and 1000-grain weight are also markedly reduced. Kobata *et al.* (2000) reported that when rice is shaded during the early grain filling stage, shade does not affect the grain dry matter increment, nevertheless, if adequate assimilates are not available during the remainder of the grain filling stage, the final grain weight will be reduced significantly.

4.10. Soil Moisture and High Temperature Stress in Combination

Drought and heat stress often occur simultaneously, but they can have very different effects on various physiological, growth, developmental, and yield processes. The few studies that examined the impact of the combined effects of drought and heat stress suggested that the combination of drought and heat stress had a significantly higher detrimental effect on growth and productivity of crops than when each stress was applied individually (Savin and Nicolas, 1996). In addition, the combination of drought and heat stress was found to alter physiological processes such as photosynthesis, accumulation of lipids, and transcript expression (Jian and Huang, 2001; Rizhysky *et al.* 2004). Responses of crop or plant species to drought and/or heat stress are highly variable.

Drought stress induces several changes in various physiological, biochemical, and molecular components of photosynthesis. Drought can influence photosynthesis either through pathway regulation by stomatal closure and decreasing flow of CO_2 into mesophyll tissue (Chaves *et al.* 2003; Flexas *et al.* 2004) or by directly impairing metabolic activities (Farquhar *et al.* 1989). The main metabolic changes are declines in regeneration of ribulose bisphosphate (RuBP) and ribulose 1,5-bisphosphate carboxylase/oxygenase (Rubisco) protein content (Bota *et al.* 2004), decreased Rubisco activity (Parry *et al.* 2002), impairment of ATP synthesis, and photophosphorylation or decreased inorganic phosphorus. In general, during the initial onset of drought stress, decreased conductance through stomata is the primary cause of decline in photosynthesis (Cornic, 2000). At later stages with increasing severity, drought stress causes tissue dehydration, leading to metabolic impairment. Drought stress has been shown to cause increases in internal CO_2 concentration (Siddique *et al.* 1999). Studies showed both diffusive limitation (through stomatal closure) and nonstomatal limitation (such as oxidative damage to chloroplast) is responsible for decline in photosynthesis under drought stress (Zhou *et al.* 2007).

The response of photosynthesis to heat stress is related to temperature dependence of Rubisco to the two substrates, carbon dioxide and oxygen. At high temperatures, the solubility of oxygen is decreased to a lesser extent than CO_2, resulting in increased photorespiration and lower photosynthesis (Lea and Leegood, 1999). In addition, the activation and activity of Rubisco are also decreased at high temperatures (Prasad *et al.* 2004). The photosynthesis apparatus, photosystem II (PSII), plays a key role in the response of leaf photosynthesis to environmental stresses. Photosystem II is relatively more tolerant to drought stress than heat stress (Havaux, 1992). Drought stress resulting in relative water content (RWC) and leaf water potential of 40% and "4 MPa, respectively, did not affect PSII functioning in dark- and light adapted leaves (Havaux, 1992). In contrast, PSII is most sensitive to heat stress. There are two main factors which make the PSII electron transport most sensitive to heat stress. First, the fluidity of thylakoid membranes increases at high temperatures; this leads to dislodging of PSII light harvesting complexes from thylakoid membrane. Second, the PSII integrity is dependent on electron dynamics. Havaux (1992) investigated the impact of drought, heat, and strong light applied separately and in combination on PSII activity and found that drought stress enhances the resistance of PSII to heat and light stress. Although Rubisco activation was more closely correlated with photosynthesis than the maximum quantum yield of photochemistry of PSII, both processes could be acclimated to heat stress by gradually increasing the leaf temperatures (Law and Crafts-Brandner, 1999).

The regulation of respiration under drought or heat stress conditions is relatively less understood. Mitochondrial respiration plays a pivotal role in determining the growth and survival of plants (Gifford, 2003). Temperature is one of the most important environmental parameters influencing mitochondrial respiration. Respiration exponentially increases with increasing temperatures from 0 to 35 or 40^{0}C, reaching plateau at 40 to 50^{0}C. At temperature above 50^{0}C, respiration decreases because of damage to respiratory mechanism. Drought stress can result in decreases in leaf and root respiration in the short term (Byrla *et al.* 2001). Temperature quotient (Q10, the relative change in a process with a 10^{0}C temperature increase) for both root and leaf respiration also decreases with increasing temperatures. However, under field conditions, the relationship between temperature and root respiration is often complicated because of the occurrence of increased soil temperature with drought. In a greenhouse study under ambient and constant soil temperatures, root respiration rates decreased under drought stress conditions (Byrla *et al.* 2001). In addition, it was also observed that drought-induced decrease in root respiration were greater in warmer soils than in cooler soils. The responses of respiration to drought and/or heat stress can vary among crop species and also with age of the organs as shown by different Q10 values (Paulsen, 1994). Mitochondria are very stable to heat stress and their activity increases over most of the range in which plants are grown. Increased respiratory losses by seeds (grains or kernels) can offset the increased influx of assimilate and can account for greater yield losses under heat stress (Wardlaw *et al.* 1980). Thus, increasing the efficiency of respiration and its resistance to heat stress could improve tolerance to growth and yield. The thermal effects of photosynthesis and respiration are related to membrane function and membrane integrity. In general, heat stress influences membrane fluidity, induces membrane leakiness, and influences the integrity of protein and membranes. Thylakoid membranes are especially sensitive to drought and heat stress; hence, disturbances in photosynthesis are among the first indicators of drought and heat stress. Under drought stress, photosynthesis decreases before the decrease of respiration, resulting in decrease in the ratio of photosynthesis and respiration and also increase in photorespiration (Prasad *et al.* 2008b).

Photosynthesis is relatively more tolerant to heat stress compared with drought stress. This differential sensitivity of photosynthesis and respiration to drought and heat stress suggests differential interaction effects. The combination of both drought and heat stress may therefore be additive or multiplicative. The limited transpirational cooling under drought stress can exacerbate the effects of already higher air temperatures. Some studies suggest that drought stress influences the thermal tolerance of photosynthesis (Lu and Zhang, 1999). In contrast, some studies have reported that drought greatly exacerbates the effects of heat stress on plant growth and photosynthesis (Xu and Zhou, 2005, 2006).

Roots and root-soil interactions

For a given water potential gradient between the soil and the roots, water flux will be driven by the root hydraulic conductance. Variation of root system hydraulic conductance with time and environmental stresses will give rise to root 'hydraulic' plasticity and acclimatization. Hydraulic conductance varies along the root according to tissue age and among root types (Doussan *et al.* 1998) and will vary with growth of the root system and its plasticity. Hydraulic conductance can be modulated by cell membrane permeability and aquaporines (water channels). In the short term, with ongoing water deficit, an increase followed by a decrease in hydraulic conductance is observed and ascribed to aquaporine activity and regulation (Maurel *et al.* 2010). In the long terms of water deficit, a further decrease in hydraulic conductance is observed due to increased suberization of root endodermis/exodermis (Vandeleur and Mayo, 2009). The decrease in hydraulic conductance reduces water flux into the plant, but also prevents water losses from the plant to the dry soil. At a further longer time scale of drought, Hydraulic conductance can be further reduced in the plant by xylem embolism, a process by which air is sucked into the xylem vessels, interrupting the sap flow (Cruiziat and Cochard, 2002). Drought resistance can be related to a greater resistance to embolism (Li *et al.* 2009). The increase in temperature increases hydraulic conductance (rather the cell membrane permeability) in roots, but up to a deleterious point (harmful for plant functions) (Lipiec *et al.* 2013).

During water shortage, the interactions between the soil and the root system will affect the level and dynamics of water stress. The first level of interaction originates from root system architecture and soil transfers: at a local scale, an increase in root clumping will decrease the efficiency of water uptake (Beudez *et al.* 2013); at the root system scale, vertical heterogeneity of soil water availability induced by water uptake combined with water transfer in the soil and in the plant leads to a water extraction front propagating downwards (Garrigues and Doussan, 2006). The extension and speed of this front can be modulated by the variation in hydraulic conductance (aquaporines, suberization) of roots and help in compensating the lower uptake in drier zones by an increase in the wetter zones. This effect can be further increased due to a decrease in soil-root hydraulic conductance related to root shrinkage, which severely hampers water flow to roots (Carminati *et al.* 2009). Recent observations point to a rhizospheric effect onto the water relations of this soil-root interface, involving mucilages, root exudates and possibly solute accumulation (Carminati and Vetterlein, 2013; McCully *et al.* 2009), which would modulate soil-root contact and water uptake with variations in dry or moist soil (White and Kirkegaard, 2010).

At longer time scales, not only plasticity in water relations but also in root growth will occur during water deficit, with a decrease in root length (reduced growth, increased mortality) in drier parts and an increase in wetter parts (Sekhon *et al.* 2010). If an increase in root growth can be observed at the onset of water stress, the continuing drought will reduce the overall root growth, resulting from uncoupling between carbon production in leaves and use in root sinks (root apex) (Muller *et al.* 2011). The influence of soil water on root growth and function is closely related to the plant species and rooting depth (Vadez *et al.* 2012). In general, shallow-rooted crops such as potatoes are less drought tolerant than deep-rooted species such as alfalfa or maize. Under water stress, some plants develop short suberized roots, as the top soil becomes dry (Gliñski and Lipiec, 1990), which helps surviving drought by reducing water loss from plant roots. A recent study has shown that in dry environments root cation exchange capacity and nutrient uptake can be significantly reduced, and the relative uptake of polyvalent cations (aluminium or heavy metals) may induce additional toxicity (Lukowska and Józefaciuk, 2013).

Root plasticity can be modulated by soil compaction and associated mechanical impedance. The negative effects of a heavily compacted subsoil layer on water uptake were partly compensated by increased uptake from looser top soil layers and significant contribution of thicker roots in water uptake. (Nosalewicz and Lipiec, 2014) Morphological and anatomical responses of the roots in dry and strong soil were related to the general shape of roots (circular or flattened) due to the spatial distribution of soil strength around the roots (Lipiec *et al.* 2012). Whalley and Clark (2011) reported that increases in soil strength sufficiently large to impede root elongation can occur after only a moderate degree of soil drying. For soils with little continuous macro- porosity, this can decrease root elongation and the maximum rooting depth attained, restraining further subsoil access to water and nutrients, and increase drought (Bengough, 1997). Water scarcity and increased soil temperature substantially affect the formation, duration, and activity of pea nodules. In the study of Siczek and Lipiec (2011), improved soil water relations due to mulching significantly increased symbiotic nitrogen fixation as measured by nitrogenase activity, nodule diameter and dry weight, and seed yield.

Phenological changes

Crop adaptation to drought and high temperature is a function of the interaction of phenology with the pattern of water use (Sekhon *et al.* 2010; Wahid *et al.* 2007). Limited shoot growth by a decreased number of tillers in response to water limitation is considered as a strategy to reduce water use under stress (El Soda *et al.* 2010). Lower rates of plant water use under good water supply at first growth phases can maintain transpiration for longer periods, with significant consequences on later responses to water deficit. Such water-sparing behaviour should yield more water available for water uptake by roots at key stages like the grain-filling period (Vadez *et al.* 2011). It was observed that earlier heading in response to high temperature conditions is advantageous in retention of more green leaves at anthesis, leading to increased evapotranspiration and smaller reduction in yield (Tewolde *et al.* 2006; Vadez *et al.* 2011). In general, short-duration varieties generally perform better under the stress conditions than long-duration ones, which could be due to their different root system (Singh *et al.* 2010). Studies under controlled growth conditions with various plants showed that high temperature is most harmful at gametogenesis (8-9 days before anthesis), anthesis, and fertilization (Wahid *et al.* 2007). It should be emphasized that plant mechanisms protecting against stress such as reduced plant size or decreased stomatal conductance may be responsible for reduced productivity (Deikman *et al.* 2012).

Water and nutrient use

Water use efficiency, defined as the amount of biomass or grain produced per unit of water used, provides a quick and simple measure of how well available water can be converted into grain and thereby is the basic indicator for measuring the effectiveness of water-saving agriculture (Sekhon *et al.* 2010). Water use efficiency is often equated with drought resistance and the improvement of crop yield under stress (Blum, 2005). Due to a decreasing amount of water available for agriculture, it is essential to maximize water use efficiency, *ie* the amount of crop per drop (Vadez *et al.* 2011). In Australia, water use efficiency can be significantly improved by reduced soil evaporation using relevant genotypes and/or agronomic practices that stimulate earlier-developing canopies during winter and minimize in spring (Siddique *et al.* 2001). Another approach to increase crop WUE and root WUE is partial root drying, an approach using split root techniques, with one drying and one well-watered root half (Davies and Hartung, 2004).

Water stress is also of great importance in the mineral nutrition of plants since most of the nutrients are provided with water. Due to this, many studies showed that application of fertilizer has no significant effects in water stress conditions, while it significantly increased yield components at optimum soil moisture content. An example of interactive effects of soil moisture content and fertilizer level impacts on crop yield is given in Abayomi and Adefila (2008). Application of potassium, known for regulating stomatal opening and closure, allows faster reopening of leaf stomata following drought-induced closure (Hu *et al.* 2011). A decrease in the concentration of potassium ions results in membrane damage and distortion of ionic homeostasis (Seyed *et al.* 2012). Deficiency of water may cause 50% decrease in the calcium concentration, which plays an important role in maintaining the integrity of cell membranes and other structures in maize leaves and roots. Moreover, deficiency of water affects metabolism of nutrients *e.g.,* inhibition of nitrate reductase activity and glutamine synthetase involved in intracellular assimilation o

Whole-Plant Responses

For most crop plants, the seed is the starting point of the growth cycle. Water uptake and imbibitions of water by seed is dependent on the soil water availability. Drought delays imbibition and thus can lead to decreased germination rates and total germination percentage. The rate of germination or seedling emergence can be calculated as the reciprocal of time to complete germination or emergence; this commonly has a linear response to temperature (Roberts, 1988), as do other developmental events such as leaf appearance and flowering (Roberts and Summerfield, 1987). At suboptimal constant temperatures, there is a positive linear relation between rates of development (*e.g.*, seed germination rate and flowering) from the base temperature (*T*b), at which the rate is zero, to the optimal temperature (*T*o) at which development occurs most rapidly. At supraoptimal temperatures, there is a negative linear relation between the optimal temperature and the ceiling temperature (*T*c), when the development rate is again zero (Roberts, 1988). At constant soil moisture conditions, percentage seed germination increases with increasing temperature above *T*b, reaching maximum at *T*o and decreasing at supraoptimal temperatures. Increasing temperature between base and optimum temperatures increase not only the rate of germination but also total percentage germination, but temperatures above optimum temperature decreases total percentage germination (Prasad *et al.* 2006c). Unlike temperature response, response to drought does not follow a bell shaped curve; rather, as soil dries (drought progresses), most of the growth and developmental events respond negatively until the developmental or growth processes cease completely.

Growth Processes

Leaf area expansion is often limited under drought stress, such that the expansion and development of the transpiration surface is drastically decreased. Leaf expansion is among the most sensitive growth processes to drought (Alves and Setter, 2004). This sensitivity is expressed in terms of smaller cells and reductions in the number of cells produced by leaf meristems (Tardieu *et al.* 2000). Alves and Setter (2004) showed that both cell expansion and production of cells contributed to a loss in leaf area depending on the developmental stage at which the leaf was stressed. In leaves that were no longer engaged in cell division, diminished cell expansion affected leaf area by reducing mature cell size, whereas, in younger leaves, inhibition of cell division resulted in fewer cells per leaf (Alves and Setter, 2004). Both cell division and cell expansion were able to recover fully when stress occurred at early phases of leaf development, but in leaves at the final phase of either cell division or cell expansion, these processes did not resume long enough to generate full size leaves (Alves and Setter, 2004). The general effects of mild drought on leaves are a reduction in leaf numbers, rate of expansion, and final leaf size. Under severe stress, the rate of leaf elongation decreases and leaf growth can cease. Drought stress can also influence total leaf area through its effect on initiation of new leaves, which is decreased under drought stress.

In contrast to drought, heat stress can stimulate cell division and cell elongation rates. High temperatures generally increase leaf appearance rates. Leaf-elongation rates increase at high temperatures, while decreasing leaf-elongation duration (Bos *et al.* 2000). The impact of heat stress on leaf area expansion and dynamics are relatively less understood and need attention. Heat stress resulted in significant increases in leaf numbers, particularly when reproductive development was arrested without any decrease in leaf photosynthetic rates (Prasad *et al.* 2006a). The importance of the leaf development and duration of crop growth is reflected in the amount of solar radiation that can be intercepted and used to accumulate crop biomass (Sinclair, 1994). Comparing the effects of drought and heat stress on leaf elongation, it was shown that within the leaf, drought decreased relative elongation rates at all the positions of leaf by nearly similar extent, except in the zone closest to the leaf insertion point, causing reduction in length of the zone of elongation (Tardieu *et al.* 2000). In contrast, temperature stress affected relative elongation rates at all positions by a similar extent; consequently, the length of zone with tissue elongation was not affected by temperature. Because of differential processes and mechanisms influencing the leaf expansion, the combination of drought and heat stress was additive (Tardieu *et al.* 2000). As early stages of development and leaf area expansion largely determine the rate of crop growth, a better understanding of these processes under various combinations of environmental conditions is very crucial for modeling the dry matter production and thus yield.

Studies have shown very strong correlations between leaf elongation rates and various physiological components (Welcker *et al.* 2007) that can be indicative of drought and/or heat stress. These relations include (i) positive correlation between the leaf elongation rates and leaf temperatures, (ii) strong linear negative correlations between leaf elongation rates and vapor pressure deficit (difference between saturated vapor pressure at leaf temperature and ambient vapor pressure), and (iii) a strong negative linear relations between leaf elongation rates and predawn leaf water potential. In other words, the response of leaf elongation rate to meristem temperature, evaporative demand, and soil water status were all linear and highly repeatable and thus could be modeled. Drought and heat stress often decrease stem growth and plant height. When plants experience drought stress, stem diameter shrinks in response to changes in internal water status (Simonneau *et al.* 1993). Changes in stem diameter were well correlated with predawn leaf water potential under prolonged drought (Katerji *et al.* 1994). Severe heat stress decreases stem growth resulting in decreased plant height (Prasad *et al.* 2006a). Root growth is very sensitive to water and heat stresses. Heat stress often decreases root growth, and it has been shown that root growth has a very narrow optimum temperature range when compared with other growth processes (Porter and Gawith, 1999). Heat stress reduced root number as well as root length and root diameter.

Developmental Processes

Drought and heat stress alter the initiation and duration of developmental phases. In most cases, the length of time from floral initiation (panicle initiation) to anthesis (panicle exsertion) is decreased by moderate drought and/or temperature stress but is increased by severe stress. Drought stress during panicle development inhibits the conversion of vegetative to reproductive phase and plants remain vegetative until the stress is relieved. Panicle initiation in sorghum was delayed by as many as 2 to 25 days and flowering by 1 to 59 days under drought stress, with more severe effects when drought was imposed both at early and late stage of panicle development (Craufurd *et al.* 1993). Drought and heat stress can delay the panicle initiation but also can cause the cessation of panicle development at any stages between panicle initiation and flowering. Severe drought or heat stress inhibits panicle exsertion and also delays flowering (Prasad *et al.* 2006a). Drought stress or heat stress during flowering and anthesis can lead to failure of fertilization because of decreasing pollen or ovule function. Drought stress or heat stress inhibits pollen development and causes sterility. Drought and/or heat stress also shortens the spike development duration (period during which potential kernel or seed numbers are determined) and the grain-filling duration (during which the grain or seed weight are determined). Drought stress during later stages of panicle or flower development decreases seed numbers and can also increase the duration from seed-set to full seed growth. Similar responses were also observed under heat stress, where the time from flowering to seed-set was increased under heat stress (Wheeler *et al.* 2000). Long duration of spikelet development and high spike weight at anthesis was positively correlated with final grain yield in wheat under drought and heat stress conditions (Bindraban *et al.* 1998). For cereal crops, longer periods of vegetative and reproductive development are often necessary to improve reproductive potential (number of productive tillers and kernels) and also leaves and tillers to provide assimilate supply during the grain filling. Studies have also shown that decreased leaf area due to drought before anthesis is correlated with reductions in the number of kernels per spike (Frederick and Camberato, 1995). Grain- or seed-filling duration is the time from seed-set to physiological maturity. For most crop species, particularly those where there is a physical restriction for growth of seeds as in case of rice, *Oryza sativa* L., (which has fixed pericarp) and legumes such as peanut, *Arachis hypogaea* L., or soybean, *Glycine max* (L.) Merr, (which has fixed locule size for development of the seed), yield capacity is mainly a function of seed numbers per unit area and seed-filling duration. Both drought (de Souza *et al.* 1997) and heat stress (Prasad *et al.* 2006a) decreases the seed-filling duration, leading to smaller seed size. Drought following flowering is known to have little effect on seed-filling rates, but seed-filling duration is shortened leading to small seed size or seed yield (Wardlaw and Willenbrink, 2000). The impact of heat stress on seed-filling rates and seed-filling duration are similar to that of drought. However, there may be a slight increase in seed-filling rate but a large decrease in seed-filling duration under heat stress. The increase in seed-filling rate does not compensate for loss of duration, thus resulting in smaller seed size and seed yields (Tashiro and Wardlaw, 1989).

Studies on interaction effects between drought and heat stress during grain filling showed that reducing kernel (seed) number did not alter the effect of heat stress following anthesis on dry weight of remaining seeds at maturity, but reducing the number of seed did result in a greater dry weight of the remaining seeds in drought stressed plants. The relationship between the response to drought and seed number was confounded by a reduction in the extent of drought stress associated with seed removal. It has been suggested that where heat stress and drought occur concurrently after anthesis, there may be a degree of drought escape associated with heat stress because of the reduction in the duration of seed filling, even though the rate of water use may be enhanced by heat stress. Under drought stress, the duration of grain filling may be controlled by the increased rate of leaf senescence, which in turn may be regulated by the nitrogen status of the plant (de Souza *et al.* 1997). Drought stress during grain filling generally decreases nitrogen accumulation of new plant tissues. Therefore, the accumulation of nitrogen in the seed during the linear seed-filling period can be met either by direct uptake of nitrogen or from remobilization of nitrogen from vegetative tissues (stems, leaves, or petioles). There is also evidence that a decrease in the seed-filling duration under drought conditions can often be compensated by increased seed-filling rate, particularly when there is access to carbohydrates either directly from the leaf photosynthesis or from those prestored in stems or leaves. Altered hormonal balance in the seeds by drought stress during seed filling, especially a decrease in gibberellic acid and an increase in abscisic acid (ABA), enhances the remobilization of prestored carbohydrates to seed (Yang *et al.* 2001). The utilization and/or remobilization of stored reserves from leaves or stems may be strongly tied to the enzymes related to carbohydrate metabolism. Acid invertase activity plays an important role in assimilate utilization (Zinselmeier *et al.* 2000), and drought (Zinselmeier *et al.* 1995) and heat stress (Cheikh and Jones, 1995) decreases acid invertase activity that could influence seed growth. However, the exact mechanisms responsible for decreased expression of invertase activity under drought and/or heat stress are unclear.

Reproductive Processes

Crop developmental stages are differentially sensitive to stress conditions. Stress just before anthesis and at anthesis caused significant increase in floral abortion and lower seed numbers in peanut (Prasad *et al.* 1999a), wheat, *Triticum aestivum* L., (Saini and Aspinall, 1981), and rice (Matsui *et al.* 2001). Most of the reproductive abortion in legumes occurs after fertilization during the early stages of embryo development. Drought stress during early stages of embryo development increased the rate of abortion (Westgate and Peterson, 1993). It is important to know if the abortion is caused directly by decreased water potential in the floral tissues (pollen or ovary) or is a result of decreased carbohydrate or nitrogen flux supply, or if it is related to whole-plant signaling system involving hormones (particularly ABA). The response of floral parts might be different than those of developing embryos because of additional connections (*e.g.*, placenta or chalaza) involved to link the embryo inside the ear or pod. Under drought stress, even though the leaf water potential was decreased, the embryos did not respond in similar fashion and had normal water potentials (Westgate *et al.* 1996). Drought imposed at flowering can also decrease photosynthetic rates and thus decrease the aMount of photosynthates allocated to floral organs, causing increased abortion (Raper and Kramer, 1987). However, the demand for photosynthates by the small embryo is low, particularly during the very early stages of development, and the sink strength of these is much lower than in other tissues (such as vegetative tissues) to experience shortage of photosynthates. This strongly suggests that additional signaling systems must be involved to link the developmental responses which can result in early embryo abortion.

Exposure to heat stress during flowering results in pollen sterility and loss of seed-set in legumes (*Arachis hypogeal;* Prasad *et al.* 2000b; *Phaseolus vulgaris* L., Prasad *et al.* 2002; *Vigna unguiculata* (L.) Walp.; Ahmed *et al.* 1992; *Glycin max*; Salem *et al.* 2007) and cereals (rice; Jagadish *et al.* 2007; wheat, Saini *et al.* 1983). Lower seed-set under heat stress can be caused either by poor anther dehiscence, hence low numbers of germinating pollen grains on the stigma (Matsui *et al.* 2000; Prasad *et al.* 2006b; Jagadish *et al.* 2007) or because of decreased pollen viability (Prasad *et al.* 2000b, 2002, 2006a, 2006b) or ovule function (Gross and Kigel, 1994). However, in some crops [*e.g.*, corn, sorghum, and millet, *Pennisetum glaucum* (L.) R. Br.] which produce large amounts of pollen grains, the ability of pollen to germinate or growth of pollen tube inside the style are more sensitive to environmental stresses. In species producing large amounts of pollen grains, loss of pollen viability under heat or drought stress would only decrease seed-set if the amount of pollen was also limited and/or if anther dehiscence was influenced by stress. Both microsporogenesis (pollen development) and megasporogenesis (stigma development) are injured under heat stress, resulting in lower seed-set (Cross *et al.* 2003; Young *et al.* 2004). Pollen is known to be relatively more sensitive to heat stress conditions.

Regarding the mechanisms responsible for pollen sterility, lower seed-set or early embryo abortion under heat stress there are several hypotheses that are proposed as possible mechanisms responsible to decreased pollen viability under drought and heat stress, some of which include: (i) developmental abnormalities in anthers leading to dislocation of microspores prematurely (Saini *et al.* 1984); (ii) dysfunction of tapetal cells because of abnormal vacuolization

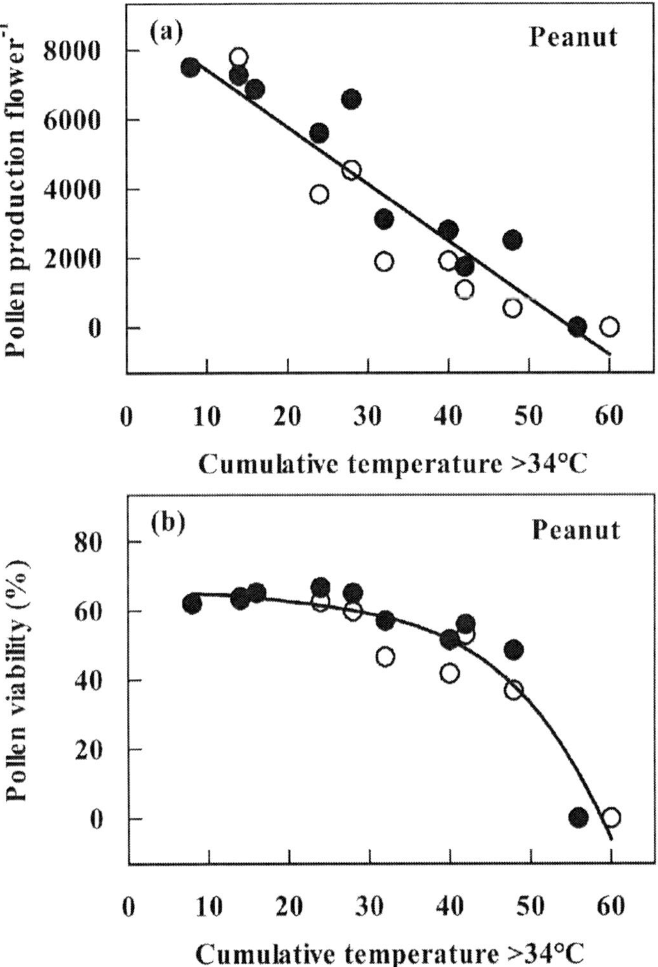

Fig. : Relationship between accumulated hourly temperature and (a) pollen production and (b) pollen viability in peanut. Redrawn with permission from Prasad *et al.*, (1999b)

(Lalonde *et al.* 1997); (iii) premature degeneration of tapetal cells and lack of endothecial development (Ahmed *et al.* 1992); (iv) altered carbohydrate accumulation and metabolism (Jain *et al.* 2007; Saini, 1997); and (v) oxygen starvation in the developing microspores which could lead to loss of gametophyte viability.

Studies suggest that there is a quantitative response to heat stress (temperature) between fertility and tissue temperature for both rice (Jagadish *et al.* 2007) and peanut (Prasad *et al.* 1999b) above a particular threshold temperature. There was a strong negative linear relation between pollen production and cumulative temperature >34°C in peanut (Fig.; Prasad *et al.* 1999b) and between spikelet fertility and cumulative temperature >33°C in rice (Jagadish *et al.* 2007). Pollen germination and rate of pollen tube growth were also highly sensitive to heat stress (Prasad *et al.* 2000b; Kakani *et al.* 2002). A modified bilinear model best described the response of pollen germination and pollen tube growth to temperature in peanut (Kakani *et al.* 2002), cotton, *Gossypium* spp., (Kakani *et al.* 2005), and soybean (Salem *et al.* 2007). Such quantitative responses of pollen production, spikelet fertility, and response of pollen germination and pollen tube growth suggest a method to model the temperature responses and the interactions between the temperature and duration of heat stress on reproductive processes at more mechanistic way, where necessary. Short periods of heat stress can also influence pollen viability, seed-set, and grain growth (Prasad *et al.* 2000b; Stone and Nicolas, 1998). Exposure to as short as 1 hour to temperature >37°C during flowering decreases seed-set (Matsui *et al.* 2000). Similarly exposure to temperature >33°C for first half of the day (6 hour after anthesis) was enough to decrease pollen viability and thus seed-set in peanut (Prasad *et al.* 2000b) and yield (Prasad *et al.* 2000a). As such, short durations of temperature stress can cause sterility; the timing of the episode of the high temperature relative to peak flowering will be very critical to quantify the impact of heat stress (Prasad *et al.* 2008b).

References

Abayomi, Y.A. and Adefila, O.E. 2008. Interactive effects of soil moisture content and fertilizer level on growth and achene yield of sunflower (*Helianthus annuus* L.). *J. Agronomy*, **7:** 182-186.

Acevedo, E., E. Fereres, T.C. Hsiao, and D.W. Henderson. 1979. Diurnal growth trends, water potential, and osmotic adjustment of maize and sorghum leaves in the field. *Plant Physiol.,* **64:** 476-480.

Ackerson, R.C., D.R. Krieg, T.D. Miller, and R.E. Zartman. 1977. Water relations of field grown cotton and sorghum: Temporal and diurnal changes in leaf water, osmotic, and turgor potentials. *Crop Sci.*, **17:** 76.

Ackerson, R.C., D.R. Krieg, C.L. Haring, and N. Chang. 1977a. Effects of plant water status on stomatal activity, photosynthesis, and nitrate reductase activity of field grown cotton. *Crop Sci.*, **17:** 81-84.

Ackerson, R.C. 1981. Osmoregulation in cotton in response to water stress. II. Leaf carbohydrate status in relation to osmotic adjustment. *Plant Physiol.*, **67:** 476-480.

Ackerson, R. C., and R.R. Hebert. 1981. Osmoregulation in cotton in response to water stress. I. Alterations in photosynthesis, leaf conductance, translocation, and ultrastructure. *Plant Physiol.,* **67:** 484-488.

Adolf, V.I., Jacobsen, S.E. and Shabala, S. 2013. Salt tolerance mechanisms in quinoa (*Chenopodium quinoa* Willd.). *Environmental and Experimental Botany,* **92:** 43-54.

Adriano, D.C., Chlopecka, A., Kapland, D.I., Clijsters, H. and Vangrosvelt, J. 1995. Soil contamination and remediation philosophy, science and technology. In," *Contaminated Soils*", Ed. R. Prost, INRA, Paris, pp. 466–504.

Ahmed, F.E., Hall, A.E. and DeMason, D.A. 1992. Heat injury during floral development in cowpea (*Vigna unguiculata*, Fabaceae). *American Journal of Botany,* **79:** 784–791.

Airaki, M., Leterrier, M., Mateos, R.M., Valderrama, R., Chaki, M., Barroso, J.B., Del Rio, L.A.,Palma, J.M. and Corpas, F.J. 2011. Metabolism of reactive oxygen species and reactive nitrogen species in pepper (*Capsicum annuum* L.) plants under low temperature stress.*Plant Cell Environ*, **35:** 281–295.

Akhtar, I. and Nazir, N. 2013. Effect of waterlogging and drought stress in plants. *International Journal of Water Resources and Environmental Sciences,* **2(2):** 34-40,

Allen, J.A., Chambers, J.L. and Stine, M. 1994. Prospects for increasing the salt tolerance of forest trees: a review. *Tree Physiology,* **14:** 843-853.

Allen Jr., L.H. and Boote, K.J. 2000.Crop ecosystem responses to climate change: soybean. In, "*Climate Change and Global Crop Productivity",* Eds. K.R. Reddy, and H.F. Hodges, CABI Publishing, Oxon, pp. 133-160.

Alfaro, E.J., Gershunov, A. and Cayan, D. 2006. Prediction of summer maximum and minimum temperature over the central and western United States: the roles of soil moisture and sea surface temperature. *J. Clim.*, **19:** 1407–1421.

Alam, I., Lee, D.G., Kim, K.H., Park, C.H., Akhtar, S., Lee, H., Oh, K., Yunand, B.W. and Lee, B.H. 2010.

Proteome analysis of soybean roots under waterlogging stress at an early vegetative stage. *J. Biosci.*, **35(1):** 49-62.

Ali, H., Khan, E. and Sajad M.A. 2013. Phytoremediation of heavy metals—concepts and application. *Chemosphere*. **91(7):** 869–881.

Aloni, B., Peet, M., Pharr, M. and Karni, L. 2001. The effect of high temperature and high atmospheric CO_2 on carbohydrate changes in bell pepper (*Capsicum annuum*) pollen in relation to its germination. *Physiologia Plantarum*, **112:** 505-512.

Alves, A.A.C. and Setter, T.L. 2004. Response of cassava leaf area expansion to water deficit: Cell proliferation, cell expansion and delayed development. *Ann. Bot. (London)*, **94:** 605–613.

Amarasinghe, V. and Watson, L. 1989. Variation in salt secretory activity of microhairs in grasses. *Australian Journal of Plant Physiology,* **16:** 219-229.

Amirjani, M.R. 2011. Effect of salinity stress on growth, sugar content, pigments and enzyme activity on rice. *International Journal of Botany,* **7:** 73-81.

Amthor, J.S. 1991. Respiration in a future, higher-CO_2 world: opinion. *Plant Cell Environ.*, **14:** 13-20.

Amthor, J.S. 2001.Effects of atmospheric CO_2 concentration on wheat yield: Review of results from experiments using various approaches to control CO_2 concentration. *Field Crops Research*, **73:**1-34.

Andalo, C., Godelle, B., LeFranc, M., Mousseau, M. and Till-Bottraud, I. 1996. Elevated CO_2 decreases seed germination in *Arabidopsis thaliana. Global Change Biol.*, **2:** 129—135.

Andalo, C., Raquin, C., Machon, N., Godelle, B. and Mousseau, M. 1998. Direct and maternal effects of elevated CO_2 on early root growth of germinating *Arabidopsis thaliana* seedlings. *Ann. Bot.*, **81:** 405—411.

Angadi, S.V., Cutforth, H.W., Miller, P.R., McConkey, B.G., Entz, H., Brandt, S.A. and Volkmar, K.M. 2000. Seeding management to reduce temperature stress in *Brassica* species during reproductive growth. *Canadian Journal of Plant Science,* **80(4):** 693-701.

Apel, K. and Hirt, H. 2004. Reactive oxygen species: metabolism, oxidative stress, and signal transduction. *Annual Review of Plant Biology*, **55:** 373-399.

Araus, J.L., Amaro, T., Casadesus, J., Asbati, A. and Nachit, M.M. 1998. Relationships between ash content, carbon isotope discrimination and yield in durum wheat. *Aust. J. Plant Phys.,* **25:** 835–842.

Araus, J.L., Slafer, G.A., Reynolds, M.P. and Royo, C. 2002. Plant breeding in C3 cereals: what should we look for? *Ann. Bot.*, **89:** 925–940.

Araus, J.L., Villegas, D., Aparicio, L.F., García del Moral, S., El Hani, S., Rharrabti, Y., Ferrio, J.P. and Royo, C. 2003. Environmental factors determining carbon isotope discrimination and yield in durum wheat under Mediterranean conditions. *Crop Sci.* **43:** 170–180.

Araus, J.L., Slafer, G.A., Royo, C. and Serret, M.D. 2008. Breeding for yield potential and stress adaptation in cereals. *Crit. Rev. Plant Sci.*, **27:** 377–412.

Ashraf, M. and Harris, P.J.C. 2004. Potential biochemical indicators of salinity tolerance in plants. *Plant Science,* **166:** 3-16.

Ashraf, M. and Foolad, M.R. 2007. Role of glycinebetaine and proline in improving abiotic stress esistance. *Environmental and Experimental Botany*, **59:** 206–216.

Ashraf, M.A.M. 2012. Waterlogging stress in plants. AJAR, **7(13):** 1976-1981.

Atkin, O.K., B. Botman, and H. Lambers. 1996. The causes of inherently slow growth in alpine plants; an analysis based on the underlying carbon economies of alpine and lowland *Poa* species. *Funct. Ecol.*, **10:** 698-707.

Atkin, O.K., I. Scheurwater, and T.L. Pons. 2007. Respiration as a percentage of daily photosynthesis in whole plants is homeostatic at moderate, but not high growth temperatures. *New Phytol.*, **174:** 367-380.

Atkin, O.K., and D. Macherel. 2009. The crucial role of plant mitochondria in orchestrating drought tolerance. *Ann. Bot.*, **103:** 581-597.

Baena-Gonzalez, E., Gray, J.C., Tyystjarvi, E., Aro, E.M. and Maenpaa, P. 2001. Abnormal regulation of photosynthetic electron transport in a chloroplast ycf9 inactivation mutant. *J Biol Chem,* **276:** 20795–20802.

Bailey-Serres, J., Fukao, T., Ronald, P., Ismail, A., Heuer, S. and Mackill, D: 2010. Submergence tolerant rice*: SUB1* 's journey from landrace to modern cultivar. *Rice,* **3:** 138–147.

Baker, A., Mc Grath, S., Reeves, R. and Smith, J. 2000. Metal hyperaccumulator plants: a review of the ecology and physiology of a biological resource for phytoremediation of metal-polluted soils. In, "*Phytoremediation of Contaminated Soils* ", Eds. N. Terry, G. Banuelos, Lewis Publisher, London, pp. 85–107.

Balasubramanian, S., Sureshkumar, S., Lempe, J. and Weigel, D. 2006. Potent induction of *Arabidopsis thaliana* flowering by elevated growth temperature. *Plos Genetics*, **2***:* 980-989.

Baldwin, I.T., Sims, C.L. and Kean, S.E. 1990. The reproductive consequences associated with inducible alkaloidal responses in wild tobacco. *Ecology*, **71:** 252–262.

Balestrasse, K.B., Tomaro, M.L., Batlle, A. and Noriega, G.O. 2010. The role of 5 aminolevulinic acid in the response to cold stress in soybean plants. *Phytochemistry*, **71:** 2038-2045.

Ball, R.A., D.M. Oosterhuis, and A. MaroMoustakos. 1994. Growth dynamics of the cotton plant during water-deficit stress. *Agron. J.*, **86:** 788-795.

Barnabás, B., Jager, K. and Feher, A. 2008. The effect of drought and heat stress on reproductive processes in cereals. *Plant, Cell and Environment*, **31:** 11-38.

Barrett-Lennanr, E.G. 2003. Theinteraction between water logging and salinity in higher plants causes, consequences and implications. *Plant and Soil,* **253:** 35-54.:

Bartoli, C.G., Simontacchi, M., Tambussi, E., Beltrano, J., Montaldi, E. and Puntarulo, S. 1999. Drought and watering-dependent oxidative stress: effect on antioxidant content in *Triticum aestivum* L. leaves. *J Exp Bot.*, **50:** 375–385.

Bastlová, D. and Kvét, J. 2002. Differences in dry weight partitioning and flowering phenology between native and non-native plants of purple loosestrife (*Lythrum salicaria* L.). *Flora*, **197:** 332–340.

Bagnall, D., Wolfe, J.O.E. and King, R.W. 1983. Chill-induced wilting and hydraulic recovery in mung bean plants. *Plant Cell and Environment*, **6(6):** 457-464.

Bengough, A.G. 1997. Modelling rooting depth and soil strength in a drying soil profile. *J. Theoret. Biol.*, **186(3):** 327-338.

Bengough, A.G., Bransby, M.F., Hans, J., McKenna, S.J., Roberts, T.J. and Valentine, T.A. 2006. Root responses to soil physical conditions; growth dynamics from field to cell. *Journal of Experimental Botany*, **57:** 437-447.

Beudez, N.and Doussan, C. 2013. Influence of three root spatial arrangement on soil water flow and uptake. Results from an explicit and an equivalent upscaled, model. *Procedia Environ. Sci.*, **19:** 37-46.

Blaylock, M.J., Elles, M.P., Nuttal, C.Y., Zdimal, K.L. and Lee, C.R. 2003. Treatment of As contaminated soil and water using Pteris vittata. Proc VI ICOBTE, Uppsala, Sv. Brooks, R.R. 1998.

Bazzaz, F.A., Ackerly, D.D., Woodward, F.I. and Rochefort, L. 1992. CO_2 enrichment and dependence of reproduction on density in an annual plant and a simulation of its population dynamics. *J. Ecol.*, **80:** 643—651.

Beppu, K., Ikeda, T. and Kataoka, I. 2001. Effect of high temperature exposure time during flower bud formation on the occurrence of double pistils in''- satohnishiki' sweet cherry. *Sci. Horticul.*, **87:** 77–84.

Bernacchi, C.J., Coleman, J.S. Bazzaz, F.A. and McConnaughay, K.D.M. 2000. Biomass allocation in old-field annual species grown in elevated CO_2 environments: No evidence for optimal partitioning. *Glob. Change Biol.*, **6:** 855–863.

Bhattacharya, A. and Singh, D.N. 1997. Physiology of seed yield under soil moisture stress and non-stress conditions in chickpea. *Indian Journal of Pulses Research Research*, **10(2):** 190-194.

Bhattacharya, A. and Desh Raj, 1997. Physiology of drought tolerance index in lentil. *Legume Research*, **20(1):** 57-51.

Bhattacharya, A. and Chandra, S. 1997. Physiological analysis of lentil seed yield under moisture stress and non-stress conditions. *Legume Research*, **20(1):** 52-56.

Bhattacharya, A. 1999. Lentil yield as affected by yield components under irrigated and non irrigated conditions.. *Legume Research*, **22(4):** 222-226.

Bhattacharya, A. and Singh, D.N. 1999. Physiological studies in chickpea genotypes: photosynthesis and allied parameters under normal and late sowing. *Indian Journal of Pulses Research*, **12(2):** 211-220.

Bhattacharya, A. and Pandey, P.S. 1999. Physiological studies in chickpea varieties: effect of temperature and time of sowing. *Indian Journal of Pulses Research*, **12(1):** 57-64.

Bhattacharya, A., Ganguly, S.B. and Singh, D.N. 1999. Late sown chickpea yield: Effect of physio-chemical traits of developing grains. *Legume Research*, **22(2):** 83-88.

Bini, C., 2005. Plants growing on abandoned mine soils: a chance in phytoremediation. Proc. III EGU Conf., Wien (CD-rom).

Bini, C., 2010. From soil contamination to land restoration. In, "*Contaminated Soils: Environmental Impact, Disposal and Treatment*", Ed. R.V. Steinberg, Nova Science Publisher, New York. ISBN: 978-1-60741-791-0.

Bini, C., Casaril, S. and Pavoni, B. 2000a. Fertility gain and heavy metal accumulation in plants and soils. *Environmental Toxicology and Chemistry*, **77:** 131–142.

Bini, C., Maleci, L., Gabbrielli, L. and Paolillo, A. 2000b. Biological perspectives in soil remedation with reference to chromium. In, "*Bioremediation of Contaminated Soils*", Ed. D. Wise, Marcel Dekker Inc., New York, pp. 663–675.

Bita, C.E. and Gerats, T. 2013. Plant tolerance to high temperature in a changing environment: scientific fundamentals and production of heat stress-tolerant crops. *Front. Plant Sci.*, **4:** 1–18.

Blum, A. 2005. Drought resistance, water-use efficiency, and yield potential – are they compatible, dissonant, or mutually exclusive? *Aust. J. Agric. Res.*, **56:** 1159-1168.

Blum, A. 2009. Effective use of water (EUW) and not water-use efficiency (WUE) is the target of crop yield improvement under drought stress. *Field Crop. Res.*, **112:** 119–123.

Bohnert, H.J., Nelson, D.E. and Jensen, R.G. 1995. Adaptations to environmental stresses. *Plant Cell*, **7:** 1099-1111.

Bohnert, H.J., Gong, Q.Q., Li, P.H. and Ma, S.S. 2006. Unraveling abiotic stress tolerance mechanisms – getting genomics going. *Curr Opin Plant Biol.*, **9:** 180–188.

Bokszczanin, K.L. and Fragkostefanakis, S. 2013. Perspectives on deciphering mechanisms underlying plant heat stress response and thermotolerance. *Front Plant Sci.*, **4:** 315–335.

Books, R.R. 1998. Phytochemistry of hyperaccumulator. In, "*Plants that Hyper accumulate Heavy Metals:* Their Role in Phytoremediation, Microbiology, Archaeology, Mineral Exploration and *Phytomining", Ed. R.* Brooks, CAB International, U.K., pp. 15–53.

Bos, H.J., Tijani-Eniola, T. and Struik, P.C. 2000. Morphological analysis of leaf growth of maize: Responses to temperature and light intensity. Neth. J. Agric. Sci. **48:**181–198.

Brown, P.W. and Zeiher, C.A. 1998. *Development of an effective screen for identifying cotton cultivars tolerant to elevated night temperatures during the monsoon* (Final Report: Project 96-342AZ). Raleigh, NC: Cotton.

Briggs, K.G. and Aytenfisu, A. 1980. Relationship between morphological characters above the flag leaf node and grain yield in spring wheat. *Crop Science*, **20:** 350-354.

Brooks, R.R. and Robinson, B.H. 1998. The potential use of hyperaccumulators and other plants for phytomining. In, *"Plants that Hyper accumulate Heavy Metals: Their Role in Phytoremediation, Microbiology, Archaeology, Mineral Exploration and Phytomining", Ed. R.* Brooks, CAB International, U.K., pp. 327–356.

Bruce, R. R. and Romkens. M.J.M. 1965. Fruiting and growth characteristics of cotton in relation to soil moisture tension. *Agron. J.*, **57:** 135-140.

Bohnert, H.J. and Jensen, R.G. 1996. Strategies for engineering water stress tolerance in plants. *Trends in Biotechnology,* **14:** 89-97.

Bossdorf, O., Auge, H., Lafuma, L., Rogers, W.E., Siemann, E. and Prati, D. 2005. Phenotypic and genetics differentiation in native versus introduced plant populations. *Oecologia*, **144:** 1–11.

Bota, J., Medrano, H. and Flexas, J. 2004. Is photosynthesis limited by decreased Rubisco acivity and RuBP content under progressive water stress? *New Phytol.*, **162:** 671–681.

Boyer, J.S. 1970. Leaf enlargement and metabolic rates in corn, soybean, and sunflower at various leaf water potentials. *Plant Physiol.*, **46:** 233-235.

Boyer, J.S. 1971. Non-stomatal inhibition of photosynthesis in sunflower at low leaf water potentials and high light intensities. *Plant Physiol.*, **48:** 532-536.

Boyer, J.S., and McPherson, H.G. 1975. Physiology of water deficits in cereal crops. *Adv.Agron.*, **27:** 1-23.

Boyer, J.S. 1982. Plant productivity and environment. *Science.* **218:** 443-448.

Boyer, J.S. and Westgate, M.E. 2004. Grain yield with limited water. *J. Exp. Bot.* **55:** 2385–2394.

Bradford, K.J. 1983. Involvement of plant growth substances in the alteration of leaf gas exchange of flooded tomato plants. *Plant Physiol.* **73:** 480-483.

Bravo, L.A., Close, T.J., Corcuera, L.J. and Guy, C.L. 1999. Characterization of an 80kDa dehydrin-like protein in barley responsive to cold acclimation. *Physiol Plant.* **106:** 177–183.

Bunce, J. 1992. Stomatal conductance, photosynthesis and respiration of temperate deciduous tree seedlings grown outdoors at an elevated concentration of carbon dioxide. *Plant Cell Environ.,* **5:** 541—549.

Bruce, K.A., Cameron, G.N., Harcombe, P.A. and Jubinsky, G. 1997. Introduction, impact on native habitats, and management of a wood invader, the Chinese tallow tree, *Sapium sebiferum* (L.) Roxb. *Natural Areas Journal*, **17:** 255–260.

Bundy, J.G., Kille, P., Liebeke, M. and Spurgeon, D.J. 2014. Metallothioneins may not be enough—the role of phytochelatins in invertebrate metal detoxification. *Environmental Science and Technology*. **48(2):** 885–886.

Buschmann, H., Edwards, P.J. and Dietz, H. 2005. Variation in growth pattern and response to slug damage among native and invasive provenances of four perennial *Brassicaceae* species. *Journal of Ecology*, **93:** 322–334.

Byrla, D.R., Bouma, T.J.,Hartmond, U. and Eissenstat, D.M. 2001. Influence of temperature and soil drying on respiration of individual roots in citrus, integrating green observations into a predictive model for the field. *Plant Cell Environ.*, **24:** 781–790.

Callaway, R.M. and Aschehoug, E.T. 2000. Invasive plants versus their new and old neighbors: a mechanism for exotic invasion. *Science*, **290:** 521–523.

Campitelli, B.E. and Stinchcombe, J.R. 2013. Testing potential selective agents acting on leaf shape *in Ipomoea hederacea: Predictions based on an adaptive leaf shape cline. Ecol Evol.,* **3:** 2409–2423.

Cao, S.Q., Zhai, H.Q., Yang, T.N., Zhang, R.X, and Kuang, T. Y. 2001. Studies on photosynthetic rate and function duration of rice germplasm resources. *Chin J Rice Sci*, **15(1):** 29–34. (in Chinese with English abstract)

Cao, J.H., Li, X.B., Lin, W.F., Xie, G.S. and Chen, J.M. 2009. Survey on resistance of 12 species oil palm cultivated in the field, *Chin. J. Trop. Agric.* **29:** 1-6.

Carminati, A. and Vetterlein, D. 2009. When roots lose contact. *Vadose Zone J.*, **8(3):** 805-809.

Carminati, A. and Vetterlein, D. 2013. Plasticity of rhizosphere hydraulic properties as a key for efficient utilization of scarce resources. *Ann. Botany*, **112(2):** 277-290.

Carter, D.R. and Peterson, K.M. 1983. Effects of a CO_2-enriched atmosphere on the growth and competitive interaction of a C3 and C4 grass. Oecologia **58:**188—193.

Case, A.L., Curtis, P.S. and Snow, A.A. 1998. Heritable variation in stomatal responses to elevated CO_2 in wild radish, *Raphanus raphanistrum* (Brassicacceae). *Am. J. Bot.,* **85:** 253—258.

Castilla, L.A., Pineda, D., Ospina, J., Echeverry, J., Perafan, R., Garcés, G., Sierra, J. and Díaz. A. 2010. Cambio climático y producción de arroz. *Rev. Arroz*, **58:** 4-11.

Chakraborty, U. and Pradhan, D. 2011. High temperature-induced oxidative stress in *Lens culinaris*, role of antioxidants and amelioration of stress by chemical pre-treatments. *J Plant Interact.,* **6:** 43–52.

Chartzoulakis, K. and Klapaki, G. 2000. Response of two greenhouse pepper hybrids to NaCl salinity during different growth stages. *Scientia Horticulturae,* **86:** 247-260.

Chaves, M.M., Pereira, J.S., Maroco, J., Rodriguez, M.L., Ricardo, C.P.P., Osorio, M.L., Carvalho, I., Faria, T., and Pinheiro, C. 2002. How plants cope with water stress in the field.

Photosynthesis and growth. *Ann. Bot.*, **89:** 907-916.

Chaves, M.M., Maroco, J.P., and Pereira, J. 2003. Understanding plant responses to drought–from genes to the whole plant. *Funct. Plant Biol.*, **30:** 239–264.

Chaves, M.M. and Oliveira, M.M. 2004. Mechanisms underlying plant resilience to water deficits: prospects for water-saving agriculture. *J. Exp. Bot.* **55:** 365-384.

Caviness, C.E. and Fagala, B.L. 1973. Influence of temperature on partially male-sterile soybean strain. *Crop Sci.*, **13:** 503–504.

Chen, S. 1991. Injury of membrane lipid peroxidation to plant cell. *Plant Physiol Commu.*, **27:** 84–90.

Chen, S., Lin, G., Huang, J. and Jenerette, G.D. 2009. Dependence of carbon sequestration on the differential responses of ecosystem photosynthesis and respiration to rain pulses in a semiarid steppe. *Global Change Biol.*, **15:** 2450–2461.

Chen, H.T. and Murata, N. 2011. Glyinebetaine protects plants against abiotic stress: Mechanisms and biotechnological application. *Plant, Cell and Environment*, **34:** 1–20.

Chhetri, N. and Chaudhary, P. 2011. Green Revolution: pathways to food security in an era of climate variability and change? *J. Disaster Res.* **6:** 486–497.

Chinnusamy, V., Zhu, J.H. and Zhu, J.K. 2007. Cold stress regulation of gene expression in plants. *Trends in Plant Sci.* **12:** 444-451.

Christensen, J.H. and Christensen, O.B. 2007. A summary of the PRUDENCE model projections of changes in European climate by the end of this century. *Clim. Change,* **81:** 7–30.

Chudhary, S. and Srivastava, M. 2015. Response of nacl on morpho-physiological parameters and nitrate reductase activity in finger millet (*Eleusine coracana*). *Global Journal of Bioscience and Biotechnology,* **4(1):** 33-36.

Clark, L.J., Whalley, W.R. and Barraclough, P.B. 2003. How do roots penetrate strong soil? *Plant and Soil,* **255:** 93-104.

Clark, L.J., Gowing, D.J., Lark, R.M., Leeds-Harrison, P.B., Miller, A.J., Wells, D.M., Whalley, W.R. and Whitmore, A.P. 2005. Sensing the physical and nutritional status of the root environment in the field: a review of progress and opportunities. *Journal of Agricultural Science, Cambridge*, **143:** 347-358.

Cornic, G. 2000. Drought stress inhibits photosynthesis by decreasing stomatal aperture: Not by affecting ATP synthesis. *Trend Plant Sci.*, **5:** 187–188.

Craufurd, P.Q. and Wheeler, T.R. 2009. Climate change and the flowering time of annual crops. *Journal of Experimental Botany*, **60:** 2529-2539.

Cole, P. and McCloud, P. 1985. Salinity and climatic effects on the yields of citrus. *Aust. J. Exp. Agric.*, **25:** 711–717.

Colmer, T.D., Cox, M.C.H. and Voesenek, L.A.C.J. 2006. Root aeration in rice (*Oryza sativa*): evaluation of oxygen, carbon dioxide, and ethylene as possible regulators of root acclimatizations. *New Phytologist,* **170:** 767–778.

Colmer, T.D. and Flowers, T.J. 2008. Flooding tolerance in halophytes. *New Phytologist,* **179:** 964–974.

Commuri, P.D. and Jones, R.D. 2001. High temperatures during endosperm cell division in maize: a genotypic comparison under in vitro and field conditions. *Crop Sci.*, **41:** 1122–1130.

Condon, A.G., Richards, R.A., Rebetzke, G.J. and Farquhar, G.D. 2004. Breeding for high water-use efficiency. *J. Exp. Bot.*, **55:** 2447–2460.

Corradi, M., Bianchi, A. and Albasini, A. 1993. Chromium toxicity in *Salvia sclarea.* Effects of hexavalent chromium on seed germination and seedling development. *Environmental and Experimental Botany,* **33(3):** 405–413.

Cox, P.M., Betts, R.A., Jones, C.D., Spall, S.A. and Totterdell, I.J. 2000. Acceleration of Global warming due to carbon-cycle feedbacks in a Coupled Climate Model. *Nature*, **408:** 184-187.

Crawford, R.M.M. 2003. Seasonal differences of plant response to flooding and anoxia. *Can. J. Bot.,* **81:**1224-1246.

Crifo, T., Puglisi, I., Petrone, G., Recupero, G.R. and Lo Piero, A.R. 2011. Expression analysis in response to low temperature stress in blood oranges: implication of the flavonoid biosynthetic pathway. *Gene*, **478:** 1–9.

Croser, J.S., Clarke, H.J., Siddique, K.H.M. and Khan, T.N. 2003. Low-temperature stress: implications for chickpea (*Cicer arietinum L.*) improvement. *Critical Reviews in Plant Sciences*, **22:** 185-219.

Cruiziat, P. and Cochard, H. 2002. Hydraulic architecture of trees: main concepts and results. *Ann. Forest Sci.*, **59(7):** 723-752.

Curtis, D.L. 1968. The relation between the date of heading of Nigerian sorghums and the duration of the growing season. *Journal of Applied Ecology*, **5:** 215-226.

Curtis, P.S., Snow, A.A. and Miller, A.S. 1994. Genotype-specific effects of elevated CO_2 on fecundity in wild radish (*Raphanus raphanistrum*). *Oecologia*, **97:** 100—105.

Curtis, P.S., Klus, D.J. Kalisz, S. and Tonsor, S.J. 1996. Intraspecific variation in CO_2 responses in *Raphanus raphanistrum* and *Plantago lanceolata*: assessing the potential for evolutionary change with rising atmospheric CO_2. In, *"Carbon Dioxide, Populations, and Communities*", Eds. C. Körner and F.A. Bazzaz. Academic Press, New York, pp 13—22.

Curtis, P.S. 1996. A meta-analysis of leaf gas exchange and nitrogen in trees grown under elevated carbon dioxide. *Plant Cell Environ.* **19:** 127–137.

Curtis, P.S. and Wang, X. 1998. A meta-analysis of elevated CO_2 effects on woody plant mass, form, and physiology. *Oecologia*, **113:** 299—313.

Cutler, J.M., Shahan, K.W. and Steponkus, P.L. 1980. Dynamics of osmotic adjustment in rice. *Crop Sci.*, **20:** 310-314.

D'Amico, M.L., Izzo, R., Navari-Izzo, F., Tognoni, F. and Pardossi, A. 2003. Sea water irrigation: antioxidants and quality of tomato berries (*Lycopersicon esculentum* Mill.). *Acta Horticulture,* (ISHS), **609:** 59-65.

Dalvi, A.A. and Bhalerao, S.A. 2013. Response of plants towards heavy metal toxicity: an overview of avoidance, tolerance and uptake mechanism. *Annals of Plant Sciences.* **2(9):** 362–368.

Dash, S. and Mohanty, N. 2001. Evaluation of assays for the analysis of thermotolerance and recovery potentials of seedlings of wheat (*Triticum aestivum* L.) cultivars. *Journal of Plant Physiology*. **158:** 153-1165.

Davletova, S., Schlauch, K., Coutu, J. and Mittler, R. 2011. The Zinc-Finger Protein Zat12 Plays a Central Role in Reactive Oxygen and Abiotic Stress Signaling in Arabidopsis. *Plant Physiol.*, **139:** 847-856.

Davis, M.A., Grime, J.P. and Thompson, K. 2000. Fluctuating resources in plant communities: a general theory of invisibility. *Journal of Ecology*, **88:** 528–534.

Davies, W.J., Bacon, M.A., Thompson, D.S., Sobeih, W. and Rodríguez, L.G. 2000. Regulation of leaf and fruit growth in plants growing in drying soil: exploitation of the plants' chemical signalling system and hydraulic architecture to increase the efficiency of water use in agriculture. *Journal of Experimental Botany*, **51:** 1617-1626.

Davies, W.J., Willkinson, S. and Loveys, B. 2002. Stomatal control by chemical signalling and the exploitation of this mechanism to increase the water use efficiency in agriculture. *New Phytologist*, **153:** 449-460.

Davies, W.J and Hartung, W., 2004. Has extrapolation from biochemistry to crop functioning worked to sustain plant production under water scarcity? New directions for a diverse planet. Proc. 4th Int. Crop Sci. Congr., September 26 –October 1, Brisbane, Australia.

Degenhardt, B. and Gimmler, H. 2000. Cell wall adaptations to multiple environment stresses in maize root. *Journal of Experimental Botany,* **51:** 595-603.

Deikman, J., Petracek, M. and Heard, J.E. 2012. Drought tolerance through biotechnology: improving translation from the laboratory to farmers' fields. *Curr. Opin. Biotechnol.*, **23:** 243-250.

de Luca, M., García Seffino, L., Grumberg, K., Salgado, M., Córdoba, A., Luna, C., Ortega, L., Rodríguez, A., Castagnaro, A. and Taleisnik, E. 2001. Physiological causes for decreased productivity under high salinity in Boma, a tetraploid *Chloris gayana* cultivar. *Australian Journal of Agricultural Research,* **52:** 903-910.

del Pozo, A., Castillo, D., Inostroza, L., Matus, I., Méndez, A.M. AND Morcuende, R. 2012. Physiological and yield responses of recombinant chromosome substitution lines of barley to terminal drought in a Mediterranean-type environment. *Ann. Appl. Biol.*, **160:** 157–167.

del Pozo, A., Matus, I., Serret, M.D., Araus, J.L. 2014. Agronomic and physiological traits associated with breeding advances of wheat under high-productive Mediterranean conditions. The case of Chile. *Env. Exp. Bot.*, **103:** 180–189.

Delauney, A. and Verma, D.P.S. 1993. Proline biosynthesis and osmo regulation in plants. *Plant J.,* **4:** 215–223.

Devasirvatham, V., Tan, D.K.Y., Trethowan, R.M., Gaur, P.M. and Mallikarjuna, N. 2010. Impact of high temperature on the reproductive stage of chickpea.. In, "*Food Security from Sustainable Agriculture Proceedings of the 15th Australian Society of Agronomy Conference* ", Eds. H. Dove, and R.A. Culvenor, Lincoln, New Zealand, pp. 15–18.

Devasirvatham, V., Tan, D.K.Y., Gaur, P.M., Raju, T.N. and Trethowan, R.M. 2012. High temperature tolerance in chickpea and its implications for plant improvement. *Crop and Pasture Science,* **63:** 419–428.

Devasirvatham, V., Gaur, P.M., Mallikarjuna, N., Raju, T.N., Trethowan, R.M. and Tan, D.K. 2013. Reproductive biology of chickpea response to heat stress in the field is associated with the performance in controlled environments. *Field Crops Research,* **142:** 9–19.

De Vries, F.W.T., Witlage, J.M. and Kremer, D. 1979. Rates of respiration and of increase in structural dry matter in young wheat, ryegrass and maize plants in relation to temperature, to water stress and to their sugar content. *Ann. Bot.*, **44:** 595-609.

DeWalt, S.J., Denslow, J.S. and Hamrick, J.L. 2004. Biomass allocation, growth, and photosynthesis of population types from the native and introduced ranges of the tropical shrub *Clidemia hirta. Oecologia*, **138:** 521–531.

Din, J., Khan, S.U., Ali, I. and Gurmani, A.R. 2011. Physiological and agronomic response of canola varieties to drought stress. *J. Anim. Plant Sci.*, **21:** 78-82.

Ding, S.B., Zhu, B.Y., Wu, D.Y. and Zhang, L. 2004. Effect of temperature and light on senescence of flag leaf and grain-filling after rice heading. *J South China Norm Univ: Nat Sci*, **46(1)**: 117–128. (in Chinese with English abstract)

Dippery, J.K., Tissue, D.T., Thomas, R.B. and Strain, B.R. 1995. Effects of low and elevated CO_2 on C3 and C4 annuals. I. Growth and biomass allocation. *Oecologia*, **101:** 13—20.

Djanaguiraman, M., Prasad, P.V.V., Boyle, D.L. and Schapaugh, W.T. 2013. Soybean pollen anatomy, viability and pod set under high temperature stress. *Journal of Agronomy and Crop Science,* **199:** 171–177.

Dodd, I.C. 2005. Root-to-shoot signalling: assessing the roles of 'up' in the up and down world of long-distance signalling in planta. *Plant Soil*, **274:** 251-270.

Doussan, C., Vercambre, G. and Pages, L. 1998. Modelling of the hydraulic architecture of root systems:an integrated approach to water absorption: distribution of axial and radial conductances in maize. *Ann. Botany*, **81:** 225-232.

Drew, M.C. 1997. Oxygen deficiency and root metabolism: Injury and acclimation under hypoxia and anoxia. *Ann. Rev. Plant Physiol Plant Mol Biol.* **48:** 223-250.

Du, J., Yang, J.-L. and Li C.-H. 2012. Advances in metallotionein studies in forest trees. *Plant OMICS,* **5(1):** 46–51.

Dufault, R.J., Ward, B. and Hassell, R.L. 2009. Dynamic relationships between field temperature and romaine lettuce yield and head quality. *Sci. Hortic.*, **120:** 452–459.

Dupuis, I. and Dumas, C. 1990. Influence of temperature stress on *in vitro* fertilization and heat-shock protein-synthesis in maize (*Zea mays* L.) reproductive tissues. *Plant Physiology,* **94:** 665-670.

Duquesnay, A., Bréda, N., Stievenard, M. and Dupouey, J.L. 1998. Changes of tree-ring d13C and water-use efficiency of beech (*Fagus sylvatica* L.) in north-eastern France during the past century. *Plant Cell Environ.*, **21:** 565—572.

Ehleringer, J.R. and Björkman, O. 1977. Quantum yields for CO_2 uptake in C3 and C4 plants. *Plant Physiol.*, **59:** 86—90.

El Soda, M., Nadakuduti, S.S., Pillen, K. and Uptmoor, R. 2010. Stability parameter and genotype mean estimates for drought stress effects on root and shoot growth of wild barley pre-introgression lines. *Molecular Breeding*, **26:** 583-593.

Ellsworth, D.S., Oren, R. Huang, C, Phillips, N. and Hendrey, G.R. 1995. Leaf and canopy responses to elevated CO_2 in a pine forest under free-air CO_2 enrichment. *Oecologia,* **104:** 139—146.

Endo, M., Tsuchiya, T., Hamada, K., Kawamura, S., Yano, K., Ohshima, M., Hegashitani, A., Watanabe, M. and Kawagishi-Kobayashi, M. 2009. High temperatures cause male sterility in rice plants with transcriptional alterations during pollen development. *Plant Cell Physiology,* **50:** 1911–1922.

Ephrath, J. E., Marani, A. and Bravdo. B.A. 1990. Effects of moisture stress on stomatal resistance and photosynthetic rate in cotton (*Gossypium hirsutum* L.) I. Controlled levels of stress. *Field Crops Res.*, **23:** 117-131.

Erfmeier, A. and Bruelheide, H. 2004. Comparison of native and invasive *Rhododendron ponticum* populations: growth, reproduction and morphology under field conditions. *Flora*, **199:** 120–133.

Erfmeier, A. and Bruelheide, H. 2005. Invasive and native *Rhododendron ponticum* populations: is there evidence for genotypic differences in germination and growth? *Ecography*, **28:** 417–428.

Evans, L.T. 1993. Crop evolution, adaptation and yield. Cambridge: Cambridge University Press;

Evans, R. and von Caemmerer, S. 2000. Would C4 rice produce more biomass than C3 rice? In, "*Redesigning rice photosynthesis to increase yield*", Ed. B. Hardy, International Rice Research Institute and Elsevier Science BV, Amsterdam, Netherlands, pp. 53-71.

Ewert, F., Rodriguez, D., Jamieson, P., Semenov, M.A., Mitchell, R.A.C., Goudriaan, J., Porter, J.R., Kimball, B.A., Pinter Jr., P.J., Manderscheid, R., Weigel, H.J., Fangmeier, A., Fereres, E. and Villalobos, F. 2002. Effects of elevated CO_2 and drought on wheat: testing crop simulation models for different experimental and climatic conditions. *Agriculture, Ecosystems and Environment*, **93:** 249-266.

Fageria, N.K. 2007. Yield physiology of rice. *J. Plant Nutr.*, **30:** 843-879.

Fageria, N.K., Stone, L.F. and Santos, A.B.D. 2012. Breeding for salinity tolerance. In, "*Plant breeding for abiotic stress tolerance*", Eds. R. Fritsche-Neto and A. Borém, Springer-Verlag, Berlin, Germany, pp. 103-122.

Fan, W.J., Zhang, M., Zhang, H.X. and Zhang, P. 2012. Improved tolerance to various abiotic stresses in transgenic sweet potato (*Ipomoea batatas*) expressing spinach betaine aldehyde dehydrogenase. *PLoS One.*, **7(5):** e37344.

FAO [Food and Agriculture Organization of the United Nations], 2007. Land and nutrition management service.

FAO [Food and Agriculture Organization of the United Nations]. 2013. *FAO Statistical Yearbook*. World food and agriculture. Rome, Italy.

Farid, M., Shakoor, M.B., Ehsan, A., Ali, S., Zubair, M. and Hanif, M.S. 2013. Morphological, physiological and biochemical responses of different plant species to Cd stress. *International Journal of Chemical and Biochemical Sciences*.**3:** 53–60.

Faver, K.L., Gerik, T.J., Thaxton, P.M. and El-Zik, K.M. 1996. Late season water stress in cotton:II. Leaf gas exchange and assimilation capacity. *Crop Sci.*, **36:** 922-928.

Ferreira, J.L., Coelho, C.H.M., Megalhães, P.C., Gamas, E.F.G. and Borem, A. 2007. Genetic variability and morphological modifications in flooding tolerance in maize varielty BRS-4154. *Crop Breeding and Applied Botechnology,* **7:** 314-320.

Ferris, R., and Taylor. G. 1993. Contrasting effects of elevated on the root and shoot growth of four native herbs commonly found in chalk grassland. *New Phytol.*, **125:** 855–866.

Farquhar, G.D., Hubick, K.T., Codon, A.G. and Richards, R.A. 1989. Carbon isotope fractionation and plant water use efficiency. In, "Stable isotopes in ecological research", Ed. P.W. Rundell, E.H. Ehleringer and Nagy, K.A. Springer-Verlag, New York, pp. 220-240.

Fischer, R.A. and Edmeades, G.O. 2010. Breeding and cereal yield progress. *Crop Sci.*, **50:** S85–S98.

Fischlin, A., Midgley, G.F., Price, J.T., Leemans, R., Gopal, B., Turley, C., Rounsevell, M.D.A., Dube, O.P., Tarazona, J. and Velichko, A.A. 2007. In, "*Ecosystems, their properties, goods and* services. Climate Change 2007: Impacts, Adaptation and Vulnerability. Contribution of Working *Group II to the Fourth Assessment Report of the Intergovernmental Panel on Climate Change*", Eds. M.L. Parry, O.F. Canziani, J.P. Palutikof, P.J. van der Linden, C.E. Hanson, Cambridge: Cambridge University Press, pp. 211–272.

Flexas, J., and Medrano, H. 2002. Drought-inhibition of photosynthesis in C3 plants: Stomatal and non-stomatal limitations revisited. *Ann. Bot.*, **89:** 183-189.

Flexas, J., Bota, J. Loreto, F., Cornic, G. and Sharkey, T.D. 2004. Diffusive and metabolic limitations to photosynthesis under drought and salinity in C3 plants. *Plant. Biol.*, **6:** 269-279.

Flexas, J., J. Bota, J. Galmes, H. Medrano, and M. Ribas-Carbo. 2006. Keeping a positive carbon balance under adverse conditions: responses of photosynthesis and respiration to water stress. *Physiol. Plant.* **127:** 343-352.

Flora, S.J.S., Mittal, M. and Mehta, A. 2008. Heavy metal induced oxidative stress and its possible reversal by chelation therapy. *Indian Journal of Medical Research,* **128(4):** 201-523.

Flower, D.J. 1996. Physiological and morphological features determining the performance of the sorghum landraces of northern Nigeria. *Experimental Agriculture*, **32:** 129-141.

Fonseca, A.E. and Westgate, M.E. 2005. Relationship between desiccation and viability of maize pollen. *Field Crops Res*., **94:** 114–125.

Fontana, S., Wahsha, M. and Bini, C. 2010. Preliminary observations on heavy metal contamination in soils and plants of an abandoned mine in Imperina Valley (Italy). *Agrochimica*, **54(4):** 218–231.

Fowler, J.E. and Quatrano, R.S. 1997. Plant cell morphogenesis: Plasma membrane interactions with the *cytoskeleton and cell wall. Annu Rev Cell Dev Biol.*, **13:** 697–743.

Garbutt, K. and Bazzaz., F.A. 1984. The effects of elevated CO_2 on plants. III. Flower, fruit and seed production and abortion. *New Phytol.*, **98:** 433—446.

Ganguly, S.B., Bhattacharya, A. and Singh, D.N. 1999. Physiological studies on chickpea varieties: physiologica; determinants and other relationships to productivity under late seeding. *Indian Journal of Pulses Research*, **12(1):** 65-74.

Garcia-Sanchez, F., Jifon, J.L., Carvajal, M. and Syvertsen, J.P. 2002. Gas exchange, chlorophyll, nutrient contents in relation to Na+ and Cl- accumulation in *Sunburst mandarin* grafted in different rootstocks. *Plant Sci.*, **162:** 705-712.

Garcés, G., Garcés, P. and Diago, M. 2005. Resultados de monitoreo de cosecha 2004 Sur del Cesar. pp. 23-26. In: Compendio resultados de investigación 2003-2005. Federación Nacional de Arroceros; Fondo Nacional del Arroz, Bogota.

Gardner, W.R., and Gardner, H.R. 1983. Principles of water management under drought conditions. *Agric. Water Manage.* **7:** 143-155.

Garg, N. and Manchanda, G. 2009. ROS generation in plants: boon or bane? *Plant Biosyst.*, **143:** 81–96.

Garrigues, E. and Doussan, C. 2006. Water uptake by plant roots: I - Formation and propagation of a water extraction front in mature root systems as evidenced by 2D light transmission imaging. *Plant Soil*, **283(1-2):** 83-98.

Gebauer, R.L.E., Reynolds, J.F. and Strain, B.R. 1996. Allometric relations and growth in *Pinus taeda*: the effect of elevated CO_2 and changing N availability. *New Phytol.*, **134:** 85—93.

Geber, M.A. and Dawson, T.E. 1990. Genetic variation in and covariation between leaf gas exchange, morphology and development in *Polygonum arenastrum*, an annual plant. *Oecologia.*, **85:** 153–158.

Geebelen, W., Vangronsveld. J., Adriano, D.C., Carleer, R. and Clijsters, H. 2002. Amendment-induced immobilization of lead in a lead-spiked soil: evidence from phytotoxicity studies. *Water Air Soil Pollut.*, **140:** 261-277.

Gerik, T.J., Faver, K.L., Thaxton, P.M. and El-Zik, K.M. 1996. Late season water stress in cotton: I. Plant growth, water use and yield. *Crop Sci.*, **36:** 914-921.

Ghannoum, O., Evans, J.R. and von Caemmerer, S. 2011. Nitrogen and water use efficiency of C4 plants. In, "*C4 Photosynthesis and related CO_2 concentrating mechanism*", Eds. A.S. Raghvendra and R.S. Sage, Springer Science+Business Media B.V., Dordrecht, The Netherlands, pp. 129-146.

Ghashgaie, J., Duranceau, M., Badeck, F.W., Cornic, G., Adeline, M.T. and Deleens, E. 2001. $\delta^{13}C$ of CO_2 respired in the dark in relation to $\delta^{13}C$ of leaf metabolites: comparison between *Nicotiana sylvestris* and *Helianthus annuus* under drought. *Plant Cell Environ.*, **24:** 505-515.

Ghobadi, M.E. and Gobadi, M. 2010. Extent of anoxia and root growth and grain yield of wheat cultivars. *World Academy of Science, Engineering and Technology*, **70:** 85-88.

Giardi, M., Masojidek, J. and Godde, D. 1997. Effects of abiotic stresses on the turnover of the D1 reaction centre II protein. *Plant Physiol.*, **101:** 635-642.

Gibon, Y., Sulpice, R. and Larher, F. 2000. Proline accumulation in canola leaf discs subjected to osmotic stress is related to the loss of chlorophylls and to the decrease of mitochondrial activity. *Physiol. Plant*, **4:** 469-476.

Gifford, R.M. 2003. Plant respiration in productivity models: Conceptualization, representation and issues for global terrestrial carbon-cycle research. *Funct. Plant Biol.*, **30:** 171–186.

Giorno, F., Wolters-Arts, M., Mariani, C. and Rieu, I. 2013. Ensuring reproduction at high temperatures: The heat stress response during anther and pollen development. *Plants*, **2:** 489-606.

Giri, B. and Mukerji, K. 1999. Improved growth and productivity of *Sesbania grandiflora* Pers. under salinity stress through mycorrhizal technology. *J. Phytol. Res.*, **12:** 35–38

Giuliani, C., Pellegrino, F., Tirillini, B. and Maleci, L. 2008. Micromorphological and chemical characterization of *Stachys recta* subsp *serpentini* (Fiori) Arrigoni in comparison to *S. recta* subsp. *recta* (Lamiaceae). *Flora*, **203:** 376–385.

Gliñski, J and Lipiec, J. 1990. Soil Physical Conditions and Plant Roots. CRC Press, Boca Raton, FL, USA.

Gote, G.N. and Padghan, P.R. 2009. Studies on different thermal regimes and thermal sensitivity analysis of tomato genotypes. *Asian J. Environ. Sci.*, **3:** 158–161.

Gonzàlez-Meler, M.A., Ribas-Carbó, M., Siedow, J.N. and Drake, B.G. 1996. Direct inhibition of plant mitochondrial respiration by elevated CO_2. *Plant Physiol.* **112:** 1349-1355.

Gonzàlez-Meler, M.A. and Siedow, J.N. 1999. Direct inhibition of mitochondrial respiratory enzymes by elevated CO_2: does it matter at the tissue or whole-plant level? *Tree Physiol.*, **19:** 253—259.

Goulas, E., Schubert, M., Kieselbach, T., Kleczkowski, L.A., Gardestrom, P., Schroder, W. and Hurry, V. 2006. The chloroplast lumen and stromal proteomes of *Arabidopsis thaliana* show differential sensitivity to short- and long-term exposure to low temperature. *Plant J*, **47:** 720–734.

Gregory, P.J. 2006. Roots, rhizosphere and soil: the route to a better understanding of soil science? *European Journal of Soil Science*, **57:** 2-12.

Grennan, A.K. 2011. Metallothioneins, a diverse protein family. *Plant Physiology,* **155(4):** 1750–1751.

Grewal, H.S. 2010. Response of wheat to subsoil salinity and temporary water stress at different stages of the reproductive phase. *Plant Soil,* **330:** 103-113

Grieve, A.M., Prior, D.L. and Bevington, B.K. 2007. Longterm effect of saline irrigation water on growth, yield, and fruit quality of Valencia orange trees. *Aus. J. Agri. Res.*, **58:** 342-348.

Griffin, K.L. and Seemann, J.R. 1996. Plants, CO_2 and photosynthesis in the 21st century. *Chem. Biol.*, **3:** 245—254.

Griffin, K.L., Sims, D.A. and Seemann, J.R. 1999. Altered night-time CO_2 concentration affects the growth, physiology and biochemistry of soybean. *Plant Cell Environ.*, **22:** 91-99.

Gross, Y. and Kigel, J. 1994. Differential sensitivity to high temperature of stages in the reproductive development of common bean (*Phaseolus vulgaris* L.). *Field Crops Research,* **36:** 201–212.

Grover, A.; ChandraMouli, A.; Agarwal, S., Katiyar-Agarwal, S., Agarwal, M. and Sahi, C. 2009. Transgenic rice for tolerance against abiotic stresses. In, "*Rice Improvement in the Genomic Era*", Ed. S.K. Datta, Hawarth Press, USA, pp. 237-267.

Gupta, D., Vandenhove, H. and Inouhe, M. 2013. Role of phytochelatins in heavy metal stress and detoxification mechanisms in plants. In, "*Heavy Metal Stress in Plants",* Berlin, Germany: Springer; 2013.; pp. 73–94.

Güsewell, S., Jakobs, G. and Weber, E. 2006. Native and introduced populations of *Solidago gigantea* differ in shoot production but not in leaf traits or litter decomposition. *Functional Ecology*, **20:** 575–584.

Hall, A.E. 1992. Breeding for heat tolerance. In, "*Plant breeding reviewers", Ed.* J. Janick, New York, NY, Wiley, pp. 129–168.

Hall, J.L. 2002. Cellular mechanisms for heavy metal detoxification and tolerance. *Journal of Experimental Botany*. **53(366):** 1–11.

Halliwell, B. 1987. Oxidative damage, lipid peroxidation and antiox-idant protection in chloroplasts. *Chem. Phys. Lipids*, **44:** 327-340.

Hameed, A., Goher, M. and Iqbal, N. 2012. Heat stress-induced cell death, changes in antioxidants, lipid peroxidation and protease activity in wheat leaves. *J Plant Growth Regul.*, **31:** 283–291.

Han, F.; Chen, H.; Li, X.J., Yang, M.F., Liu, G.S. and Shen, S.H. 2009. A comparative proteomic analysis of rice seedlings under various high-temperature stresses. *Biochim Biophys Acta*, **1794:** 1625-1634.

Han, H., Gao, S., Li, B., Dong, X.C., Feng, H.L. and Meng, Q.W. 2010. Overexpression of violaxanthin de-epoxidase gene alleviates photoinhibition of PSII and PSI in tomato during high light and chilling stress. *J Plant Physiol*, **167:** 176–183.

Hanley, M., Trodmov, S. and Taylor, G. 2004. Species level effects more important than functional group level responses to elevated CO_2: Evidence from simulated turves. *Funct. Ecol.*, **18:** 304–313.

Hansen, W.J., Sato, M., Ruedy, R., Lacis, A. and Oinas, V. 2000. Global warming in the twenty first century: An Alternative Scenario. *Proceedings of National Academy of Science*, **97:** 9875-9880.

Hardy, R.W.F. and Havelka, U.D.K. 1975. Symbiotic N2 fixation: Multi-fold enhancement by CO_2-enrichment of field-grown soybeans. *Plant Physiol.*, **48(Suppl.):** 35.

Hardham, A.R. and Gunning, B.E. 1978. Structure of cortical microtubule arrays in plant cells. J Cell Biol., **77:** 14–34.

Hatfield, J.L., Boote, K.J., Fay, P., Hahn, L., Izaurralde, R.C., Kimball, B.A., Mader, T., Morgan, J., Ort, D.,Polley, W., Thomson, A. and Wolfe, D. 2008. Agriculture. In, "*The Effects of Climate Change on Agriculture, Land Resources, Water Resources, and Biodiversity in the United States*. Eds. J.L.

Hatfield, K.J. Boote, P. Fay, L. Hahn, R.C. Izaurralde, B.A. Kimball, T. Mader, J. Morgan, D. Ort, W. Polley, A. Thomson, A., D. Wolfe, D. Hatfield, J.L., Boote, K.J., Kimball, B.A., Ziska, L.H., Izaurralde, R.C., Ort, D., Thomson, A.M. and Wolfe, D.W. 2011. Climate impacts on agriculture: implications for crop production. *Agron. J.* **103:** 351–370

Hatfield, J.L., Takle, G., Grothjahn, R., Holden, P., Izaurralde, R.C., Mader, T., Marshall, E., Liverman, D. 2014. In, "*Agriculture. Climate Change Impacts in the United States: The Third National Climate Assessment.*", Eds. J.M. Melillo, T.C. Richmond, G.W. Yohe, pp. 150–174.

Hattori, Y., Nagai, K. and Ashikari, M. 2011. Rice growth adapting to deepwater, Current Opinion in Plant Biology, **14:** 100–105.

Hauagge, R. 2010. ''IPR julieta', a new early low chill requirement apple cultivar. *Acta Hortic. (ISHS)*, **872:** 193–196.

Hauser, M.T., Morikami. A. and Benfey, P.N. 1995. Conditional root expansion mutants of *Arabidopsis. Development,* **121:** 1237-1252.

Havaux, M. 1992. Stress tolerance of photosystem II in vivo: Antagonistic effects of water, heat and photoinhibition stresses. *Plant Physiol.*, **100:** 424–432.

Hawkesford, M., Araus, J., Park, R., Calderini, D., Miralles, D., Shen, T., Zhang, J. and Parry, M.A. 2013. Prospects of doubling global wheat yields. *Food Energy Secur.*, **2:** 34–48.

Hazer Jaribi, E., Jabarov, K., Sobouri, A. and Kavandi, R. 2013. Evaluation of genotypic x environments interactions using path coefficienr analysis in soybean. *Caspian Journal of Applied Sciences Research,* **2(7):** 46-54.

Hearn, A.B. 1994. The principles of cotton water relations and their application in management. In, "*Challenging the Future*", Eds. Constable G.A. and N.W. Forrester, Proc. World Cotton Conf. Brisbane, Australia. pp. 66-92.

Hebbara, M., Rajakumar, G.R., Ravishankar, G. and Raghavaiah, C.V. 2003. Effect of salinity stress on seed yield through physiological parameters in sunflower genotypes. *Helia,* **26:**155-160.

Hedhly, A., Hormaza, J.I. and Herrero, M. 2005. The effect of temperature on pollen germination pollen tube growth, and stigmatic receptivity in peach. *Plant Biology*, **7:** 476-483.

Hedhly, A., Hormaza, J.I. and Herrero, M. 2007. Warm temperatures at bloom reduce fruit set in sweet cherry. *Journal of Applied Botany Food Quality,* **81:** 158–164.

Hedhly, A., Hormaza, J.I. and Herrero, M. 2008. Global warming and plant sexual reproduction. *Trends in Plant Science*, **14:** 30-36.

Heinemann, A.B., Maia Ad, H.N., Dourado-Neto, D., Ingram, K.T. and Hoogenboom. G. 2006. Soybean [*Glycine max* (L.) Merr.] growth and development response to CO_2 enrichment under different temperature regimes. *Eur. J. Agron.*, **24:** 52–61.

Heitholt, J.J. and Schmidt, J.H. 1994. Receptacle and ovary assimilate concentrations and subsequent boll retention in cotton. *Crop Sci.*, **34:** 125-131.

Helios-Rybicka, E., 1996. Impact of mining and metallurgical industries on the environment in Poland. *Applied Geochemistry*, **11(1–2):** 3–11.

Hemantaranjan, A., Bhanu, A.N., Singh, M.N., Yadav, D.K., Patel, P.K., Singh, R. and Katiyar. D. 2014.

Heat stress responses and thermotolerance. *Adv. Plants Agri. Res.*, **3:** 1–10.

Henson, I.E., Mahalakshmi, V., Bidinger, F.R. and Alagarswamy, G. 1982. Osmotic adjustment to water stress in pearl millet under field conditions. *Plant Cell Environ*, **5:** 147-154.

Herrero, M.P. and Johnson, R.R. 1980. High-temperature stress and pollen viability of maize. *Crop Science,* **20:**, 796–800.

Herrero, M. and Arbeloa, A. 1989. Influence of the pistil on pollen-tube kinetics in peach (*Prunus persica*). *American Journal of Botany*, **76:** 1441-1447.

Herrero, M. 2003. Male and female synchrony and the regulation of mating in flowering plants. Philosophical *Transactions of the Royal Society of London Series B-Biological Sciences*, **358:** 1019-1024.

Hidema, J., Makino, A., Mae, T. and Ojima, K. 1991. Photosynthetic characteristics of rice leaves aged under different irradiances from full expansion through senescence. *Plant Physiol.*, **97:** 1287-1293.

Hilal, M., Zenoff, A.M., Ponessa, G., Moreno, H. and Massa, E.M. 1998. Saline stress alters the temporal patterns of xylem differentiation and alternative oxidase expression in developing soybean roots. *Plant Physiology,* **117:** 695-701.

Hinsinger, P., Bengough, A.G., Vetterlein, D. and Young, I.M. 2009. Rhizosphere: biophysics, biogeochemistry and ecological relevance. *Plant and Soil*, **321:** 117.

Hoad, S.P., Russell, G., Lucas, M.E. and Bingham, I.J. 2001. The management of wheat, barley, and oat root systems. *Advances in Agronomy*, **74:** 193-254.

Hoffmann, A.A. and Merilä, J. 1999. Heritable variation and evolution under favourable and unfavourable conditions. *Trends Ecol Evol.*, **14:** 96–101.

Hossain, M.A., Piyatida, P., da Silva, J.A.T. and Fujita, M. 2012. Molecular mechanism of heavy metal toxicity and tolerance in plants: central role of glutathione in detoxification of reactive oxygen species and methylglyoxal and in heavy metal chelation. *Journal of Botany*. **2012:** 37.

Hsiao, T.C. 1973. Plant responses to water stress. *Annu. Rev. Plant Physiol.* **24:** 519-570.

Hsiao, T.C., E. Acevedo, E. Fereres, and D.W. Henderson. 1976. Stress metabolism: Water stress, growth and osmotic adjustment. *Phil. Trans. R. Soc. Lond. B*. **273:** 479-500.

Hsu, F.S., Lin, J.B. and Chang, S.R. 2000. Effets of waterlogging on seed germination, electrical conductivity, of seed leakage and development of hypocotyls and radical in sudangrasses. *Bot Acad Bul Sin,* **41:** 267-273.

Hu, L., Wang, Z. and Huang, B. 2011. Effects of cytokinin and potassium on stomatal and photosynthetic recovery of Kentucky Bluegrass from drought stress. *Crop Sci*. **53:** 221–231.

Hume, D.J. and Jackson, A.K.H. 1981. Frost tolerance in soybeans. *Crop Science*, **21:** 689-692.

Hungate, B.A., Holland, E.A., Jackson, R.B., Chapin, F.S. III, Mooney, H.A. and Field, C.B. 1997. The fate of carbon in grasslands under carbon dioxide enrichment. *Nature*, **388:** 576—579.

Hughes, M.A. and Dunn, M.A. 1996. The molecular physiology of plant acclimation to temperature. *J. Exp. Bot.*, **47:** 291-305.

Hutton, R.J. and Landsberg, J.J. 2000. Temperature sums experienced before harvest partially determine the post-maturation juicing quality of oranges grown in the Murrumbidgee Irrigation Areas (MIA) of New South Wales. *J. Sci. Food Agric.*, **80:** 275–283.

Intergovernmental Panel Climate Change (IPCC). 2007. Climate Change 2007: Impacts, Adaptation and Vulnerability: Contribution of Working Group II to the Fourth Assessment Report of the Intergovernmental Panel on Climate Change. Cambridge University Press, Cambridge, U.K. and New York, NY.

Ishikawa, H. and Evans, M.L. 1995. Specialized zones of development in roots. *Plant Physiology,* **109:** 725-727.

Ito, Y., Nakanomyo, I., Motose, H., Iwamoto, K., Sawa, S., Dohmae, N. and Fukuda, H. 2006. Dodeca-CLE peptides as suppressors of plant stem cell differentiation. *Science,* **313:** 842-845.

Ivanov, A.G., Sane, P.V., Krol, M., Gray, G.R., Balseris, A., Savitch, L.V., Oquist, G. and Huner, N.P. 2006. Acclimation to temperature and irradiance modulates PSII charge recombination. *FEBS Lett,* **580:** 2797–2802.

Izanloo, A., Condon, A.G., Langridge, P., Tester, M. and Schnurbusch, T. 2008. Different mechanisms of adaptation to cyclic water stress in two South Australian bread wheat cultivars. *J Exp Bot.*, **59:** 3327–3346.

Jach, M.E. and Ceulemans. R. 1999. Effects of elevated atmospheric CO_2 on phenology, growth and crown structure of Scots pine (*Pinus sylvestris*) seedlings after two years of exposure in the field. *Tree Physiol.*, **19:** 289—300.

Jackson, M.B. 1990. Hormones and developmental change in plants subjected to submergence or soil water logging. *Aquat Bot*, **38:** 49-72.

Jackson, R.B., Sperry, J.S. and Dawson, T.E. 2000. Root water uptake and transport: using physiological processes in global predictions. *Trends Plant Sci.*, **5:** 482–488.

Jagadish, S.V.K., Craufurd, P.Q. and Wheeler, T.R. 2008. Phenotyping parents of mapping populations of rice for heat tolerance during anthesis. *Crop Science*, **48:** 1140-1146.

Kaushal, N., Bhandari, K., Siddique, K.H.M. and Nayyar, H. 2016. Food crops face rising temperatures: An overview of responses, adaptive mechanisms, and approaches to improve heat tolerance. *Cogent Food and Agriculture,* **2(1):** 1134380

Jenabiyan, M., Pirdashi, H. and Yaghoubiab, Y. 2014. The combined effect of cold and light intensity stress on some morphological and physiological parameters in two soybean (*Glycin max* L.) cultivars. *International Journal of Biosciences*, **5(3):** 189-197.

Jewell, M.C., Campbell, B.C. and Ian, D. 2010. Transgenic plants for abiotic stress resistance. In, "*Transgenic Crop Plant*", Eds. C. Kole, C.H. Michler, A.G. Abbott, and T.C. Hall, Springer. Berlin 67-132.

Jian, Y.and Huang. B. 2001. Drought and heat stress injury to two cool-season turf grass in relation to antioxidant and lipid peroxidation. *Crop Sci.*, **41:** 436–442.

Jain, M., Prasad, P.V.V., Boote, K.J., Hartwell, A.L. and Chourey, P.S. 2007. Effects of season-long high temperature growth conditions on sugar-to-starch metabolism in developing microspores of grain sorghum (*Sorghum bicolor* L. Moench). *Planta*, **227:** 67-79.

Jiang, Y. and Huang, B. 2001. Drought and heat stress injury to two cool-season turfgrass in relation to antioxidant metabolism and lipid peroxidation. *Crop Sci.*, **41:** 436-442.

Jiao, D. and Li, X. 2001. Cultivar differences in photosynthetic tolerance to photooxidation and shading in rice (*Oryza sativa* L.). *Photosynthetica*, **39(2):** 167–175.

John, R., Ahmad, P., Gadgil, K. and Sharma, S. 2009. Heavy metal toxicity: effect on plant growth, biochemical parameters and metal accumulation by *Brassica juncea* L. *International Journal of Plant Production*. **3(3):** 65–76.

Johnson, R.R. and Moss, D.N. 1976. Effect of water stress on $^{14}CO_2$ fixation and translocation in wheat during grain filling. *Crop Sci.*, **16:** 697-701.

Jones, M.M., and Turner, N.C. 1978. Osmotic adjustment in leaves of sorghum in response to water deficits. *Plant Physiol.*, **61:** 122-126.

Juan, A., González, M. Gallardo, A. Hilal, Rosa, M. and Prado, F.E. 2009. Physiological responses of quinoa (*Chenopodium quinoa* Willd.) to drought and waterlogging stresses: drymatter partitioning *Botanical Studies*, **50:** 35-42.

Kadir, S., Sidhu, G. and Al-Khatib, K., 2006. Strawberry (Fragaria ananassa duch.) growth and productivity as affected by temperature. *Hort Science*, **41:** 1423–1430.

Kamal, A., Qureshi, M.S., Ashraf, M.Y. and Hussain, M. 2003. Salinity induced changes in some growth and physiochemical aspects of two soybeans [*Glycine max* (L.) Merr.] genotypes. *Pak. J. Bot.*, **35:** 93-97.

Karimzadeh, G., Sharifi-Sirchi, G.R., Jalali-Javaran, M., Dehghani, H. and Francis, D. 2006. Soluble proteins induced by low temperature treatment in the leaves of spring and winter wheat cultivars. *Pak J Bot.*, **38:** 1015–1026.

Kariola, T.; Brader, G.; Helenius, E., Li, J., Heino, P. and Palva, T. 2006. Early responsive to dehydration 15, a negative regulator of abscisic acid responses in *Arabidopsis, Plant Physiol,* **142:** 1559- 1573.

Katerji, N., Tardieu, F., Bethenod, O. and Quetin. P. 1994. Behavior of maize stem diameter during drying cycles: Comparison of two methods for detecting water stress. *Crop Sci.*, **34:** 165–169.

Kaushal, N., Awasthi, R., Gupta, K., Gaur, P., Siddique, K.H.M. and Nayyar, H. 2013. Heat-stress-induced reproductive failures in chickpea (*Cicer arietinum*) are associated with impaired sucrose metabolism in leaves and anthers. *Functional Plant Biology,* **40:** 1334–1349.

Kawai, M., Samarajeewa, P.K., Barrero, R.A., Nishiguchi, M. and Uchimiya, H. 1998. Cellular dissection of the degradation pattern of cortical cell death during aerenchyma formation of rice roots. *Planta,* **204:** 277–287.

Kaya, Y., Topal, R., Gonulal, A.E. and Arisoy, R.Z. 2002. Factor analyses of yield traits in genotypes of durum wheat (*Triticum durum*). *Indian Journal of Agricultural Science* **72:** 301-303.

Keane, B., Collier, M.H., Shann, J.R. and Rogstad, S.H. 2001. Metal content of dandelion (*Taraxacum officinale*) leaves in relation to soil contamination and airborne particulate matter. *The Science of the Total Environment,* **281:** 63–78.

Keller, F. and Ludlow, M.M. 1993. Carbohydrate metabolism in drought-stressed leaves of pigeonpea (*Cajanus cajan*). *J. Exp. Bot.*, **44:** 1351-1359.

Keles, Y. and Oncel, I. 2004. Growth and solute composition in two wheat species experiencing Combined influence of stress conditions. *Russ. J. Plant Physiol.*, **51:** 228-233.

Kennedy, R.A., Rumpho, M.E and Fox, T.C. 1992. Anaerobic metabolism in plants. *Plant Physiol*, **100:** 1-6.

Keskitalo, J., Bergquist, G., Gardeström, P. and Jansson, S. 2005. A cellular timetable of autumn senescence. *Plant Physiol,* **139:** 1635–1648.

Keutgen, A.J. and Pawelzik, E. 2008. Quality and nutritional value of strawberry fruit under long term salt stress. *Food Chemistry,* **107:** 1413-1420.

Khalili, M., Naghavi, M.R., Pour Aboughadareh, A. and Talebzadeh, S.J. 2012. Evaluating of drought stress tolerance based on selection indices in spring canola cultivars (*Brassica napus* L.). *J. Agric. Sci.*, **4:** 78-85.

Kidd, P., Barcelo, J., Bernal, M.P., Navari-Izzo, F., Poschenrieder, C., Shilev, S., Clemente, R. and Monterroso, C. 2009. Trace element behaviour at the root–soil interface: implications in phytoremediation. *Environmental and Experimental Botany*, **67:** 243–259.

Sarret, G., Vangronsveld, J., Roux, M., Coves, J. and Manceau, A. 2001. Bioaccumulation of metal in plants and microorganisms studied by electron microscopy and EXAFS spectroscopy. Proc. VI ICOBTE, Guelph, p. 555 (abstract).

Kim, H.Y., Horie, T., Nakagawa, H., and Wada, K. 1996. Effects of elevated CO_2 concentration and high temperature on growth and yield of rice. II. The effect of yield and its component of Akihikari rice. *Jpn. J. Crop Sci.*, **65:** 644–651.

Kimball, B.A. 2011. Lessons from FACE: CO_2 effects and interactions with water, nitrogen andtemperature. In, "Handbook of climate change and agroecosystems: Impacts, adaptation, and mitigation", Eds. D. Hillel and C. Rosenzweig, Imperial College Press, Hackensack, NJ. pp. 87–107.

King, J.S., Thomas, R.B. and Strain, B.R. 1996. Growth and carbon accumulation in root systems of *Pinus taeda* L. and *Pinus ponderosa* Dougl. *ex* Laws. seedlings as affected by varying CO_2, temperature, and nitrogen. *Tree Physiol.*, **16:** 635—642.

Kiran kumara, K.S.P., Sridevi, V. and Chandana Lakshmi, M.V.V. 2012: Studies on effect of salt stress on some medicinal plants. *International Journal of Computational Engineering Research,* **2:** 143-149.

Klein, J.A., Harte, J. and Zhao, X.-Q. 2007. Experimental warming, not grazing, decreases rangeland quality on the Tibetan plateau. *Ecol. Appl.*, **17:** 541–557.

Knapp, A.K., Briggs, J.M. and Koelliker, J.K. 2001. Frequency and extent of water limitation to primary production in a mesic temperate grassland. *Ecosystems.*, **4:** 19–28.

Knight, H., Brandt, S. and Knight, M.R. 1998. A history of stress alters drought calcium signalling pathways in *Arabidopsis. Plant J,* **16:** 681–687.

Knowles, N., Dettinger, M.D. and Cayan, D.R. 2006. Trends in snowfall versus rainfall in the western United States. *J. Clim.*, **19:** 4545–4559.

Kramer, P.J. 1980. Drought stress and the origin of adaptation. In, *"Adaptation of plants to water and high temperature stress*", Eds. N.C. Turner, and P.J. Kramer, John Wiley, New York, pp. 7-20.

Kramer, P.J. 1983. Water deficits and plant growth. pp: 342-389. In, "Water relations of plants", Ed. P.J. Kramer. Academic Press, New York.

Krastanov, A. 2010. Metabolomics - The state of art. *Biotechnology and Biotechnological equipment.* **24:** 1537-1543.

Krieg, D.R., and Sung, F.J.M. 1986. Source-sink relationships as affected by water stress. In, "*Cotton Physiology",* Eds. J.R. Mauney and J.M. Stewart, The Cotton Foundation, Memphis, Tenn., pp. 73-79.

Kobayashi, H., Masaoka, Y., Takahashi, Y., Ide, Y. and Sato, S. 2007. Ability of salt glands in Rhodes Grass (*Chloris gayana* Kunth) to secrete Na+ and K+. *Soil Science and Plant Nutrition,* **53:** 764-771.

Kobata, T., Sugawara, M. and Takatu, S. 2000. Shading during the early grain filling period does not affect potential grain dry matter increase in rice. *Agron J,* **92(3):** 411–417.

Kozela, C. and Regan, S. 2003. How plants make tubes. *Trends in Plant Science,* **8:** 159–164.

Krishnamurthy, R. and Bhagwa, K.A. 1995. Effect of NaCl on enzymes in salt tolerance and salt sensitive rice cultivars. *Acta Agron. Hung.* **43:** 51-57.

Królak, E. 2003. Accumulation of Zn, Cu, Pb and Cd by dandelion (T*araxacum officinale* Web.) in environments with various degrees of metallic contamination. *Polish Journal of Environmental Studies,* **12(6):** 713–721.

Krômer, S. 1995. Respiration during photosynthesis. *Annu. Rev. Plant Physiol. and Plant. Mol.Biol.,* **46:** 45-70.

Kumar, A., Singh, P., Singh, D.P., Singh, H. and Sharma, H.C. 1984. Differences in Osmoregulation in *Brassica* Species. *Annals Bot.*, **54:** 537-541.

Kumar, S., Thakur, P., Kaushal, N., Malik, J.A., Gaur, P. and Nayyar, H. 2013. Effect of varying high temperatures during reproductive growth on reproductive function, oxidative stress and seed yield in chickpea genotypes differing in heat sensitivity. *Archieves of Agronomyand Soil Science,* **59:** 823–843.

Kumudini, S., Andrade, F.H., Boote, K.J., Brown, G.A., Dzotsi, K.A., Edmeades, G.O., Gocken, T., Goodwin, M., Halter, A.L., Hammer, G.L., Hatfield, J.L., Jones, J.W., Kemanian, A.R., Kim, S.-H., Kiniry, J., Lizaso, J.I., Nendel, C., Nielsen, R.L., Parent, B., Stçckle, C.O., Tardieu, F., Thomison, P.R., Timlin, D.J., Vyn, T.J., Wallach, D., Yang, H.S. and Tollenaar, M. 2014. Predicting maize phenology: intercomparison of functions for developmental response to temperature. *Agron. J.* **106:** 2087–2097.

Lacape, M.J., Wery, J. and Annerose, D.J.M. 1998. Relationships between plant and soil water status in five field-grown cotton cultivars. *Field Crops Res.*, **57:** 29-43.

Larcher W. 1995. Physiological plant ecology. Springer-Verlag, New York, pp. 506.

Larkindale, J.; Hall, J.D.; Knight, M.R. and Vierling, E. 2005. Heat stress phenotypes of *Arabidopsis mutants* implicate multiple signaling pathways in the acquisition of Thermotolerance. *Plant Physiol*, **138:** 882-97.

Lakshmi-Prada, M., Vanangamudi, M. and Thandapi, V. 2004. Effect of low light on yield and physiological attributes of rice. *IRRN*, **29(2):** 71-73.

Lara, M.V. and Andrea, C.S. 2005. Photosynthesis in non typical C4 species. In, "*Handbook of photosynthesis*", Ed. M. Pessarakli, Second Edition, CRC Press, Tylor and Francis Group, Boca Ratôn, FL, USA, pp. 391-421.

Larson, R.A. 1988. The antioxidants of higher plants, *Phytochemistry,* **27:** 969-978.

Laugier, E., Tarrago, L., Vieira Dos Santos, C., Eymery, F., Havaux, M. and Rey, P. 2010. *Arabidopsis thaliana* plastidial methionine sulfoxide reductases B, MSRBs, account for most leaf peptide MSR activity and are essential for growth under environmental constraints through a role in the preservation of photosystem antennae. *Plant J*, **61:** 271–282.

Law, D.R., and Crafts-Brandner, S.J. 1999. Inhibition and acclimation of photosynthesis to heat stress is closely correlated with activation of ribulose-1,5-bisphosphate carboxylase / oxygenase. *Plant Physiol.*, **120:** 173–181.

Lavania, D., Dhingra, A., Siddiqui, M.H., Al-Whaibi, M.H. and Grover, A. 2015a. Current status of the production of high temperature tolerant transgenic crops for cultivation in warmer climates. *Plant Physiol. Biochem.*, **86:** 100–108.

Lavania, D., Dhingra, A., Siddiqui, M.H., Al-Whaibi, M.H. and Grover, A. 2015b. Genetic approaches for breeding heat stress tolerance in faba bean (*Vicia faba* L.). *Acta Physiol. Plant*, **37:** 1737

Lawlor, D. W., and H. Fock. 1977. Water stress induced changes in the aMounts of some photosynthetic assimilation products and respiratory metabolites of sunflower leaves. *J. Exp.Bot.,* **28:** 329-337.

Lawlor, M.M., and G. Cornic. 2002. Photosynthetic carbon assimilation and associated metabolism in relation to water deficits in higher plants. *Plant Cell Environ.* **25:** 275-294.

Lawlor, D.W. 2002. Limitation to photosynthesis in water stressed leaves: stomata vs. metabolism and the role of ATP. *Ann. Bot.*, **89:** 871-885.

Le Houerou, H.N. 1996. Climate changes, drought and desertification. *J. Arid. Environ.* **34:** 133-185.

Lea, P.J. and Leegood, R.C. 1999. Plant Biochemistry and Molecular Biology. John Wiley, Chichester.

Leidi, E.O., Lopez, J.M., Lopez, M. and Gutierrez, J.C. 1993. Searching for tolerance to water stress in cotton genotypes: photosynthesis, stomatal conductance and transpiration. *Photosynthetica.* **28:** 383-390.

Levy, Y. and Syvertsen, J.P. 2004. Irrigation water quality and salinity effects in citrus trees. *Hortic. Rev.*, **30:** 37-82.

Lewis, A.M. and Boose, E.R. 1995. Estimating volume flow rates through xylem conduits. *American Journal of Botany* **82** : 1112-1116.

Lewis, J.D., Olszyk, D., and Tingey, D.T. 1999. Seasonal patterns of photosynthetic light response in Douglas-fir seedlings subjected to elevated atmospheric CO_2 and temperature. *Tree Physiol.*,**19:** 243-252.

Li, J.-H., Dijkstra, P. Hinkle, R. Wheeler, R.M. and Drake, B.G. 1999. Photosynthetic acclimation to elevated atmospheric CO_2 concentration in the Florida scrub-oak species *Quercus geminata* and *Quercus myrtifolia* growing in their native environment. *Tree Physiol.*, **19:** 229—234.

Li, Y.Y., Sperry, J.S. and Shao, M. 2009. Hydraulic conductance and vulnerability to cavitation in corn (*Zea mays* L.) hybrids of differing drought resistance. *Environ. Exp. Botany*, **66(2):** 341-346.

Liang, L.H., Mei, X., Lin. F., Xia. J., Liu, S.J. and Wang, J.H. 2009. Effect of low temperature stress on tissue structure and physiological index of cashew young leaves. *Ecol. Environ. Sci.* **18:** 317-320.

Liao, C.T. and Lin, C.H. 2000. Physiological adaptation of crop plants to flooding stress. *Proc. Natl. Sci. Counc. ROC(B)*, **25(3):** 148-157.

Lim, P.O., Kim, H.J. and Nam, H.G. 2007. Leaf senescence. *Annu Rev Plant Biol*, **58:** 115–136.

Lipiec, J., Horn, R., Pietrusiewicz, J. and Siczek, A. 2012. Effects of soil compaction on root elongation and anatomy of different cereal plant species. *Soil Till. Res.*, **121:** 74-81.

Lipiec, J., Doussan, C., Nosalewicz, A. and Kondracka, K. 2013. Effect of drought and heat stresses on plant growth and yield: a review. *Int. Agrophys*, **27:** 463-477.

Liu, F., Jensen, C.R. and Andersen, M.N.. 2004. Drought stress effect on carbohydrate concentration in soybean leaves and pods during early reproductive development: its implication in altering pod set. *Field Crops Res.*, **86:** 1-13.

Liu, Q H, Zhou X B, Yang L Q, Li T, Zhang J J. 2009. Effects of early growth stage shading on rice flag leaf physiological characters and grain growth at grain-filling stage. *Chin J Appl Ecol*, **20(9):** 2135–2141. (in Chinese with English abstract)

Liu, Q., Wu, X., Chen, B., Ma, J. and Jia-qing, Gao, J. 2014. Effects of low light on agronomic and physiological characteristics of rice including grain yield and quality. *Rice Science*, **21(5):** 243-251.

Lizaso, J.I. and Ritchie, J.T. 1997. Maize root and shoot response to root zone saturation during vegetative growth. *Agronomy Journal*, **89:** 125-134.

Lobell, D.B., Burke, M.B., Tebaldi, C., Mastrandrea, M.D., Falcon, W.P. and Naylor, R.L. 2008. Priorizing climate change adaptation needs for food security in 2030. *Science*, **319:** 607–610.

Lobell, D.B., Schlenker, W. and Costa-Roberts, J. 2011. Climate trends and global crop production since 1980. *Science*, **333:** 616–620.

Lobell, D.B. and Gourdji, S.M. 2012. The influence of climate change on global crop productivity. *Plant Physiol.*, **160:** 1686–1697.

Lone, A.A. and Warsi, M.Z.K. 2009. Response of maize (*Zea mays* L.) to excess soil moisture (ESM) tolerance at different stages of life cycle. *Botany Research International*, **2:** 211-217.

Long, S.P. 1991. Modification of the response of photosynthesis productivity to rising temperature by atmospheric CO_2 concentrations: has its importance been under estimated? *Plant Cell and Environment*, **14:** 729-739.

Long, S.P. and Drake, B.G. 1991. Effect of the long-term elevation of CO_2 concentration in the field on the quantum yield of photosynthesis of the C3 sedge, *Scirpus olneyi*. *Plant Physiol.*, **96:** 221—226.

Long, S.P., Baker, N.R. and Raines, C.A. 1993. Analyzing the responses of photosynthetic CO_2 assimilation to long-term elevation of atmospheric CO_2 concentration. *Vegetatio*, **104/105:** 33—45.

Long, S.P. 1999.. Environmental responses. In, "*C4 Plant Biology*", Eds. R.F. Sage and R.K. Monson, Academic Press, San Diego, USA, pp. 215-249.

Lopareva-Pohu, A., Verdin, A., Garçon, G., Sahraoui, A.L., Pourrut, B., Debiane, D.,Waterlot, C., Laruelle, F., Bidar, G., Douay, F. and Shirali, P. 2011. Influence of fly ash aided phytostabilisation of Pb, Cd and Zn highly contaminated soils on *Lolium perenne* and *Trifolium repens* metal transfer and physiological stress. *Environmental Pollution,* **159:** 1721–1729.

Lopes, M.S. and Reynolds, M.P. 2012. Stay-green in spring wheat can be determined by spectral reflectance measurements (normalized difference vegetation index) independently from phenology. *J Exp Bot*, **63:** 3789–3798.

Lou, Q., Chen, L., Sun, Z., Xing, Y., Li, J., Xu, X., Mei, M., Luo, J. and Luo, M.L. 2007. A major QTL associated with cold tolerance at seedling stage in rice (*Oryza sativa L.*). *Euphytica,* **158:** 87-94.

Lu, C. and Zhang, J. 1999. Effects of water stress on Photosystem II photochemistry and its thermostability in wheat plants. *J. Exp. Bot.*, **50:** 1199–1206.

Ludlow, M.M. and Muchow, R.C. 1990. A critical evaluation of traits for improving crop yields in water-limited environments. *Advances in Agronomy*, **43:** 107-149.

Luedeling, E., Zhang, M. and Girvetz, E.H. 2009. Climate changes lead to declining winter chill for fruit and nut trees in California during 1950-2009. *PLOS One* **4:** e6166.

Lukowska, M. and Józefaciuk, G. 2013. Unknown mechanism of plants response to drought: low soil moisture and osmotic stresses induce severe decrease in CEC and increase in acidity of barley roots. *J. Agric. Sci.*, **5(10):** 204-213.

Madhu, M. and Hatfield, J.L. 2015. Elevated carbon dioxide and soil moisture on early growth response of soybean. *Agricultural Sciences*, **6:** 263-278.

Malekshahi, F., Dehghani, H. and Alizadeh, B. 2009. A study of drought tolerance indices in Canola (*Brassica napus* L.) genotypes. *J. Sci. Tech. Agric. Nat. Res.*, **13:** 77-90.

Malkowski, E., Kurtyka, R. and Kita. A. 2005. Accumulation of Pb and Cd and its effect on Ca distribution in maize seedlings (*Zea mays* L.). *Polish J. Environ. Stud.*, **14:** 203-207.

Macovei, A., Ventura, L., Donà, M., Faè, M., Balestrazzi, A. and Carbonera, D. 2010. Effects of heavy metal treatments on metallothionein expression profiles in white poplar (*Populus albaL.*) cell suspension cultures. *Analele Universitãtii din Oradea-Fascicula Biologie,* **18(2):** 274–279.

Mahajan, S. and Tujeta, N. 2005. Cold, salinity and drought stresses:An overview. *Archives of Biochemistry and Biophysics*, **444:** 139–158.

Maherali, H., Caruso, C.M., Sherrard, M.E. and Latta, R.G. 2010. Adaptive value and costs of physiological plasticity to soil moisture limitation in recombinant inbred lines of A*vena barbata. Am Nat.*, **175:** 211–224.

Malawska, M. and Wilkomirski, B. 2001. An analysis of soil and plant (*Taraxacum officinale*) contamination with heavy metals and polycyclic aromatic hydrocarbons (PAHs) in the area of the railway junction Ilawa Glòwna, Poland. *Water, Air, and Soil Pollution,* **127:** 339–349.

Maleci, L., Gentili, L., Pinetti, A., Bellesia, F. and Servettaz, O. 1999. Morphological and phytochemical characters of *Thymus striatus* Vahl growing in Italy. *Plant Biosystems,* **133(2):** 137–144.

Maleci, L., Bini, C. and Paolillo, A. 2001. Chromium (III) uptake by *Calendula arvensis* L. and related phytotoxicity. Proc. VI ICOBTE, Guelph, On. , p. 384 (abstract).

Malgorzata, H. and Andzej, N. 2005. Monitoring of bioremediation of soil polluted with diesel fuel applying bioassays. *Elect. J. Polish Agric. Univ.*, **8:** 1-2

Malik, R.S., Dhankar, J.S. and Turner, N.C. 1979. Influence of soil water deficits on root growth of cotton seedlings. *Plant Soil.* **53:** 109-115.

Manara, A. 2012. Plant responses to heavy metal toxicity. In, "*Plants and Heavy Metals", Ed. A.* Furini, Dordrecht, Netherlands: Springer; pp. 27–53.

Mangabeira, P., Almeida, A.A., Mielke, M., Gomes, F.P., Mushrifah, I., Escaig, F., Laffray, D., Severo, M.I., Oliveira, A.H. and Galle, P. 2001. Ultrastructural investigations and electron probe X-ray microanalysis of chromium-treated plants. Proc. VI ICOBTE, Guelph, p. 555 (abstract).

Mano, Y., Omori, F., Takamizo, T., Kindiger, B., Bird, R.M. and Loaisiga, C.H. 2006. Variation in root aerenchyma formation in flooded and non flooded maize and teosinate seedlings. *Plant and Soil*, **281:** 269-279.

Massaci, A., Nabiev, S.M., Petrosanti, L., Nematov, S.K., Chernikova, T.N., Thor, K. and Leipner. J. 2008. Response of the photosynthetic apparatus of cotton (*Gossypium hirsutum* L.) to the onset of drought stress under field conditions studied by gas-exchange analysis and chlorophyll fluorescence imaging. *Plant Physiol. Biochem.* **46:** 189-195.

Mathews, K. L.; Malosetti, M.; Chapman, S., McIntyre, L., Reynolds, M., Shorter, R. and van Eeuwik, F. 2008. Multi-environment QTL mixed models for drought stress adaptation in wheat. *Theor Appl Genet*, **117:** 1077-1091.

Matsuda, K., and Riazi. A. 1981. Stress-induced osmotic adjustment in growing regions of barley leaves. *Plant Physiol.*, **68:** 571-576.

Matsui, T. and Omasa, K. 2002. Rice (*Oryza sativa* L.) cultivars tolerant to high temperature at flowering: anther characteristics. *Annals of Botany,* **89:** 683-687.

Matteucci, M., D'Angeli, S. Errico, S., Lamanna, R., Perrotta, G. and Altamura, M.M. 2011. Cold affects the transcription of fatty acid desaturases and oil quality in the fruit of *Olea europaea* L. genotypes with different cold hardiness. *J Exp Bot*, **62:** 3403–3420.

Matus, I., Mellado, M., Pinares, M., Madariaga, R. and del Pozo, A. 2012. Genetic progress in winter wheat cultivars released in Chile from 1920 and 2000. *Chil. J. Agr. Res.*, **72:** 303–308.

Marchiol, L., Sacco, P., Assolari, S. and Zerbi, G. 2004. Reclamation of polluted soils:phytoremediation potential of crop-related Brassica species. *Water, Air, and Soil Pollution,* **158(1):** 345–356.

Maurel, C., Simonneau, T. and Sutka, M., 2010. The significance of roots as hydraulic rheostats. *J. Exp. Botany*, **61(12):** 3191-3198.

McCully, M.E and Cann, M.J. 2009. Cryo-scanning electron microscopy (CSEM) in the advancement of functional plant biology. Morphological and anatomical applications. *Functional Plant Biol.*, **36(2):** 97-124.

McCree, K.J., Kallsen, C.E. and Richardson. S.G. 1984. Carbon balance of sorghum plants during osmotic adjustment to water stress. *Plant Physiol.*, **76:** 898-902.

Mc Grath, S. 1998. Phytoextraction for soil remediation. In, "*Plants that Hyperaccumulate Heavy Metals:* Their Role in Phytoremediation, Microbiology, Archaeology, Mineral Exploration and *Phytomining*", Ed. R. Brooks, CAB International, U.K., pp. 261–287.

McMichael, B.L.,Jordan, W.R. and Powel, R.D. 1972. An effect of water stress on ethylene production by intact cotton petioles. *Plant Physiol.* **49:** 658-660.

McMichael, B.L., Jordan, W.R. and Powell, R.D. 1973. Abscission processes in cotton: Induction by plant water deficit. *Agron. J.,* **65:** 202-204.

McMichael, B.L., and Hesketh, J.D. 1982. Field investigations of the response of cotton to water deficits. *Field Crops Res.,* **5:** 319- 333.

McMichael, B.L., and Quisenberry, J.E. 1991. Genetic variation for root-shoot relationship among cotton germplasm. *Environ. Exp. Bot.*, **31:** 461-470.

Meehl, G.A., Stocker, T.F., Collins, W.D., Gaye, A.J., Gregory, J.M., Kitoh, A., Knutti, R., Murphy, J.M., Noda, A., Raper, S.C.B., Watterson, J.G., Weaver, A.J. and Zhao, Z. 2007. In, "*Global Climate Projections*", Eds. S. Solomon, D. Qin, M. Manning, Z. Chen, M. Marquis, K.B. Averyt, M. Tignor, H.L. Miller, Cambridge University Press, Cambridge, U.K. and New York, NY.

Mengel, K., Kirkby, E.A., Kosegarten, H. and Appel, T. 2001. Principles of plant nutrition. Kluwer, Dordrecht.

Miller, G., Suzuki, N., Citci-Yilmaz, S. and Mittler, R. 2010. Reactive oxygen species homeostasis and signalling during drought and salinity stresses. *Plant Cell Envir.* ,**33(4):** 453–467.

Mittler, R., Finka, A. and Goloubinoff, P. 2012. How do plants feel the heat? *Trends Biochem. Sci.*, **37(3):** 118–125.

Mishra, S. and Dubey, R.S. 2006. Heavy metal uptake and detoxification mechanisms in plants. *International Journal of Agricultural Research,* **1(2):** 122–141.

Moghbeli, E., Fathollahi, S., Salari, H., Ahmadi, G., Saliqehdar, F., Safari, A. and Grouh, M.S.H. 2012. Effects of salinity stress on growth and yield of *Aloe vera* L. *Journal of Medicinal Plants Research,* **6:** 3272-3277.

Mohammadi-Nikpoor, A. and Keshavarzi, A. 1995. Study of effect of sowing date and plant density on yield and yield components of safflower in Mashad, Iran. M.Sc. Thesis on Agriculture, Ferdowsi University of Mashad, Mashad, Iran.

Mooney, H.A., Canadell, J., Chapin, F.S., Ehleringer, J., Körner, C., McMurtrie, R., Parton, W.J., Pitelka, L. and Schulze, E.-D. 1999. The terrestrial biosphere and global change: Ecosystem physiology responses to global change. In, "*Implications of Global Change for Natural and Managed Ecosystems: A Synthesis of GCTE and Related Research*", Eds. B.H. Walker, J. Canadell, and J.S.I. Ingram. Cambridge University Press, Cambridge.

Mooney, H.A. and Cleland, E.E. 2001. The evolutionary impact of invasive species. *Proceedings of National Academy of Sciences, USA*, **98:** 5446–5451.

Montillet, J.L., Chamnongpol, S., Rustérucci, C., Dat, J., van de Cotte, B., Agnel, J.P., Battesti, C., Inzé, D., Van Breusegem, F. and Triantaphylides C. 2005. Fatty acid hydroperoxides and H_2O_2in the execution of hypersensitive cell death in tobacco leaves. *Plant Physiol.*, **138:** 1516–1526.

Morgan, J.M. 1987. Differences in osmoregulation between wheat genotypes. *Nature*. **270:** 234-235.

Morrison, M.J. and Stewart, D.W. 2002. Heat stress during flowering in summer Brassica. *Crop Science*, **42:**797-803.

Moosavi, S.G.H.R. 2007. Effect of plant density and planting pattern on agronomic traits and yield of forage sorghum. Final report of agricultural design in Islamic Azad University, Birjand Branch, Iran.

Morgan, J.M. 1984. Osmoregulation and water stress in higher plants. *Ann Rev Plant Physiol.,* **35:** 299–319.

Morrison, M.J. and Stewart, D.W. 2002. Heat stress during flowering in summer *Brassica. Crop Science,* **42:** 797–803.

Moula, G. 2009. Effect of shade on yield of rice crops. *Pakistan J. Agric. Res.*, **22:** 24-27.

Müller, J., Menzel, D. and Šamaj, J. 2007. Cell-type-specific disruption and recovery of the cytoskeleton *in Arabidopsis thaliana epidermal root cells upon heat shock stress. Protoplasma,* **230:** 231–242.

Mukhtar, S., Bakar, J.L. and Kanwar, R.S. 1990. Corn growth as affected by excess soil water. *Transaction of American Society of Agricultural Engineering*, **33:** 437-442.

Munns, R. 2002. Comparative physiology of salt and water stress. *Plant, Cell and Environment,* **25:** 239- 250.

Murchie, E.H., Pinto, M. and Horton, P. 2009. Agriculture and the new challenges for photosynthesis research. *New Phytol*, **181:** 532–552.

Mukhopadhyay, A., Vij, S. and Tyagi, A.K. 2004. Over expression of a zinc-finger protein gene from rice confers tolerance to cold, dehydration, and salt stress in transgenic tobacco. *Academic Sci. USA*, **101:** 6309-6314.

Muller, B., Pantin, F., Génard, M., Turc, O., Freixes, S., Piques, M. and Gibon Y. 2011. Water deficits uncouple growth from photosynthesis, increase C content, and modify the relationships between C and growth in sink organs. *J. Exp. Bot.*, **62:** 1715-1729.

Munns, R. and Tester, M. 2008. Mechanisms of salinity tolerance. *Annual Review of plant Biology,* **59:** 651-681

Myers, D.A., Thomas, R.B. and DeLucia, E.H. 1999. Photosynthetic responses of loblolly pine (*Pinus taeda*) needles to experimental reduction in sink demand. *Tree Physiol.*, **19:** 235-242.

Nakano, H., Kobayashi, M. and Terauchi, T. 1998. Sensitive stages to heat stress in pod setting of common bean (*Phaseolus vulgaris* L.). *Japanese Journal of Tropical Agriculture,* **42:**78–84.

Neumann, P. 1995. Inhibition of root growth by salinity stress: toxicity or an adaptive biophysical response. In, "*Structure and function of roots*", Eds. F. Baluska, M. Ciamporová, M.O.

Gasparikova, and P.W. Barlow, Academic Kluwer Publishers, Dordrecht, pp. 299-304. Nieto-Sotelo, J.; Kannan, K. B.; Martinez, L. M. and Segal, C. 1999. Characterization of a maize heat-shock protein 101 gene, HSP101, encoding a ClpB/Hsp100 protein homologue. *Gene*, **230:** 187-195.

Nishiuchi, S., Yamauchi, T., Takahashi, H., Kotula, L. and Nakazono, M. 2012. Mechanism for cropping with submergence and waterlogging in rice. *Rice*, **5:** 2.

Nonami, H. 1998. Plant water relations and control of cell elongation at low water potentials, *J. Plant Res*, **111:** 373- 382.

Norby, R.J., Gunderson, C.A., Wullschleger, S.D., O'Neill, E.G. and McCracken. M.K. 1992. Productivity and compensatory responses of yellow-poplar trees in elevated. *Nature (London),* **357:** 322 332.

Norton, L.R., Firbank, L.G. and Watkinson, A.R. 1995. Ecotypic differentiation of response to enhanced CO_2 and temperature levels in *Arabidopsis thaliana. Oecologia*, **104:** 394-396.

Nosalewicz, A.and Lipiec, J. 2014. The effect of compacted soil layers on vertical root distribution and water uptake by wheat. *Plant and Soil*, **375(1):** 229-240..

Nouri, A., Etminan, A., Teixeira da Silva, J.A. and Mohammadi, R. 2011. Assessment of yield, yield-related traits and drought tolerance of durum wheat genotypes (*Triticum turjidum* var. durum Desf.). *Aust. J. Crop Sci.*, **5:** 8–16.

Obrist, D. and Arnone, J.A. 2003. Increasing CO_2 accelerates root growth and enhances water acquisition during early stages of development in *Larrea tridentate. New Phytol.*, **159:** 175–184.

Oechel, W.C. and Strain, B.R. 1985. Native species responses to increased atmospheric carbon dioxide concentration. In, *"Direct Effects of Increasing Carbon Dioxide on Vegetation"*, Eds. B.R. Strain and J.D. Cure. Washington, D.C: U.S. Department of Energy, Office of Basic Energy Sciences, Carbon Dioxide Research Division, Springfield, VA, pp 117-154.

Oechel, W.C., Cowles, S., Grulke, N., Hastings, S.J., Lawrence, B., Prudhomme, T., Riechers, G., Strain, B., Tissue D. and Vourlitis, G. 1994. Transient nature of CO_2 fertilization in Arctic tundra. *Nature*, **371:** 500—503.

Okada, K. and Katoh, S. 1998. Two long-term effects of light that control the stability of proteins related to photosynthesis during senescence of rice leaves. *Plant Cell Physiol*, **39(4):** 394–404.

Oliver, S.N., Van Dongen, J.T., Alfred, S.C., Ezaza, A. Mamum, E.A., Zhao, X., Saini, H.S., Fernandes, S.F. Blancard, C.L., Sutton, B.G., Geigenberger, P., Dennis, E.S. and Dolferus, R. 2005. Cold-induced repression of the rice anther-specific cell wall invertase gene *OSINV4* is correlated with sucrose accumulation and pollen sterility. *Plant, Cell and Environment*, **28:** 1534-1551.

Oliver, S.N., Dennis, E.S. and Dolferus, R. 2007. ABA regulates apoplastic sugar transport and is a potential signal for cold-induced pollen sterility in rice. *Plant and Cell Physiology,* **48:** 1319-1330.

Oosterhuis, D.M., Wullschleger, S.D., and Stewart, J.M. 1987. Osmotic adjustment in commercial cultivars and wild types of cotton. *Agronomy Abstracts.* p. 97.

Oosterhuis, D.M., and S.D. Wullschleger. 1987. Osmotic adjustment in cotton (*Gossypium hisrsutum* L.) leaves and roots in response to water stress. *Plant Physiol.*, **84:** 1154-1157.

Orsini, F., Alnayef, M., Bona, S., Maggio, A. and Gianquinto, G. 2012. Low stomatal density and reduced transpiration facilitate strawberry adaptation to salinity. *Environmental and Experimental Botany,* **81:** 1-10.

Ortega, L., Fry, S.C. and Taleisnik, E. 2006. Why are *Chloris gayana* leaves shorter in salt-affected plants? Analyses in the elongation zone. *Journal of experimental Botany,* **57:** 3945-3952.

O'Toole, J/C/, Chang. T/T/ and Singh, T.N. 1979. Leaf rolling and transpiration. *Plant Science*, **16:** 111-114.

Pace, P.F., Crale, H.T., El-Halawany, S.H.M., Cothren, J.T. and Senseman, S.A. 1999. Drought induced changes in shoot and root growth of young cotton plants. *J. Cotton Sci.*, **3:** 183-187.

Parida, A.K., Dagaonkar, V.S., Phalak, M.S., Umalkar, G.V. and Aurangabadkar, L.P. 2007. Alterations in photosynthetic pigments, protein and osmotic components in cotton genotypes subjected to short-term drought stress followed by recovery. *Plant Biot. Rep.*, **1:** 37-48.

Parry, M.A.J., Androlojc, P.J., Khan, S., Lea, P.J. and Keys, A.J. 2002. Rubisco activity: Effects of drought stress. *Ann. Bot. (London)*, **89:** 833–839.

Parry, M.A.J., Reynolds, M., Salvucci, M.E., Raines, C., Andralojc, P.J., Zhu, X.-G., Price, G.D., Condon, A.G. and Furbank, R.T. 2011. Raising yield potential of wheat. II. Increasing photosynthetic capacity and efficiency. *J Exp Bot*, **62:** 453–467.

Passarini, F., Wientjes, E., Hienerwadel, R. and Croce, R. 2009. Molecular basis of light harvesting and photoprotection in CP24: unique features of the most recent antenna complex. *J Biol Chem,* **284:** 29536–29546.

Patra, M., Bhowmik, N., Bandopadhyay, B. and Sharma, A. 2004. Comparison of mercury, lead and arsenic with respect to genotoxic effects on plant systems and the development of genetic tolerance. *Environmental and Experimental Botany*. **52(3):** 199–223.

Parvaiz, A. and Satyawati, B. 2008. Salt stress and phyto-biochemical responses of plants – a review. *Plant, Soil and Environment*, **54:** 89–99.

Pataki, D.E., Oren, R. and Tissue, D.T. 1998. Elevated carbon dioxide does not affect average canopy stomatal conductance of *Pinus taeda* L. *Oecologia*, **117:** 47—52.

Paulsen, G.M. 1994. High temperature responses of crop plants. In, "*Physiiology and Determination of Yield*", Ed. K.J. Boote, ASA, Madison, WI, pp. 365-389.

Pearcy, R.W., Björkman, O., Caldwell, M.M., Keely, J.E., Monson, R.K. and Strain, B.R. 1987. Carbon gain by plants in natural environments. *BioScience*, **37:** 21—29.

Peet, M., Sato, S. and Clément, C.P. 2003. Heat stress increases sensitivity of pollen, fruit and seed production in tomatoes (*Lycopersicon esculentum* Mill.) to non-optimal vapor pressure deficits. *Acta Hort. (ISHS)*, **618:** 209–215.

Peñuelas, J., Munné-Bosch, S., Llusià, J. and Filella, I. 2004. Leaf reflectance and photo- and antioxidant protection in field-grown summer-stressed *Phillyrea angustifolia*. Optical signals of oxidative stress? *New Phytol.*, **162:** 115–124.

Perata, P. and Alpi, A. 1993. Plant responses to anaerobiosis. *Plant Science*, **93(1-2):** 1-17.

Perata, P., Armstrong, W. and Voesenek, L.M.C. 2011. Plant and flooding stress. *New Phytology,* **190:** 269-273.

Pettigrew, W.T. 2004. Moisture deficit effects on cotton lint yield, yield components, and boll distribution. *Agron. J.*, **96:** 377-383.

Peyrano, G., Taleisnik, E., Quiroga, M., de Forchetti, S.M. and Tigier, H. 1997. Salinity effects on hydraulic conductance, lignin and peroxidase activity in tomato roots. *Plant Physiology and Biochemistry,* **35:** 387-393.

Pickering, N.B., Allen Jr., F.L.H., Albrecht, S.L., Jones, P., Jones, J.W. and Baker, J.T. 1994. Environmental plant chambers: controls and measurements using cr-10t data loggers. In, "*Computers in Agriculture*: *Proceedings of the 5th International Conference",* Eds. *D.G.* Watson, F.S. Zuzueta, and T.V. Harrison, American Society of Agricultural Engineers, Orlando, pp. 29-35.

Pilumwong, J., Senthong, C., Srichuwong, S. and Ingram. K.T. 2007. Effects of temperature and elevated CO_2 on shoot and root growth of Peanut (*Arachis hypogaea* (L.)) grown in controlled environment chambers. *Sci. Asia,* **33:** 79–87.

Polley, W.H., Johnson, H.B., Mayeux, H.S., Tischler, C.R. and Brown, D.A. 1996. Carbon dioxide enrichment improves growth, water relations, and survival of droughted honey mesquite (*Prosopis glandulosa*) seedlings. *Tree Physiol.*, **16:** 817—823.

Polowick, P.L. and Sawhney, V.K. 1988. High-temperature-induced male and female sterility in canola (*Brassica napus* L.). *Annals of Botany,* **62:** 83–86.

Porter, J.R. and Gawith, M. 1999. Temperature and the growth and development of wheat: a Review. *Eur J Agron*, **10:** 23-36.

Poorter, H., Berkel, Y.V., Baxter, R., Den Hertog, J., Dijkstra, P., Gifford, R.M., Griffin, K.L., Roumet, C., Roy, J. and Wong, S.C. 1997. The effect of elevated CO_2 on the chemical composition and construction costs of leaves of 27 C3 species. *Plant, Cell and Environment*, **20**: 472-482.

Prasad, P.V.V., Craufurd, P.Q. and Summerfield, R.J. 1999a. Sensitivity of peanut to timing of heat stress during reproductive development. *Crop Sci.*, **39:**1352–1357.

Prasad, P.V.V., Craufurd, P.Q. and Summerfield, R.J. 1999b. Fruit number in relation to pollen production and viability in groundnut exposed to short episodes of heat stress. *Ann. Bot. (Lond.),* **84:** 381–386.

Prasad, P.V.V., Craufurd, P.Q. and Summerfield, R.J. 2000a. Effect of high air and soil temperature on dry matter production, pod yield and yield components of groundnut. *Plant Soil*, **222:** 231–239.

Prasad, P.V.V., Craufurd, P.Q., Summerfield, R.J. and Wheeler, T.R. 2000b. Effects of short episodes of heat stress on flower production and fruit-set of groundnut (*Arachis hypogaea* L.). *J. Exp. Bot.* **51:**777–784.

Prasad, P.V.V., Craufurd, P.Q., Kakani, V.G., Wheeler, T.R. and Boote, K.J. 2001. Influence of high temperature during pre- and post-anthesis stages of floral development on fruit-set and pollen germination in peanut. *Aust. J. Plant Physiol.*, **28:** 233–240.

Prasad, P.V.V., Boote, K.J., Allen, Jr., L.H. and Thomas, J.M.G. 2002. Effects of elevated temperature and carbon dioxide on seed-set and yield of kidney bean (*Phaseolus vulgaris* L.). *Glob. Change Biol.*, **8:** 710–721.

Prasad, P.V.V., Boote, K.J., Allen, Jr., L.H. and Thomas, J.M.G. 2003. Super-optimal temperatures are detrimental to reproductive processes and yield of peanut under both ambient and elevated carbon dioxide. *Glob. Change Biol.*, **9:** 1775–1787.

Prasad, M.N.V. 2004. Phytoremediation of metals in the environment for sustainable development. *Proceedings of the Indian National Science Academy*. **70(1):** 71–98.

Prasad, P.V.V., Boote, K.J., Vu, J.V.V. and Allen, Jr. L.H. 2004. The carbohydrate metabolism enzymes sucrose-P synthase and ADG-pyrophosphorylase in phaseolus bean leaves are up-regulated at elevated growth carbon dioxide and temperature. *Plant Sci.,***166:** 1565-1573.

Prasad, P.V.V., Boote, K.J. and Allen, Jr. L.H. 2006a. Adverse high temperature effects on pollen viability, seed-set, seed yield and harvest index of grain sorghum [*Sorghum bicolor* (L.) Moench] are more severe at elevated carbon dioxide due to higher tissue temperatures. *Agric. For. Meteorol.*, **139:** 237–251.

Prasad, P.V.V., Boote, K.J., Allen, Jr., L.H., Sheehy, J.E. and Thomas. J.M.G. 2006b. Species, ecotype and cultivar differences in spikelet fertility and harvest index of rice in response to high temperature stress. *Field Crops Res.*, **95:** 398–411.

Prasad, P.V.V., Boote, K.J., Thomas, J.M.G., Allen, Jr., L.H. and Gorbet, D.W. 2006c. Influence of soil temperature on seedling emergence and early growth of peanut cultivars in field conditions. *J. Agron. Crop Sci.*, **192:** 168–177.

Prasad, P.V.V., Pisipati, S.R. Mutava, R.N. and Tuinstra, M.R. 2008a. Sensitivity of sorghum to high temperature stress during reproductive development. *Crop Sci.*, **48(5):** 1911-1917.

Prasad, P.V.V., Staggenborg, S.A. and Ristic, Z. 2008b. Impacts of drought and/or heat stress on physiological, developmental, growth, and yield processes of crop plants. In, "*Response of crops* to limited water: Understanding and modeling water stress effects on plant growth processes. *Advances in Agricultural Systems Modeling Series 1*. Madison, WI 53711, USA. pp. 301-355.

Preeti, P. and Tripathi, A.K., 2011. Effect of heavy metals on morphological and biochemical characteristics of *Albizia procera* (Roxb.) Benth. seedlings. *International Journal of Environmental Science,* **1:** 5.

Pritchard, S.G. and Rogers. H.H. 2000. Spatial and temporal deployment of crop roots in CO_2-enriched environments. *New Phytol.*, **147:** 55–71.

Putterill, J., Laurie, R. and Macknight, R. 2004. It's time to flower: the genetic control of flowering time. *Bioessays*, **26:** 363-373.

Qin, L.; He, J.; Lee, S. K. and Dodd, I.C. 2007. An assessment of the role of ethylene in mediating lettuce (*Lactuca sativa*) root growth at high temperatures. *J Exp Bot*, **58:** 3017-3024.

Rahi, T.S., Singh, K. and Singh, B. 2013. Screening of sodicity tolerance in *Aloe vera*: An industrial crop for utilization of sodic lands. *Industrial Crops and Products,* **44:** 528-533.

Rahimi-Dehgolan, R., Sarvestani, T.Z., Rezazadeh, S.A. and Dolata-badian, A. 2012. Morphological and physiological characters of *Aloe vera* subjected to saline water irrigation. *Journal of Herbs, Spices and Medicinal Plants,* **18:** 222-230.

Ram, P.C., Singh, A.K., Singh, B.B., Singh, V.K., Singh, H.P., Setter, T.L., Singh, V.P. and Singh, R.K. *1999.* Environmental characterization of floodwater in eastern India: relevance to submergence tolerance of lowland rice. *Experimental Agriculture,* **35:** 141–152.

Ramirez, P. and Kelly, J.D. 1998. Traits related to drought resistance in common bean. *Euphytica.* **99:** 127-136.

Ramos, J., Perreta. M.G., Tivano. J.C. and Vegetti, A.C. 2004. Variaciones anatómicas en la raíz de Pappophorum philippianum inducidas por salinidad. *Phyton (USA)*: **73:**103-109.

Rampino, P.; Pataleo, S.; Gerardi, C.; Mita, G. and Perrotta, C. 2006. Drought stress response in wheat: physiological and molecular analysis of resistant and sensitive genotypes., *Plant Cell Environ*, **29(12):** 2143-2152.

Rana, R.S., Bhosale, A.B., Sood, R., Sharma, R. and Chander, N. 2011. Simulating impact of climate change on mustard (*Brassica juncea*) production in Himachal Pradesh. *J. Agromet.* **13(2):** 104-109.

Ranathunge, K., Steudle, E. and Lafitte, R. 2003. Control of water uptake by rice (*Oryza sativa* L.): role of the outer part of the root. *Planta,* **217:** 193–205.

Rao, V.U.M., Rao, B.B., Nair, L., Singh, D., Sekhar, C. and Venkateswa, B. 2011. Thermal sensitivity of mustard (*Brassica juncea* L.) crop in Haryana. *J. Agromet.* **13(2):** 131-134.

Rascio, N. and Navari-Izzo, F. 2010. Heavy metal hyperaccumulating plants: how and why do they do it? And what makes them so interesting? *Plant Science*, **180(2):** 169–181.

Rathore, T.R., Warsi, M.Z.K., Singh, N.N. and Vasal, S.K. 1998. Production of maize under excess soil moisture (water logging) conditions. 2nd Asian Regional Maize Workshop PCCARD, Los Banos, Philippines, Feb 23-27, 1998.

Rauf, S. 2008. Breeding sunflower (*Helianthus annuus* L.) for drought tolerance. *Commun. Biometry Crop Sci.* **3:** 29-44.

Raveh, E. 2005. Methods to assess potential chloride stress in citrus: analysis of leaves, fruit, stem-xylem sap and roots. *Hort. Technol.*, **15:** 104-108.

Raveh, E. and Levy, Y. 2005. Analysis of xylem water as an indicator of current chloride uptake status in citrus trees. *Sci.Hortic.*, **103:** 317-327.

Rebetzke, G.J., López-Castañeda, C., Botwright Acuña, T.L., Condon, A.G. and Richards, R.A. 2008. Inheritance of coleoptile tiller appearance and size in wheat. *Aust J Agric Res*, **59:** 863–873.

Reddy, K.R., Kakani, V.G., Zhao, D., Koti, S. and Gao, W. 2004. Interactive effects of ultraviolet-bradiation and temperature on cotton physiology, growth, development and hyperspectral reflectance. *Photochemistry and Photobiology*, **79:** 416-427.

Reekie, E.G. and Bazzaz, F.A. 1991. Phenology and growth in four annual species grown in ambient and elevated CO_2. *Can. J. Bot.*, **69:** 2475—2481.

Reginato, G.H., Callejas, R.H., Sapiaín, R.A. and García-de-Cortázar, V. 2010. Rest completion and growth of' 'thompson seedless' grapes as a function of temperatures. *Acta Hortic. (ISHS)*, **872:** 427–430.

Reichard, S.H. and Hamilton, C.W. 1997. Predicting invasions of woody plants introduced into North America. *Conservation Biology*, **11:** 193–203.

Reinhardt, D.H. and Rost, T.L. 1995. Salinity accelerates endodermal development and induces an exodermis in cotton seedlings roots. *Environmental and experimental Botany,* **35:** 563-574.

Ren, W.J., Yang, W.Y., Xu, J.W., Fan, G.Q., Wang, L.Y and Guan, H. 2002. Impact of low-light stress on leaves characteristics of rice after heading. *J Sichuan Agric Univ*, **20**(3): 205–208. (in Chinese with English abstract)

Ren, W.J., Yang, W.Y., Fan, G.Q., Zhu, X.M., Ma, Z.H. and Xu, J.W. 2003a. Effect of low light on dry matter accumulation and yield of rice. *J Sichuan Agric Univ*, **29**(4): 292–296. (in Chinese with English abstract)

Ren, W.J., Yang, W.Y., Zhang, G.Z., Zhu, X., Fan, G.Q, and Xu, J.W. 2003b. Effect of low-light stress on nitrogen accumulation, distribution and grains protein content of indica hybrid. *Plant Nutr Fert Sci*, **9**(3): 288–293. (in Chinese with English abstract)

Rengasamy, P. 2002. Transient salinity and subsoil constraints in dry land farming in Australian sodic soils: An overview. *Aust. J Expt Res.,* **42:** 351-361.

Restrepo, H. and Garcés, G. 2013. Evaluation of low light intensity at three phenological stages in the agronomic and physiological responses of two rice (*Oryza sativa* L.) cultivars. *Agron. colomb.*, **31(2):** 195-200.

Reynolds, M.P., Rajaram, S. and Sayre, K.D. 1999. Physiological and genetic changes of irrigated wheat in the post-green revolution period and approaches for meeting projected global demand. *Crop Science*, **39:** 1611-1621.

Reynolds, J.F., Kemp, P.R., Ogle, K. and Fernández, R.J. 2004. Modifying the 'pulse-reserve' paradigm for deserts of North America: precipitation pulses, soil water and plant responses. *Oecologia.*, **141:** 194–210.

Reynolds, M.P., Pierre, C.S., Saad, A.S.I., Vargas, M. and Condon, A.G. 2007. Evaluating potential genetic gains in wheat associated with stress-adaptive trait expression in elite genetic resources under drought and heat stress. *Crop Sci*, **47:** S-172–S-189.

Reynolds, M.P. and Tuberosa, R. 2008. Translational research impacting on crop productivity in drought prone environments. *Curr. Opin. Plant Biol.*, **11:** 171-179.

Ribas-Carbo, M., Taylor, N.L., Giles, L., Busquets, S., Finnegan, P.M. and Day. D.A. 2005. Effects of water stress on respiration in soybean leaves. *Plant Physiol.*, **139:** 466-473.

Ricard, B., Couée, I., Raymond, P., Saglio, P.H., Saint Ges, V. and Pradet, A. 1994 Plant metabolism under hypoxia and anoxia. *Plant Physiol Biochem*, **32:** 1-10.

Richards, R.A. and Lukacs, Z. 2002. Seedling vigour in wheat: sources of variation for genetic and agronomic improvement. *Aust J Agric Res*, **53:** 41–50.

Richards, R.A. 2006. Physiological traits used in the breeding of new cultivars for water-scarce environments. *Agric. Water Manag.*, **80:** 197–211.

Rivero, R.M., Ruiz, J.M., Garcý´a, P.C., Lopez-Lefebre, L.R., Sa´nchez, E. and Romero, L. 2001. Resistance to cold and heat stress: accumulation of phenolic compounds in tomato and watermelon plants. *Plant Science*, **160:** 315 – 321.

Rivero, H., Huertas, M., Francini, R., Vila, L. and Darre, L.E. 2006. Concentrations of As, Cd, Co, Cr, Cu, Fe, Hg, K, Mg, Mn, Mo, Na, Ni, Pb and Zn in Uruguayan rice determined by atomic absorption spectrometry. *Atomic Spectroscopy,* **27:** 48-55.

Rivero, R.M., Kojima, M., Gepstein, A., Sakakibara, H., Mittler, R., Gepstein, S., and Blumwald, E. 2007. Delayed leaf senescence induces extreme drought tolerance in a flowering plant. *Proc. Natl. Acad. Sci. U.S.A.,* **104:** 19631–19636.

Rizhysky, L. Liang, H., Shuman, J., Shulaev, V., Davletova, S. and Mittler. R. 2004. When defense pathways collide: The response of Arabidopsis to a combination of drought and heat stress. *PlantPhysiol.*, **134:** 1683–1696.

Roberts, E.H. and Summerfield, R.J. 1987. Measurement and prediction of flowering in annual crops. In, "*Manipulation of flowering*", Ed. J.G. Atherton, Butterworth, London, pp. 17–50.

Roberts, E.H. 1988. Temperature and seed germination. In, "*Plants and temperature*", Ed. S.P. Long and F.I. Woodward, Society of Experimental Biology, Cambridge, UK. Pp. 109-132.

Roberts, E.H., Qi, A., Ellis, R., Summerfield, R.J., Lawn, R.J. and Shanmugasundaram, S. 1996. Use of field observations to characterise genotypic flowering responses to photoperiod and temperature: a soyabean exemplar. *Theoretical and Applied Genetics* , **93:** 519-533.

Rochefort, L. and Bazzaz, F.A. 1992. Growth response to elevated CO_2 in seedlings of four co-occurring birch species. *Can. J. For. Res.*, **22:** 1583—1587.

Rodrigo, J. and Herrero, M. 2002. Effects of pre-blossom temperatures on flower development and fruit set in apricot. *Scientia Horticulturae,* **92:** 125–135.

Rogers, W.E. and Siemann, E. 2004. Invasive ecotypes tolerate herbivory more effectively than native ecotypes of the Chinese tallow tree *Sapium sebiferum. Journal of Applied Ecology*, **41:** 561–570.

Rogers, W.E. and Siemann, E. 2005. Herbivory tolerance and compensatory differences in native and invasive ecotypes of Chinese tallow tree (*Sapium sebiferum*). *Plant Ecology*, **181:** 57–68.

Rosegrant, M.W. and Cline, S.A. 2003. Global food security: challenges and policies. *Science,* **302:** 1917–1919.

Rosselli, W., Rossi, M. and Sasu, I. 2006. Cd, Cu and Zn contents in the leaves of *Taraxacum officinale*. Swiss Federal Institute for Forest. *Snow and Landscape Research*, **80(3):** 361–366.

Ruelland, E. and Zachowski, A 2010. How plants sense temperature. *Environ Exp Bot*, **69:** 225–232.

Rymen, B., Fiorani, F., Kartal, F., Vandepoele, K., Inze, D. and Beemster, G.T. 2007. Cold nights impair leaf growth and cell cycle progression in maize through transcriptional changes of cell cycle genes. *Plant Physiology*, **143(3):** 1429-1438.

Sachs, M.M., Freeling, M. and Okimoto, R. 1980. The anaerobic proteins of maize. *Cell*, **20:** 761-767

Sacks, W.J. and Kucharik, C.J. 2011. Crop management and phenology trends in the U.S. corn belt: Impacts on yields, evapotranspiration and energy balance. *Agric. For. Meteor.* **151:** 882–894.

Sage, R.F. 1994. Acclimation of photosynthesis to increasing atmospheric CO_2: the gas exchange perspective. *Photosynth. Res.*, **39:** 351-368.

Sage, R.F. 2002. Variation in the k(cat) of Rubisco in C(3) and C(4) plants and some implications for photosynthetic performance at high and low temperature. *J Exp Bot,* **53:** 609–620.

Sage, R.F. and McKown, A.D. 2006. Is C4 photosynthesis less phenotypicaly plastic than C3 photosynthesis? *Journal of Experimental Botany*, **57(2):** 303-317.

Sage, R.F., Kocacinar, F. and Kubien, D.S. 2011. C4 photosynthesis and temperature. In, "*C4 Photosynthesis and related CO_2 concentrating mechanism*", Eds. A.S. Raghvendra and R.S. Sage, Springer Science+Business Media B.V., Dordrecht, The Netherlands, pp. 161-195.

Saini, H.S., Sedgley, M. and Aspinall, D. 1983. Effect of heat-stress during floral development on pollen-tube growth and ovary anatomy in wheat (*Triticum aestivum* L.). *Australian Journal of Plant Physiology,* **10:**137-144.

Salsman, K.J., Jordan, D.N., Smith, S.D. and Neuman, D.S. 1999. Effect of atmospheric enrichment on root growth and carbohydrate allocation of *Phaseolus* spp. *Int. J. Plant Sci.,* **160:** 1075–1081.

Salt, D.E., Blaylock, M., Kumar, N., Dushenkov, V., Ensley, B.D., Chet, I. and Raskin, I. 1995. Phytoremediation: a novel strategy for the removal of toxic metals from the environment using plants. *Biotechnology*, **13:** 468–474.

Salvucci, M.E. and Crafts-Brandner, S.J. 2004. Inhibition of photosynthesis by heat stress: The activation *state of Rubisco as a limiting factor in photosynthesis. Physiol Plant.,* **120:** 179-186.

Samson, D.A. and Werk, K.S. 1986. Size-dependent effects in the analysis of reproductive effort in plants. *Am. Nat.*, **127:** 667—680.

Sánchez-García, M., Royo, C., Aparicio, N., Martín-Sánchez, A. and Álvaro, F. 2013. Genetic improvement of bread wheat yield and associated traits in Spain during the 20th century. *J. Agric. Sci.,***151:** 105–118.

Sanderson, M., Stair, D. and Hussey, M. 1997. Physiological and morphological responses of perennial forages to stress. *Advances in Agronomy,* **59:** 171-224.

Sangwan, V.; Orvar, B. L.; Beyerly, J.; Hirt, H. and Dhindsa, R.S. 2002. Opposite changes in membrane fluidity mimic cold and heat stress activation of distinct plant MAP kinase pathways. *Plant J,* **31:** 629-638.

Sataka, T. and Hayase, H. 1970. Male sterility caused by cooling treatment at the young microspore stage in rice plants. V. Estimation of pollen developmental stage and the most sensitive stage to coolness. *Proceedings of the Crop Science Society of Japan*, **39:** 468-473.

Satake, T. and Yoshida, S. 1978. High temperature-induced sterility in indica rice at flowering. *Jpn. J. Crop Sci.*, **47:** 6–17.

Sato, S., Peet, M.M. and Thomas, J.F. 2000. Physiological factors limit fruit set of tomato (*Lycopersicon esculentum* Mill.) under chronic, mild heat stress. *Plant Cell Environ.*, **23:** 719–726.

Sato, S., Peet, M.M. and Thomas, J.F. 2002. Determining critical pre- and post-anthesis periods and physiological processes in *Lycopersicon esculentum* Mill. exposed to moderately elevated temperatures. *Journal of Experimental Botany*, **53:** 1187-1195.

Sato, S. 2006. The effects of moderately elevated temperature stress due to global warming on the yield and the male reproductive development of tomato (*Lycopersicon esculentum* Mill.). *Hort Research*, **60:** 85–89.

Savin, R., and Nicolas, M.E. 1996. Effect of short episodes of drought and high temperature on grain growth and starch accumulation of two malting barley cultivars. *Aust. J. Plant Physiol.,* **23:** 201–210.

Seyed, Y., Lisar, S., Motafakkerazad, R., Hossain, M.M. and Ismail, Rahman, M. 2012. Water stress in plants: causes, effects and responses, water stress. Ed. Ismail, Md. Mofizur Rahman, Intech Europe, Rijeka, Croatia.

Schlenker, W. and Roberts, M.J. 2009. Nonlinear temperature effects indicate severe damages to U.S. crop yields under climate change. *Proc. Natl. Acad. Sci.* **106:** 15594–15598.

Schlesinger. 1997. Biogeochemistry: An analysis of global change. Academic Press, New York, pp. 588.

Schmid, B., Birrer, A. and Lavigne, C. 1996. Genetic variation in the response of plant populations to elevated CO_2 in a nutrient-poor, calcareous grassland. *In, "Carbon Dioxide, Populations, and Communities",* Eds. C. Körner and F.A. Bazzaz. Academic Press, New York, pp 31—50.

Schild, L.N., Parfitt, J.M.B., Porto M.B. and Silva, C.A. 1999. Comportamento, do milho, en planossolo, sob condcôes de excess hìdrico. I – Desempenho agronôpecãria Clima Temperado, **2:** 97-109.

Schützendübel, A. and Polle, A. 2002. Plant responses to abiotic stresses: heavy metal-induced oxidative stress and protection by mycorrhization. *The Journal of Experimental Botany,* **53(372):** 1351–1365.

Schwarz, D., Rouphael, Y., Colla, G. and Venema, H.J. 2010. Grafting as a tool to improve tolerance of vegetables to abiotic stresses:Thermal stress, water stress and organic pollutants. *Scientia Horticulturae*, **127:** 162-171.

Seif, S. and Gruppe, W. 1985. Chilling requirements of sweet cherries (*Prunus avium*) and interspecific cherry hybrids (*Prunus* x ssp.). *Acta Hortic. (ISHS)*, **169:** 289–294.

Sekhon, H.S., Singh, G., Sharma, P. and Bains, T.S. 2010. Water use efficiency under stress environments. In, "*Climate change and management of cool season grain legume crops",* Eds. S.S. Yadav, D.L. Mc Neil, R. Redden, and S.A. Patil, Springer Press, Dordrecht-Heidelberg-London-New York.

Sepehri, A. and Golparvar, A.R. 2011. The effect of drought stress on water relations, chlorophyll content and leaf area in canola cultivars (*Brassica napus* L.). *Electronic J. Biol.,* **7:** 49-53.

Setter, T.L. and Waters, I. 2003. Review of prospects for germplasm improvement for waterlogging tolerance in wheat, barley and oats. *Plant and Soil,* **253:** 1–34.

Singh, B.N., Rane, J., McDonald, G., Khabaz-Saveri, H., Biddulph, T.B., Wilson, R., Barclay, I. McLean, R. and Cakir, M. 2009. Review of wheat improvement for waterlogging tolerance in Australia and India: the importance of anaerobiosis and element toxicities associated with different soils. *Annals of Botany,* **103:** 221–235.

Shah, F., Huang, J., Cui, K., Nie, L., Shah, T., Chen, C. and Wang, K. 2011. Impact of high temperature stress on rice plant and its traits related to tolerance. *J. Agric. Sci.*, **149:** 545–556.

Shah, N.S., Srivastava, J.P., Jaime, A., da Silva, T. and Shahi, J.P. 2012. Morphological and yield response of maize (*Zea mays* L.) genotypes subj ected to root zone excess soil moisture stress. *Plant Stress,* **6(1):** 59-72.

Shannon, M.C. 1998. Adaptation of plants to salinity. *Adv Agron*, **60:** 75-120.

Sharma, N., Cram, D., Huebert, T., Zhou, N. and Parkin. I.A 2007. Exploiting the wild crucifer *Thlaspi arvense* to identify conserved and novel genes expressed during a plant's response to cold stress. *Plant Mol Biol*, **63:** 171–184.

Sheehy, J.E., Elmido, A., Centeno, G. and Pablico, P. 2005. Searching for new plants for climate change. *J. Agric. Meterol.* **60:** 463–468.

Shiono, K., Ogawa, S., Yamazaki, S., Isoda, H., Fujimura, T., Nakazono, M. and Colmer, T.D. 2011. Contrasting dynamics of radial O_2-loss barrier induction and aerenchyma formation in rice roots of two lengths. *Annals of Botany,* **107:** 89–99.

Seo, P.J., Kim, M.J., Park, J.Y., Kim, S.Y., Jeon, J., Lee, Y.H., Kim, J. and Park, C.M. 2010. Cold activation of a plasma membrane-tethered NAC transcription factor induces a pathogen resistance response in *Arabidopsis. Plant J,* **61:** 661–671.

Sibomana, I.C., Aguyoh, J.N. and Opiyo, A.M. 2013. Water stress affects growth and yield of container grown tomato (*Lycopersicon esculentum* Mill) plants. *Global Journal of Bioscience and Biotechnology,* **2(4):** 461-466.

Siddique, M.R.B., Hamid, A. and Islam, M.S. 1999. Drought stress effects on photosynthetic rates and leaf gas exchange of wheat. *Bot. Bull. Acad. Sin.*, **40:** 141–145.

Siddique, K.H.M., Regan, K.L., Tennant, D. and Thomson, B.D. 2001. Water use and water use efficiency of cool season grain legumes in low rainfall mediterranean-type environments. *Eur. J. Agron.*, **15:** 267-280.

Simonneau, T., Habib, R., Goutouly, J.P. and Buguet. J.G. 1993. Diurnal changes in stem diameter depend upon variation in water content: Direct evidence from peach trees. *J. Exp. Bot.,* **44:** 615–621.

Sing, R.and Ghildyal, B.P. 1980. Soil submergence effects on nutrient uptake, growth and of five corn cultivars. *Agronomy Journal,* **77:** 737-741.

Sherrard, M.E., Maherali, H. and Latta, R.G. 2009. Water stress alters the genetic architecture of functional traits associated with drought adaptation in *Avena barbata. Evolution,* **63:** 702-715.

Shi, X.D., Wen, Z.Q., Liu, Y.F. and Wang, W.W. 2006. Research on the effect of different light stresses on crop growth. *J Anhui Agric Sci,* **34(17):** 4216–4218. (in Chinese with English abstract)

Shinozaki, K., Shinozaki, Y.K. and Seki, M. 2003. Regulatory network of gene expression in the drought and cold stress responses. *Plant Biol.,* **6:** 410-417.

Shirani-rad, A.H. and Abbasian, A. 2011. Evaluation of drought tolerance in rapeseed genotypes under non stress and drought stress conditions. *Not. Bot. Horti. Agrobo.*, **39:** 164-171.

Shukla, D., Tiwari, M., Tripathi, R.D., Nath, P. and Trivedi, P.K. 2013. Synthetic phytochelatins complement a phytochelatin-deficient *Arabidopsis* mutant and enhance the accumulation of heavy metal(loid)s. *Biochemical and Biophysical Research Communication,* **434(3):** 664–669.

Siczek, A. and Lipiec, J. 2011. Soybean nodulation and nitrogen fixation in response to soil compaction and surface straw mulching. *Soil Till. Res.*, **114:** 50-56.

Silva, H., Sagardia, S., Sequel, O., Torres, C., Tapia, C., Franck, N. and Cardemil, Z. 2010. Effect of water availability on growth and water use efficiency for biomass and gel production in *Aloe vera* (*Aloe barbadensis* M.). *Industrial Crops and Products,* **31:** 20-27.

Sheen, J. 1990. Metabolic repression of transcription in higher plants. *Plant Cell,* **2:** 1027-1038.

Sims, D.A., Luo, Y. and Seemann, J.R. 1998. Comparison of photosynthetic acclimation to elevated CO_2 and limited nitrogen supply in soybean. *Plant Cell Environ.*, **21:** 945—952.

Sinclair, T.R. 1994. Limits to crop yield. In, "*Physiology and determination of Yield*", Ed. K.J. Boote, A.S.A. Madison, WI, p. 509–532.

Singh, D.P., Singh, P., Kumar, A. and Sharma, H.C. 1985. Transpirational cooling as a screening technique for drought tolerance in oilseed *Brassicas. Ann. Bot.*, **56:** 815-820.

Singh, V., van Oosterom, E.J., Jordan, D.R., Messina, C.D., Cooper, M. and Hammer,,G.L. 2010. Morphological and architectural deve- lopment of root systems in sorghum and maize. *Plant Soil,* **333:** 287-299.

Singh, V., Nguyen, C.T., van Oosterom, E.J., Chapman, S.C., Jordan, D.R. and Hammer, G. L. 2015. Sorghum genotypes differ in high temperature responses for seed set. *Field Crops Res.,* **171:** 32–40.

Smith, P. and Olesen, J.E. 2010. Synergies between the mitigation of, and adaptation to, climate change in agriculture. *J Agric Sci, Cambridge.* **148:** 543–552.

Snider, J.L., Oosterhuis, D.M., Skulman, B.W. and Kawakami, E.M. 2009. Heat stress-induced limitations to reproductive success in. *Gossypium hirsutum. Physiologia Plantarum,* **137:** 125-138.

Solomon. S., Qin, D., Manning, M., Marquis, M., Averyt, K., Tignor, M.M.B., Miller, HL and Cheng, Z. 2007. Climate change 2007: the physical science basis. Contribution of Working Group I to the Fourth Assessment Report of the Intergovernmental Panel on Climate Change. New York, Cambridge University Press; 2007

Sønsteby, A. and Heide, O.M. 2008. Temperature responses, flowering and fruit yield of the June-bearing strawberry cultivars florence, frida and korona. *Sci. Hortic.*, **119:** 49–54.

Song, Y.H., Ito, S. and Imaizumi, T. 2013. Flowering time regulation: photoperiod-and temperature-sensing in leaves. *Trends Plant Sci.*, **18(10):** 575–583.

Song, W.-Y., Mendoza-Cózatl, D.G., Lee, Y., Schroeder, J.I., Ahn, S.N., Lee, H.S., Wicker, T. and Martinoia, E. 2014. Phytochelatin-metal(loid) transport into vacuoles shows different substrate preferences in barley and *Arabidopsis. Plant, Cell and Environment.* **37(5):** 1192–1201.

Sowinski, P., Rudzinska-Langwald, A., Adamczyk, J., Kubica, I., and Fronk, J. 2005. Recovery of maize seedling growth, development and photosynthetic efficiency after initial growth at low temperature. *Journal of Plant Physiology*, **162:** 67-80.

Srinivasan, A., Saxena, N.P. and Johansen, C. 1999. Cold tolerance during early reproductive growth of chickpea (*Cicer arietinum* L.): genetic variation in gamete development and function. *Field Crops Research,* **60:** 209-222.

Srivastava, A.K., Adak, T. and Chakravarty, N.V.K. 2011. Quantification of growth and yield of oilseed *Brassica* using thermal indices under semi-arid environment. *J. Agromet.* **13(2):** 135-140.

Srivastava, G.C. 2011. Crop physiology. Biotech Books, New Delhi. St. Omer, L. and Horvath, S.M. 1983. Elevated carbon dioxide concentrations and whole plant senescence. *Ecology,* **64:** 1311—1314.

Stanfield, B., Ormrod, D.P., and Fletcher, H.F. 1966. Response of peas to environment: II. Effects of temperature in controlled-environment cabinets. *Canadian Journal of Plant Science,* **46:** 195–203.

Steponkus, P.L., Uermura, M. and Webb, M.S. 1993. A contrast of the cryostability of the plasma membrane of winter rye and spring oat-two species that widely differ in their freezing tolerance and plasma membrane lipid composition. In, "*Advances in Low-Temperature Biology*", Ed. P.L. Steponkus, JAI Press. London **2:** 211-312.

Stitt, M. 1991. Rising CO_2 levels and their potential significance for carbon flow in photosynthetic cells. *Plant Cell Environ.*, **14:** 741-762.

Strain, B.R. 1985. Physiological and ecological controls on carbon sequestering in terrestrial ecosystems. *Biogeochemistry*, **1:** 219—232.

Strain, B.R. 1991. Possible genetic effects of continually increasing atmospheric CO_2. In, *"Ecological Genetics and Air Pollution*", Eds. G.E. Taylor, Jr., L.F. Pitelka and M.T. Clegg. Springer-Verlag, New York, pp 237—244.

Strain, B.R. and Thomas, R.B. 1995. Anticipated effects of elevated CO_2 and climate change on plants from Mediterranean-type ecosystems utilizing results of studies in other ecosystems. In, *"Anticipated Effects of a Changing Global Environment on Mediterranean-Type Ecosystems*", Eds. J.M. Moreno and W.W. Oechel. Springer-Verlag, New York, pp 121-139.

Sukhvibul, N., Whiley, A.W., Smith, M.K., Hetherington, S.E. and Vithanage, V. 1999. Effect of temperature on inflorescence and floral development in four mango (*Mangifera indica*L.) cultivars. *Scientia Horticulturae,* **82:** 67–84.

Sun, S.B., Shen, Q.R., Wan, J.M. and Liu, Z.P. 2003. Induced expression of the gene for NADP-malic enzyme in leaves of *Aloe vera* L. under salt stress. *Acta Biochimica et Biophysica Sinica,* **35:** 423-429.

Sun, Y.Y., Sun, Y.J., Chen, L., Xu, H. and Ma, J. 2012. Effects of different sowing dates and low-light stress at heading stage on the physiological characteristics and grain yield of hybrid rice. *Chin J Appl Ecol*, **23(10):** 2737–2744. (in Chinese with English abstract)

Suter, D., Frehner, M., Fischer, B.U., Nosberger, J. and Luscher, A. 2002. Elevated CO_2 increases carbon allocation to the roots of *Lolium perenne* under freeair CO_2 enrichment but not in a controlled environment. *New Phytol.*, **154:** 65–75.

Suzuki, K., Tsukaguchi, T., Takeda, H. and Egawa, Y. 2001. Decrease of pollen stainability of green bean at high temperatures and relationship to heat tolerance. *Journal of the American Society for Horticultural Science,* **126:** 571–574.

Suzuki, N., Koussevitzky, S., Mittler, R. and Miller, G. 2011. ROS and redox signaling in the response of plants to abiotic stress. *Plant Cell Environ*, **35:** 259–270.

Swemmer, A.M., Knapp, A.K. and Snyman, H.A. 2007. Intra-seasonal precipitation patterns and above-ground productivity in three perennial grasslands. *J Ecol.*, **95:** 780–788.

Sytar, O., Kumar, A., Latowski, D., Kuczynska, P., Strzaka, K. and Prasad, M.N.V. 2013. Heavy metal-induced oxidative damage, defense reactions, and detoxification mechanisms in plants. *Acta Physiologiae Plantarum.* **35(4):** 985–999.

Taiz, L. and Zeiger, E. 2006. Plant Physiology. Sinauer Associates Inc. Publishers, Sunderland, Massachusetts, **4:** 312-315.

Takahashi, D., Li, B., Nakayama, T., Kawamura, Y. and Uemura, M. 2013. Plant plasma membrane proteomics for improving cold tolerance. *Front Plant Sci.*, **4:** Article ID 90.

Taleisnik, E., Peyrano, G. and Arias, C. 1997. Response of *Chloris gayana* cultivars to salinity. Germination and early vegetative growth. *Tropical grassland,* **31:** 232-240.

Tangahu, B.V., Abdullah, S.R.S., Basri, H., Idris, M., Anuar, N. and Mukhlisin, M. 2011. A review on heavy metals (As, Pb, and Hg) uptake by plants through phytoremediation. *International Journal of Chemical Engineering.* **2011:** 31.

Tardieu, F., Reymond, M., Hamard, P., Granier, C. and Muller, B. 2000. Spatial distribution of expansion rate, cell division rate and cell size in maize leaves: A synthesis of the effects of soil water status, evaporative demand and temperature. *J. Exp. Bot.*, **51:** 1505–1514.

Tawfik, K.M., Sheteawi, S.A. and El-Gawad, Z.A. 2001. Growth and aloin production of *Aloe vera* and *Aloe eru* under different ecological conditions. Proceedings of the First International Conference (Egyptian British Biological Society, EBB Soc.) *Egyptian Journal of Biology,* **3:** 149-159.

Teixeira, E.I., Fischer, G., van Velthuizen, H., Walter, C. and Ewert, F. 2013. Global hot-spots of heat stress on agricultural crops due to climate change. *Agric. Forest Meteorol.*, **170:** 206–215.

Tepwadee, C., Shigenori, M., Abe, J., Kaori, I., Ryosuke, T. and Anan, P. 2008. Root at seedling stage of hree cordage fiber crops. *Plant Prod. Sci.*, **1(2):** 232-237.

Tesfaendrias, M.T., McDonald, M.R. and Warland, J. 2010. Consistency of long-term marketable yield of carrot and onion cultivars in muck (organic) soil in relation to seasonal weather. *Can. J. Plant Sci.*, **90:** 755–765.

Tewolde, H., Fernandez, C.J. and Erickson, C.A. 2006. Wheat cultivars adapted to post-heading high temperature stress. *J. Agron. Crop Sci.*, **192:** 111-120.

Thakur, P., Kumar, S., Malik, J.A., Berger, J.D. and Nayyar, H. 2010. Cold stress effects on reproductive development in grain crops: an overview. *Environmental and Experimental Botany,* **67:** 429-443.

Tian J., Belanger F. C. and Huang, B. 2009. Identification of heat stress-responsive genes in heat-adapted thermal *Agrostis scabra* by suppression subtractive hybridization. *J. Plant Physiol.,* **166:** 588–601.

Tian, Z., Wang, F., Zhang, W., Liu, C. and Zhao, X. 2012. Antioxidant mechanism and lipid peroxidation patterns in leaves and petals of marigold in response to drought stress. *Hort Environ Biotechnol.*, **53(3)**:183–192.

Tickoo, J. L., Gajraj, R. M. and Manji, C. 1996. Plant type in mungbean (*Vigna radiata* L.) Wilczek). In, "*Proceedings of recent advances in Mungbean research", Eds.* A.N. Asthana and D.H. Kim, Kanpur, India: Indian Society of Pulses Research and Development, Indian Institute of Pulses Research, pp. 197-213.

Timpa, J.D., Burke, J.J., Quisenberry, J.E. and Wendt. C.W. 1986. Effects of water stress on the organic acid and carbohydrate composition of cotton plants. *Plant Physiol.*, **82:** 724-728.

Tissue, D.T. and Oechel, W.C. 1987. Responses of *Eriophorum vaginatum* to elevated CO_2 and temperature in the Alaskan tussock tundra. *Ecology*, **68:** 401—410.

Tissue, D.T., Megonigal, J.P. and Thomas, R.B. 1997*a*. Nitrogenase activity and N_2 fixation are stimulated by elevated CO_2 in a tropical N_2-fixing tree. *Oecologia*, **109:** 28—33.

Tissue, D.T., Griffin, K.L., Thomas, R.B. and Strain, B.R. 1995. Effects of low and elevated CO_2 on C3 and C4 annuals. II. Photosynthesis and leaf biochemistry. *Oecologia,* **101:** 21-28.

Tissue, D.T., Griffin, K.L. and Ball, J.T. 1999. Photosynthetic adjustments in field-grown ponderosa pine trees after six years of exposure to elevated CO_2. *Tree Physiol.*, **19:** 221-228.

Tissue, D.T., Thomas, R.B. and Strain, B.R. 1997*b*. Atmospheric CO2 enrichment increases growth and photosynthesis of *Pinus taeda*: a 4 year experiment in the field. *Plant Cell Environ.,* **20:** 1123-1134.

Thomas, R.B. and Griffin, K.L. 1994. Direct and indirect effects of atmospheric carbon dioxide enrichment on leaf respiration of *Glycine max* (L.) Merr. *Plant Physiol.*, **104:** 355—361.

Tognetti, R., Longobucco, A., Miglietta, F. and Raschi, A. 1999*a*. Water relations, stomatal response and transpiration of *Quercus pubescens* trees during summer in a Mediterranean carbon dioxide spring. *Tree Physiol.*, **19:** 261—270.

Tognetti, R., Longobucco, A. and Raschi, A. 1999*b*. Seasonal embolism and xylem vulnerability in deciduous and evergreen Mediterranean trees influenced by proximity to a carbon dioxide spring. *Tree Physiol.*, **19:** 271—277.

Tolley, L.C. and Strain, B.R. 1985. Effects of CO_2 enrichment and water stress on gas exchange of *Liquidambar styraciflua* and *Pinus taeda* seedlings grown under different irradiance levels. *Oecologia*, **65:** 166—172.

Tonsor, S.J., Scott, C., Boumaza, I., Liss, T.R., Brodsky, J.L. and Vierling, E. 2008. Heat shock protein 101 effects in *A. thaliana*: genetic variation, fitness and pleiotropy in controlled temperature conditions. *Molecular Ecology*, **17:** 1614-1626.

Tripathi, A. and Tripathi, S. 1999. Changes in some physiological and biochemical characters in *Alizia* as bio-indicator of heavy metal toxicity. *J. Environ. Biol.*, **20:** 93-98.

Triapthi, S. 2000. Studies on screening of inbred lines for waterlogging tolerance in maize. M. Sc. (Ag.) Thesis, G.B. Pant Unversity of Agriculture and Technology, Pantnagar, India, pp. 97.

Trolinder, N.L., B.L. McMichael, and D.R. Upchurch. 1993. Water relations of cotton flower petals and fruits. *Plant Cell Environ.*, **16:** 755-760.

Tuberosa, R. 2012. Phenotyping for drought tolerance of crops in the genomics era. *Frontiers Physiol.,* **3:** 347-356.

Turner, N.C., J.E. Begg, and M.L. Tonnett. 1978. Osmotic adjustment of sorghum and sunflower crops in response to water deficits and its influence on the water potential at which stomata close. *Aust. J. Plant Physiol.*, **5:** 597-608.

Turuga, L., Albulescu, M., Popovici, H. and Puscas, A. 2008. *Taraxacum officinale* in phytoremediation of contaminated soils by industrial activities. *Annals of West University of Timisoara, Series Chemistry,* **17(2):** 39–44.

Tyagi, N.K. and Mitra, P.S. 1998. Agricultural salinity management in India. In, "*Agricultural salinity management in India*", Eds. N.K. Tyagi, P.S. Mitra,, Central Soil Salinity Research Institute, Kernal, India. 43

Uemura, M. and Steponkus, P.L. 1999. Cold acclimation in plants: relationship between the lipid composition and the cryostability of the plasma membrane. *J Plant Res,* **112:** 245–254.

USDA-ARS. 2008. Research database bibliography on salt tolerance. George, E., Brown, Jr. Salinity Lab. US Dept. Agric. Res. Serv. Riverside, CA.

Ushio, A., Mae, T. and Makino, A. 2008. Effects of temperature on photosynthesis and plant growth in the assimilation shoots of a rose. *Soil Sci. Plant Nutr.* **54:** 253-258.

Vacca, R.A.; de Pinto, M.C.; Valenti, D., Passarella, S., Marra, E. and De Gara, L. 2004. Production of reactive oxygen species, alteration of cytosolic ascorbate peroxidase, and impairment of mitochondrial metabolism are early events in heat shock-induced programmed cell death in tobacco Bright-Yellow 2 cells. *Plant Physiol,* **134:** 1100-1112.

Vadez, V., Kholova, J., Choudhary, S., Zindy, P., Terrier, M., Krishnamurth, L., Kumar, P.R. and Turner, N.C. 2011. Whole plant response to drought under climate change. In, "*Crop adaptation to climate change*", Eds. S.S. Yadav, R. Redden, J.L. Hatfield, H. Lotze-Campen, A.E. Hall. Chichester-Wiley-Blackwell.

Vadez, V., Berger, J.D., Warkentin, T., Asseng, S., Ratnakumar, P., Rao, K.P.C., Gaur, P.M., Munier-Jolain, N., Larmure, A., Voisin, A.S., Sharma, H.C., Pande, S., Sharma, M., Krishnamurthy, L. and Zaman, M.A. 2012. Adaptation of grain legumes to climate change: a review. *Agron. Sustain. Dev.*, **32:** 31-44.

Vanaja, M., Raghuram Reddy, P., Jyothi Lakshmi, N., Maheswari, M., Vagheera, P., Ratnakumar, P., Jyothi, K., Yadav, S.K. and Venkatswarlu, B. 2007. Effect of elevated atmospheric CO_2concentrations on growth and yield of black gram (*Vigna mungo* (L.) Hepper)– A rainfed pulse crop. *Plant Soil Environ.* **53(2):** 81–88.

Vandeleur, R.K.and Mayo, G. 2009. The role of plasma membrane intrinsic protein aquaporins in water transport through roots: diurnal and drought stress responses reveal different strategies between isohydric and anisohydric cultivars of grapevine. *Plant Physiol.*, **149(1):** 445-460.

Van Iersel, M.W., and D.M. Oosterhuis. 1996. Drought effects of the water relations of cotton fruits , bracts and leaves during ontogeny. *Exp. Environ. Bot.*, **36:** 51-59.

van Oosten, J.-J. and Besford, R.T. 1995. Some relationships between the gas exchange, biochemistry and molecular biology of photosynthesis during leaf development of tomato plants after transfer to different carbon dioxide concentrations. *Plant Cell Environ.*, **18:** 1253—1266.

Vara Prasad, P.V., Craufurd, P.Q., Summerfield, R.J. and Wheeler, T.R. 2000. Effects of short episodes of heat stress on flower production and fruit-set of groundnut (*Arachis hypogaea* L.). *Journal of Experimental Botany*, **51:** 777-784.

Vasquez, M.D., Poschenrieder, C. and Barcelo, J. 1991. Ultrastructural effects and localization of low cadmium concentrations in bean roots. *The New Phytologist*, **120:** 215–226.

Verkleij, J.A.C., Sneller, F.E.C. and Schat, H. 2003. Metallothioneins and phytochelatins: ecophysiological aspects. In, "*Sulphur in Plants*", Eds. Y.P. Abrol, and A. Ahmad, Dordrecht, Netherlands: pringer; pp. 163–176.

Viehweger, K. 2014. How plants cope with heavy metals. *Botanical Studies.* **55(35):** 1–12.

Villegas, D., Garcia del Moral, L.F., Rharrabti, Y., Martos, V. and Royo, C. 2006. Morphological traits above the flag leaf nodes as indicators of drought susceptibility index in durum wheat. *Journal of Agronomy and Crop Science,* **193(2):**103-116.

Vijayalakshmi, K., Fritz, A., Paulsen, G., Bai, G., Pandravada, S. and Gill, B. 2010. Modeling and mapping QTL for senescence-related traits in winter wheat under high temperature. *Mol Breed*, **26:** 163–175.

Vijaylaxmi, 2013. Biochemical changes in cotyledons of germinating mungbean seeds from summer and rainy season. *Indian Journal of Plant Physiology*, **18(4):** 377-380.

Voesenek, L.A.C.J., Van der Sman, A.J.M., Harren. F.J.M. and Blom, C.W.P.M. 1992. An amalgamation between hormone physiology and plant ecology: a review on flooding resistance and ethylene. *J Plant Growth Regul,* **11:** 171-188.

Vogg, G., Heim, R., Gotschy, B., Beck, E. and Hansen, J. 1998. Frost hardening and photosynthetic performance of Scots pine (*Pinus sylvestris* L.). II. Seasonal changes in the fluidity of thylakoid membranes. *Planta*, **204:** 201–206.

Vollenweider, P. and Günthardt-Goerg, M.S. 2005. Diagnosis of abiotic and biotic stress factors using the visible symptoms in foliage. *Environ. Pollut.* **137:** 455–465.

Wagstaffe, A. and Battey, N.H. 2006. The optimum temperature for long-season cropping in the ever bearing strawberry ′everest′. *Acta Hortic. (ISHS)*, **708:** 45–50.

Wahid, A. 2007. Physiological implications of metabolite biosynthesis for net assimilation and heat-stress tolerance of sugarcane (*Saccharum officinarum*) sprouts. *Journal of Plant Research,* **120:** 219–228.

Wahid, A.; Gelani, S.; Ashraf, M. and Foolad, M. R. 2007. Heat tolerance in plants: An overview. *Environmental and Experimental Botany*, **6:** 199- 223.

Wang, Z. and Huang, B. 2004. Physiological recovery of kentucky bluegrass from simultaneous drought and heat stress. *Crop Sci.* **44:** 1729-1736.

Wang, H.-C., Wu, J.-S., Chia, J.-C., Yang, C.-C., Wu, Y.-J. and Juang, R.-H. 2009. Phytochelatin synthase is regulated by protein phosphorylation at a threonine residue near its catalytic site. *Journal of Agricultural and Food Chemistry*. **57(16):** 7348–7355.

Wang, Z. 2011. Plant Physiology. Beijing: Chinese Agriculture Press: 130–131. (in Chinese)

Wahid, A., Gelani, S., Ashraf, M., Foolad, M.R. 2007. Heat tolerance in plants: an overview. *Environmental and Experimental Botany*, **61:** 199-223.

Warrag, M.O.A. and Hall, A.E. 1984. Reproductive responses of cowpea (*Vigna nguiculata* (L.) Walp) to heat stress. II. Responses to night air temperature. *Field Crops Research,* **8:** 17–33.

Ward, J.K. and Strain, B.R. 1997. Effects of low and elevated CO_2 partial pressure on growth and reproduction of *Arabidopsis thaliana* from different elevations. *Plant Cell Environ.*, **20:** 254—260.

Ward, J.K. and Strain, B.R. 1999. Elevated CO_2 studies: Past, Present and Future. *Tree Physiol.* **19:** 211-220.

Ward, J. T.; Lahner, B.; Yakubova, E.; Salt, D. E. and Raghothama, K. G. 2008. The effect of iron on the primary root elongation of *Arabidopsis* during phosphate deficiency. *Plant Physiol*, **147:** 1181-1191.

Wardlaw, I.F., Sofield, I. and Cartwright, P.M. 1980. Factors limiting the rate of dry matter accumulation in the grain of wheat grown at high temperatures. *Aust. J. Plant Physiol.*, **7:** 387–400.

Warrington, I.J., Fulton, T.A., Halligan, E.A. and de Silva, H.N. 1999. Apple fruit growth and maturity are affected by early season temperatures. *J. Am. Soc. Hortic. Sci.*, **124:** 468–477.

Warren, C.R., N.J. Livingston, and D.H. Turpin. 2004. Water stress decreases the transfer conductance of Douglas-fir (*Pseudotsuga menziensii*). *Tree Phys.*, **24:** 971-979.

Wassmann, R., Jagadish, S., Sumfleth, K., Pathak, H., Howell, G., Ismail, A., Seeraj, R., Redona, E. Singh, R.K. and Heuer, S. 2009. Regional vulnerability of climate change impacts on Asian rice production and scope for adaptation. *Advances in Agronomy,* **102:** 91–133.

Weber, H., Heim, U., Golombek, S., Borisjuk, L. and Wobus, U. 1998. Assimilate uptake and the regulation of seed development. *Seed Sci. Res.*, **8:** 331-345.

Welch, J.R., Vincent, J.R., Auffhammer, M., Moya, P.F., Dobermann, A. and Dawe, D. 2010. Rice yields in tropical/subtropical Asia exhibit large but opposing sensitivities to minimum and maximum temperatures. *Proc. Natl. Acad. Sci.*, **107:** 14562–14567.

Welcker, C., Boussuge, B., Bencivenni, C., Ribaut, J.M. and Tardieu. F. 2007. Are source and sink strengths genetically linked in maize plants subjected to water deficit? A QTL study of the response of leaf growth and of anthesis-silking interval to water deficit. *J. Exp. Bot.*, **58:** 339–349.

Wen, J., Li, J. and Guo, Y. 2002. Use carbonyl content of proteins in evaluating antioxidative health food. *J. Food Hygien.*, **14:** 13-17.

Wiebbecke, C.F., Graham, M.A., Cianzio, S.R. and Palmer, R.G. 2012. Day temperature influences the male-sterile locus ms9 in soybean. *Crop Sci.*, **52:** 1503–1510.

Wilsey, B.J. and Polley, H.W. 2006. Aboveground productivity and root–shoot allocation differ between native and introduced grass species. *Oecologia*, **150:** 300–309.

Wilson, R.F., Burke, J.J. and Quisenberry, J.E.. 1987. Plant morphological and biochemical responses to field water deficits. II. Responses of leaf glycerolipid composition in cotton. *Plant Physiol*, **84:** 251-254.

Wilson, J.W., Hand. D.W. and Hannah, M.A. 1992. Light interception and photosynthesis efficiency in some glasshouse crops. *J Exp Bot*, **43(3):** 363–373.

Whalley, W.R. and Clark, L.J. 2011.Drought stress, effect on soil mechanical impedance and root (crop) growth. In, "*Encyclopedia of Agrophysics*", Eds. J. Gliñski, J. Horabik, J. Lipiec, Springer Press, Dordrecht-Heidelberg-London-New York.

White, R.G.and Kirkegaard, J.A. 2010. The distribution and abundance of wheat roots in a dense, structured subsoilimplications for water uptake. *Plant Cell Environ.*, **33:** 133-148.

Whittle, C.A., Otto, S.P., Johnston, M.O. and Krochko, J.E. 2009. Adaptive epigenetic memory of ancestral temperature regime in. *Arabidopsis thaliana. Botany-Botanique*, **87:** 650-657.

Wong, H.L., Sakamoto, T., Kawasaki, T., Umemura, K. and Shimamoto, K. 2004. Down-regulation of metallothionein, a reactive oxygen scavenger, by the small GTPase OsRac1 in rice. *Plant Physiology*. **135(3):** 1447–1456.

Wuana, R. A. and Okieimen, F.E. 2011. Heavy metals in contaminated soils: a review of sources, chemistry, risks and best available strategies for remediation. *ISRN Ecology*. **2011:** 20.

Xin, Z. and Browse, J. 2000. Cold comfort farm: the acclimation of plants to freezing temperatures. *Plant Cell Environ.*, **23:** 893-902.

Xu, Z.Z. and Zhou. G.S. 2005. Effects of water stress and nocturnal temperature on carbon allocation in the perennial grass, *Leymus chinensis*. *Physiol. Plant.*, **123:** 272–280.

Xu, C.X., Liu, Y.L., Zheng, Q.S. and Liu, Z.P. 2006. Silicate improves growth and ion absorption and distribution in *Aloe vera* under salt stress. *Journal of Plant Physiology and Molecular Biology*, **32:** 73-78.

Xu, S., Li, J., Zhang, X., Wei, H. and Cui, L. 2006. Effects of heat acclimation pretreatment on changes of membrane lipid peroxidation, antioxidant metabolites, and ultrastructure of chloroplasts in two cool-season turfgrass species under heat stress. *Environ Exp Bot.*, **56:** 274–285.

Xu, Z.Z., Zhou, G.S. and Shimizu, H. 2009. Are plant growth and photosynthesis limited by pre-drought following rewatering in grass? *J Exp Bot.*, **60:** 3737–3749.

Xu, Z.Z., Zhou, G.S. and Shimizu, H. 2009. Effects of soil drought with nocturnal warming on leaf stomatal traits and mesophyll cell ultrastructure of a perennial grass. *Crop Sci.*, **49:** 1843–1851.

Yadav, D.K. 2010. Diurnal and temporal variations in sone reactive oxygen species scavenging enzymes in maize (*Zea mays* L.) genotypes under water logged conditions. M. Sc. Thesis, Benaras Hindu University, Varanasi, India, pp. 50.

Yadav, S.K. 2010a. Cold stress tolerance mechanisms in plants. *Journal Agronomy for Sustainable*, **30(3):** 515-527.

Yamada, M., Morishita, H., Urano, K., Shinozaki, N., Yamaguchi-Shinozaki, K., Shinozaki, K. and Yoshiba, Y. 2005. Effects of free proline accumulation in *Petunia* under drought stress. *Experimental Botany*, **56:** 1975–1981.

Yamamoto, Y., Kurokawa, H., Nitta, Y. and Yoshida T. 1995. Varietal differences of tillering response to shading and nitrogen levels in rice plant: Comparison between high tillering semidwaft indica and low tillering japonica. *Jpn J Crop Sci*, **64(2):** 227–234. (in Japanese with English abstract) Yamauchi, T., Rajhi, I. and Nakazono, M. 2011. Lysigenous aerenchyma formation in maize root is confined to cortical cells by regulation of genes related to generation and scavenging of reactive oxygen species. Plant Signaling and Behavior, **6:** 759–761.

Yarnia, M., Arabifard, N., Rahimzadeh Khoei, F. and Zandi, P. 2011. Evaluation of drought tolerance indices among some winter rapeseed cultivars. *Afr. J. Biotechnol.*, **10:** 10914-10922.

Yang, H.G. and Lin, W.F. 2008. Effect of low temperature stress on some physiological characteristics of oil palm (*Elaeis guineensis* Jacq.). *Seedlings*, **29:** 326-332.

Yang, Z., and Chu, C. 2011. Towards understanding plant response to heavy metal stress. In, *"Abiotic Stress in Plants—Mechanisms and Adaptations"*, Shanghai, China: Tech; **2011:** 59–78.

Yang, D., Duan, L.S., Xie, H.A., Li, Z.H. and Huang, T.X. 2011. Effect of pre-flowering light deficiency on biomass accumulation and physiological characteristics of rice. *Chin J Eco-Agric*, **19(2):** 347–352. (in Chinese with English abstract)

Yang, A., Dai, X. and Zhang, W.H. 2012. A R2R3-type MYB gene, OsMYB2, is involved in salt, cold, and dehydration tolerance in rice. *J. Exp. Bot.*, **63:** 2541-2556.

Yoshida, R., Kanno, A. and Kameya, T. 1996. Cool temperature-induced chlorosis in rice plants. II. Effects of cool temperature on the expression of plastid-encoded genes during shoot growth in darkness. *Plant Physiology*, **112:** 585-590.

Young, L.W., Wilen, R.W. and Bonham-Smith, P.C. 2004. High temperature stress of *Brassica napus* during flowering reduces micro- and megagametophyte fertility, induces fruit abortion, and disrupts seed production. *Journal of Experimental Botany*, **55:** 485-495.

Zadeh, H.M. and Naeini, M.B. 2007. Effects of salinity stress on the morphology and yield of two cultivars of canola (*Brassica napus* L.). *Journal of Agronomy,* **6:** 409-414.

Zagorchev, L., Seal, C.E., Kranner, I. and Odjakova, M. 2013. A central role for thiols in plant tolerance to abiotic stress. *International Journal of Molecular Science,* **14(4):** 7405-7432.

Zaidi, P.H. and Singh, N.N. 2001. Effect of water logging on growth, biochemical composition and reproduction in maize (*Zea Mays* L.). *Journal of Plant Biology,* **28:** 61-70.

Zaidi, P.H., Rafique, S. and Singh, N.N. 2003. Response of maize genotypes to excess moisture stress: morpho-physiological effects and basis of tolerance. *Eur J Agron*, **19:** 383-399.

Zaidi, P.H., Rafique, S., Rai, P.K., Singh, N.N. and Srinivasan, G. 2004. Tolerance to excess moisture in maize (*Zea mays* L.): Susceptible crop stages and identification of tolerant genotypes. *Field Crops Research*, **90:** 189-202.

Zaidi, P.H., Maniselvan, P., Sultana, R., Yadav, M., Singh, R.P., Singh, S.B., Das, S. and Srinivasan, G. 2007. Importance of secondary traits in improvements of maize (*Zea mays* L.) for enhancing tolerance to excessive soil moisture stress. *Cereal Research Communications*, **35:** 1427-1435.

Zekri, M. 2001. Salinity effect on seedling emergence, nitrogen and chloride concentrations, and growth of citrus rootstocks. *Proceedings of the Florida State Horticultural Society*, **114:** 79-82.

Zekri, M. 2004. Strategies to manage salinity problem in citrus. *Proceedings of International Society of Citriculture*, 639-643.

Zeller, S., and Feller, U. 1999. Long-distance transport of cobalt and nickel in maturing wheat. *Eur. J. Agronomy*, **10:** 91-98.

Zengin, F.K. and Munzuroglu, O. 2005. Effects of some heavy metals on content of chlorophyll, proline and some antioxidant chemicals in bean (*Phaseolus vulgaris* L.) seedlings. *Acta Biologica Cracoviensia Series Botanica.* **47(2):** 157–164.

Zhai, F.-F., Han, L., Liu, J.-X., Qian, Y.-Q., Ju, G.-S., Li, W., Zhang, S.-W. and Sun, Z.-Y. 2013. Assessing cold resistance of mutagenic strains of perennial ryegrass under artificial low-temperature stress. *Acta Prata Sin.*, **22:** 268–279.

Zhang, K. and Lin, Y. 1994. Chinese Tallow tree. China Forestry Press, Beijing (in Chinese).

Zhang, J. and Lechowicz, M.J. 1995. Responses to CO_2 enrichment by two genotypes of *Arabidopsis thaliana* differing in their sensitivity to nutrient availability. *Ann. Bot.*, **75:** 491—499.

Zhang, S., Jiang, H., Peng, S., Korpelainen, H. and Li, C. 2011. Sex-related differences in morphological, physiological, and ultrastructural responses of *Populus cathayana* to chilling. *J Exp Bot* , **62:** 675–686.

Zhong, H. and Lauchli, A 1993. Changes of cell wall composition and polimer size in primary roots of cotton seedlings under high salinity. *Journal of experimental Botany,* **44:** 773-778.

Zhou, Y., He, Z.H., Chen, X.M., Wang, D.S., Yan, J., Xia, X.C. and Zhang, Y. 2007. Genetic improvement of wheat yield potential in North China, in Wheat Production in Stressed Environments, Eds, H.T. Buck, J.E. Nisi, N. Salomón, Springrt-Verlag, New York, 583–589.

Zhou, Y. Lam, H.M. and J. Zhang, J. 2007. Inhibition of photosynthesis and energy dissipation induced by water ad high light stresses in rice. *J. Exp. Bot.*, **58:** 1207–1217.

Zhou, B., Yao, W., Wang, S., Wang, X., and Jiang, T. 2014. The metallothionein gene, *TaMT3*, from *Tamarix androssowii* confers Cd^{2+} tolerance in Tobacco. *International Journal of Molecular Sciences.* **15(6):** 10398–10409.

Zhu, J., Talbott, L.D., Jin, X. and Zeiger, E. 1998. The stomatal response to CO_2 is linked to changes in guard cell zeaxanthin. *Plant Cell Environ.*, **21:** 813—820.

Zhu, J.K. 2002. Salt and drought stress signal transduction in plants. *Annu Rev Plant physiol Plant Mol Biol.*, **53;** 247-273.

Zhu, J.H., Dong, C.H. and Zhu, J.K. 2007. Interplay between cold-responsive gene regulation,metabolism and RNA processing during plant cold acclimation. *Curr Opin Plant Biol*, **10:** 290–295.

Zhu, P, Yang S.M., Ma. J., Li, S.X. and Chen, Y. 2008. Effect of shading on the photosynthetic characteristics and yield at later growth stage of hybrid rice combination. *Acta Agron Sin*, **34(11):** 2003–2009. (in Chinese with English abstract)

Zidan, I, Azaizah, H. and Neumann, P.M. 1990. Does salinity reduce growth in maize root epidermal cells inhibiting their capacity for cell wall acidification? *Plant Physiology,* **93:** 7-11.

Zinn, K.E., Tunc-Ozdemir M. and Harper, J.F. 2010. Temperature stress and plant sexual reproduction: uncovering the weakest links. *J. Exp. Bot.*, **61:** 1959–1968.

Zinselmeier, C., J.R. Schussler, M.E. Westgate, and R.J. Jones. 1995. Low water potential disrupts carbohydrate metabolism in maize ovaries. *Plant Physiol.*, **107:** 385-391.

Zinselmeier, C., B.R. Jeong, and J.S. Boyer. 1999. Starch and the control of kernel number in maize at low water potentials. *Plant Physiol.*, **121:** 25-35.

Zou, J.W., Rogers, W.E., DeWalt, S.J. and Siemann, E. 2006. The effect of Chinese tallow tree (*Sapium sebiferum*) ecotype on soil-plant system carbon and nitrogen processes. *Oecologia*, **150:** 272–281.

Zupan, M., Hudnik, V., Lobnik, F. and Kadunc, V. 1995. Accumulation of Pb, Cd, Zn from contaminated soil to various plants and evaluation of soil remediation with indicator plant (*Plantago lanceolata*). In, "*Contaminated Soils*", Ed. R. Prost, INRA, Paris, pp. 325–335.

Zupan, M., Kralj, T., Grcman, H., Hudnik, V. and Lobnik, F. 2003. The accumulation of Cd, Zn, Pb in Taraxacum officinale and Plantago lanceolata from contaminated soils. Proc VII ICOBTE, Uppsalȧ Sv.

5

Abiotic Stress and Heat Shock Proteins (HSPs) in Plants

A. Bhattacharya

The stresses, drought, salinity, chemicals, cold and hot temperatures, have interconnected effects on plants, resulting in the disruption of protein homeostasis. Maintenance of proteins in their functional native conformations and preventing aggregation of non-native proteins are important for cell survival under stress. The heat shock proteins buffer this environmental variation and are therefore important factors for the maintenance of homeostasis across environmental regimes. Heat shock proteins are known to be expressed in plants experiencing high-temperature stress. Heat shock proteins are important in relation to stress resistance and adaptation to the environment. Their molecular function might be membrane or protein stabilization, but, for now, for most heat shock proteins the molecular function remains to be elucidated. The general observation of the response and induction of heat shock proteins following heat stress suggests that they play a fundamental role in cellular homeostasis. It is possible that heat shock proteins may be involved with the phenomenon of acquired thermotolerance. Heat stress disturbs cellular homeostasis and can lead to severe retardation in growth and development, and even death. Temperature stress can have also a devastating effect on both the plant anabolism and catabolism, initially acting on protein complexes, being the quaternary structure the first folding that is lost during heat stress or heating. Understanding the role of HSPs in relation to stress resistance in a more applied perspective as a potential indicator of stress is important. New technological developments make it possible to investigate the role of genes coding for heat shock proteins in greater detail. A combination of genomics and proteonomics will further elucidate the effects of stress on expression patterns at the DNA, RNA and protein level.

Agriculture production must continue to meet the demands of the growing population. In the past decades agricultural production increased due to higher yields and by bringing more land under agricultural production. The scarcity of productive agricultural land may force us to grow agricultural crops in harsher environments. Temperature is an important environmental factor affecting crop productivity. Crops have high productivity when grown under optimum environmental conditions, particularly temperature optimum for various growth and metabolic processes. Temperature higher than the optimum, decrease both the rate and duration of metabolic processes and thus decrease the yield.

Stress is generally defined as all effects that have negative impact on plants (Taiz and Zaiger, 1988). Active plant growth in mesophile organisms is in between 10-40 °C. Any temperature which is higher and/or below that range creates heat stress on metabolic activities of plants (Treshow, 1970). Plant growth is a function of biochemical reactions and those reactions are controlled by enzymes. Most of the chemical reactions increase as two fold by every 10 °C increase in between 20-30 °C. Temperatures above that range, reaction speed decreases since enzymes step by step denatured or inactivated. In addition, reactions catalyzed by enzymes depend on completeness of tertiary structure of enzymes. Increasing molecular collisions as temperature increase, harms tertiary structures and reduce enzyme activities and reaction rates. Faster increase in temperature brings faster denaturation. When an enzyme is inactivated, growth processes related to that enzyme is depressed (Treshow, 1970). Photosynthesis is an important reaction affected by heat. Photosynthesis is performed in between 10-30 °C range in temperate climate plants. Above 10°C photosynthesis increases as temperature increases, it decreases above 30°C (Treshow, 1970). Respiration generally adapt a normal course at 5-25 °C, it is accelerated up to 30-35 °C and decelerated above 35 °C (Kadýoðlu, 2004). It was reported that heat stress disrupts water, ion and organic solute movement across plant membranes (Ebrahim and Quick, 2001), reduces chemical reactions, gas solubility, mineral absorption and water take up (Taiz and Zaiger, 1988), impairs photosynthetic electron transport system, and increases oxidative degeneration of membrane lipids (Dash and Mohanty, 2002). Toxic material formation has been detected as a result of high temperature. It was suggested that heat injury is due to the toxic effect of NH_3 produced at high temperatures and that this effect is counteracted by the respiratory production of organic acids. This may help to explain the heat tolerance of succulents, since they posses a highly acid metabolism. On the other hand, their acidity is at a maximum when danger of heat is minimal (at night) and at a minimum when the danger is maximal (in the afternoon) (Levitt, 1980). Heat stress is a major factor limiting the productivity and adaptation of crops, especially when temperature extremes coincide with critical stages of plant development. The rate of temperature change and the duration and degree of high temperatures all contribute to the intensity of heat stress. Where heat stress occurs, it is important that plants possess a certain degree of heat tolerance to survive the stress period. In addition, plant response to heat stress depends on the thermal adaptation, the duration of the exposure and the stage of growth of the exposed tissue (Chen *et al.* 1982).

5.1. Abiotic Stress and Plants

Basic stresses such as drought, salinity, temperature, and chemical pollutants are simultaneously acting on the plants causing cell injury and producing secondary stresses such as osmotic and oxidative ones (Wang *et al.* 2003). Plants could not change their sites to avoid such stresses, but have different ways and morphological adaptations to tolerate these stresses. The ways of defense at the molecular level are very important for the survival and growth of plants. Plants show a series of molecular responses to these stresses. The physiological processing basis for these molecular responses will not be covered here as it has been reviewed in depth lately (Shao *et al.* 2007a). Heat stress as well as other stresses can trigger some mechanisms of defense such as the obvious gene expression that was not expressed under "normal" conditions (Morimoto *et al.* 1993; Feder, 2006). In fact, this response to stresses on the molecular level is found in all living things, especially the sudden changes in genotypic expression resulting in an increase in the synthesis of protein groups. These groups are called "heat-shock proteins", "Stress-induced proteins" or "Stress proteins" (Lindquist and Craig, 1988; Morimoto *et al.* 1994; Gupta *et al.* 2010). Almost all kinds of stresses induce gene expression and synthesis of heat-shock proteins in cells that are subjected to stress (De Maio, 1999). In *Arabidopsis* and some other plant species low temperature, osmotic, salinity, oxidative, desiccation, high intensity irradiations, wounding, and heavy metals stresses were found to induce the synthesis of heat shock proteins (Swindell *et al.* 2007). However, stressing agents lead to an immediate block of every important metabolic process, including DNA replication, transcription, mRNA export, and translation, until the cells recover (Biamonti and Caceres, 2009).

It was known a long time ago that the most damage to crop plants in fields occurs when two or more stresses are prevailing (Mittler, 2006). Hence, in order to study the plant tolerance, it is very necessary to mimic the natural conditions in a specific area. Studies indicate that the plant responses to two or more factors are unique and differ from the response to one factor only. For example, subjecting the plants to drought only leads to high content of proline, but subjecting the same species to drought combined with high temperature leads to high content of sucrose and other sugars, but not proline (Rizhsky *et al.* 2004). Hence, Mittler (2006) studying all prevailing abiotic factors has suggested to treat this situation as a new stress condition that he called "Stress combination". The mechanisms of plant tolerance to a combination of diverse stress conditions, particularly those that mimic the field environment, have gained interest particularly for the biotechnologists (Chen and Zhu, 2004; Al-Babili and Beyer, 2005; Luo *et al.* 2005; Muns, 2005; Shao *et al.* 2007b).

Heat stress – high temperature – affects the metabolism and structure of plants, especially cell membranes and many basic physiological processes such as photosynthesis, respiration, and water relations (Wahid *et al.* 2007). On the molecular level, this effect of heat stress reflects the temperature dependence of Michaelis–Menton constant (K_m) of every enzyme participating in the process (Mitra and Bhatia, 2008). Plants must cope with heat stress for survival, so they developed different mechanisms including the maintenance of cell membrane stability, capturing the reactive oxygen species (ROS), synthesis of antioxidants, accumulation and osmoregulation of osmoticum, induction of some kinases that respond to stress, Ca-dependent kinase proteins, and enhancing the transcription and signal transfer of chaperones (Wahid *et al.* 2007). The induction and synthesis of heat-shock proteins due to high temperature exposure are common phenomena in all living organisms from bacteria to human beings (Parsell and Lindquist, 1993; Vierling, 1991; Gupta *et al.* 2010). It seems that the synthesis of these proteins is costly energy wise that is reflected on the yield of the organism.

Heat shock proteins are a family of proteins that are produced by cells in response to exposure to stressful conditions. They were first described in relation to heat shock, but are now known to also be expressed during other stresses including exposure to cold, UV light, and during wound healing or tissue remodeling. Many members of this group perform chaperone function by stabilizing new proteins to ensure correct folding or by helping to refold proteins that were damaged by the cell stress. Heat shock proteins are found in virtually all living organisms, from bacteria to humans. Heat-shock proteins are named according to their molecular weight. For example, HSP60, HSP70, and HSP90 refer to families of heat shock proteins on the order of 60, 70, and 90 kilo daltons (kDa) in size, respectively. Heat shock proteins play an important role in protein-protein interactions such as folding and assisting in the establishment of proper protein conformation (shape) and prevention of unwanted protein aggregation. By helping to stabilize partially unfolded proteins, heat shock proteins aid in transporting proteins across membranes within the cell. Some members of the heat shock proteins family are expressed at low to moderate levels in all organisms because of their essential role in protein maintenance.

The grass family Poaceae is comprised of nearly 10,000 species, including annual species cultivated as major grain crops and perennial species cultivated as forage for livestock or turf grasses on home lawns, commercial landscapes, roadsides, parks, athletic fields, and golf courses. Based on the ranges of temperature and precipitation that grasses adapt to, they are classified into warm-season and cool season categories. Cool-season and warm-season grasses have distinctive photosynthetic pathways, referred as C3 and C4 pathways, respectively. C3 or cool-season grass species grow most actively at temperatures ranging from 18^{0}C–24^{0}C while C4 or warm-season grass species have an optimum growth temperature between 30^{0}C–35^{0}C (Fry and Huang, 2004). Heat stress is particularly detrimental to cool-season grass species. Temperature rises beyond a threshold (5–10^{0}C above ambient) may cause irreversible damages to plant function and development or alteration of metabolism, resulting in reduction in growth and yield (Porter, 2005). The extent to which heat stress causes damage on plants is a complex issue. It depends on the intensity, duration and rate of increase in temperature, as well as other environmental conditions, such as when the high temperature occurs (during the day or the night) and where it occurs (in the air or the soil) (Sung *et al.* 2003). Since predicted global warming has become a serious threat for sustainable agriculture worldwide, an increasing challenge has been imposed to improve grass tolerance to high temperature.

Prokaryotic and eukaryotic organisms respond potentially harmful stimulations such as high and low temperature, radiation (UV), heavy metals, pesticide, salinity, bacterial and viral infections, parasitism, drought, oxidative stress, senescence, harmful mutations and reduction of cellular energy by inducing the synthesis of protein family called "stress proteins". Heat shock proteins and other stress proteins have been known to protect cells against deleterious effects of stress (Blumenthal *et al.* 1990; Feder and Hofman, 1999: Iba, 2002; Lindquist and Craig, 1988; Sioransen *et al.* 2003; Young and Elliott, 2002). Homologs of heat shock proteins were found in every species that have been studied (Feder and Hofman, 1999). Heat shock proteins are also expressed in some cells either constitutively or under cell cycle or developmental control (Vierling, 1997; Waters *et al.* 1996). Heat shock proteins and their cognates are found in every organism at ordinary growth temperature and play an important role in cellular functions related with growth (Lindquist and Craig, 1988; Waters *et al.* 1996). The major stress proteins occur at low to moderate levels in cells that have not been stressed but accumulate to very high levels in stressed cells (Weng and Nguyen, 1992).

Heat shock proteins, or stress proteins, are highly conserved and present in all plants and animals (Vierling, 1991). Some heat shock proteins also referred to as molecular chaperones, play an important role in protein stabilization such as assembling of multi-protein complexes, folding or unfolding of proteins, transport or sorting of proteins into correct compartments at sub-cellular level, control of cell-cycle and signaling, as well as cell protection against stress or apoptosis (Lindquist and Craig, 1988). When plants were exposed to high temperatures a new expression pattern was observed by the biosynthesis of heat shock proteins (Lindquist and Craig, 1988). The heat-shock response in plants is similar to that of other organisms. Activation of heat shock factors regulate the induction of heat shock genes (Young, 2010). Heat shock factors act in the promoter region of the heat shock proteins genes through a well-defined and highly conserved heat shock element.

Heat shock proteins are essential components of cells and developmental processes under normal physiological conditions besides its role in heat stress (Pratt *et al.* 2001). Earlier, it was revealed that most heat shock proteins serve as molecular chaperones (Young, 2010). Heat stress is one of the main abiotic stresses and affects the production of various crops (De Souza, 2012). High temperatures alter several metabolic processes reducing photosynthetic activity that results mainly in grain yield losses. Plants response to heat shock leads to changes in the cellular membrane structure, protein metabolism, level of enzymes and photosynthesis activity. It has been reported that high temperature changed the properties of membranes of nucleus, endoplasmic reticulum, mitochondria and chloroplasts of *Oryza sativa* (Shah *et al.* 2011). Lipids in the thylakoid membranes of the chloroplast are very important to improve photosynthesis and hence stress tolerance (Singh *et al.* 2007). Heat stress together with some stresses will cause defiance mechanisms such as the clear expression of genes that are expressed under the heat condition but not under favorable conditions (Feder, 2006; Morimoto, 1993). These genes are referred to as either "heat-shock proteins", "stress-induced proteins" or "stress proteins" (Lindquist and Craig, 1988; Gupta *et al.* 2010). Almost all kinds of stresses induce gene expression and synthesis of heat-shock proteins in cells that are subjected to stress (Iqbal *et al.* 2010; Westerheide *et al.* 2009). In *Arabidopsis* and some other plant species low temperature, osmotic, salinity, oxidative, desiccation, high intensity irradiations, wounding, and heavy metals stresses were found to induce the synthesis of heat shock proteins (Swindell *et al.* 2007). Heat stress "high temperature" affects the metabolism and structure of plants, especially cell membranes and many basic physiological processes such as photosynthesis, respiration, and water relations (Wahid *et al.* 2007). Plants living in extreme environments require a cell physiology able to compensate for stressful conditions and membranes and proteins are among the most affected cell components during stress (Huang and Xu, 2008). Changes in these components can alter several processes, such as uptake of water and ions, translocation of solutes, photosynthesis and respiration, and produce inactivation of enzymes, accumulation of unprocessed peptides and proteolysis. Thus, metabolic damage is produced and growth is, therefore, reduced (Lyons, 2012; Ferguson *et al.* 1990). One common protective response in most organisms to temperature stress is the induction of genes that code for het shock proteins (Lindquist and Craig, 1988; Essemine *et al.* 2010; Babiychuk *et al.* 2011; Guy and Li, 1998). It seems that the synthesis of these proteins is costly energy wise that is reflected on the yield of the organism. Thus, there is the need to shed more light on the various mechanisms on how plants respond to stress, such as heat, and the role they play with regards to stress tolerance (Wang *et al.* 2004).

Heat shock proteins can be said to be proteins which are usually expressed when responding to stress conditions (Lindquist and Craig, 1988; Craig and Gross, 1991; Banilas *et al.* 2012; Saidi *et al.* 2009; Krivokapich *et al.* 2006; Tsan and Gao, 2006; Gorski *et al.* 2003). Heat shock proteins have been studied widely in the fruit fly (*Drosophila melanogaster*) where the idea was first identified. Both at cellular and organismic level exposure, heat shock proteins are highly conserved (Lindquist and Craig, 1988; Saidi *et al.* 2009). For a cell to survive under stress it is quite important to maintain the proteins in their functional conformation and aggregation of non-native proteins should be prevented. Plants, as sessile organisms, rely on proteomic plasticity to remodel themselves during periods of developmental change, to endure varying environmental conditions, and to respond to biotic and abiotic stresses. The latter are the basic reason for the loss of crops all over the world, which reduces the average yields of virtually all major crop plants by more than fifty percent (Swindell *et al.* 2007). Abiotic stresses usually cause protein dysfunction (Wang *et al.* 2004). Different families of proteins are known to be connected with the ability of the plant to respond to physiological stress either by synthesis, accumulation or reduction (Åkerfelt *et al.* 2010). Among other things, these proteins are involved in signaling, carbohydrate metabolism and amino acid metabolism. In fact, proteins are now understood to mediate signaling, translation, host-defense mechanisms, carbohydrate metabolism and amino acid metabolism by playing a significant function in controlling the genome and ultimately features that are obvious such as in xerophytes or coming across stressors directly such as antioxidant enzymes and chaperonins or not directly like a key enzyme in the synthesis of osmolyte (Li and Mark, 2009). The most studied of these proteins have molecular weights of 100, 90, 70, 60 and 12-40 kDa and referred to as HSP100, HSP90, HSP70, HSP60 and sHSPs respectively (Whaibi, 2011). There is a difference in the number of heat shock proteins between organisms and even within an organism among the different cell types. To whatever degree, all organisms show the expression of heat shock proteins that belongs to the family HSP70 and the molecular weight of the family falls between 68-78kDa (Lindquist and Craig, 1988; Saidi *et al.* 2009). The family of the HSP70, out the other families of heat shock proteins, has a strong correlation with resistance to heat, either permanent or transient (Neznanov *et al.* 2011; Tang *et al.* 2005; Daugaard *et al.* 2007). The heat shock response in plants has been well characterized primarily in two crop plants, maize and soybean (Ding *et al.* 2009; Ma *et al.* 2006; Wu *et al.* 2010), cotton (Cottee *et al.* 2014; De Ronde *et al.* 1993), and to a lesser extent in a few other systems (Snider *et al.* 2009; Grigorova *et al.* 2011; Süle *et al.* 2004; Zubo *et al.* 2008; Senthil-Kumar *et al.* 2007). There have been reports on the HSR of a cell culture of a wild species of tomato, *Lycopersicon peruvianum* (Weber *et al.* 2008; Sun and McRae, 2005) and a report on the tissue specific expression of an HSP70 cognate in cultivated tomato (Jovanoviæ *et al.* 2011).

5.2. Heat Shock Proteins: Classification and Major Functions

Organisms have evolved a wide array of mechanisms for adapting to stressful environments. One of the most closely studied of these is the induction of heat shock proteins, which comprise several evolutionarily conserved protein families. All of the major heat shock proteins (that is, those expressed in very high amounts in response to heat and other stresses) have related functions: they ameliorate problems caused by protein misfolding and aggregation. However, each major heat shock protein family has a unique mechanism of action. Some promote the degradation of misfolded proteins, others bind to different types of folding intermediates and prevent them from aggregating, and still another (HSP100) promotes the reactivation of proteins that have already aggregated (Parsell and Lindquist, 1993, 1994).

Heat shock proteins are generally categorized into 5 evolutionarily conserved families: HSP100, 90, 70, 60, and the small heat shock proteins (HSPs) (Krishna, 2003). Some of the heat shock proteins act as molecular chaperones. Most, but not all, heat shock proteins are molecular chaperones, which bind and stabilize proteins at intermediate stages of folding, assembly, degradation, and translocation across membranes.

The family of HSP70

HSP70 proteins compose a large family of highly conserved molecular chaperones widely found in almost all organisms (Boorstein *et al.* 1994). The sequence identity between bacterial and eukaryotic HSP70s is about 50%, suggesting its critical functions in various life forms (Boorstein *et al.* 1994; Boston *et al.* 1996). Most eukaryotic organisms have multiple HSP70 homologs, located in diverse cell compartments including cytosol, mitochondrion, chloroplast and endoplasmic reticulum (Boston *et al.* 1996). In addition to their known function in preventing protein aggregation and assisting refolding of nonnative proteins in unfavorable environments, many HSP70 proteins also play essential roles in housekeeping activities under normal conditions. As a good example, in addition to the stress-inducible HSP70s, some HSP70 homologs, which are so called heat shock cognate 70 (HSC70), are constitutively expressed in the eukaryotic cytosol. HSC70 stabilizes nascent proteins being released from ribosomes, preventing possible misfolding and aggregation of partially synthesized polypeptide chains before the end of protein expression (Boston *et al.* 1996; Hartl, 1999; Fink, 1999). A comprehensive expression profile analysis of the *Arabidopsis thaliana* HSP70 gene family detected 2-20-fold induction of eleven HSP70 genes while the expression of another two HSP70 genes was not enhanced by heat shock treatment (Jiang *et al.* 2005).

The family of heat shock protein with molecular weight of 70kDa is wide and acts as molecular chaperones and is virtually found in almost all plant and animal species (Boorstein *et al.* 1994). The sequence arrangement identity between bacterial and eukaryotic HSP70 identity is around fifty percent, indicating its important and necessary functions in various life forms (Boorstein *et al.* 1994; Tompa and Kovacs, 2010). Regardless of their function in preventing aggregation of proteins and helping again in folding of proteins that are not native in environments that is not favorable, several HSP70 also play a crucial role in housekeeping activities under favorable conditions (Tompa and Kovacs, 2010). Some HSP70 homologues are constitutively expressed in the cytosol of eukaryotic organisms. These homologues are referred to as heat shock cognate 70 (HSC70) (Fink, 1999). They help to stabilize newly developing proteins being released from ribosome, preventing newly synthesized polypeptide chains from any possible misfolding and aggregation before the expression of the protein is completed (Fink, 1999; Hartl, 1996). A comprehensive analysis of the expression profile of *Arabidopsis thaliana*, HSP70 gene family founded two-twenty-fold induction of 11 HSP70 genes while another two HSP70 genes was expressed and not increased by heat shock treatment (Sung *et al.* 2001). In addition, some members of HSP70 are involved in controlling the biological activity of folded regulatory proteins, and might act as negative repressors of HSF mediated transcription (Tompa and Kovacs, 2010). In addition to its general chaperone functions (*i.e.* preventing accumulation and assisting in the folding again of proteins that are not native under the conditions of stress), HSP70 also plays a regulatory role in other stress associated gene expression (Barrientos *et al.* 2010). It was suggested that the interaction between HSP70 and heat shock factor prevents the trimerization and binding of heat shock factor to heat shock element, thereby blocking the transcriptional activation of heat-shock genes by their heat shock factors (Kim and Schöffl, 2002). HSP70 is also involved in the modulation of signal transducers such as protein kinase A, protein kinase C and protein phosphatase (Vigh *et al.* 2007).

The family of HSP90

Despite the versatile functions, all HSP70 proteins in higher eukaryotes including plants share similar structures. As represented by a bovine HSC70, the typical structure of HSP70 homolog proteins is composed of an N-terminal ATPase domain about 45 kDa and a C-terminal substrate-binding domain about 25 kDa, which are joined by an inter domain linker (Jiang *et al.* 2005). The ATPase domain resembles the structure of actin (Flaherty *et al.* 1990) and shares about 64% sequence identity among all eukaryotic HSP70s (Hartl, 1999). The substrate binding domain has relatively low-sequence conservation (~43% identity) but generally binds short stretches of hydrophobic peptides, which are normally buried inside the folded proteins (Hartl, 1999). The affinity for substrate peptide binding is mediated by different nucleotide-binding states of the ATPase domain. ATP-bound HSP70 binds and releases substrates at fast rates. As the ATP is hydrolyzed to ADP, an allosteric conformational change occurs between the two domains of HSP70, resulting in a higher substrate binding affinity (Jiang *et al.* 2005; Schmid *et al.* 1994; Chang *et al.* 2008). The switch of HSP70 nucleotide states is facilitated by J-domain cochaperones (HSP40) (Fink, 1999). This process enables HSP70 to go through cycles of substrate binding and releasing in an ATP-dependent manner, which stabilizes the exposed hydrophobic segments of nonnative proteins, prevents aggregation, and assists the correct folding (Hartl, 1999; Szabo, 1994).

HSP90 is one of the most abundant proteins in the cytoplasm, where it constitutes 1–2% of total protein levels (Czar *et al.* 1997). The essential eukaryotic chaperone HSP90 facilitates the maturation of a variety of meta-stable protein substrates, many of which are central regulators of biological circuits. Due to HSP90s central position in numerous pathways regulating growth and development, the chaperone has emerged as a principal focus of investigation in diverse fields (García-Cardeña *et al.* 1998). The HSP90s are highly conserved and abundant cytosolic proteins in eukaryotes (Prasinos *et al.* 2005). Unlike other heat shock proteins, the chaperone activity of the HSP90 is exerted on a number of target substrates or client proteins including steroid hormone receptors, cell cycle kinases, signal transduction pathway components, proteolytic machinery, and microtubule dynamics (Holt *et al.* 1991; Nathan *et al.* 1997). However, the high abundance of HSP90 compared with that of its client proteins, the high levels of expression in response to stress, and the existence of endoplasmic reticulum (ER), chloroplasts, and mitochondria homologues suggests that this is a very narrow view of the HSP90 cellular activity and further indicates that it may contribute to additional functions in the cytosol and organelles under physiological or stress conditions (Beradini *et al.* 2001). HSP90 acts as part of a multi-chaperone machine together with HSP70 and co-operates with a cohort of co-chaperones, including Hip (HSP70 interacting protein), Hop (HSP70 / HSP90 organizing protein), p23 and HSP40 (Ludwig-Müller *et al.* 2000). HSP90 regulates diverse cellular functions and exerts marked effects on normal biology, disease and evolutionary processes (Taipel *et al.* 2010). HSP90s mediate plant abiotic stress signal pathways (Liu *et al.* 2006a), but again the mechanisms are unclear. Under normal conditions, HSP90.2 in the cytoplasm negatively regulates transcription of heat-induced genes by suppression of HSF (Yamada *et al.* 2007). HSP90.2 is instantaneously inactivated under heat shock, and HSF is activated to induce expression of downstream genes containing HSF elements (Yamada *et al.* 2007). It was subsequently confirmed that over-expression of HSP90.2 suppresses HsfA2 transcription, and that HsfA2 is induced under inhibition of HSP90.2 in order to adapt to oxidative stress (Nishizawa-Yoki *et al.* 2010).

The family of HSP60

This class HSP60 is called in some of the literatures as chaperonins and it is generally agreed that they are important in assisting plastid proteins such as Rubisco (Wang *et al.* 2004). Some studies pointed out that this class might participate in folding and aggregation of many proteins that were transported to organelles such as chloroplasts and mitochondria (Ahsan and Komatsu, 2009). These HSP60 bind different types of proteins after their transcription and before folding to prevent their aggregation (Kalmar and Greensmith, 2009). Functionally, plant chaperonins are limited and the general idea is that stromal chaperones (HSP70 and HSP60) are involved in attaining functional conformation of newly imported proteins to the chloroplast (Jackson-Constan *et al.* 2001). Functional characterization of plant chaperonins is limited. It is generally agreed that they are important in assisting plastid proteins such as Rubisco (Young *et al.* 2004; Tompa and Kovacs, 2010). It has been reported that a mutated species of *Arabidopsis* chloroplast Cpn60a exhibits defects in chloroplast development and, subsequently, in the proper development of the plant embryo and seedling (Apuya *et al.* 2001). Antisense Cpn60b transgenic tobacco plants showed drastic phenotypic alterations, including slow growth, delayed flowering, stunting and leaf chlorosis (Zabaleta *et al.* 1994). Chloroplast Cpn60 is a homologue of the bacterial HSP60-type chaperone GroEL (Mayer, 2010; Bukau *et al.* 2006). Such chaperones assemble into two stacked heptameric rings, and cooperate with a cochaperone termed Cpn10 (an HSP10/GroES homologue), which may form heptameric caps at either end of the Cpn60 tetradecamer. Client proteins enter a central cavity in the Cpn60 complex, providing a protected environment in which folding can take place, which is controlled by the nucleotide-dependent cycling of Cpn60 subunits between binding-active and folding-active states

(Mayer, 2010; Bukau *et al.* 2006). The role of the Cpn60/Cpn10 chaperonin system in the folding of Rubisco subunits has been well documented (Boston *et al.* 1996). It has been suggested that stromal HSP70 and Cpn60 act sequentially to assist the maturation of imported proteins; particularly those destined for the thylakoid membranes. Another factor that facilitates the passage of new proteins through the stroma to the thylakoids is cpSRP (chloroplast Signal Recognition Particle), the clients of which are light-harvesting chlorophyll-binding proteins. In fact, one of the cpSRP components, a protein termed cpSRP43, was recently reported to have unique chaperone-like properties (Jaru-Ampornpan *et al.* 2010; Falk and Sinning, 2010).

The best-characterized HSP60 protein is GroEL from *E. coli*. GroEL-like HSP60 homologs have been found in mitochondrion and chloroplast of plant cells but not in cytosol (Fink, 1999). Under heat-shock conditions, expression of mitochondrion HSP60 is induced and protects preexisting proteins in the organelle from denaturation or inactivation (Martin *et al.* 1992), whereas expression of chloroplast HSP60 (chHSP60) is constitutively produced with only modest increase (Vitanen *et al.* 1995). Due to the high sequence similarity between plant HSP60 and GroEL, the structure of GroEL reported by Xu *et al.* (1997) to represent the typical assembly of HSP60. GroEL forms a huge homo-oligomer that is composed of two stacked rings, with each ring containing 7 monomers. Each GroEL monomer is about 58 kDa and can be divided into three separate domains: a nucleotide-binding equatorial domain, a flexible apical domain, and a hinge-like intermediate domain (Braig *et al.* 1994). The unique structure of GroEL-like HSP60 homo-oligomeric complex creates a hydrophobic cavity about 50 °A in the center of each stacked ring, allowing the accommodation of unfolded polypeptides with size ranging from 10 to 60 kDa (Hartl, 1999; Horwich, 1993). Following the entry of substrates (non-native proteins), GroEL binds ATP and associates with a 10 kDa co-chaperone and GroES (chaperonin 10) (Xu *et al.* 1997). GroES forms a heptametric ring and interacts with the apical domain of GroEL, acting like a dome-shaped lid that closes the central cavity in the chaperonin. The GroEL/GroES association traps nonnative proteins in an enclosed hydrophobic environment that is amenable to proper folding. Subsequently, the hydrolysis of ATP leads to GroEL conformational change and dissociation of GroES. This results in the release of encapsulated substrates and initiates the next cycle of substrate binding and folding (Hartl, 1999). A primary mechanistic distinction between HSP60 and HSP70 is that HSP60 is capable of binding an entire domain or complete protein, unlike HSP70 that recognizes only short peptide segments. It is worth noting that chHSP60 is also named as "Rubisco large subunit binding protein" as it plays an essential role in the folding and assembly of Rubisco large subunits (Hemmingsen *et al.* 1988). Despite the overall sequence and structural similarity (40%~50% sequence identity) to GroEL, chHSP60 is a hetero-olgiomeric protein complex consisting of α and β subunits (Dickson *et al.* 2000). It has been found that functional Rubisco protein can only be expressed in *E. coli* in the presence of GroEL/GroES system (Boston *et al.* 1996). This suggests that chHSP60 probably facilitates Rubisco folding in a similar manner as its probacterial homolog.

The family of HSO90 and HSP100

In accordance to HSP70 and HSP60 families, HSP90 proteins are also ATP-dependent molecule chaperones widely expressed in most organisms (Pearl and Prodromou, 2006). However, HSP90 features unique substrate specificity. Instead of binding a wide spectrum of unfolded proteins, HSP90 only interacts with relatively well-folded proteins involved in transcription regulation and signal transduction pathways (Majoul *et al.* 2003; Zhao and Houry, 2005). Furthermore, the function of HSP90 requires the formation of large protein complexes involving multiple co-chaperones, including HSP70 and HSP40, which indicates close cooperation between different molecule chaperone families (Zhao and Houry, 2005). Although structural information about plant HSP90 is sparse, high-sequence conservations of the protein family across the phylogeny suggests a similar functional mechanism. Typically, a HSP90 protein is composed of an N-terminal ATPase domain, a middle substrate protein-binding domain, and a C-terminal dimerization domain (Zhao and Houry, 2005). It is proposed that the ATP binding and hydrolysis regulate different conformational states of HSP90, which make the dimeric molecule chaperone to bind and release substrate proteins like a molecule clamp. However, dissection of HSP90 function is still restricted by the limited understanding of full-length HSP90 structure and its interaction with co-chaperones (Xu *et al.* 2011).

Although all organisms synthesize heat shock proteins in response to heat, the balance of proteins synthesized and the relative importance of individual heat shock protein families in stress tolerance vary greatly among organisms. For example, in yeast, a member of the HSP100 (ClpB/C) family, HSP104, is strongly expressed in the nuclear-cytoplasmic compartment in response to stress and plays a particularly pivotal role in tolerance to extreme conditions (Sanchez *et al.* 1992; Parsell *et al.* 1994). Yeast cells expressing HSP104 survive exposure to high temperatures or high concentrations of ethanol 1000- to 10,000-fold better than do cells not expressing HSP104. Members of the HSP100 family also play critical roles in the stress tolerance of bacterial cells (Schirmer *et al.* 1996), including photosynthetic cyanobacteria (Eriksson and Clarke, 1996). In contrast, the fruit fly *Drosophila* makes no protein of this type in response to stress; instead, the induction of HSP70 plays the central role in stress tolerance in this organism (Solomom *et al.* 1991; Welte *et al.* 1993).

HSP100 (a.k.a. Clp proteins) is another class of ATP dependent molecular chaperones. The unique feature of HSP100 family is their capacity to solubilize aggregated proteins and involvement in protein degradations (Boston *et al.* 1996; Honwich, 1995). The best characterized HSP100 proteins are ClpA from *E. coli* and HSP104 from *S. cerevisiae.* Both ClpA and HSP104 assemble as a hexameric ring with a narrow central pore (Fink, 1999; Guo *et al.* 2002). This conserved structural organization of HSP100 proteins implies similar functional mechanisms in different organisms. HSP100 plays an essential role in plant survival of severe heat stress (hong and Vierling, 2000), but it is absent in some other organisms (*e.g., Drosophila* and vertebrates) that rely on HSP70 and other HSPs to prevent aggregation and accommodate refolding under severe heat stress (Gurley, 2000).

The family of HSP100, also act as molecular chaperones, is among the members of the super family AAA ATPase with a wide range of different functional properties (Agarwal *et al.* 2001; Schirmer *et al.* 1996). Beside playing a role in protein accumulation and misfolding, the HSP100 family plays a significant position in the disaggregation of proteins and/or degradation. The removal of polypeptides that are not functioning and are potentially harmful as a result of misfolding, accumulation or denaturation is crucial for the maintenance of homeostasis in the cells. Another unique function of this class is the reactivation of aggregated proteins (Kalmar and Greensmith, 2009) by resolubilization of non-functional protein aggregates and also helping to degrade irreversibly damaged polypeptides (Kim and Schöffl, 2002; Bösl *et al.* 2006). HSP100 have been studied in several plant species such as *Arabidopsis*, *Glycine max*, *Nicotiana tabacum*, *Oryza sativa*, maize (*Zea mays*), Lima bean (*Phaseolus lunatus*) and *Triticum sativum (*Agarwal *et al.* 2001; Adam *et al.* 2001*; Keeler et al.* 2000). HSP100 like other molecular chaperones are often expressed in plants, but the environmental stresses such as high and low temperatures, dehydration, high salt or dark-induced etiolating are known to be the agents that induce them and the HSP100 expression is developmental (Agarwal *et al.* 2001; Keeler *et al.* 2000; Queitsch *et al.* 2000). Agarwal *et al* (2003) indicated that in rice HSP100 protein production was correlated with the disappearance of protein granules in yeast cells. It was further shown that the dissolution of electron-dense granules by HSP101 takes place during the post-stress phase, implying a role for HSP100 in the recovery of cell stress. Given the high sequence similarity of HSP101 with yeast HSP104 and their conserved functions in thermotolerance, one can reasonably propose that HSP101 in Arabidopsis is also acting to facilitate reactivation of proteins denatured by heat (Queitsch *et al.* 2000). The finding that HSP101 plays a crucial role in thermotolerance in plants, together with the conserved function of HSP101, suggests that engineering plants to express increased HSP101 may improve survival during periods of acute environmental stress (Queitsch *et al.* 2000).

HSP100 family members, which play such a major role in the stress tolerance of bacteria and fungi, have also been identified in higher plants (Lee *et al.* 1994; Boston *et al.* 1996; Schirmer *et al.* 1996; Wells *et al.* 1998). Similar to many other heat shock protein families, the HSP100 protein family comprises both heat-inducible and constitutive members. Among plants, bacteria, and yeast, these heat-inducible members are more closely related to each other than they are to their own constitutively expressed relatives (Schirmer *et al.* 1996). Their sequence homology and similar patterns of induction suggest a related function in stress tolerance. Moreover, the *Arabidopsis*, soybean, wheat, and tobacco HSP100 homologs can at least partially restore thermotolerance to yeast cells carrying an HSP104 deletion (Lee *et al.* 1994; Schirmer *et al.* 1996; Wells *et al.* 1998).

Although all organisms synthesize heat shock proteins in response to heat, the balance of proteins synthesized and the relative importance of individual heat shock protein families in stress tolerance vary greatly among organisms. For example, in yeast, a member of the HSP100 (ClpB/C) family, HSP104, is strongly expressed in the nuclear-cytoplasmic compartment in response to stress and plays a particularly pivotal role in tolerance to extreme conditions (Parsell *et al.* 1994). Yeast cells expressing HSP104 survive exposure to high temperatures or high concentrations of ethanol 1000- to 10,000-fold better than do cells not expressing HSP104. Members of the HSP100 family also play critical roles in the stress tolerance of bacterial cells (Schirmer *et al.* 1996), including photosynthetic cyanobacteria (Ericksson and Clarke, 1996). In contrast, the fruit fly *Drosophila* makes no protein of this type in response to stress; instead, the induction of HSP70 plays the central role in stress tolerance in this organism (Solomon *et al.* 1991; Welte *et al.* 1993).

Determining which proteins play the most crucial roles in stress tolerance in different types of organisms requires genetic analysis. Among organisms amenable to such analysis, higher plants present a particularly interesting subject. First, their immobility limits the range of their behavioral responses to stress and places a strong emphasis on cellular and physiologic mechanisms of protection. Second, their natural environments subject them to wide variations in temperature, both seasonally and diurnally. Third, they are developmentally complex, and the nature of the stresses to which they are exposed as well as their responses to stress are likely to vary in different tissues. Even for a particular organ—among leaves, for example—temperatures can vary dramatically with position on the plant (sun exposure) and can change abruptly with a shift in shading. Finally, the ability to withstand heat stress, especially in combination with water stress, may be of great importance in agricultural productivity (Levitt, 1980; Frova, 1997).

Surprisingly, the critical factors conferring temperature tolerance in higher plants are still poorly understood. Much indirect evidence suggests that heat shock proteins, as a general class, are likely to play some role. Several studies have correlated the induction of heat shock proteins by mild heat stress with the induction of tolerance to much more severe stress (Vierling, 1991; Howarth and Skot, 1994). In addition, over expression of certain transcriptional regulators of heat shock protein expression, HSF1 and HSF3, causes plants to constitutively express at least some HSPs and produces somewhat higher basal thermotolerance (Lee *et al.* 1995; Prändl *et al.* 1998). The only direct evidence for the function of an individual heat shock protein in stress tolerance in plants comes from transgenic carrot culture cells and plants. Changes in the expression of HSP17.7 cause modest changes in growth rates of tissue culture cells and electrolyte leakage of leaves after heat stress (Malik *et al.* 1999).

HSP100 family members, which play such a major role in the stress tolerance of bacteria and fungi, have also been identified in higher plants (Lee *et al.* 1994; Boston *et al.* 1996; Schirmer *et al.* 1996; Wells *et al.* 1998). Similar to many other heat shock protein families, the HSP100 protein family comprises both heat-inducible and constitutive members. Among plants, bacteria, and yeast, these heat-inducible members are more closely related to each other than they are to their own constitutively expressed relatives (Schirmer *et al.* 1996). Their sequence homology and similar patterns of induction suggest a related function in stress tolerance. Moreover, the *Arabidopsis*, soybean, wheat, and tobacco HSP100 homologs can at least partially restore thermotolerance to yeast cells carrying an HSP104 deletion (Lee *et al.* 1994; Schirmer *et al.* 1994; Wells *et al.* 1998).

The small heat shock protein family (smHSPs)

The family of small heat shock proteins (smHSPs) range in different sizes from 15 to 42 kDa and are characterized by a conserved sequence at their C terminus (Sun *et al.* 2002). The most prominent and diverse group of heat induced proteins in plants are sHSPs, a super-family of chaperones that are defined by a conserved carboxy-terminal domain of about 90 amino acids, referred to as the α-crystallin domain (de Jong *et al.* 1998). Higher plants possess at least 20 types of sHSP, but some species could contain even 40 types (Vierling, 1991). The six nuclear gene families encoding sHSPs have been recognized in higher plants (Scharf *et al.* 2001). Each gene family encodes proteins of a specific class that are found in a distinct cellular compartment, including cytosol/nucleus (classes CI, CII, and CIII), plastids (class P), mitochondria (class M) and endoplasmic reticulum (class ER) (Sun *et al.* 2002). The role of sHSPs during heat stress involves the formation of complexes with heat-denatured proteins. According to the model proposed by Lee *et al.* (1997), sHSP oligomers interact with unfolded proteins, prevent protein thermal aggregation, maintain proteins in a state competent for refolding and serve as a reservoir of the unfolded substrate for HSP70 chaperones. So far, the cooperation between the pea small HSP18.1 and HSP70 system has been observed *in vitro* (Lee and Vierling, 2000). The molecular chaperone activity of tomato HSP17.7 class I and HSP17.3 class II sHSPs was confirmed *in vivo*, as shown by their ability to prevent the thermal inactivation of firefly luciferase in a transgenic *Arabidopsis thaliana* cell culture (Löw *et al.* 2000). In addition to the chaperone function, sHSPs may have other roles. Malik *et al.* (1999) have suggested that carrot HSP17.7 CI may be involved in translational control upon heat stress. Recent findings indicate that sHSPs may play an important role in membrane quality control and thereby potentially contribute to the maintenance of membrane integrity under stress conditions (Nakamoto and Vigh, 2007). In many higher plants, sHSPs are the most abundant group of proteins produced under HS; for instance, quantitative analyses of soybean seedling proteins under HS have revealed that sHSPs accumulate to over 1% of total leaf proteins (Hsieh *et al.* 1992).

Sequence analysis of sHSPs shows that members of this protein family include an evolutionarily divergent N-terminal part and a short C-terminal tail (de Jong *et al.* 1993). The number of sHSP genes increases along the evolutionary scale (Haslbesck *et al.* 2005). Single-celled organisms such as bacteria have only one or two sHSP, whereas multi cellular organisms have many sHSP in the genome. Particularly, ten separate families of sHSPs have been recognized to be conserved in both monocot and dicot plants, indicating the potential for diversity in sHSP mechanisms (Siddique *et al.* 2008; Scharf *et al.* 2001). sHSPs encoded by four of these families localize to the cytoplasm, and those encoded by the other six families localize to cellular organelles including nucleus, chloroplasts, mitochondria, endoplasmic reticulum, and peroxisomes (Basha *et al.* 2010).

The current model for sHSP chaperone activity was defined based on studies of a cytosolic sHSP family named as Class I sHSPs (sHSP-CI), which represent the most abundant sHSP in plants. The model suggests that sHSP assembles into a large homo-oligomer, which binds denatured proteins in an ATP-independent manner, keeping them in a folding competent state. Then, it cooperates with ATP-dependent molecular chaperones, such as HSP70 and HSP90, to refold those proteins. Notably, sHSP has a much larger binding stoichiometry than other molecular chaperones, which has led to the speculation that sHSP functions as a reservoir to stabilize the flood of denatured proteins in response to stress (Guan *et al.* 2004; Lee and Vierling, 2000). It has been proposed that heat-induced oligomer dissociation is a major mechanism by which plant sHSPs can expose normally inaccessible, hydrophobic client binding surfaces. Nevertheless, the details about the interactions between sHSP and nonnative proteins and how these nonnative proteins are subsequently refolded are still lacking. This is partially due to limited knowledge on the molecular structure of sHSPs (Haslbeck *et al.* 2005). Among the few solved crystallographic structures of sHSPs is a wheat TaHSP16.9-CI (wHSP16.9, PDB Id: 1GME) (Montfort *et al*, 2001). The basic building block of wHSP16.9 is a dimer, which further assembles as a 12-mer consisting of two trimers of dimers. In solution, wHSP16.9 can dissociate into smaller oligomeric states in a temperature dependent manner (Montfort *et al.* 2001). On the basis of this observation, it is likely that heat-induced dissociation of sHSP oligomers may expose the hydrophobic patches buried in the oligomeric interface, resulting in binding and stabilization of denatured proteins (Montfort *et al.* 2001; Haslbeck *et al.* 2005).

Furthermore, the sHSPs of *Arabidopsis thaliana* and *Lycopersicon esculentum* are divided into three subclasses (Scharf *et al.* 2001; Siddiqui *et al.* 2003). These included:

- Subclass CI represented by six proteins in *A. thaliana* and five proteins in *L. esculentum*
- Subclass CII represented by two genes in both plants,
- Subclass CIII represented by one gene in both plants.

Siddiqui *et al.* (2006) reported the presence of other groups in the cytoplasm of *A. thaliana* cells and could be categorized into subclasses: CIV, CV, CVI, and CVII. Each subclass has its own distinct characteristics and role.

There are six groups of genes that encode for the sHSPs. The grouping is based on the sequence similarity and the location of these proteins in the cell. There are two classes of proteins (Class I and Class II) in the cytoplasm encoded by two groups of genes. Other locations are chloroplasts, endoplasmic reticulum, mitochondria, and membranes (Vierling, 1991; Waters *et al.,* 1996). The expression of genes for these sHSPs is limited in the absence of environmental stress and occurs in some stages of growth and development of plants such as embryogenesis, germination, development of pollen grains, and fruit ripening (Sun *et al.,* 2002).

The transcription of these genes is controlled by regulatory proteins called heat stress transcription factors (Hsfs) located in the cytoplasm in an inactive state. So these factors are considered as transcriptional activators for heat shock (Baniwal *et al.* 2004; Hu *et al.* 2009). Each factor has one carboxylic terminal (C-terminal) and three amino terminal (N-terminal) and has the amino acid leucine (Schuetz *et al.* 1991). Plants are characterized by a large number of transcriptional factors [at least 21 (Nover and Baniwal, 2006)]. These factors have been classified (Tripp *et al.* 2009) into three classes according to the structural differences in their aggregation in triples, *i.e.* oligomerization domains as follows:

- Plant HsfA such as HsfA1 and HsfA2 in *L. esculentum*
- Plant HsfB such as HsfB1 in *L. esculentum*
- Plant HsfC

Each factor has its role in the regulatory network in plants. However, all cooperate in regulating many functions and different stages of response to periodical heat stress (triggering, maintenance, and recovery). This role is represented in tomato system where HsfA1a is the master regulator that is responsible for the induced-stress gene expression including the synthesis of both HsfB1 and HsfA2 as these factors are found after the induction by heat treatment. These three factors are necessary for plant acquisition of heat tolerance (Baniwal *et al.* 2004). Mishra *et al.* (2002) have reported that there is an Acquired Thermotolerance (AT) phenomenon which is supported by a study on *Arabidopsis thaliana* that indicated the participation of HsfA2 (Chang *et al.* 2007). Furthermore, HsfA2 was finely regulated with HSP17-CII during anther development of a heat-tolerant tomato genotype and was further induced under both short and prolonged heat stress conditions (Giomo *et al.* 2010). The study data suggested that HsfA2 may be directly involved in the activation of protection mechanisms in the tomato anther during heat stress and, thereby, may contribute to tomato fruit set under adverse temperatures.

The adverse effects of high temperatures on potato growth and yield are numerous and include the acceleration of stem growth with assimilate partitioned more toward the stem; the reduction of photosynthesis and increase of respiration; reduction of root growth; inhibition of tuber initiation and growth; frequent tuber disorders; reduction of tuber dry matter and increase of glycoalkaloid level (Struik, 2007). Because of global warming, the global mean air surface temperatures have increased by approximately 0.7 ^{0}C in the past century, with a projected further rise of 1.1- 6.4 ^{0}C by the end of 21st century (Christensen *et al.* 2007). Using simulation model-based predictions of global warming over the next 60 years, Hijmans (2003) predicted potato yield losses in the range of 18 to 32%. However, these losses can be reduced to 9-18% with adaptations to production methods, such as terms of planting time and use of heat-tolerant cultivars.

Some potato cultivars have already been developed using conventional breeding to enhance heat tolerance for warmer climate cultivation (Veilleux *et al.* 1997; Minhas *et al.* 2001). Although traditional breeding is one of the most frequently used practices for developing heat tolerance in plants, it is challenged by some limitations. Nowak and Colborne (1989) reported on the possibility of *in vitro* selection for heat-tolerance in the potato based on the microtuberization behavior of two heat-tolerant and two heat-susceptible varieties under HS conditions. *In vitro* culture provides an opportunity to perform HS experiments under strictly controlled conditions with a variation of just one factor – temperature. Also, the physiological status of the potato cultivars, which may differ in plant maturation time, can be synchronized *in vitro*. A combination of *in vitro* cultivation and fast-screening methods based on effective selection markers for the detection of heat-tolerant genotypes can significantly reduce some of the limitations of field-carried searching for heat tolerance. Small HSPs could be used as markers for detecting heat tolerant (HT) genotypes; a tight correlation between the synthesis of sHSPs and the development of heat tolerance has been reported in a number of plant species, including soybean (Lin *et al.* 1984), *Sorghum bicolor* L. (Howarth and Skot, 1994), apple (Bowen *et al.* 2002), rice (Murakami *et al.* 2004), tomato (Mamedov and Shono, 2008) and durum wheat (Rampino *et al.* 2009). Based on differential expression observed in heat-tolerant and heat-sensitive cultivars, the employment of sHSPs as potential heat tolerance markers has been proposed, so far, for barley (Sule *et al.* 2004) and wheat (Yildiz *et al.* 2008).

The α-crystalline-related smHSPs are ubiquitous in nature, but 20–40 kDa smHSPs are unusually abundant in higher plants. Small HSPs are divided into six classes based on their sequence alignments, immunological cross reactivity, and cellular compartmentalization. Three classes (I, II and III) are localized in the cytoplasm or the nucleus and the other three are localized in the chloroplast, the endoplasmic reticulum, and the mitochondria (Sun *et al.* 2002; Vierling, 1991). Although smHSPs' precise functional mechanism is still unclear, *in vivo* studies have shown that smHSPs act as molecular chaperones (Jaya *et al.* 2009; Löw *et al.* 2000; Yu *et al.* 2005). Plants synthesize significant amounts of smHSPs when exposed to high temperatures (Ahn and Zimmerman, 2006; Charng *et al.* 2006; Volkov *et al.* 2006) drought stress (Sato and Yokoya, 2008; Zahur *et al.* 2009), oxidative stress (Sato and Yokoya, 2008; Neta-Sharir, 2005), cold acclimation (Guo *et al.* 2007), salts (Sun *et al.* 2001), and ABA treatment (Yu *et al.* 2005; Zahur *et al.* 2009; Timperio *et al.* 2008). These results suggest that smHSPs play an important part in plant endurance to abiotic stress. Because each smHSPs gene has a unique sequence with the potential for a special function, the unusual abundance of sHSPs (20–40 kDa) in plants could reflect their functional diversity. The smHSPs cannot refold non-native proteins, but they can bind to partially folded or denatured substrates proteins, preventing irreversible unfolding or wrong protein aggregation (Mogk *et al.* 2003). It has been shown that the smHSPs18.1 isolated from *Pisum sativum*, as well as the sHSPs16.6 from *Synechocystis* sp. PCC6803 under *in vitro* conditions, binds to unfolded proteins and allows further refolding by HSP70/HSP100 complexes (Mogk *et al.* 2003). It was noticed that there was a positive qualitative relation between the accumulation of smHSPs in the plastids and thermotolerance of heat shock (from 28 to 40ºC) in six divergent *Anthophyta* species, including C3, C4, CAM, monocotyledonous and dicotyledonous species. Similar results were obtained separately with four non *Anthophyta* species (Downs and Heckathorn, 1998). Downs and Heckathorn (1998) had indicated that the mitochondrial smHSPs protected NADH: ubiquinone oxidoreductase (complex I) during heat stress in apple fruit of *Pyrus pumila* (P. Mill) K. Koch var. McIntosh. This information might indicate some role of these proteins in adaptation of plants to heat stress. Snyman and Cronje, (2008) had concluded that there were some indications that smHSPs family plays a significant role in the quality membrane control and hence has the potential contribution in the membrane integrity maintenance especially under the conditions of stress. Liming et al (2008) showed that transforming plants with HSP24 from *Trichoderma harzianum* was found to confer significantly higher resistance to heat stress when constitutively expressed in *Saccharomyces cerevisia.* Some studies also support that there is a positive correlation between the heat shock protein level in the cell and respective stress tolerance (Sun *et al.* 2001; Wang *et al.* 2005). Though it is not very much clear that how HSPs confer heat stress tolerance, a recent investigation focused on the in vivo function of thermo protection, governed by sHSPs, is achieved through its assembly into functional stress granules or heat shock granules (Miroshnichenko *et al.* 2005).

Functioning as molecular chaperones *in vitro* and *in vivo*, smHSPs can prevent irreversible protein aggregation and maintain denatured proteins in a folding-competent state under abiotic stress conditions (Lee *et al.* 1997; Low *et al.* 2000; Sun *et al.* 2001; Basha *et al.* 2006; Ahman *et al.* 2007; Mu *et al.* 2011). In the absence of such stresses, however, smHSPs can also be produced specifically in reproductive organs at certain developmental stages, including seed maturation and germination, pollen development, and fruit maturation (Waters *et al.* 1996; Neta-Sharir *et al.* 2005; Volkov *et al.* 2005). Three smHSPs – OsHSP17.0, OsHSP26.7, and OsHSP24.1– are predominantly expressed in the spikes and/or imbibed seed embryos, indicating certain roles for them in pollen development and seed germination (Zou *et al.* 2009). Therefore, many smHSPs can function in either the presence or absence of abiotic stress (Mu *et al.* 2013). For example, AtHSP17.6, an *Arabidopsis* cytoplasmic smHSP, is expressed in heat-shocked leaves but not in untreated control leaves of *Arabidopsis*. However, *Arabidopsis* embryos show high constitutive expression of AtHSP17.6 in meristematic and pro-vascular tissues (Prandl *et al.* 1995).

Research on smHSPs has generally emphasized plant tolerances to various environmental stresses, *e.g.*, high temperature (Malik *et al.* 1999), salt (Hamilton and Heckathorn, 2001), osmotic pressure (Sun *et al.* 2001) and oxidation (Lee *et al.* 2000). Other studies have used genetic modifications to focus on transgenic plants that over-express smHSPs, leading to improved agronomic traits with respect to basal thermotolerance (Parez *et al.* 2009; Sanmiya *et al.* 2004), seed longevity (Prieto-Dapena *et al.* 2006), and tolerances to osmotic stress (Sun *et al.* 2001) and chilling (Guo *et al.* 2007).

5.3. Mechanisms for Heat Tolerance

Plants may survive heat stress through heat-avoidance or heat-tolerance mechanisms (Levitt, 1972). Heat avoidance is the ability of plants to maintain internal temperatures below lethal stress levels, including transpirational cooling, changes in leaf orientation, reflection of solar radiation, leaf shading of tissues that are sensitive to sunburn, and extensive rooting (Bonos and Murphy, 1999). Heat tolerance is the ability of plants to survive high internal tissue temperatures. Heat-tolerant cultivars can be resistant in both humid and arid conditions. Plant tolerance to high temperature may be achieved through various mechanisms, including changes at the molecular, cellular, biochemical, physiological, and whole-plant levels (Sung *et al.* 2003; Wahid *et al.* 2007). Typically, heat tolerant grass species and cultivars exhibit higher activity in the photosynthetic apparatus (Huang *et al.* 1998; Ristic *et al.* 2007) and higher carbon allocation and nitrogen uptake rates (Xu and Huang, 2006; Cui *et al.* 2006) when exposed to supra-optimal temperature. Heat stress was found to induce oxidative stress in grasses so that species and cultivars variations in the activities of antioxidant enzymes were associated with differences in heat tolerance (Dash and Mohanty, 2002; Sairam *et al.* 2000; Liu and Huang, 2000; Jang and Huang, 2001). Major hormones such as cytokinins and ethylene are also found to play regulatory roles in heat tolerance of grasses (Veselov *et al.* 1998; Musatenko *et al.* 2003; Xu and Huang, 2009; Zhang and Erwin, 2008). Heat stress has significant effects on protein metabolism, including degradation of proteins, inhibition of protein accumulation, and induction of certain protein synthesis, depending on the level and duration of heat stress (Monjardino *et al.* 2005; He and Huang,

2007). Moderate heat response involves down regulation of proteins functioning in lipid biogenesis, cytoskeleton structure, sulfate assimilation, amino acid biosynthesis, nuclear transport and antioxidant response (Ferreira *et al.* 2006; Demirevska-Kepova *et al.* 2005). While synthesis of most normal proteins and mRNAs is inhibited in heat stress conditions, the transcription and translation of a small set of proteins, called heat shock proteins (HSPs), may be induced or enhanced when plants are exposed to elevated temperatures (Blumenthal *et al.* 1990; Vierling, 1990; Krishnan *et al.* 1989).

Induction of HSPs due to heat stress

Heat shock, the stressor that has been studied greatly, together with a wide range of other metabolic stressors is known to cause the synthesis of heat responsive proteins (Bösl *et al.* 2006). The heat shock response causes the enhanced expression of heat stress genes, multi-gene families encoding molecular chaperones (Berardini *et al.* 2001; Tompa and Kovacs, 2010; Snyman and Cronje, 2008; Bukau *et al.* 2006; Hartl and Hayer-Hartl, 2009; Morimoto, 2008). Fender and O'Connell, (1990) had showed that heat shock proteins are more expressed *in vivo* and *in vitro* with the increase in temperature for incubation or room temperature respectively. Besides, heat shock proteins expression is reduced at 47°C and ceased at 49°C due to the inability of the cells to survive at higher temperature (Haddadi, 2009). During heat shock, at temperatures above 35°C, in tomato, like other plants, very few proteins other than those induced by heat are detected following *in vivo* labeling (Fender and O'Connell, 1990). They also studied that in addition to the induction of typical HSPs at temperatures approximately 10°C higher than standard growth temperatures, an additional set of heat induced proteins were observed at temperatures 15 to 20°C higher than standard growth temperatures. Persistent heat shock induces persistent heat shock protein synthesis in tomato, until no protein synthesis can be detected. In their analysis of the duration of heat shock protein synthesis in tomato, they investigated the response at two temperatures, 35 and 37°C, the temperatures of induction and of maximal heat shock protein synthesis. Different heat shock proteins have been identified in many crop plants associated with heat stress (Table 1). A more complete identification and comparison of heat responsive proteins in the two leguminous plants contrasting in heat tolerance were achieved through proteomic analysis (Ortiz and Cardemil, 2001). They observed that accumulation of HSP70 is higher in *P. chilensis* than in soybean at 20min of treatment at 40°C as well as after 60 and 90min of heat shock, indicating that there is also a better protection and for a longer time against heat shock stress.

Table 1: Expression of heat shock protein major families in some crop plants responding to heat stress

Major Class of HSP	Crop Species	Responsible function	References
HSP100	Wheat, Tomato, Fescue *Arabidopsis*, Rice	Facilitate reactivation of proteins denatured by heat	Queitsch *et al.*, 2000; Fender and O'Connell, 1990; Campbell *et al.*, 2001; Batra *et al.*, 2007
HSP90	Maize, Populus, Tomato, Wheat	Acts as part of a multi-chaperone machine together with HSP70 and co-operates with a cohort of co-chaperones	Fender and O'Connell, 1990; Zhang and Denlinger, 2010; Zhang *et al.*, 2013; Kumar *et al.*, 2012
HSP70	Whet, Tomato, Creeping soybean	Chaperone function under heat stress, modulating the expression of many downstream genes in signal transduction pathways during stress	Fender and O'Connell, 1990; Duan *et al.*, 2011; Xing *et al.*, 2009
HSP60	Mize, bentgrass, Barley, Rye, Wheat, Creeping grass	Participate in folding and aggregation of many proteins	Zhang *et al.*, 2010; Duan *et al.*, 2011; Xu *et al.*, 2011
sHSP	Barley, Mize, Rice, Wheat, Tomato, Pearl millet, Sorghum	Bind to partially folded or denatured substrates proteins, preventing irreversible unfolding or wrong protein aggregation	Zhang and Denlinger, 2010; Duan *et al.*, 2011; Süle *et al.*, 2004; Lee *et al.*, 2007

Expression of heat shock proteins as well as other proteins is a strategy for adaptation to high temperatures, and HSP induction may be correlated with thermotolerance (Wahid *et al.* 2007). An example can be enhanced expression of HSP70 under heat stress in cells of many plant species (Neumann *et al.* 1993). It is worth adding that expression of some heat shock proteins (*e.g.* HSP70) can be enhanced by abscisic acid (Pareek *et al.* 1998b). Expression of genes inducing heat shock proteins can be an important mechanism of increasing stress tolerance (thermotolerance) (Wahid *et al.* 2007) through getting better photosynthesis and water and nutrient use efficiency (Camejo *et al.* 2005) and cellular membrane stability (Ahn and Zimmerman, 2006) or hydration of cellular structures (Wahid and Close, 2007). In a review, Ahuja *et al.* (2010) have indicated that during drought stress some genes *e.g* ,dehydrin genes (in wheat) and superoxide dismutases (in alfalfa) were induced or upregulated whereas the protein concentration (in black poplar) decreased. The proteins have been involved in glycolysis and gluconeogenesis and proposed as putative biomarkers to define physiological effects at the molecular level and as targets for improving drought resistance in wheat. Systems biology approaches based on plant molecular stress responses reveal the contribution of different signalling pathways defining plant '-omic' architectural responses in relation to changes in environmental stress factors (Ahuja *et al.* 2010; Ashraf and Harris, 2013). To increase plant tolerance to abiotic stresses and maintain a high relative water content, plants may accumulate compounds of low molecular mass such as proline and gibberellins (Kavi Kishor *et al.* 2005; Zlatev and Lidon, 2012), possibly through buffering the cellular redox potential (Wahid and Close, 2007). The accumulation capacity of the compounds protects protein structures as cells dehydrates and is linked to genetic variability of plants for moisture stress tolerance (Monica *et al.* 2007; Zlatev and Lidon, 2012). Some authors indicated that the relationship between turgor and proline accumulation could be a useful drought-injury sensor (Iannucci *et al.* 2000), since during stress the proline level can be as much as 100-fold higher than in normal conditions (Bellinger and Larher, 1987).

5.4. Physiological Function of HSPs

The function of any protein is determined by its formation and folding into three dimensional structure (Levitt *et al.* 1997). Formation of three dimensional structures requires 50% of principle amino acids sequence (Dobson *et al.* 1998). That is where the role of heat shock proteins in the folding of other proteins is important. Morimoto and Santoro (1998) indicated that heat shock proteins protect cells from injury and facilitate recovery and survival after a return to normal growth conditions. On the other hand, Timperio *et al.* (2008) specified that upon heat stress, the role of heat shock proteinsHSP as molecular

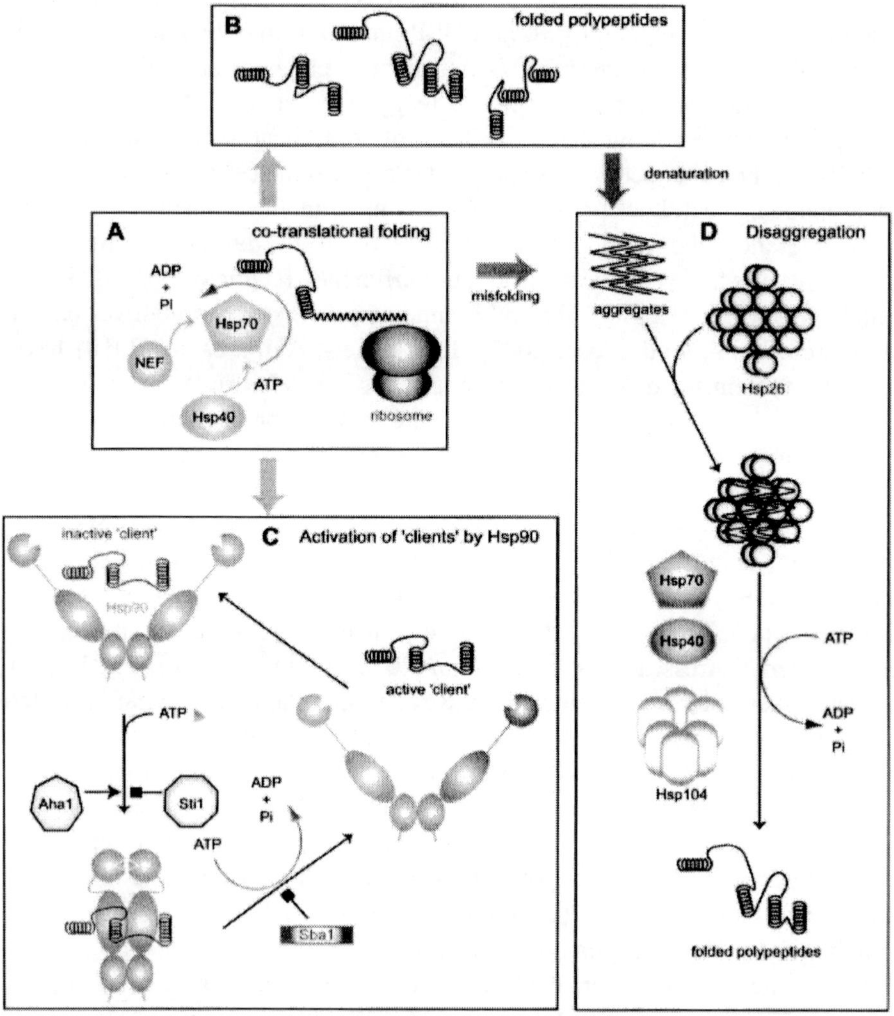

Fig. 1: Simple illustration of part of the chaperone machines that operate in the cytosol: (A) Folding of proteins by HSP70 is co-translational, nucleotide exchange factors (NEFs) and HSP40 facilitate this process. (B) Once protein synthesis is complete. Homologues of HSP70 promote folding in other cellular comparetments. (C) Certain proteins are presented in a largely folded though inactive state, to the HSP90 chaperosome, the ATP-dependent action of which leads to activation of the substrate protein. Co-chaperones act as adaptors between HSP70 and HSP90, with specific co-chaperones acting as inhibitors (e.g. Sti1) or stimulators (e.g. Aha1) of the HSP90 ATPase. (D) Misfolding and cellular stress lead to aberrant protein conformations, which can lead to aggregation. HSP104 catalyses disaggregation, a process facilitated by HSP's 70, 40 and 26. (From Panaretou and Zhai, 2008).

chaperones is without doubt, their function in non-thermal stress could be different: unfolding of proteins is not the main effect and protection from damage could occur in an alternative way apart from ensuring the maintenance of correct protein structure. It has been suggested that HSPs general role is to act as molecular chaperones (Figure 1) regulating the folding and accumulation of proteins as well as localization and degradation in all plants and animal species (Feder and Hofmann, 1999; Schulze-Lefert, 2004; Panaretou and Zhai, 2008; Hu *et al.* 2009; Gupta *et al.* 2010). These proteins, as chaperones, prevent the irreversible aggregation of other proteins and participate in refolding proteins during heat stress conditions (Tripp *et al.* 2009). Each group of these HSPs has a unique mechanism and the role of each is briefed (Al-Whaibi, 2011).

The induction of heat shock proteins is characteristic of an emergency response, *i.e.*, they are extremely rapid and very strong. For example, in *Glycine max* seedlings, heat shock protein mRNAs are observed within 5 minutes of heat shock, and up to 20,000 fold induction of heat shock proteins occurs (28). Secondly, the induction temperature reflects stress conditions for the organism (Table 2).

Tble 2: Heat shock protein induction temperature in organisms

Organism heat shock protein maximum	Growth temperature optimum (^{0}C)	Induction temperature (^{0}C)
I. Archaebacteria		
i. Extreme thermophile *Pyrodicim occultum*	102	108
ii. Prokaryote: *E. coli*	38	42
II. Eukaryotes		
a). Archtic fishes	0	5-10
b). *Drosophila*	25	33-38
c). Birds	25-35	43-45
d). Mammals		45
e). Snow fungus *Fusarium nivale*	12	25
f). Antarctic algae *Plocamium cartilagineum*	0	5
g). Yeast *Saccharomyces cervisiae*	25-28	37
h). Higher plants		
i. *Lolium temulentum*	25-30	35
ii. *Triticum* sp.	20-25	32-40
iii. *Glycine max*	28	40
iv. *Sorghum bicolor*	35	43-45
v. Maize	35	43-45
vi. *Pennisetum glaucum*	35	45
vii. Tomato	25	35-37
viii. Cotton	30	40
ix. *Arabidopsis thaliana*	22	35

The plants species adapted to temperate environment (soybean, maize, pea, and wheat) begin to synthesize heat shock proteins when tissue temperature exceeds 32-33 ^{0}C (Vierling, 1991). Thus heat shock protein-inducing temperature of these organisms reflects the thermal characteristics of the environment in which these organisms are growing. Under field conditions, soil water deficit enhanced midday canopy temperature of 40 ^{0}C induced heat shock proteins in cotton (Bruke *et al.* 1985). Even in irrigated wheat, heat shock proteins were expressed in field condition when the flag leaf temperature reached 32-35 ^{0}C and heat shock proteins expression was correlated with the thermo tolerance (Nguyen *et al.* 1994). Hence heat shock proteins are expressed in these organisms, when their temperature increases above the normal, rather than at a universal temperature threshold. The kinetics of heat shock protein induction and acquired thermo tolerance are tightly coupled and highly conserved among the evolutionary diverse organisms. Moreover, heat shock proteins show highly similarity in nucleotide and amino acid sequences among eukaryotes, and in some cases also with prokaryotes. These evolutionary conservations clearly suggest the importance of heat shock proteins in heat stress tolerance of organisms (Viswanathan, Khanna and Chopra, 1996).

Heat shock proteins may bind with proteins that are sensitive to heat in a cell under heat shock to protect them from degradation and as well preventing any havoc following precipitation and permanently influencing the viability of the cells (Young, 1990; Gilliham *et al.* 2011). Studies on protein purification and function, molecular cloning both *in vitro* and *in vivo* have figure-out clearly the position of HSPs in the sorting of proteins, folding as well as assembly of proteins. As a result HSPs have been called 'molecular chaperones'. It is observed that HSPs prevents the cells from the damage effect caused by high temperature (Tang *et al.* 2005). Function of any protein is determined by its formation and folding into three dimensional structures (Levitt *et al.* 1997). Formation of three dimensional structures requires 50% of principle amino acids sequence (Nakamoto and Vigh, 2007). That is why the role of HSPs in the folding of other proteins is important. Morimoto and Santoro (1998) had indicated that HSPs protect cells from injury and facilitate recovery and survival after a return to normal growth conditions. It has been observed that upon heat stress, the role of HSPs as molecular chaperones is without doubt, their function in non-thermal stress could be different: unfolding of proteins is not the main effect and protection from damage could occur in an alternative way apart from ensuring the maintenance of correct protein structure. It has been concluded that the role of HSPs generally is acting as molecular chaperones by controlling the folding/unfolding and protein accumulation as well as localization and breakdown in both plant and animal species (Gupta *et al.* 2010; Guy and Li, 1998; Hu *et al.* 2009; Panaretou and Zhai, 2008; Schulze-Lefert, 2004). They prevent the irreversible accumulation of other proteins and are involve in the refolding of proteins under heat stress conditions (Tripp *et al.* 2009). Several evidences now are accumulating showing that HSPs are required to get native structures of monomeric and oligomeric proteins after they have been produced on the ribosomes or after been moved across the membranes (Haddadi, 2009; Miler and Qureshi, 1992). Cells having been subjected to a non-harmful heat shock develop an increased resistance to prolong heat stress subsequently (Åkerfelt *et al.* 2010; Wahid and Close, 2007). The expression of HSPs group has a relationship with tolerance to heat. They observed that when cell are exposed to heat, proteins synthesis ceased rapidly and it takes five to six hours to recover. On the contrary, Calderwood (2010) indicated that after six to eight hours the development of thermotolerance and induction of protein are completed. Finally, heat tolerance reaches its highest level and the synthesis of heat shock protein is returned to normal.

During photosynthesis, reactive oxygen species (ROS) generation occurs via electron transport reactions in the chloroplasts, such as the Mehler reaction, which generates superoxide anion that is converted to H_2O_2. ROS are also produced in the chloroplasts through the photoreduction of the herbicide methyl viologen (MV), which is a superoxide anion propagator. The relatively low reactivity of H_2O_2 suggests that it could diffuse from the chloroplast to initiate signaling events. Alternatively, the accumulation of H_2O_2 could be limited to chloroplasts initiating distinct signaling cascade (Mullineaux *et al.*

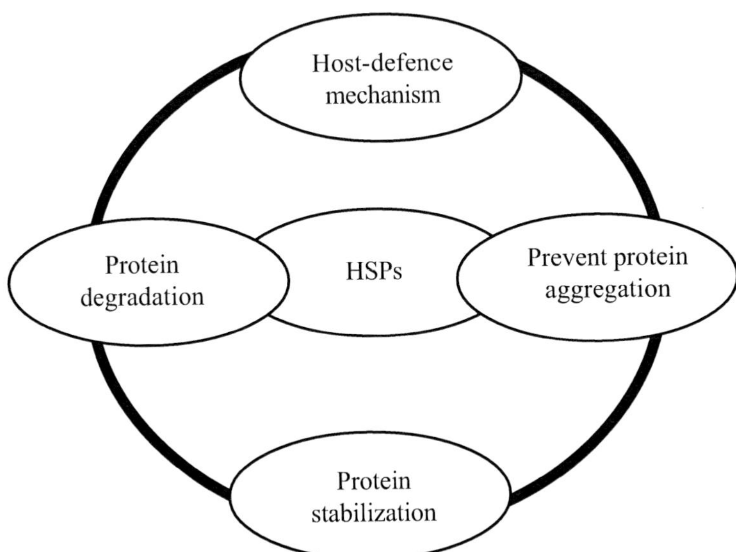

Fig. 2: Function of heat shock protein under heat stress

2006). Environmental stresses like high light intensity, drought, extreme temperatures, heavy metals and UV radiations all enhance photosynthetic ROS generation. In the case of heat shock, there is emerging evidence that there is a cross-talk between heat and oxidative stress signaling. A burst of H_2O_2 was reported to occur after very short periods at high temperature, apparently as a result of NADPH oxidase activity (Desikan *et al.* 2004). This burst has been correlated with the induction of heat responsive genes, a process assumed to be mediated through direct sensing of H_2O_2 by heat shock transcription factors (Hsfs) (Scarpeci *et al.* 2008; Miller and Mittler, 2006).

Recent studies have revealed that several HSPs are redox regulated. The best studied is the prokaryotic HSP33, which is induced by oxidative stress. Under reducing conditions HSP33 is present in the cytosol as an inactive monomer. Oxidative stress will cause the intracellular redox environment to become more electropositive, favouring HSP33 activation by dimerization through the formation of two intramolecular disulfide bonds. In plants, little is known about the redox regulation of HSPs. The deduced protein sequences of HSPs genes analyzed here are quite different in Cys contents. Thus, the cytosolic HSP70 contains 7 Cys residues while they are absent in HSP17.6 B-CI, suggesting that only HSP70 could suffer redox regulation. Further analysis of HSP17.6B-CI subcellular location using Target P prediction showed that it contains a putative chloroplastic transit peptide of 33 residues with a 0.479 score. Another sHSP, the tomato chloroplast HSP21, protects PSII from temperature-dependent oxidative stress. It is important to note that the expression of these two HSP genes (At3g12580 and At2g29500) were not affected by apoplastic H_2O_2 or H_2O_2 produced by the plasma membrane, indicating that chloroplastic ROS contribute to the signaling cascade of these genes. The redox regulation of HSPs is matter of current research.

Response of heat shock proteins to heat stress in plants

The Heat Shock Response in plants and mammals is regulated by a set of highly conserved proteins known as heat shock proteins, and expression of this protein is governed by heat shock factors (Snyman and Cronje, 2008). Interestingly, the number of heat shock factors in plants far exceeds those found in mammalian cells, most probably to equip plants with the ability to withstand various forms of external stressors at any time during their lifespan. There are 15 known heat shock factors in *Arabidopsis thaliana* (Nover *et al.* 2001) and >21 in *Solanum lycopersicum* (formerly *Lycopersicon esculentum*), and all are thought to play key roles in stress responses (Scharf *et al.* 1998). In tomato, only two of these heat shock factors, HSFA2 and HSFB1, are heat inducible (Scharf *et al.* 1990), but their expression is controlled by HSFA1, referred to as the master regulator of the heat shock response (Mishra *et al.* 2002). HSFA2 is seen as the "work horse" of the stress response and becomes the most dominant heat shock factors during a heat stress (Mishra *et al.* 2002). When cells are exposed to various stress conditions, such as heat stress, the heat shock factors that reside in the cytosol dissociate from the heat shock protein (*e.g.* HSP70), are activated, and undergo trimerization. These heat shock factor trimers are phosphorylated and translocated to the nucleus where they bind to the heat shock element, which is located in the promoter region on the heat shock protein genes. Within the diverse heat shock protein family, HSP70 is the most widely studied member, and a highly conserved 70 kDa protein that plays a key role in the stress response in plants as it does in mammals. Heat shock proteins are hypothesized to prevent and/or repair stress induced damage (Lindquist and Craig, 1988). Exposure to elevated temperatures leads to activation of the cellular heat shock response, which is conserved throughout all kingdoms of life. The heat shock response causes the enhanced expression of heat stress genes, multigene families encoding molecular chaperones (Tompa and Kovacs, 2010; Richter *et al.* 2010; Bösl *et al.* 2006; Bukau *et al.* 2006; Hartl and Hayer-Hartl, 2009; Morimoto, 2008). Among these, HSP70 is one of the most abundant heat shock proteins in eukaryotic cells. HSP70 binds in an ATP-dependent manner to hydrophobic patches of partially unfolded proteins and prevent protein aggregation (Mayer and Bukau, 2005). While HSP70s accumulate during heat shock, their constitutively expressed cognates (HSC70) are essential for general cellular functions due to their involvement in the control of protein homeostasis. They assist the folding of nascent polypeptides released from the ribosome (Hartl, and Hayer-Hartl, 2002), sorting of proteins to cell organelles by interaction with mitochondrial and chloroplast protein import complexes (Mirus and Schleiff, 2009; Zhang and Glaser, 2002), and form a link to the ubiquitin-mediated proteasomal degradation pathway (Ballinge *et al.* 1991; Lüders *et al.* 2000).

Correlation between heat shock protein expression and thermotolerance

Acquired thermo tolerance is correlated with heat shock protein synthesis in many organisms, including higher plants (Nover *et al.* 1989; Vierling, 1991; Lindquist, 1986). Using etiolated soybean seedlings Lin *et al.* (1984) have shown that the rate of synthesis of the low molecular weight heat shock proteins is correlated with acquired thermo tolerance. Soybean seedlings are able acquire thermo tolerance by a pre-treatment of 2 hours at 40 °C or 10 minutes at 45 °C followed by 2 hours at 25 °C. Thus, the pre-heat shock treatment which induced heat shock protein synthesis resulted in the tolerance of seedlings at 45 °C, 2 hour heat shock. The 40 °C pre-heat shock not only induced the heat shock protein synthesis in soybean seedlings, but also resulted in localization and stable association of heat shock proteins with cell organelle fractions (nuclei, mitochondria and ribosomes). Similarly in other crops like wheat (Abernathy *et al.* 1989; Krishnan *et al.* 1989; Blumenthal *et al.* 1990; McElwain and Spiker, 1992), sorghum, pearl millet (Howarth, 1990; Howrath and Skot, 1994), and maize (Ristic *et al.*1991) the seedling thermo tolerance and otherwise lethal temperature is correlated with the kinetics of heat shock protein synthesis. These studies strongly suggest that the accumulation of heat shock proteins is important for protection from thermal killing. If so, quantity or the quality of heat shock protein is important for providing thermo tolerance. *Triticum monococcum* L. cultivars, M_3 and M_9, which differ in thermo tolerance, did not differ in the quality of heat shock protein induced at 37 ^{0}C. But Northern analysis using heat shock protein cDNAs as probes revealed that the tolerant genotypes M_3 was able to accumulate higher steady state mRNA levels of 16.9, 26 and 70 kDa heat shock proteins than the susceptible M_9 during heat hardening (Vierling and Nguyen, 1992).

In *Triticum aestivum*, the heat tolerant variety Mustang maintained its cell viability up to 80% when pre-heat shocked (at 34^0 C) wheat leaves were exposed to 50^0 C for 1 hour, while the susceptible genotypes Sturdy could maintain only 40% cell viability. Mustang also maintained its capacity to synthesize a small subunit of Rubisco at 34^0 C, while Sturdy could not. Two dimensional gel electrophoresis analysis revealed the presence of three unique heat shock proteins (16, 17 and 26 kDa) in Mustang (Krishnan *et al.* 1989). In case of maize also, the drought and heat tolerant genotype ZPBL1304 synthesized an unique 42 kDa heat shock protein, which was absent in susceptible line ZPL389 (Ristic *et al.* 1991). Thus not only are the heat shock protein synthesis and acquired thermo tolerance tightly coupled, but also the interspecific differences in quantity, quality and the rate of accumulation of heat shock proteins are highly correlated with thermo tolerance (Viswanathan and Khanna-Chopra, 1996).

Function of Plant HSFs in Other Abiotic Stress Responses

Under natural conditions, plants frequently suffer from various abiotic stresses simultaneously; HS is compounded by additional abiotic stresses such as drought and salt stress (Bita and Gerats, 2013). The response of plant cells encountering a single stress condition can not reflect the real conditions in the field (Nishizawa *et al.* 2006). Gene manipulation of *HSFs* in plants is a significant approach to ameliorate the effects of combined HS and other abiotic stresses. Characterization of the functional HSFs involved in various abiotic stresses is necessary. The *Arabidopsis* HSFA1s are involved in response and tolerance to salt, osmotic, and oxidative stresses during seedling establishment (Liu *et al.*

2011). Especially, *Arabidopsis* HSFA1b controls a developmental component to drought tolerance and water productivity, however, the effect of *HSFA1b* over-expression on drought/dehydration tolerance does not involve changes in the expression of DREB2A or many other ABA- or dehydration-responsive genes (Bechtold *et al.* 2013). Given that *Arabidopsis* HSFA3 is regulated by DREB2A as part of drought stress signaling pathway (Scharf *et al.* 2012), it is tempting to speculate that *Arabidopsis* HSFA1b and *A3* involve in different signal pathways to enhance the tolerance to drought stress. In addition, over-expression of chickpea CarHSFB2 in *Arabidopsis* can increase the transcript levels of some stress-responsive genes (RD22, RD26, and RD29A) at seedling stage under drought stress conditions, thus improving their drought-tolerance (Ma *et al.* 2016), co-overexpression of sunflower *HaHSFA4a* and *A9* in transgenic tobacco results in synergistic effects on seedling tolerance to severe dehydration and oxidative stress (Personat *et al.* 2014). As the dominant *HSF* in thermotolerant cells, HSFA2 also enhances tolerance to various other abiotic stresses, including salt/osmotic stress (Ogawa *et al.* 2007; Yokotani *et al.* 2008), anoxia stress (Banti *et al.* 2010), and combined high-light (HL) and HS stresses (Nishizawa *et al.* 2006). Unlike the above active regulation factors, tomato SlHSFA3 and *V. pseudoreticulata* VpHSF1 play negative roles in salt and osmotic stress, respectively (Li *et al.* 2013; Peng *et al.* 2013). These results suggest that the complex family of plant HSFs presents a functional diversity under different abiotic stress conditions.

5.5. Cellular Localization of HSPs: Heat Shock Dependent

Heat shock proteins are located in both the cytoplasm and organelles, such as the nucleus, mitochondria, chloroplasts and endoplasmatic reticulum. The positive correlation between acquisition of thermo tolerance and heat shock proteins appears to be depend not only upon synthesis of heat shock proteins but also on their cellular localization (Lindquist, 1986; Velanquez and Lindquist, 1984; Lin *et al.* 1985). In soybean, heat hardening, which induced seedling thermo tolerance, also induced synthesis and selective localization of heat shock proteins. Cell fractionation studies revealed that low molecular weight 15-18 kDa heat shock proteins selectively localized and associated with nuclei, mitochondria and ribosomes, a lesser amount of 68-70 and 90 KDa heat shock proteins localized in these organelles. While some 22-24 kDa heat shock proteins remained soluble in cytosol, they remained organelle associated during a chase at 40^0 C, but dissociated gradually during a chase at 28^0 C. If again 10 minutes heat shock at 45^0 C was given, the localization occurred within 15 minutes. Arsenite-induced heat shock proteins did not localize at 28^0 C, but they became organnele associated during subsequent heat stress (Lin *et al.* 1984).

Similarly pea heat shock protein 22 was strongly associated with chloroplast thylakoids when the temperature was raised above 38^0 C, at high light intensities (Kloppstech *et al.* 1985; Glaczinski and Kloppstech, 1988; Chen *et al.* 1990). Studies conducted with *Chlamydomonas* showed association of low molecular weight heat shock proteins wirg PS II which protected the PS II from photoinhibition (Schuster *et al.* 1988). In higher plants too, the localization and association of heat shock proteins with organelles demonstrated the protective role of heat shock proteins during heat stress. In soybean seedlings, at 38^0 C, all heat shock proteins were synthesized, but their organelle localization occurred at 42.5^0 C. The 15-18 kDa heat shock protein and 70 kDa heat shock heat shock proteins were associated with mitochondria at 42.5^0 C and dissociated during 4 hour recovery period at 20^0 C. This association protected the mitochondrial phosphorylation at non-permissive temperature (Chou *et al.* 1989).

Genomic response

In terms of gene expression, the response of plant to heat stress was also studied by analysis of Arabidopsis transcriptomes (Schramm *et al.* 2006). Some genes appeared up-regulated. Heat shock provokes a rapid reprogramming of gene expression to favor translation of heat shock proteins in which translation factors could play a role. In general, the expression of heat shock proteins and its factors, heat shock factors, was induced largely by heat and some other stresses. Presence of heat shock proteins in higher plants was discovered by some researchers in tobacco and soybean using cell culture technique. When soybean was subjected to 40ºC for four hours, ten new proteins were found, but disappeared after 3 h treatment at 28 ºC (Ma *et al.* 2006). Studying the gene expression of HSP90 in rice plant (*Oryza sativa*) indicated that the HSP87 was present after 2 hours of heat shock (from 28 to 45 ºC), and its quantity was high and stable even after long heat stress (4 hours) and then return to normal conditions *i.e* condition before stress. According to (Fender and O'Connell, 1990) during heat shock at temperatures above 35 ºC, in tomato, like other plants, very few proteins other than those induced by heat are detected following *in vivo* labeling. The transcription levels of all 9 OsHSP genes studied were significantly enhanced under heat stress, which showed they may play roles in heat stress (Ye *et al.* 2012). Expression profiles of PtHSP90 genes obtained by RNAseq was carried out by (Zhang *et al.* 2013), using qRT-PCR of seven selected PtHSP90 genes on three different tissues of *Populus* plant under heat stress.

The transcription of Heat shock protein encoding genes is controlled by regulatory proteins called heat stress transcription factors (Hsfs), which exist as inactive proteins mostly found in the cytoplasm (Beniwal *et al.* 2004). Thus, the heat stress response is controlled by HSFs, which are activated by the presence of heat shock protein genes, acting by binding to the highly conserved heat shock elements in the promoters of target genes (HSPs), activating expression of them (Flowers and Yeo, 1995). Heat shock transcription factors and the promoter heat shock elements are among the most highly conserved transcriptional regulatory elements in nature (Hayhn *et al.* 2004). In addition to mediating a relatively large part of the defense response of eukaryotes to heat stress, HSFs are also thought to be involved in different pathological conditions, cellular responses to oxidative stress, heavy metals, amino acid analogs and metabolic inhibitors, and certain developmental and differentiation processes (Miller and Mittler, 2006; Hahn *et al.* 2004; Scharf *et al.* 2012).

Several studies have been reported about the role of HSFs in heat stress. In the tomato, HSFA2 was up-regulated early in development (Giomo *et al.* 2010). HSFA2 also improved heat and osmotic tolerance in wild *Arabidopsis* (*Hu et al.* 2009). Also in *Arabidopsis*, HSFs are induced by all major abiotic stresses: heat, cold, osmosis and salt (Swindell *et al.* 2007). There are reports about HSFs regulating other HSFs as well. HSFA1d and HSFA1e are key regulators of HSFA2 in *Arabidopsis* under heat and high light stress (Nishizawa *et al.* 2011). Also, attempts to increase thermotolerance by overexpression of a single Hsf or heat shock protein gene have had limited impact because of the genetic complexity of the heat stress response (Vinocur and Altman, 2005). Hahn *et al.* (2011) found a versatile regulatory mechanism in the tomato, where HSP70 and HSP90, together regulate different HSFs. Some reports also showed a role for HSFs under other abiotic stresses. HSFA2 enhances the anoxia tolerance in wild *Arabidopsis*, besides the heat tolerance (Banti *et al.* 2010). HSFA2 also is induced under salt and drought stress, as reported in rice (Chauhan *et al.* 2011).The main studies on plant genetic transformation with heat shock protein genes have investigated mostly heat stress and thermo-tolerance. Positive correlations between the expression levels of several heat shock proteins and stress tolerance have been described extensively by functional genomics and proteomics in different plant species. Comparison of expression data under variable conditions, for example from different tissue types, developmental stages, growth conditions, or applications and durations of stress treatments, shows related patterns of transcript accumulation, with the expression of about 2% of the genome being affected (Vinocur and Altman, 2005; Kotak *et al.* 2007). In the tomato, for instance, several HSPs are induced by heat stress (Frank *et al.* 2009). Nine heat shock proteins were reported as up-regulated in rice under heat stress (Ye *et al.* 2012). In *Porphyra seriata*, PsHSP70 enhanced heat stress tolerance (Park *et al.* 2012). Sung and Guy (2003) also related an altered expression in these proteins in *Arabidopsis*. They noticed an increased expression of some HSP70 under heat stress. Overexpression of HSP101 from *Arabidopsis* in rice plants, that are sensitive to heat stress, resulted in a significant improvement of growth performance during their recovery (Katiyar-Agarwal *et al.* 2003). sHSPs were also up-regulated under heat treatment in *Arabidopsis (Hu et al. 2009)*. HSP17-CII is activated early in the development under heat stress (Giomo *et al.* 2010). HSP90 was greatly induced by heat stress, but not by other abiotic stresses (Hu *et al.* 2009). In a proteomic analysis, Neilson *et al.* (2010) found various HSPs up-regulated by heat stress in many plant species.

Proteomic response

When plants are exposed to high temperatures they synthesize both high molecular mass heat shock proteins (from 60 to 110 kDa) and small heat shock proteins (from 15 to 45 kDa) (Memyk, 1997; Renaut *et al.* 2006). A total of 48 differentially expressed proteins in samples taken after 12 or 24 h of heat exposure compared to controls were revealed, 18 of which were heat shock proteins (HSP70, dnak-type molecular chaperone BiP, HSP100, CPN60, small heat shock proteins). Studies carried out by Polenta *et al.* (2007) regarding isolation and characterization of relevant heat shock proteins from plant tissues induced in tomato pericarp by thermal treatment, highlighted the presence of two major (HSPC1, HSPC2) and two minor (HSPC3 and HSPC4) class I smHSPs, identified and characterized by using mono specific polyclonal antiserum and MS/MS analysis of tryptic peptides. In this case the authors claimed that smHSPs accumulation is an appropriate parameter for monitoring treatment intensities and consequently for preventing chilling injury. According to Lee *et al.* (2007), they identified a total of 18 proteins as HSPs in the rice leaves studied under heat stress. It is interesting that all of these heat shock proteins were identified from Polyethylene Glycol (PEG) pellet fraction samples, which suggest that PEG pellet fractions may be used as a potential fractionation tool in the analysis of multimeric proteins.

Tolerant plant response

Heat shock proteins are known to be induced not only in response to short-term stress, but their production is a necessary step in plant heat acclimation. It is important to note, in fact, the early production of HSP101 Chaperone by plants exposed to heat stress, but there are also many other pathways involved in adaptation to higher temperatures. The studies of HSPs expression made by (Larkindale and Vierling, 2008) demonstrated that a better acclimation occurs when plants are gradually exposed to increasing temperatures, without changing parameters suddenly. They also examined clusters of genes whose expression increased during heat stress; in particular, cluster 45 contains 18 HSPs (HSP101, 14 small HSPs and 3 HSP70). Variation in heat shock protein production among closely related species from thermally contrasting habitats has been observed even when these species are grown and heat stressed under identical conditions (Efeoðlu, 2009). This variation in heat shock protein production is often correlated with organismal thermotolerance (Downs *et al.* 1998). There are many studies showing quantitative variations for their tolerance degrees among genotypes. The acquisition of thermotolerance appears to depend on not only upon the synthesis of heat shock proteins but also on their selective cellular localization. Krishnan *et al.* (1989) reported that in the soybean seedlings, several heat shock proteins become selectively localized in or associated with nuclei, mitochondria and ribosomes at 40 °C, they were absent in other parts. The selective localization of heat shock proteins is temperature dependent. Genotypic variations have been observed for thermo-sensitivity in studies up to date. Heat shock sensitive genotypes synthesized heat shock proteins earlier than tolerant genotypes (Finka *et al.* 2012). Especially very early detected (in a few minutes in heat shock) specific heat shock proteins transcripts might have effect on order of heat shock protein gene expression and cause different tolerance levels among species (Weng and Nguyen, 1992.). Tolerance to heat shock cannot be controlled by only one gene. The heat shock tolerance is determined by different gene sets in different stages of life cycle and in different tissues. HSP70 was shown to be causally involved in the capacity to acquire thermotolerance in Arabidopsis and HSP27 in another thermo-sensitive Arabidopsis mutant. HSP17 was identified as a factor of acquired thermo tolerance in the study of transgenic cells (Maestri *et al.* 2002).

5.6. Identification and Characterization of HSPs Associated with Heat Tolerance in Grasses

The presence and role of heat shock proteins in heat tolerance has been examined in various annual grasses cultivated as cereal crops, most of which belong to the genera of rice (*Oryza* sp.), wheat (*Triticum* sp.), maize (*Zea* sp.), sorghum (*Sorghum* sp.), rye (*Secale* sp.), barley (*Hordeum* sp.), and oat (*Avena* sp.). The involvement of heat shock proteins in thermal tolerance has been studied in only a few perennial species such as creeping bentgrass (*Agrostis stolonifera*), fescues (*Festuca* sp.), and orchard grass (*Dactylis glomerata*). Table 3 summarizes the heat shock proteins reported in the grass family and their tissue specificity, which may play a crucial role in defending each type of tissues against heat stress (Ahsan *et al.* 2010).

Table 3: Tissue-specific expression of five families of HSPs in cereal and forage and turf grasses responding to heat stress.

HSP Family	Tissue	Cereal species
HSP100	Leaf	Wheat (Cambell *et al.* 2001)
	Root	Wheat (Cambell *et al.* 2001)
	Seed	Maize, Rice, and Wheat (Singla *et al.* 1998)
HSP90	Seed	Maize (Cooper *et al.* 1984)
HSP70	Leaf	Wheat (Necchi *et al.* 1987)
	Root	Maize (Cooper *et al.* 1984), Wheat (Necchi *et al.* 1987)
	Seed	Wheat (Maestri *et al.* 2002)
HSP60	Leaf	Maize (Cooper *et al.* 1984)
	Root	Barley, Rye, and Wheat (Necchi *et al.* 1987), Maize (Cooper *et al.* 1984)
	Seed	Maize (Cooper *et al.* 1984)
sHSP	Leaf	Barley (Sule *et al.* 2004), Maize (Jorgensen *et al.* 1992), Rice (Lee *et* al 2007), Wheat (Necchi *et al.* 1987)
	Root	Maize (Cooper *et al.* 1984), Wheat (Necchi *et al.* 1987)
	Seed	Pear millet and Sorghum (Howarth, 1989), Wheat (Maestri *et al.* 2002; Majoul *et al.* 2004)

After Xu et al. 2011

HSPs Identified in perennial species cultivated as forage or turf grasses

Park *et al.* (1996) first detected HSPs (97, 83, 70, 40, 25, and 18 kDa) in heat-tolerant and non-tolerant variants of creeping bentgrass, a major cool-season turf species. They also found the heat-tolerant variants synthesized two to three additional sHSP (25 kDa). Zhang *et al.* (2005) cloned four classes of HSPs (HSP100, HSP90, HSP70, and sHSPs) that are differentially expressed under heat stress between the two genotypes of fescues, which are widely used as both forage and turf grasses. Cha *et al.* (2009) characterized an endoplasmic reticulum-resident HSP90 gene from orchard grass, whose expression increased during heat stress. This protein functions as a molecular chaperone by preventing thermal aggregation of malate dehydrogenase and citrate synthase. Heat acclimation has been found to induce heat shock proteins in various plant species (Vierling, 1991), In a study (He *et al.* 2005) examining the effects of heat acclimation (gradual temperature increase) and sudden heat stress (direct temperature increase) on protein synthesis and degradation in a heat-sensitive creeping bentgrass cultivar "Penncross", it was found that both heat treatments led to the accumulation of several HSPs (23, 36, and 66 kDa); in addition, heat acclimation induced a few extra cytoplasmic HSPs (57 and 54 kDa), which were not present in the unacclimated plants under heat stress. These results suggest that up-regulation of HSPs, primarily sHSP, HSP60 or HSP70 based on their molecular weights, is a typical response of perennial grasses to heat stress. Especially, due to the fact that heat acclimation improved heat tolerance of the plants as manifested by lower electrolyte leakage in the leaves of heat-acclimated plants, induction of the two HSP60 proteins during heat acclimation could be related to enhanced thermotolerance in perennial grasses.

It is known that there exists a positive correlation between cytokinin content and heat tolerance in creeping bentgrass (Xu and Huang, 2007), and exogenous application of cytokinins improves heat tolerance (Liu and Huang, 2002; Liu *et al.* 2002). Veerasamy *et al.* (2007) further investigated the effects of exogenous applied zeatin riboside (ZR), a synthetic cytokinin, on protein metabolism associated with heat tolerance in "Penncross". Improved heat tolerance of ZR-treated plants were manifested by less heat induced degradation of ribulose-1,5-bisphosphate carboxylase proteins and lower protease activity than untreated plants. Particularly, the expression levels of a few heat shock proteins (32 and 57 kDa) were upregulated in ZR-treated plants under heat stress. These results suggest that some sHSP and HSP60 proteins are among the primary targets in cytokinin regulation of heat tolerance in cool-season perennial grass species.

In order to better understand the roles of heat shock proteins in heat tolerance, a unique C3 perennial grass species, rough bentgrass (*Agrostis scabra*) identified in Yellowstone National Park, has been investigated. The thermal *A. scabra* grows actively in the chronically hot soils (Tercek *et al.* 2003), which may have adopted both heat avoidance and tolerance strategies (Stout and Al-Niemi, 2002). The physiological traits associated with superior thermotolerance of the species were described in a few recent publications from our lab. This species could maintain the canopy photosynthesis and respiration rates responding to short-term soil temperature elevation (Lyons *et al.* 2007). Its roots tolerate high soil temperature by holding high proportion of alternative respiration (Rachmilevitch *et al.* 2007), low maintenance and ion uptake costs (Rachmilevitch *et al.* 2006a), as well as efficient expenditure and adjustment of carbon and nitrogen allocation patterns between growth and respiration (Rachmilevitch *et al.* 2006b)..

Heat-induced changes in one-dimensional protein profiles of thermal *A. scabra* were compared to those of *A. stolonifera*. In the shoots, significant protein degradation was observed at 30–45^{0}C in "Penncross" and a new heat-tolerant cultivar of creeping bentgrass "L93", but not until 40–45^{0}C in *A. scabra.* Meanwhile, expression of HSPs (23, 32, 36, and 66 kDa) was induced or enhanced at 35–45^{0}C in "L-93" and *A. scabra,* but only at 40–45^{0}C in "Penncross". Moreover, stronger expression of HSP60 and HSP70 proteins in the shoots of *A. scabra* or "L-93" than "Penncross" at 35–45°C of 3 d was revealed by immune-blotting. In the roots, heat-induced degradation of proteins including HSPs was mitigated in the thermal species, especially at the extreme temperature (45°C). Immuno-blotting detected induction of HSP60 and multiple sHSP (Class I) proteins at elevated temperatures in both species, but the induction in *A. stolonifera* was triggered later under heat stress and/or by higher temperature compared to the thermal species; HSP70 was constitutively produced during heat-shock treatment (2 and 4 h) but prolonged heat stress increased its expression level (24 and 28 h) (Huang, unpublished data). The results from both shoots and roots indicate a correlation between early induction of major heat shock proteins as well as maintenance of them under elevated temperature and better heat tolerance of cool season perennial grasses (Xu *et al.* 2011).

A more complete identification and comparison of heat responsive proteins in the two *Agrostis* grass species contrasting in heat tolerance were achieved through proteomic analysis. Among the hundreds of proteins identified in the leaves is an HSC70, the abundance of which decreased under heat stress in both species (Xu and Huang, 2010). However, the degradation ceased at 2 d in *A. scabra* but continued to 10 d in *A. stolonifera*. It suggests that maintaining production of constitutively expressed heat shock proteins such as HSC70 is important for sustaining grass plant growth under heat stress. In the roots, proteomic analysis revealed the increase of an heat shock protein Sti (stress-inducible protein) in both species under heat stress, which contains two heat shock protein binding motif, three tetratricopeptide repeat and two Sti1 domains (Xu and Huang, 2008). Sti proteins are involved in HSP90 signaling and interaction (Flom *et al.* 2006). As heat-induced accumulation of this protein was earlier and greater in the thermal species compared to heat sensitive *A. stolonifera*, it indicated that up regulation of HSP90-related proteins such as Sti may contribute to whole plant thermo tolerance in perennial grasses. The involvement of heat shock proteins in heat tolerance was also determined at the gene level. A suppression subtractive hybridization (SSH) library was constructed by Tian *et al.* (2009) to identify heat-responsive genes for thermal *A. scabra*. In this study, genes of an HSP20-like chaperone and an HSP70 were isolated. Expression of the HSP70 gene was constitutively expressed under optimum temperature but strongly up regulated under heat stress in both shoots and roots. The HSP20-like chaperone is highly homologous to an HSP20-like chaperone from clover (*Medicago truncatula*) that contains the p23 domain. As p23 is one of the cochaperones of HSP90 and stabilizes the HSP90 heterocomplex (McLaughlin *et al.* 2006; Cha *et al*, 2009), enhanced expression of this chaperone gene under heat stress also indicated that up regulation of

HSP90-related proteins are important for heat tolerance in perennial grasses, as discussed above in the proteomic study. In another study, using the sequence of the *HSP70* gene isolated by SSH in *A. scabra* and the reported sequence of a sHSP (*HSP16*) gene in *A. stolonifera*, the expression levels of the two genes were compared between heat-sensitive *A. stolonifera* and thermal *A. scabra*. The expression of *HSP16* was highly induced in both species at 45^{0}C after 24 hours, but the induction was more substantial in the thermal species, whereas, *HSP70* gene was constitutively expressed at optimum temperature but the expression was slightly up regulated at elevated temperatures in both species. The response of heat shock protein gene expression to increasing temperature is in accordance with the response of heat shock protein in abundance to elevated temperature, confirming the direct association of heat shock proteins with heat tolerance in perennial grasses.

Heat shock proteins identified in annual species cultivated as cereal crops

Expression of HSPs in cereal species was first revealed in some early works in 1980s. In a study (Necchi *et al.* 1987) examining heat shock protein metabolism in seedlings of five cereal species (common, drurm wheat, barley, rye, and triticale) responding to heat shock at 40^{0}C, inductions of 13 heat shock proteins (14-15, 35–69, 83–99 kDa) were detected. It was also reported that distinct levels of acquired thermal tolerance between wheat varieties were associated with significant quantitative differences in the synthesis of multiple HSPs (16, 17, 22, 26, 33, and 42 kDa) (Krishnan *et al.* 1989).

More thorough characterization of heat-responsive proteins including HSPs benefits from successful application of proteomic-based techniques, particularly two-dimensional gel electrophoresis coupled with mass spectrometry. Lee *et al.* (2007) identified 18 heat shock proteins in a study investigating rice leaf proteome in response to heat stress, including seven HSP70s, three HSP100s, one HSP60, and seven newly induced or highly up regulated sHSPs. Majoul *et al.* (2004) detected up regulation of five sHSPs in a study analyzing the effect of heat stress on hexaploid wheat grain proteome. Using a novel hybrid mass spectrometer (an electrospray ionization-quadrupole linear ion trap (Q-TRAP) combined with nano-HPLC), Sule *et al.* (2004) were able to distinguish six isoforms of a 16.9 kDa sHSP in a proteomic study of barley heat response.

Since each heat shock protein family generally shares high-sequence similarity across diverse cereal species, the anti-HSP antibodies could exhibit relatively broad cross-species activities. Pareek *et al.* (1995) purified and raised highly specific polyclonal antisera against two rice heat shock proteins (104 and 90 kDa), both of which accumulate in response to heat stress. Using these reagents, they detected heat-induced accumulation of the immunological homologues of both heat shock proteins in seedlings of wheat, sorghum, and maize in Western blotting experiments.

5.7. Molecular Regulation of Heat Shock Response

Threshold temperature for inducing heat shock protein synthesis is generally related to temperature at which each species ordinarily grows (Feder and Hofman, 1999). Major prokaryotic heat shock proteins are constitutively expressed by special genes at all temperatures. After switching temperature or treatment with any agent harming to proteins, expression of these genes speeds up and reaches the level that is characteristic for that species in a few minutes. Heat shock response is controlled by α32 polypeptide at transcriptional level. There is an analog situation in eukaryotes. Expression of heat shock proteins genes are primarily regulated at transcriptional level (Iba, 2002). Heat shock response is controlled by heat shock transcription factors binding to specific DNA sites (HS elements) at transcriptional level (Ang *et al.* 1991). Heat shock factors specifically binds to heat shock elements in promoters and activates the transcription of HS genes (Iba, 2002: Lohmann *et al.* 2004). Unlike animals and yeasts, which may have 4 or fewer heat shock factors, plants have been shown to have multiple copies of these genes: tomato has at least 17 and *Arabidopsis* has 21 different heat shock factor genes (Krishna, 2004). HSFA2 of *Arabidopsis* has been shown as a key regulator in response to several types of environmental stress (Nishizawa *et al.* 2006).

An important component of thermotolerance is heat tolerance of gene expression. Heat stress is known to swiftly alter the pattern of gene expression, inducing the heat shock protein complement and inhibiting expression of many genes expressed under temperature (Efeoglu, 2009). The mRNAs encoding non-heat stress-induced proteins are destabilized during heat stress. Heat stress inhibits splicing and it was hypothesized that heat shock proteins encoding mRNAs can be processed properly due to the absence of introns in the corresponding genes (Maestri *et al.* 2002). The heat shock response is connected with profound changes of the intracellular protein distribution in general. This is observed in "microscale" *e.g.* the migration of RNAPII from control genes to heat shock-activated genes or binding of the activated cytoplasmic heat shock transcription factor to nuclear heat shock promoter sites, but also in macroscale. Examples are the reorganization of the cytoskeleton, the massive accumulation of nonhistone proteins in the chromatin or the formation of heat shock granules. Restoration of the normal protein pattern of protein distribution is an essential part of the recovery period. It is improved in thermo tolerant cells and precedes the recovery of normal cellular functions (Nover, 1991). Plant sHSPs are controlled at transcriptional level in heat stress. sHSPs might protect and keep stable untranslated normal cellular mRNAs during heat stress (Sun and Montagu, 2002; Waters *et al.* 1996).

The high selectivity of heat shock protein synthesis depends on the massive synthesis of new mRNAs encoded by the heat shock protein genes and the number of alterations of the translation apparatus leading to an efficient discrimination of non-heat shock protein mRNAs. The selectivity of heat shock protein synthesis in heat shocked cells does not correspond to the composition of the mRNA fraction. Most of the "control" mRNAs, though not translated, are maintained intact and can be reactivated in the recovery period. Heat shock mRNAs rapidly accumulate during heat shock and preferentially occupy the polysomal apparatus (Nover *et al.* 1989). A hypothetic multistep model of essential translation events during heat shock can be put forward: In 0-5 min, an immediate initiation deficiency and the collapse of the cytoskeleton lead to the destruction of polysomes, 15 min. after stress, formation of Hsgs (nontranslated control mRNAs are protected from degradation by increasing deposition of HSPs); in late phase of the (heat shock) hs or in the recovery period, stored mRNAs are reactivated (Nover *et al.* 1989). Slow adaptation to heat or pre-stimulation increase translation at strong heat shock situations. Many modifications of translation process were described in microorganisms, plants and animals. Storing of transiently untranslated mRNAs is needed to efficiently recover synthesis of control proteins (Nover *et al.* 1989).

Thermal tolerance and molecular chaperones

Biological function of proteins is critically dependent on the formation/ dissolution of weak chemical bonds. This requirement irrevocably synthesizes organisms to environmental factors that affect weak bonds. If these factors are present in excess or in insufficient quantity, the result is departure from the native structure of proteins, with ultimately devastating consequences for protein functions. Non-native proteins expose regions that are satisfied in the native structure; these exposed unsatisfied regions can bind other such regions in other proteins and so lead to aggregations of proteins that at worst are cytotoxic and at best reduce the pool of functional protein in the cell. This problem is ancient and widespread as life itself- and so is one major biological solution: molecular chaperones. Molecular chaperons are a class of proteins that function to minimize the problems that arise when other proteins are in nonnative conformations. Molecular chaperons can recognize and bind to nonnative and release them in highly related fashion, allowing the bound proteins to attain/ reattain their native conformation and/or be targeted for degradation and removal from the cell (Feder, 1999). Chaperons are responsible for protein folding, assembly, translocation and degradation in many cellular processes. They stabilize proteins and membranes and can assist in protein refolding under stress conditions. Heat shock proteins function as molecular chaperones; i.e. they interact with other proteins and, in so doing, minimize the probability that these other proteins will interact inappropriately with one another. Heat shock proteins recognize and bind to other proteins when these other proteins are in nonnative conformations, whether due to protein denaturing stress or because the peptides they comprise have not yet been fully synthesized, folded, assembled or localized to an appropriate cellular compartment (Figure 2). Typically heat shock proteins function as oligomers. They are responsible for maintaining heat shock proteins partner proteins in a folding competent, folded or unfolded state; organellar localization, import, and/or export; minimizing the aggregation of non-native proteins; and targeting non-native or aggregated proteins for degradation and removal from the cell. Chaperones generally should recognize structural elements exposed by non-native proteins (Feder and Hofman, 1999; Hendrick and Hartl, 1993; Waters *et al.* 1996). Five major families of HSPs are known as chaperones: the HSP70 (DnaK), the chaperonins (GroEL and HSP60), the HSP90, the HSP100 (Clp) and the small heat shock proteins (sHSP) family. Aside from these major families, there are other proteins with chaperone functions, such as protein disulfide isomerase and calnexin / calreticulin, which assist in protein folding in the endoplasmic reticulum (Wang *et al.* 2004).

Evident differences of the thermotolerant states observed:

a) After a short and mild heat shock

b) After a moderate to severe heat shock applied to few hours

c) After a long term adaptation to growth or at least survival at elevated temperatures.

Experimental heat shock research mostly concentrates on conditions 'a' and 'b'. If both are compared two different states of thermotolerance can be defined. Long term heat shock (c) has two important effects: destruction of the chloroplast structure connected with bleaching observed in the virus infections from growing meristems of many cultural plants (Nover *et al.* 1989).

For thermotolerance induction in plants, tissues or cell cultures are either subjected to a continuous, moderate heat shock, to a preinduction procedure or they are slowly adapted to increasingly severe heat shock (Nover *et al.* 1989).

Preinduction heat shock procedures also induce tolerance to other stresses and other types of stresses induce tolerance to heat stress (Lindquist and Craig, 1988).

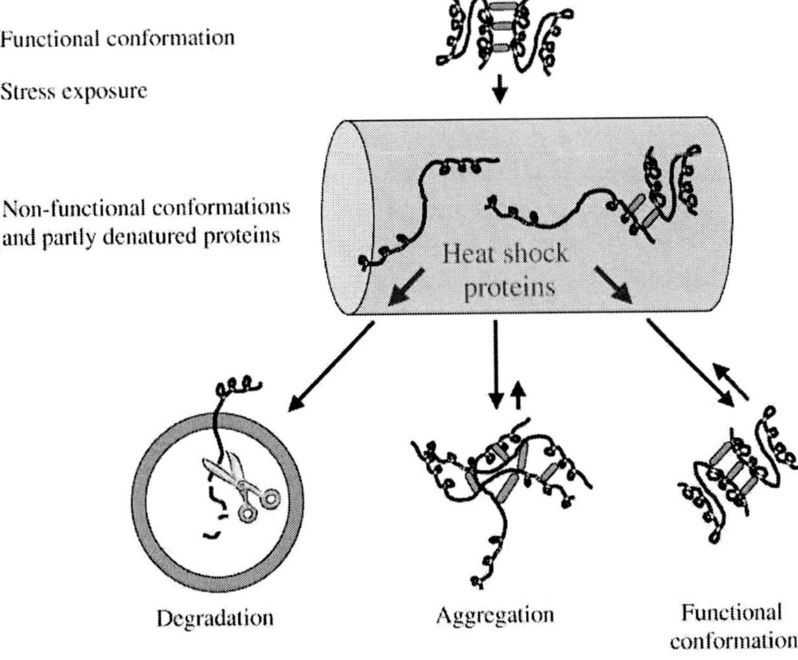

Fig. 3: Cellular function of heat shock proteins. The fate of proteins with non-functional conformations after stress exposure may be either to re-obtain the functional conformation, form aggregations with other misfolded proteins or become degraded. Hsps play a helper role in shifting the equilibrium in the direction of more functional proteins or degradation of damaged proteins (Soransen *et al.,* 2003)

Two tightly related aspects have to be considered:

1. the protection of cellular fine structure during the stress period
2. the effective repair of eventual damages with restoration of the normal metabolic and gene expression activities mainly in the recovery period (Nover *et al.* 1989).

Similar to animal systems, the 3 phases of heat shock induction (few minutes), expression (1-3 hrs) and decay (12-48 hrs) of heat shock proteins and thermotolerance are very similar also in plants. Generally it is not justified to conclude from analyses of growth of survival curves that heat shock proteins are required or not required for thermotolerance. For example, growing pollen tubes of *Tradescantia paludosa,* though unable to heat shock protein synthesis, exhibit certain aspects of induced thermotolerance simply demonstrate that other protective agents contribute as well (Nover *et al.* 1989). Variation in heat shock proteins production among closely related species from thermally contrasting habitats has been observed even when these species are grown and heat stressed under identical conditions. This variation in heat shock protein production is often correlated with organismal thermotolerance (Downs *et al.* 1998). There are many studies showing quantitative variations for their tolerance degrees among genotypes. The acquisition of thermotolerance appears to depend on not only upon the synthesis of heat shock proteins but also on their selective cellular localization. Lin *et al.* (1984), reported that in the soybean seedlings, several heat shock proteins become selectively localized in or associated with nuclei, mitochondria and ribosomes at 40 °C, they were absent in other parts. The selective localization of heat shock proteins is temperature dependent. Genotypic variations have been observed for thermosensitivity in studies up to date. Heat shock sensitive genotypes synthesized heat shock proteins earlier than tolerant genotypes (Krishnan *et al.* 1989). Especially very early detected (in a few minutes in heat shock) specific heat shock protein transcripts might have effect on order of heat shock protein gene expression and cause different tolerance levels among species (Weng and Nguyen, 1996). Tolerance to heat shock cannot be controlled by only one gene. The heat shock tolerance is determined by different gene sets in different stages of life cycle and in different tissues. HSP70 was shown to be causally involved in the capacity to acquire thermotolerance in *Arabidopsis*. HSP27 in another thermosensitive *Arabidopsis* mutant. HSP17 was identified as a factor of acquired thermo tolerance in the study of transgenic cells (Maestri *et al.* 2002).

Regulation of heat shock proteins

Heat shock proteins are present in cells under normal conditions, but are expressed at high levels when exposed to a sudden temperature jump or other stress. Heat shock proteins stabilize proteins and are involved in the folding of denatured proteins. High temperatures and other stresses, such as altered pH and oxygen deprivation, make it more difficult for proteins to form their proper structures and cause some already structured proteins to unfold. Left uncorrected, mis-folded proteins form aggregates that may eventually kill the cell. Heat shock proteins are induced rapidly at high levels to deal with this problem. Increased expression of HSPs is mediated at multiple levels: mRNA synthesis, mRNA stability, and translation efficiency.

Stress proteins also assist in the repair of denatured proteins or promote their degradation after stress or injury. They have been referred to as "molecular chaperones" because of this function.

It is thought that stress proteins are produced in response to nonlethal stress to protect organisms from subsequent severe stress that would otherwise be lethal. In the case of exposure to heat, this phenomenon has been called "thermotolerance" and has launched many experiments in which an association has been found between the heat shock response and protection against other stresses, such as hypoxia or ischemia. The addition of one type of stress may provide protection against other types of insults, which results in cross-tolerance. As examples, stress protein induction by hyperthermia may provide protection during a subsequent arterial injury or exposure to a heavy metal may provide subsequent protection against heat or ischemic injury. This thermotolerance treatment strategy has proved successful in experimental models of cardiac ischemia, arterial injury, endotoxic shock, renal and hepatic ischemia, ethanol-induced gastric ulcerations, and skeletal muscle ischemia-reperfusion (Rajoriya *et al.* 2014).

Many of the stress proteins are present continuously (constitutive expression), and expression of other proteins is increased by stress (stress inducible). Stress proteins are rapidly induced through transcription (messenger RNA production from DNA occurs in minutes) and translation (protein production from messenger RNA) mechanisms. Gene transcription is controlled by heat shock transcription factors. Different members of the heat shock transcription factor family may be activated by specific stresses. Inactive heat shock factors exist as monomers. However, once activated, they trimerize into an active form that is capable of binding to the promotor site of the stress protein gene and initiating transcription and translation.

Threshold temperature for inducing heat shock protein synthesis is generally related to temperature at which each species ordinarily grows (Feder and Hoffman, 1999). Major prokaryotic heat shock proteins are constitutively expressed by special genes at all temperatures. After switching temperature or treatment with any agent harming to proteins, expression of these genes speeds up and reaches the level that is characteristic for that species in a few minutes. Heat shock response is controlled by α 32 polypeptide (rpoH gene product) at transcriptional level. There is an analog situation in eukaryotes. Expression of heat shock protein genes are primarily regulated at transcriptional level (Iba, 2002; Weng and Nguyen, 1992). Heat shock response is controlled by heat shock transcription factors (HSF) binding to specific DNA sites (HS elements) at transcriptional level. HSF specifically binds to heat shock elements (HSEs) in promoters and activates the transcription of HS genes (Iba, 2002; Lohmann *et al.* 2004). HSFs, once plant has sensed an increase in temperature, go from a monomeric state in the cytoplasm to a trimeric state in the cell nucleus where they can bind the HSEs. HSF binding recruits other transcription components, resulting in gene expression within minutes. Unlike animals and yeasts, which may have 4 or fewer HSFs, plants have been shown to have multiple copies of these genes: tomato has at least 17 and Arabidopsis has 21 different HSF genes (Krishna, 2004). HSFA2 of *Arabidopsis* has been shown as a key regulator in response to several types of environmental stress (Nishizawa *et al.* 2006). An important component of thermotolerance is heat tolerance of gene expression. Heat stress is known to swiftly alter the pattern of gene expression, inducing the heat shock protein complement and inhibiting expression of many genes expressed under temperature. The mRNAs encoding non-heat stress-induced proteins are destabilized during heat stress. Heat stress inhibits splicing and it was hypothesized that heat shock protein encoding mRNAs can be processed properly due to the absence of introns in the corresponding genes (Maestri *et al.* 2002). The heat shock response is connected with profound changes of the intracellular protein distribution in general. This is observed in "microscale" *e.g.* the migration of RNAPII from control genes to heat shock-activated genes or binding of the activated cytoplasmic heat shock transcription factor to nuclear heat shock promotor sites, but also in macroscale. Examples are the reorganization of the cytoskeleton, the massive accumulation of nonhistone proteins in the chromatin or the formation of HSG. Restoration of the normal protein pattern of protein distribution is an essential part of the recovery period. It is improved in thermo tolerant cells and precedes the recovery of normal cellular functions (Nover, 1991). Plant sHSPs are controlled at transcriptional level in heat stress. sHSPs might protect and keep stable untranslated normal cellular mRNAs during heat stress (Sun,and Montagu, 2002; Waters *et al.* 1996). Describing some isoforms of sHSPs indicates those proteins are coded by different post-translational modifications (Skylas *et al.* 2002). The high selectivity of heat shock protein synthesis depends on the massive synthesis of new mRNAs encoded by the heat shock protein genes and the number of alterations of the translation apparatus leading to an efficient discrimination of nonHSP mRNAs. The selectivity of heat shock protein synthesis in heat shocked cells does not correspond to the composition of the mRNA fraction. Most of the "control" mRNAs, though not translated, are maintained intact and can be reactivated in the recovery period. Heat shock mRNAs rapidly accumulate during heat shock and preferentially occupy the polysomal apparatus (Nover *et al.* 1989). A hypothetic multistep model of essential translation events during heat shock can be put forward: In 0-5 min, an immediate initiation deficiency and the collapse of the cytoskeleton lead to the destruction of polysomes, 15 min. after stress, formation of Hsgs (nontranslated control mRNAs are protected from degradation by increasing deposition of heat shock proteins); in late phase of the (heat shock) hs or in the recovery period, stored mRNAs are reactivated (Nover *et al.* 1989). Slow adaptation to heat or pre-stimulation increase translation at strong heat shock situations. Many modifications of translation process were described in microorganisms, plants and animals. Storing of transiently untranslated mRNAs is needed to efficiently recover synthesis of control proteins (Nover *et al.* 1989).

5.8. Soil Salinity and Heat Shock Proteins in Plants

There are two main components of salt stress/shock: osmotic and ionic components. Salt shock immediately induces osmotic shock, when plants are suddenly exposed to large differences in osmolarity (osmotic pressure) between external solutes with a high concentration of NaCl and internal solutes in the cell cytoplasm. Osmotic stress as a first component of salt stress is characterized by relatively small differences in osmotic pressure inside and outside of the cells, and occurs several consequent times if plants are exposed to gradually increasing NaCl concentrations. Osmotic adjustment refers to the adjustment of osmotic pressure inside plant cells in response to osmotic stress, and needs to be much stronger if plants are to survive the effects of salt (osmotic) shock. The ionic component of salt stress/shock follows the osmotic response, with a time delay because the concentration of Na^+ ions has to reach a toxic level in the plant cell cytoplasm (Savrukov, 2012).

To study the effects of salt stress on plants, scientists apply NaCl or sea salts to plants grown in the laboratory, growth chamber, greenhouse, or other facilities. Hydroponics (with or without supporting agents, aerated, flood-drained, or passive) is the most popular tool for studying salt stress, because NaCl application can be easily controlled. However, some prefer to use soil mixtures (Ayarpadikannan *et al.* 2012; Chakraborty, *et al.* 2012). Experiments with salinity are extremely variable, depending on the species being studied, purpose of the research, and even personal preference. However, one point remains particularly crucial for interpreting the results of experiments. How is salt applied to the plants: gradually or suddenly, in a single step?

Generally, the application of salt stress involves gradual application of NaCl, usually of 25 or maximum 50 mM increments of NaCl, twice daily, until a final, predetermined salt concentration is reached (Munns *et al.* 2000; Munns and James, 2003; Huang *et al.* 2006; James *et al.* 2006; Sharukov *et al.* 2006, 2009; Byrt *et al.* 2007; Roshandel and Flowers, 2009). Various researchers refer to this method as 'progressive imposition' (Almansouri *et al.* 1999), 'salt acclimation' or 'gradual step acclimation' (Sanchez *et al.* 2008), or 'salt-adapting' (Baisakh *et al.* 2006). Ideally, salt application must be made as smoothly as possible, for example, 2.083 mM NaCl every hour to reach 50 mM NaCl by the end of first day (24 hours). However, this 'ideal' type of gradual salt application is technically difficult. The more gradual the application of NaCl, the more closely reactions of plants will reflect those expected for salt stress in saline field environments.

In contrast, salt shock occurs when plants are suddenly transferred from normal growth solution (without NaCl) into solution containing high concentrations of NaCl. The main component of salt shock is osmotic shock or plasmolysis, especially in root cells (Munns, 2002), when the cell protoplast shrinks and detaches from the cell wall, as has been observed in barley after transfer to 200 mM NaCl (Pritchard *et al.* 1991). Desperate attempts by the cells to maintain equilibrium between external and internal water content results in the leakage of cell solution into open spaces between the cell wall and plasma membrane. These apoplastic solutes, containing high concentrations of Na^+, can freely flow through the open spaces in root cells and be transported to the shoot with minimal control by the plant. There is consequently rapid activation of many genes, in response to osmotic shock and damaged plasma membrane in root cells and to ionic stress in shoot cells. The mechanism of osmotic shock is universal for all plant species because it results from the physical–chemical effects of loss of cell turgor as described above. Plants from different species, regardless of their level of salt tolerance, will only differ in the degree of damage to the plasma membrane during plasmolysis and in how quickly normal structure and function of affected cells is restored.

It has been reported that young wheat plants suffer osmotic shock and plasmolysis at 150mM NaCl, when salt is applied in a single step. Older plants will suffer osmotic shock at lower levels, down to 100 mM NaCl (Pritchard *et al.* 1991). Therefore, it can be assumed that, when applied in a single application, levels of NaCl higher than 100–150 mM NaCl, in general, will cause plasmolysis. Salinity levels ranging between 50 and 100 mM NaCl can be assumed as intermediate between osmotic stress and osmotic shock, while application of 50 mM NaCl or less will not cause plasmolysis but only osmotic stress, where many plants can manage with moderated adjustments of cell turgor and osmolarity.

A primary response of plants when exposed to increased soil sodium is a decrease in plant water potential, because the osmotic and water potential of the soil decreases. One well-characterized response of plants to low soil water potential is the accumulation of osmotically compatible cellular solutes (*e.g.*, proline and glycine betaine) as a mechanism to decrease plant osmotic and water potential. In native species, osmotically compatible solute accumulation is correlated with soil sodium gradients and reduced physiological stress (Briens and Larher, 1982; Hester *et al.* 1996). In several species that accumulate proline, there is a delay (3–6 days) in proline accumulation following the onset of sodium stress (Brady *et al.* 1984; Lerner, 1985; Yoshiba *et al.* 1997). Therefore, the early stages of sodium stress are critical for plant survival because plants are not capable of osmotically adjusting water potential during this period.

The proteins, which accumulate in response to salt stress, are referred to as Salt Shock Protein (SSP), which may have some relevance in stress tolerance. The salinity promotes synthesis of SSP, causes either increase or decrease in the level and total and soluble protein, depending on the plant parts studied and leads to increased activity of many enzymes (Igarashi *et al.* 1997). The SSP-23 protein, which disappeared in *B. parviflora* under salinity stress, reappeared when these salinized seedlings were desalinized. These observations suggest the possible involvement of these polypeptides for osmotic adjustment under salt stress (Parida *et al.* 2004). Gomathi and Vasantha (2006) revealed that changes in RNA and DNA content under salinity stress might be responsible for specific expression of Salt Shock Proteins (SSP) with MW of 15, 28 and 72 kDa in tolerant genotypes, while there were completely absent in susceptible. However, supplementing GA_3 under salt stress condition enhanced nucleic acid content (RNA) and thus induced expression of SSP in susceptible genotypes with low intensity (Joseph and Jini, 2010).

It is possible that prior to the accumulation of protective solutes heat shock proteins may play a protective role. Accumulating evidence suggests that HSPs play a role in tolerance to a variety of stresses (Downs *et al.* 1999). Heat shock proteins are general stress proteins involved in maintaining cell function and thus survival during stress or facilitating recovery from stress (Vierling, 1991; Parsell and Lindquist, 1994; Downs *et al.* 1997; Downs and Heckathorn, 1998; Guy and Li, 1998; Heckathorn *et al.* 1998). Of particular interest, because of their identified importance in many types of plant stress, are two classes of heat shock proteins, the HSP70 class (70 kD) and the small heat shock protein (smHSP) class (15–30 kD) (Vierling, 1991; Parsell and Lindquist, 1994). The HSP70 class of proteins are chaperones and are found in the cytosol and most organelles (Vierling, 1991; Wang *et al.* 1993; Parsell and Lindquist, 1994). The smHSP class consists of five major forms localized to the (1) cytosol (2 forms), (2) chloroplast, (3) mitochondria, and (4) endoplasmic reticulum. Functions of the chloroplast and mitochondrion localized proteins have been identified. They both protect electron transport when plants are exposed to heat and oxidative stress (and also photoinhibition in chloroplasts; Downs and Heckathorn, 1998; Heckathorn *et al.* 1998; Downs *et al.* 1999). Since salt stress increases cellular ionic concentration, which can affect protein synthesis and structure (Yoshiba *et al.* 1997; Hare *et al.* 1998), the induction of heat shock proteins by sodium is likely to be an adaptive trait, which we predict is associated with Na-tolerant species (Hamilton *et al.* 2001).

Many researchers (Flom *et al.* 2006; Chan *et al.* 2006; Guyomarch *et al.* 2004; Izhaki *et al.* 2001; Kurek *et al.* 2002) confirmed that protein *sti1* appeared to be up-regulated in response to salt stress; this protein contains two heat shock chaperonin binding motif (STI1), three Tetratrico Peptide Repeat (TPR) and two *Sti1* domains. The up-regulation of this regulatory protein may decrease the sterility of pollen during another development. It is believed that HSP90 interacts with TPR-containing proteins to modulate diverse cellular processes through protein-protein interaction. HSP70 is confirmed as biomarker of stress produced by NaCl in marine macro algae and fresh water plant species, emphasized its role in protecting plants against stress (Ireland *et al.* 2004). Dual channel imaging and warping of 2-DE protein gels were also used to visualize global changes in the protein synthesis pattern of cells in response to osmotic stress (6% NaCl). Extensive studies suggest that different chaperones follow distinct strategies to prevent protein misfolding and aggregation in the highly crowded cellular environment. For example, the monomeric HSP70 recognizes short hydrophobic peptide segments and binds at an early stage of folding (Bukau *ey al.* 2000).

In addition to the cellular adaptations to salt stress, many species that are exposed to persistently high sodium concentrations exude sodium salts onto the surface of leaves (Marcum and Murdoch, 1992; Marcum *et al.* 1998; Naidoo and Naidoo, 1998). Exudation exists and is effective at reducing the concentration Na in tissues in *Sporobulus virginicus,* a salt marsh species (Marcum and Murdoch, 1992; Naidoo and Naidoo, 1998), and other *Sporobulus* species (Lipschitz and Waisel, 1974). Exudation of sodium from the leaf allows for the uptake of water and nutrients while eliminating sodium (Lipschitz and Waisel, 1974; Naidoo and Naidoo, 1998).

Salt stress also causes an augmentation of several proteins with protective functions such as chaperones from HSP90 family in tomato roots (Manaa, 2011) and HSP70 family, Hsc70 (heat-shock cognate) proteins in *Arabidopsis thaliana (*Pang *et al.* 2010) and *P. patens* (Wang *et al.* 2008). DnaK protein, and others in salt-treated rice seedlings (Chitteti and Peng, 2007; Kim *et al.* 2005) or *Arabidopsis* cell culture (Ndimba *et al.* 2005). Increased relative abundance of several small heat shock proteins (mitochondrial small heat shock protein, chloroplast heat shock protein, 17.8 kDa class I small heat shock protein, HSP20) was found in salt-treated tomato hypocotyls (Chen *et al.* 2009) and *Aster tripolium* leaves (Geissler *et al.* 2010). Increased relative abundance of STI1 protein, a stress-responsive protein with two heat shock chaperonin-binding motifs and three tetratricopeptide repeats (TPR) in salt-treated rice panicles (Dooki *et al.* 2006) points toward a large regulatory network affected by salt stress since TPR-containing proteins have been reported as being involved in myriads of processes including HSP90 signalling, gibberellin signalling and protein mitochondrial transport. Several pathogenesis-related proteins such as PR5 protein, PR10 proteins, TSI-1 protein, elicitor peptide three precursor and β-glucosidases have been found as being induced by salinity (Pang *et al.* 2010; Jellouli *et al.* 2008; Caruso *et al.* 2008; Sugimoto and Takeda, 2009; Aghaei *et al.* 2008 and 2009; Jain *et al.* 2006; Vincent *et al.* 2007). PR10 proteins have been reported as being involved not only in defence reactions, but also in JA signaling (Hashimoto *et al.* 2004), they could reveal ribonuclease activity (Park *et al.* 2004) and could bind several ligands including cytokinins, fatty acids, flavonoids and brassinosteroids (Mogensen *et al.* 2002). β-glucosidases catalyse hydrolysis of 1,3-β-D-glucosidic linkages in 1,3-β-D-glucans and they are implicated in three processes: (1) alterations of specific â-linked polysaccharides during cell expansion in development; (2) pathogen defence response by cyanogenesis since the enzymes catalyse hydrolysis of cyanogenic glucosides after pathogen attack; (3) regulation of phytohormone activity via the release of active cytokinins, gibberellins and auxins from biologically inactive hormone-glucoside conjugates.

Other stress-responsive proteins include germin like proteins (GLP) which play an important role in plant embryogenesis. Some germin-like proteins display oxalate oxidase and superoxide dismutase activities. Increased relative abundance of germin-like proteins was observed under several abiotic and biotic stress conditions, for example in salt-stressed barley leaves (Fatehi *et al.* 2012) and *Arabidopsis* roots (Jiang *et al.* 2007). Another interesting protein group induced by salinity are lectins which are known as being involved in protein-saccharide interactions and stress signalling. Small lectins with a jacalin domain have been proposed to function in plant defence mechanisms (Zhang *et al.* 2000). An increased relative abundance of a jacalin lectin family protein was observed in salt-treated *A. thaliana* leaves (Pang *et al.* 2010) and an induction of mannose-binding RICE lectin (MRL) was reported in rice roots after a short-term salt stress (Chitteti and Peng, 2007) while a decreased relative abundance of lectin-like protein was found in salt-treated soybean root and hypocotyls (Aghaei *et al.* 2009).

Other stress-protective proteins, such as osmotin and osmotin-like proteins are associated with a plant adjustment to an enhanced osmotic stress. Increase in osmotin and osmotin-like proteins has been found in various salt-treated plants ranging from a salt-tolerant cultivar of potato (Aghaei *et al.* 2008) and hypocotyls of tomato (Chen *et al.* 2009) to roots of a halophytic mangrove plant *Bruguiera gymnorhiza (Tada and Kashimura, 2009).*

5.9. Heat Shock Proteins in Response to Heavy Metals in Plants

Heavy metal stresses usually give rise to dysfunctional protein conformations. Molecular chaperones are stress proteins and many of them were originally identified as heat shock proteins. According to current knowledge, Heat shock proteins facilitate protein refolding and stabilize polypeptides and membranes. HSP70 has essential functions in preventing aggregation and assisting refolding of nonnative proteins under stress conditions (Wang *et al.* 2003). Photosynthesis is typically decreased by increased level of heavy metals. The specific effects of heavy metal on photosynthesis vary among species (Patsikka *et al.* 2002; Vinit-Dunand *et al.* 2002). Heavy metals damages both membrane and soluble phases of chloroplasts through multiple mechanisms that include protein denaturation and oxidative damage (Hall, 2002). General response of plants to elevated levels of heavy metals appears to be increased synthesis of various heat-shock proteins (Barque *et al.* 1996; Hall, 2002). Heat-shock proteins are general stress proteins involved in protection of photosynthesis during heat, oxidative, and photoinhibitory stress, by protecting PSII or other aspects of thylakoids (Nakamoto *et al.* 2000). Chloroplast has small heat shock protein (smHSPs). Chloroplast small heat-shock proteins protect photosynthesis during heavy metal stress (Scott *et al.* 2004). Which protect photosynthesis through more than one mechanism by preventing irreversible protein aggregation, stabilizing chloroplast membranes and by site-specific antioxidants.

There have been several reports of an increase in heat shock protein expression in plants in response to heavy metal stress. It has been reported that, in rice, both heat stress and heavy metal stress increased the levels of mRNAs for low molecular mass heat shock proteins (16–20 kDa) (Tseng *et al.* 1993), while Neumann *et al.* indicated that HSP17 is expressed in roots of *Armeria maritima* plants grown on Cu rich soils (Neumann *et al.* 1995). Small heat shock proteins (*e.g.* HSP17) were also shown to increase in cell cultures of *Silene vulgaris*and *Lycopersicon peruvianum* in response to a range of heavy metal treatments (Wollgiehn and Neumann, 1999), however, no or very low amounts of heat shock proteins were found in plants growing on metalliferous soils, suggesting that heat shock proteins are not responsible for the heritable metal tolerance of *Silene*.

Working with cell cultures of *L. peruvianum*, it was shown that a larger heat shock protein (HSP70) also responds to Cd stress (Neumann *et al.* 1994). It is of interest that antibody localization showed that HSP70 was present in the nucleus and cytoplasm, but also at the plasma membrane. This suggests that HSP70 could be involved in the protection of membranes against Cd damage. Expression of HSP70 also increased in the seaweed *Enteromorpha intestinalis* after exposure to a variety of stressors including Cu (Lewis *et al.* 2001). Thus, tolerance mechanisms involving a more resistant plasma membrane or improved repair mechanisms, heat shock proteins could have an important role in this respect. Interestingly, it was reported that a short heat stress given prior to heavy metal stress induces a tolerance effect by preventing membrane damage, as judged by ultrastructural studies (Neumann *et al.* 1994). Clearly more molecular evidence is required to support such an important repair or protective role.

Heavy metals constitute a heterogeneous group of elements widely varied in their chemical properties and biological functions. Heavy metals are chemical elements with a high specific gravity. Heavy metals occur in environment from natural processes and anthropogenic activities (Connell *et al.* 1999; Franca *et al.* 2005). Heavy metal such as iron, chromium, nickel commonly known as trace elements play an important role in biological systems, yet they may become highly toxic when present in high concentrations (Ibok *et al.* 1989). Cadmium, mercury and lead are non-essential metals. They are toxic, even in trace amounts (Fernandes *et al.* 2008). Pollution by heavy metals is a world wide problem due to its persistency and continuing accumulation of metals in the environment. The fact that heavy metals cannot be destroyed through biological degradation and have the ability to accumulate in the organs of living organisms make these toxicants deleterious to living organisms and consequently to humans. Disruption of normal cellular processes may cause rapid increase in the synthesis of a group of proteins which belong to the heat shock protein families. The heat shock protein genes are highly conserved and have been characterized in a wide range of organisms. The heat shock response is an evolutionarily conserved heat shock response is an evolutionarily conserved mechanism for maintaining cellular homeostasis following sublethal noxius stimuli (Lindquist, 1986; Lindquist and Craig, 1988).

Tolerance to heavy metals in plants may be defined as the ability to survive in a soil that is toxic to other plants, and is manifested by an interaction between a genotype and its environment (Macnair *et al.* 2000), although the term is frequently used more widely in the literature to include changes that may occur experimentally in the sensitive response to heavy metals. In a number of thorough genetic studies, such adaptive metal tolerance has been shown to be governed by a small number of major genes with perhaps contributions from some more minor modifier genes (Macnair *et al.* 2000; Schat *et al.* 2000). The question of whether this means that only a single biochemical or molecular change is required to produce tolerance to a specific metal remains to be resolved. Related to this question is the occurrence of multiple tolerance and co tolerance where plants can grow on soils enriched in combinations of several heavy metals. This tolerance could result from a less specific mechanism that confers a broad resistance to several different metals (co tolerance) or may involve a series of independent metal specific mechanisms (multiple tolerance) (Schat *et al.* 2000). However, the evidence for co tolerance is not strong, suggesting that specific mechanisms are involved for each metal present at a toxic concentration (Macnair *et al.* 2000; Schat *et al.* 2000).

Heavy metals are simultaneously acting on the plants causing cell injury and producing secondary stresses such as osmotic and oxidative ones (Wang *et al.* 2003). Heat stress as well as other stresses can trigger some mechanisms of defense such as the obvious gene expression that was not expressed under "normal" conditions (Morimoto, 1993; Feder, 2006). Heat shock proteins and other stress proteins have been known to protect cells against deleterious effects of stress. Heat shock proteins are also expressed in some cells either constitutively or under cell cycle or developmental control. Heat-shock proteins are stress proteins involved in the protection, repair, and degradation of damaged cell components, especially proteins, during most abiotic stresses (Parsell and Lindquist, 1994; Hamilton and Heckathorn, 2001). Most heat shock proteins are molecular chaperones. Chaperones aid in the transport of proteins throughout the cell's various compartments.

5.10. Heat Shock Proteins in Leguminous Plants

Most cultivated species have wild relatives which exhibit high tolerance to abiotic stress (Wu and Wallner, 1983). *Prosopis chilensis* (Mol.) Stuntz (Chilean algarrobo) is a leguminous tree which grows in the arid and semi arid regions of Northern and Central Chile. This tree has been described as resistant to temperature, drought, salinity, and wounding (Rodriguez and Cardemil, 1994; Ortiz *et al.* 1995; Cazebonne *et al.* 1999). *P. chilensis* has a high capability for acquiring thermotolerance since the seedlings are able to survive at 50 ^{0}C after germination at 35 ^{0}C, or after exposure for 2 hour at the sublethal temperature of 40 ^{0}C (Medina and Cardemil, 1993). It has also been shown that, under field conditions, *P. chilensis* expresses heat shock proteins daily at the hours of higher temperatures, which is presumed to be an adaptation to its harsh natural environment (Ortiz *et al.* 1995). Thermotolerance acquisition and expression of heat shock proteins has also been found in the cultivated leguminous soybean (Hermandez and Vierling, 1993).

Since *Prosopis chilensis* could be considered a genetic resource for crop improvement in desert environments, this work attempts to evaluate and compare the heat shock responses of seedlings and plants of two legumes, *P. chilensis* and cultivated *Glycine max* (soybean), studying growth, membrane damage, fluorescence emission, and heat shock protein levels in response to heat stress. The level of thermotolerance of *P. chilensis* could be related to higher amounts of heat shock proteins, induced by thermal stress as compared with the amounts induced in soybean, another leguminous plant able to acquire thermotolerance when subjected to sublethal temperatures (Hermandez and Vierling, 1993).

Ortiz and Cardemil (2001) compared relative growth rates, basal and acclimated thermotolerance, membrane damage, fluorescence emission, and relative levels of free and conjugated ubiquitin and HSP70 were compared after 2 hours of treatment at different temperatures between *Prosopis chilensis* and *Glycine max* (soybean), cv. McCall, to evaluate if the thermotolerance of these two plants was related to levels of accumulation of heat shock proteins. It was observed that seedlings of *P. chilensis* germinated at 25 °C and at 35 °C and grown at temperatures above germination temperature showed higher relative growth than soybean seedlings treated under the same conditions. The lethal temperature of both species was 50 °C after germination at 25 °C. However, they were able to grow at 50 °C after germination at 35 °C. Membrane damage determinations in leaves showed that *P. chilensis* has an *LT*50 6 °C higher than that of soybean. There were no differences in the quantum yield of photosynthesis (*F*v/*F*m), between both plants when the temperatures were raised. *P. chilensis* showed higher relative levels of free ubiquitin, conjugated ubiquitin and HSP70 than soybean seedlings when the temperatures were raised. Time course studies of accumulation of these proteins performed at 40 °C showed that the relative accumulation rates of ubiquitin, conjugated ubiquitin and HSP70 were higher in *P. chilensis* than in soybean. In both plants, free ubiquitin decreased during the first 5 min and increased after 30 min of heat shock, conjugated ubiquitin increased after 30 min and HSP70 began to increase dramatically after 20 min of heat shock.

5.11. Phenomena of Induction of HSPs in Plants

Presence of heat shock proteins in higher plants was discovered in tobacco and soybean using cell culture technique (Barnett *et al.* 1980). When soybean was subjected to 40 °C for four hours, ten new proteins were found, but disappeared after 3 h treatment at 28 °C (Key *et al.* 1981). Studying the gene expression of HSP90 in rice plant (*Oryza sativa*) indicated that the heat-shock protein HSP87 was present after 2 hours of heat shock (from 28 to 45 °C), and its quantity was high and stable even after long heat stress (4 h) and the return to normal conditions (no stress). It was found, also, that HSP90 (HSP85 and HSP87) could be induced by other kind of stress such as salinity, drought, and cold. This study reported the accumulation of different levels of these proteins in fifteen wild species of rice (Pareek *et al.* 1998a). In another study it was indicated that HSP90 and the definition of the gene (rHSP90, GenBank Accession No. AB037681) that encodes them in rice plant, and the finding that they participate in plant tolerance of other abiotic stresses such as salinity (NaCl, $NaHCO_3$, and Na_2CO_3), desiccation (of polyethylene glycol, PEG), high pH (8.0 and 11.0), and high temperatures, viz 42 and 50 °C (Liu *et al.* 2006b).

Several studies on other plants (Singla *et al.* 1997) indicated that heat shock protein synthesis qualitatively and quantitatively was dependent on cell/tissue type and/or the degree of differentiation and development. Earlier, the presence of a cytoplasmic class of proteins (Class 1) in seeds of wild and commercial legumes was reported (Hernandez and Vierling, 1993). This indicated the expression of this class under natural environment. A further field study of the expression of this class in leaves, flowers and developing seed pods in *Medicago sativa* was carried out. Results indicated the repeated formation of these proteins in flowers and buds, even in plants that did not have these proteins expressed in their leaves (Hernandez and Vierling, 1993). During storage of beech (*Fagus sylvatica* L.) seeds, a sHSP with molecular mass of approximately 22 kDa was identified (Kalemba and Pikacka, 2008). The largest content of this protein was observed in the oldest seeds, especially in embryonic axes.

There were some studies about the presence of heat shock proteins in plants subjected to two or more stress factors mimicking natural field conditions. In the field, heat stress was usually accompanied by one or more of stress factors such as drought, high irradiation, salinity, or others, but studies of this kind are scarce. One of these studies was performed on irrigated and non irrigated cotton plants (*Gossypium hirsutum* L.) where most growth parameters decreased (80% to 85%) in non irrigated plants (Burke *et al.* 1985). The study, also, indicated reduced photosynthesis (two folds) at midday compared to irrigated plants where the temperature under the canopy of irrigated plants reached 30 °C while that was 40 °C under the canopy of non irrigated plants. These differences between the two treatments were accompanied by differences in protein content too. Plant leaves of non irrigated accumulated a steady level of proteins that have molecular weight of 100, 94, 89, 75, 60, 58, 37, and 21 kD after several weeks, and these proteins were not detected in leaves of irrigated plants. Pursuing these results, leaves of cotton plant that was grown in growth chamber were incubated at 40 °C with the labeled amino acid [^{35}S] methionine and after three hours, the same proteins but labeled appeared. The final conclusion of this study was that cotton plants accumulated heat-shock proteins under natural conditions of drought stress and 40 °C temperature.

Photoinhibition is a limiting factor in photosynthesis and in natural environment the light intensity is very high, at least in some areas. Light induces the synthesis of the HSPs and they therefore might ameliorate the damage caused by high intensity of light. This possibility was investigated by comparing the content of heat shock proteins of leaves exposed to direct sun light with shaded leaves of *Solidago altissima*, family Asteraceae in cold days and warm ones in the field (Barua and Heckathom, 2006). The results indicated that the heat shock proteins content in the leaves exposed to direct sun light and at natural heat stress was higher significantly. Both light and temperature significantly affected accumulation of heat shock proteins in the laboratory.

Another field study on the desert legume *Retama raetam* and the interaction of stress factors in arid regions indicated the presence of daily periodism of photosynthesis (Merguiol *et al.* 2002). One period was between 07:00 and 10:00, while the second was between 15:00 and 17:00. Similar periodism was reported for another desert legume *Prosopis chilensis* growing in a desert of Chile (Ortiz and Cardemi, 2001). During the reduction in photosynthesis rate (from 11.00 to 15.00) there was an induction of transcripts of enzymes participating in defense (removal of reactive oxygen intermediates) and the synthesis of HSPs. The final conclusion was that *R. raetam* used a combination of avoidance and active defense mechanisms to withstand the stressful conditions that prevail within desert land.

Plant response to abiotic stress factors is controlled by complex net of genes. In the project of gene expression database of *A. thaliana* with the title of AtGenExpress, the effect of several abiotic stresses (heat, cold, drought, salinity, osmotic, UV-B, light, and wounding) was studied under similar conditions on the seedlings of *A. thaliana* (Kilan *et al.* 2007), and the results were analyzed by the DNA Microarray Technology. This study provided the types of gene expression induced by abiotic stresses including models for the analysis of the information of gene expression in response to UV-B, drought, and cold stresses. Results indicated that the first reaction to stress on the level of transcription in this plant included a group of stress-response genes. These genes might have a crucial role in the response to different stresses as well as the main role of systemic signals generated by the tissue exposed to stress.

The interaction of different biotic and abiotic stresses with heat stress was studied and analyzed, and the information in respect of transcription of HSPs and HSFs in the plant *A. thaliana* was deposited in the database AtGenExpress Consortium (Schmid *et al.* 2005). Results of the analysis indicated that all stresses interacted in the response pathways of heat-shock proteins and their factors, but the degree of interaction was different which suggested a cross-talk in the regulating net. Hu *et al.* (2009) examined a global expression profiling with heat stressed rice seedling, and then compared their own results with the previous rice data under cold, drought, and salt stresses. They concluded that HSPs and HSFs might be important elements in cross-talk of different stress signal transduction networks.

In general, the expression of HSPs and its factors Hsfs was induced largely by heat, cold, salinity, and osmotic stresses. The response to other stress factors depended on protein class and tissue. For example, under all types of stresses, high expression response for class HSP20 was recorded with high similarity of their information. Wounding the roots of the plant stimulated (after 12 hours) the expression of several genes from the classes HSP20, HSP70, and HSP100.

High response of expression of the genes for HSPs and its factors Hsfs was observed under UV-B stress in aerial tissues (shoot), but in non aerial tissues (root system) there were no expression (Swindell *et al.* 2007).

There are some indications about the relationship between the response to heat stress and the response to oxidative stress as both stresses induce the pathways leading to the expression/ accumulation of heat shock proteins (Dat *et al.* 1998; Lee *et al.* 2000). On the other hand, heat stress induced the expression of antioxidant enzymes (Gong *et al.* 1998; Lee *et al.* 1999). One of these important enzymes is ascorbate peroxidase (APX) which uses ascorbate to nullify the toxicity of hydrogen peroxide (one of reactive oxygen species). Different isozymes of this enzyme were found in the cytoplasm of the plant cell and in some of its organelles (Panchuk *et al.* 2002). It was reported that there was an interdependence signaling for both stresses.

5.12. Other Abiotic Stresses and Heat Shock Proteins

Importantly, while heat shock proteins were so named because they are up-regulated by an acute increase in temperature (Lindquist, 1980), a variety of other stresses, including photoinhibitory high light, also induce HSPs (Feder and Hoffman, 1999; Wang *et al.* 2004). Light stress often occurs under field conditions, and photoinhibition is a major limitation for photosynthesis in the field (Long *et al.* 1994). Photoinhibition occurs when excessive radiation leads to the over-reduction of the components of photosynthetic electron transport. This results in the impairment of electron transport, and ultimately, in irreversible damage to the photosynthetic reaction centers (Ohad *et al.* 1990) and high temperatures exacerbate these effects of high light. Heat shock proteins can protect against photoinhibition (*e.g.*, by protecting photosystem II [PSII] or stabilizing thylakoid membranes), and light induction of heat shock proteins may ameliorate the damaging effects of excess light (Stapel *et al.* 1993; Downs *et al.* 1999; Schroda *et al.* 1999).

The effect of light on the induction of heat shock proteins has been demonstrated in cyanobacteria (Glatz *et al.* 1997; Hihara *et al.* 2001; Asadulghani *et al.* 2003), algae (von Gromoff *et al.* 1989; Dryzmalla *et al.* 1996), and higher plants (Stapel *et al.* 1993; Debel *et al.* 1994; Rossel *et al.* 2002). Light regulation of constitutive heat shock proteins (*i.e.*, heat-shock cognates) has also been demonstrated in spinach (Li and Guy, 2001). A regulatory pathway independent of heat shock probably mediates light induction of heat shock proteins (Kropat *et al.* 1997). These studies show that light can induce heat shock proteins independent of temperature, and in most cases, there is an additive effect of high light (high light vs. dark) and high temperature (high vs. low) on accumulation of heat shock proteins (Barua and Heckathorn, 2006).

For example, increased light levels decreased the minimum temperature required to induce mitochondrial small heat shock protein in *Chenopodium rubrum* (Knack and Kloppstech, 1992). However, these studies have been conducted under laboratory conditions, and thus, with plants that have not been acclimated to natural high light or dynamic light regimes, which is likely to alter the response of heat shock proteins to light (*e.g.*, decrease the light-responsiveness of heat shock proteins to light, due to acclimation to high or variable light).

To investigate heat and light controls on heat shock protein production in the field, accumulation of heat shock proteins in naturally occurring *Solidago altissima* (goldenrod). Representatives of two of the five major families of heat shock proteins (Wang *et al.* 2004), HSP70 and small heat shock proteins (sHSPs) were examined and it found that they differed in their accumulation in goldenrods from contrasting light microclimates (open sun vs. shade) and on cool vs. warm days. A second component of the study was conducted in the laboratory, under more controlled conditions, to confirm field results regarding light and heat interaction on heat shock protein content. Here temperature response of heat shock proteins was studied to determine temperatures required to induce heat shock proteins, temperatures of maximal heat shock protein accumulation, and shut-off temperatures for heat shock protein accumulation (Barua and Heckathorn, 2006). Effects of low or moderately elevated light levels influence these set points for heat shock proteins or the magnitude of the response. Also, to determine the effects of higher light intensities on heat shock proteins (moderate light levels of 600 μmol m^{-2} s^{-1} PPFD in the controlled-environment growth chambers was used) and to determine if high light can induce heat shock proteins independent of temperature in goldenrod, light response of heat shock protein accumulation was examined to low, moderate, and high light, at control and heat-shock temperatures. PSII function (dark-adapted *Fv*/*F*m) was monitored to assess the effects of temperature and light on plant performance, as PSII is extremely sensitive to both light and temperature stress (Ohad *et al.* 1990).

Nitrogen is the most needed mineral nutrient for plants, and it is also important to maintain good crop quality, including color, density, growth, and resistance to stress conditions (Liu *et al.* 2007). Plants fertilized with nitrogen during heat stress had greater fresh and dry weight, and significantly higher membrane thermostability than those fertilized with nitrogen before heat stress. This result had suggested being due to greater rhizospheric nitrogen availability during heat stress (Tawfik *et al.* 1996). A study reported that higher nitrogen helped to maintain higher photosynthesis and photosynthetic nitrogen-use efficiency in maize under heat stress (Wang *et al.* 2008). In heat stressed cool-season turfgrasses, additional foliar nitrogen supply was found to be beneficial (Fu and Huang, 2003; Zhao *et al.* 2008), with enhanced antioxidative response being suggested as a mechanism accounting for improved tolerance (Fu and Huang, 2003). However, other mechanisms may be important for improved heat stress response by nitrogen, such as induction and change of expression pattern of the major heat shock proteins. In addition, although annual nitrogen fertilization programs for sand-based creeping bentgrass putting greens are well developed, recommendations for optimum nitrogen application during summer heat stress periods are not well defined. For instance, Beard (2002) suggested minimizing nitrogen application during summer heat stress. He also indicated a need for nitrogen to maintain healthy turf, but no specific rates were recommended. Duble (2001) also pointed out that very little fertilizer should be used in summer on bentgrass greens with possible monthly applications of nitrogen at 12.5 kg ha^{-1} (Wang *et al.* 2014)

References

Abernathy, R.H., Thiel, D.S., Peterson, N.S. and Helm, K. 1989. Thermotolerance is developmentally dependent in germinating wheat seed. *Plant Physiol.*, **89:** 569-576.

Adam, Z., Adamska, I., Nakabayashi, K., Ostersetzer, O.,Haussuhl, K., Manuell, A. and Shinozaki, K. 2001. Chloroplast and mitochondrial proteases in arabidopsis. A proposed nomenclature. *Plant Physiol.*, **125(4):** 1912-1918.

Agarwal, M., Katiyar-Agarwal, S., Sahi, C., Gallie, D. and Grover, A. 2001. *Arabidopsis thaliana* HSP100 proteins: Kith and kin. *Cell Stress Chaperon.*, **63:** 219.

Agarwal, M., Sahi, C., Katiyar-Agarwal, S., Agarwal, S., Young, T., Gallie, D. and Grover, A. 2003. Molecular characterization of rice HSP101: Complementation of yeast HSP104 mutation by disaggregation of protein granules and differential expression in indica and japonica rice types. *Plant Mol. Biol.*, **51:** 543-553.

Aghaei, K., Ehsanpour. A.A. and Komatsu, S. 2008. Proteome analysis of potato under salt stress. *J. Proteome Res*, **7:** 4858–4868.

Aghaei, K., Ehsanpour, A.A., Shah, A.H. and Komatsu, S. 2009. Proteome analysis of soybean hypokotyl and root under salt stress. *Amino Acids*, **36:** 91–98.

Ahn, Y. and Zimmerman, J. 2006. Introduction of the carrot HSP17. 7 into potato (*Solanum tuberosum* L.) enhances cellular membrane stability and tuberization *in vitro. Plant Cell Environ.*, **29:** 95-104,

Ahrman, E., Lambert, W., Aquilina, J.A., Robinson, C.V. and Emanuelsson, C.S. 2007. Chemical cross-linking of the chloroplast localized small heat-shock protein, HSP21, and the model substrate citrate synthase. *Protein Sci.*, **16:** 1464–1478.

Ahsan, N. and Komatsu, S. 2009. Comparative analyses of the proteomes of leaves and flowers at various stages of development reveal organ specific functional differentiation of proteins in soybean. *Proteomics*, **9:** 4889-4907.

Ahsan, N., Donnart, T., Nouri, M. and Komatsu, S. 2010. Tissue-specific defense and thermo-adaptive mechanisms of soybean under heat stress revealed by proteomic approach, *Journal of Proteome Research*, **9:** 4189–4204.

Ahuja, I., De Vos, R.C., Bones. A.M. and Hall. R.D. 2010. Plant molecular stress responses face climate change. *Trends Plant Sci.*, **15:** 664-674.

Åkerfelt, M., Morimoto, R.I. and Sistonen, L. 2010. Heat shock factors: integrators of cell stress, development and lifespan. *Nature reviews Molecular cell biology*, **11:** 545-555.

Al-Babili, S. and Beyer, P. 2005. Golden rice-five years on the road-five years to go? *New Phytol.*, **10:** 565–573.

Al-Whaibi, M.H. 2011. Plant heat shock proteins: A mini review. *Journal of King Saud University – Science*, **23(2):** 139-150.

Allakhverdiev, S.I., Kreslavski, V.D., Klimov, V.V., Los, D.A., Carpentier, R. and Mohanty, P. 2008. Heat stress: an overview of molecular responses in photosynthesis. *Photosynthesis Research*, **98(1–3):** 541–550.

Almansouri, M., Kinet, J.M. and Lutts, S. 1999. Compared effects of sudden and progressive impositions of salt stress in three durum wheat (*Triticum durum*Desf.) cultivars. *Journal of Plant Physiol.*, **154:** 743–752.

Ang, D., Liberek, K., Skowyra, D., Zylicz, M. and Georgopoulos, C. 1991. Biological role and regulation of the universally conserved heat shock proteins, *The Journal of Biological Chemistry*, **266(36):** 24233-24236.

Asadulghani, Y. and Nakamoto, S.H. 2003. Light plays a key role in the modulation of heat shock response in the cyanobacterium *Synechocystis* sp. PCC 6803. *Biochem Biophy Res Comm.*, **306:** 872-879.

Ashraf, M. and Harris, P.J.C. 2013. Photosynthesis under stressful environments: An overview. *Photosynthetica*, **51(2):** 163-190.

Apuya, N., Yadegari, R., Fischer, R.L., Harada, J.J., Zimmerman, J.L. and Goldberg, R.B. 2001. The *Arabidopsis* embryo mutant schlepperless has a defect in the chaperonin-60á gene. *Plant Physiol.*, **126:** 717-730.

Ayarpadikannan, S., Chung, E., Cho, C.W., So, H.A., Kim, S.O., Jeon, J.M., Kwak, M.H., Lee, S.W. and Lee, J.H. 2012. Exploration for the salt stress tolerance genes from a salt-treated halophyte, *Suaeda asparagoides*. *Plant Cell Reports*, **31:** 35–48.

Babiychuk, E., Vandepoele, K., Wissing, J., Garcia-Diaz, M., De Rycke, R., Akbari, H., Joubes, J., Beeckman, T., Jansch, L., frentzen, M., Van Montaqu, M.C. and Kushnir, S. 2011. Plastid. gene expression and plant development require a plastidic protein of the mitochondrial transcription termination factor family. *Proceedings of the National Academy of Sciences*, **108:** 6674-6679.

Baisakh, N., Subudhi, P.K. and Parami, N.P. 2006. cDNA-AFLP analysis reveals differential gene expression in response to salt stress in a halophyte *Spartina altemiflora* Loisel *Plant Science*, **170:** 1141–1149.

Bellinger, Y. and Larher, F. 1987. Proline accumulation in higher plants: A redox buffer? *Plant Physiol. (Life Sci Adv.)*, **6:** 23-27.

Ballinge, C.A., Connell, P., Wu, Y., Hu, Z., Thompson, L.J., Yin, L. and Patterson, C. 1991. Identification of CHIP, a novel tetratricopeptide repeat-containing protein that interacts with heat shock proteins and negatively regulates chaperone functions. *Mol. Cell. Biol.*, **19:** 4535-4545.

Barnett, T., Altschuler, M., McDaniel, C.N. and Mascarenhas, J.P. 1980. Heat shock induced proteins in plant cells. *Dev. Genet.*, **1:** 331–340.

Banilas, G., Korkas, E., Englezos, V., Nisiotou, A.A. and Hatzopoulos, P. 2012. Genome wide analysis of the heat shock protein 90 gene family in grapevine (*Vitis vinifera* L.). *Australian Journal of Grape and Wine Research*, **18:** 29-38.

Banti, V., Mafessoni, F., Loreti, E., Alpi, A. and Perata, P. 2010. The heat-inducible transcription factor HsfA2 enhances anoxia tolerance in *Arabidopsis*. *Plant Physiol.*, **152:** 1471–1483.

Barrientos, M.M., Hermosa, R., Cardoza, R.E., Gutiérrez, S., Nicolás, C. and Monte, E. 2010. Transgenic expression of the *Trichoderma harzianum* HSP70 gene increases *Arabidopsis* resistance to heat and other abiotic stresses. *Journal of plant physiology*, **167:** 659-665.

Barque, J.P., Abahamid, A., Chacun, H. and Bonaly, J. 1996. Different heat-shock proteins are constitutively overexpressed in cadmium and pentachlorophenol adapted *Euglena gracilis* cells. *Biochemical and Biophysical Research Communications*, **223:** 7-11.

Basha, E., Friedrich, K.L. and Vierling, E. 2006. The N-terminal arm of small heat shock proteins is important for both chaperone activity and substrate specificity. *J Biol Chem.*, **281:** 39943–39952.

Basha, E. Jones, C., Wysocki, V. and Vierling, E. 2010. Mechanistic differences between two conserved classes of small heat shock proteins found in the plant cytosol. *Journal of Biological Chemistry,* **285(15):** 11489–11497.

Batra, G., Chauhan, V.S., Singh, A., Sarkar, N.K. and Grover, A. 2007. Complexity of rice HSP100 gene family: lessons from rice genome sequence data. *Journal of biosciences*, **32:** 611-619.

Baniwal, S.K.,Bharti, K., Chan, K.Y., Fauth, M., Ganguli, A., Kotak, S., Mishra, S.K., Nover, L., Port, M., Scharf, K., Tripp, L., Weber, C., Zielinski, D., von Koskull-Döring, P. 2004. Heat stress response in plants: a complex game with chaperones and more than 20 heat stress transcription factors. *J. Biosci.*, **29:** 471–487.

Beard, J.B. 2002. Turf Management for Golf Courses. Chelsea, MI: Ann Arbor Press.

Berardini, T.Z., Bollman, K., Sun, H. and Poethig, S.R. 2001. Regulation of vegetative phase change in *Arabidopsis thaliana* by cyclophilin 40. *Science Signaling.*, **291:** 2405.

Biamonti, G and Caceres, J.F. 2009. Cellular stress and RNA splicing. *Trends Biochem. Sci.*, **34:** 146–153.

Bita, C.E. and Gerats, T. 2013. Plant tolerance to high temperature in a changing environment: scientific fundamentals and production of heat stress-tolerant crops. *Front. Plant Sci.* **4:** 273.

Blumenthal, C. Bekes, F., Wrigley, C.W. and E. W. Barlow, E.W. 1990. The acquisition and aintenance of thermotolerance in Australian wheats. *Australian Journal of Plant Physiology,* **17(1):** 37–47.

Bonos, S.A. and J. A. Murphy, J.A. 1999. Growth responses and performance of Kentucky bluegrass under summer stress. *Crop Science*, **39(3):** 770–774.

Boorstein, W.R., Ziegelhoffer, T. and Craig, E.A. 1994. Molecular evolution of the HSP70 multi gene family. *Journal of Molecular Evolution*, **38(1):** 1–17.

Barua, D. and Heckathom, S.A. 2006. The interactive effects of light and temperature on heat-shock protein accumulation in *Solidago altissima* (Asteraceae) in the field and laboratory. *Am. J. Bot.*, **93(1):** 102-109.

Boston, R.S., Viitanen, P.V. and Vierling, E. 1996. Molecular chaperones and protein folding in plants. *Plant Mol. Biol.*, **32:** 191–222.

Bösl, B., Grimminger, V. and Walter, S. 2006. The molecular chaperone HSP104: A molecular machine for protein disaggregation. *J. Struct. Biol.*, **156:** 139-148.

Boston, R.S., Viitanen, P.V. and Vierling, E. 1996. Molecular chaperones and protein folding in plants. *Plant Molecular Biology*, **32(1-2):** 191–222.

Bowen, J., Michael, L-Y., Plummer, K.I.M. and Ferguson, I.A.N. 2002. The heat shock response is involved in thermotolerance in suspension-cultured apple fruit cells. *J. Plant Physiol.*, **159:**599–606.

Brady C.J., Gibson, T.S., Barlow, E.W.R., Speirs, J., Wyn Jones, R.G. 1984. Salt-tolerance in plants. I. Ions, compatible organic solutes and the stability of plant ribosomes. *Plant, Cell and Environment*, **7:** 571-578.

Braig, K., Otwinowski, Z., Hegde, R., Boisvert, D.C., Joachimiak, A., Horwich, A.L. and Sigler, P.B. 1994. The crystal structure of the bacterial chaperonin GroEL at 2.8 A. *Nature*, **371(6498):** 578–586.

Briens, M. and Larher, F. 1982. Osmoregulation in halophytic higher plants: a comparative study of soluble carbohydrates, polyols, betaines and free proline. *Plant, Cell and Environment,* **5:** 287-292.

Bukau, B., Deuerling, E., Pfund, C. and Craig, E. 2000. Getting newly synthesized proteins into shape. *Cell*, **101:** 119-122.

Bukau, B., Weissman, J. and Horwich, A. 2006. Molecular chaperones and protein quality control. *Cell*, **125:** 443–451.

Burke, T.J., Hartfield, J.L., Klein, R.R. and Mullet, J.E. 1985. Accumulation of heat shock proteins in field-grown cotton. *Plant Physiol.*, **78:** 394-398.

Byrt, C.S., Platten, J.D., Spielmeyer, W., James, R.A., Lagudah, E.S., Dennis, E.S., Tester, M. and Munns, R. 2007. HKT1;5-like cation transporters linked to Na^+ exclusion loci in wheat , *Nax2 and Kna1*. *Plant Physiology*, **143:** 1918–1928.

Calderwood, S.K. 2010. Heat shock proteins in breast cancer progression-A suitable case for treatment? *International Journal of Hyperthermia*, **26(7):** 681-685.

Camejo, D., Rodriguez, P., Morales, M.A., Dell'amico, J.M., Torrecillas, A., and Alarcon, J.J., 2005. High temperature effects on photosynthetic activity of two tomato cultivars with different heat susceptibility. *J. Plant Physiol.*, **162:** 281-289.

Campbell, J.L., Klueva, N.Y., Zheng, H.G., Nieto-Sotelo, J., Ho, T.H. and Nguyen, H.T. 2001. Cloning of new members of heat shock protein HSP101 gene family in wheat (*Triticum aestivum* (L.) Moench) inducible by heat, dehydration, and ABA. *Biochimica Biophysica Acta*, **1517:** 270–277.

Caruso, G., Cavaliere, C., Guarino, C., Gubbiotti, R. and Foglia, P. 2008. Lagana A. Identification of changes in *Triticum. durum* L. leaf proteome in response to salt stress by two-dimensional electrophoresis and MALDI-TOF mass spektrometry. *Anal. Bioanal. Chem.*, **391:** 381–390.

Cazebonne, C., Vega, A., Varela, D. and Cardemil, L.1999. Salinity effects on germination and growth of *Prosopis chilensis*. *Revista Chilena de Historia Natural*, **72:** 83–91.

Cha, J.Y., Ermawati, N., Jung, M.H., Suudi, M., Kim, K.Y., Han, C.D., Lee, K.H. and Son, D. 2009. Characterization of orchardgrass p23, a flowering plantHSP90 cohort protein. *Cell Stress and Chaperones*, **14(3):** 233–243.

Cha, J.Y., Jung, M.H., Ermawati, N., Suudi, M., Rho, G.J., Han, C.D., Lee, K.H. and Son, D. 2009. Functional characterization of orchardgrass endoplasmic reticulum resident HSP90 (DgHSP90) as a chaperone and an ATPase. *Plant Physiology and Biochemistry,* **47(10):** 859–866. Chakraborty, K., Sairam, R.K. and Bhattacharya, R.C. 2012. Differential expression of salt overly sensitive pathway genes determines salinity stress tolerance in *Brassica* genotypes. *Plant Physiol and Biochemistry*, **51:** 90–101.

Charng, Y., Liu, H., Liu, N., Hsu, F. and Ko, S. 2006. *Arabidopsis* HSA32, a novel heat shock protein, is essential for acquired thermotolerance during long recovery after acclimation. *Plant Physiol.,* **140:** 1297-1305.

Charng, Y.Y., Liu, H.C., Liu, N.Y., Chi, W.T., Wang, C.N., Chang, S.H. and Wang, T.T. 2007. A heat-inducible transcription factor, HSFA2, is required for extension of acquired .thermotolerance in *Arabidopsis*. *Plant Physiol.*, **143:** 251–262.

Chan, N.C., Likic, V.A., Waller, R.F., Mulhern, T.D. and Lithgow, T. 2006. The C-terminal TPR domain of Tom70 defines a family of mitochondrial protein import receptors found only in animals and fungi. *J. Mol. Biol.*, **358:** 1010-1022.

Chang, Y.W., Sun, Y.J., Wang, C. and Hsiao, C.D. 2008. Crystal structures of the 70-kDa heat shock proteins in domain disjoining conformation. *Journal of Biological Chemistry*, **283(22):** 15502–15511.

Chauhan, H., Khurana, N., Agarwal, P. and Khurana, P. 2011. Heat shock factors in rice (*Oryza sativa L.):* Genome-wide expression analysis during reproductive development and abiotic stress. *Mol. Genet. Genomics*, **286:** 171–187.

Chen, H.H., Shen, Z.Y. and Li, P.H. 1982. Adaptability of crop plants to high temperature stress. *Crop Science*, **22:** 719-725.

Chen, Q., Lauzon, L., DeRocher, A. and Veirling, E. 1990. Accumulation stability and localization of major chloroplast heat shock protein. *J Cell Biol.*, **110:** 1873-1883.

Chen, W.Q.J. and Zhu, T. 2004. Networks of transcription factors with roles in environmental stress response. *Trends Plant Sci.*, **9:** 591–596.

Chen, S., Gollop, N. and Heuer, B. 2009. Proteomic analysis of salt-stressed tomato (*Solanum lycopersicum*) seedlings: effect of genotype and exogenous application of glycinebetaine. *J. Exp. Bot.*, **60:** 2005–2019.

Chitteti, B. and Peng, Z. 2007. Proteome and phosphoproteome differential expression under salinity stress in rice (*Oryza sativa*) roots. *J. Proteome Res.*, **6:**1718–1727.

Chou, M., Chen, Y.M. and Lin, C.Y. 1989. Thermotolerance of isolated mitochondria associated with heat shock protein. *Plant Physiol.*, **89:** 617-621.

Christensen, J.H., Hewitson, B., Busuioc, A., Chen, A., Gao, X., Held, I., Jones, R., Kolli, R.K., Kwon, W.-T., Laprise, R., Magaña Rueda, V., Mearns, L., Menéndez, C.G., Räisänen, J., Rinke, A., Sarr, A. and Whetton, P. 2007. Regional Climate Projections. In, "*Climate Change 2007: The* Physical Science Basis. Contribution of Working Group I to the Fourth Assessment Report of *the 3 Intergovernmental Panel on Climate Change*", Eds. S. Solomon, Qin, D., Manning, M., Chen, Z., Marquis, M. Averyt, K.B. Tignor, M. and Miller, H.L., Cambridge University Press, Cambridge, United Kingdom and New York, NY, USA.

Connell, D., Lam, P., Richardson, B. and Wu, R. 1999. Introduction to Ecotoxicology. Oxford, UK: Blackwell Science Ltd, 71.

Cooper, O., Ho, T.D. and Hauptmann, R.M. 1984. Tissue specificity of the heat-shock response in maize. *Plant Physiology*, **75:** 431–441.

Cottee, N.S., Wilson, I.W., Tan, D.K. and Bange, M.P. 2014. Understanding the molecular events under-pinning cultivar differences in the physiological performance and heat tolerance of cotton (*Gossypium hirsutum*). *Funct. Plant Biol.*, **41:** 56-67.

Craig, E.A. and Gross, C.A. 1991. Is HSP70 the cellular thermometer? *Trends Biochem. Sci.*, **16:** 135-140. Cui, L., Cao, R., Li, J., Zhang, L. and J. Wang, J. 2006. High temperature effects on ammonium assimilation in leaves of two *Festuca arundinacea* cultivars with different heat susceptibility. *Plant Growth Regulation*, **49(2-3):** 127–136.

Czar, M.J., Galigniana, M.D., Silverstein, A.M., Pratt, W.B. and Geldanamycin, W.B. 1997. A heat shock protein 90-binding benzoquinone ansamycin, inhibits steroid-dependent translocation of the glucocorticoid receptor from the cytoplasm to the nucleus. *Biochem.*, **36:**, 7776-7785. Dash, S. and N. Mohanty, N. 2002. Response of seedlings to heatstress in cultivars of wheat: growth temperature-dependent differential modulation of photosystem 1 and 2 activity, and foliar antioxidant defense capacity. *Journal of Plant Physiology*, **159(1):** 49–59.

Daugaard, M., Rohde, M. and Jäättelä, M. 2007. The heat shock protein 70 family: Highly homologous proteins with overlapping and distinct functions. *FEBS letters*, **581:** 3702-3710.

Debel, K., Knack, G. and Kloppstech, K. 1994. *Accumulation of plastid HSP 23 of Chenopodium rubrum is controlled post-translationally by light. Plant Journal*, **6:** *79-85.* de Jong, W.W., Leunissen, J.A.M. and C. E.M.Voorter, C.E.M. 1993. Evolution of the α- crystallin/small heat-shock protein family. *Molecular Biology and Evolution,* **10(1):** 103–126.

de Jong, W.W., Caspers, G.J. and Leunissen, J.A.M. 1998. Genealogy of the á-crystallin-small heat-stress superfamily. *Int. J. Biol. Macromol.*, **22:** 151–162.

De Maio, A. 1999. Heat shock proteins: facts, thoughts, and dreams. *Shock (Augusta, Ga.)*, **11:** 1–12.

De Ronde, J.A., Van Der Mescht, A. and Cress, W.A. 1993. Heat-shock protein synthesis in cotton is cultivar dependent. *S. Afr. J. Plant Soil*, **10:** 95-97.

De Souza, M.A., Pimentel, A.J.B. and Ribeiro, G. 2012. Breeding for heat-stress tolerance. In, *"Plant breeding for abiotic stress tolerance"*. Springer, pp. 137-156.

Demirevska-Kepova, K., H¨olzer, R., Simova-Stoilova, L. and Feller, U. 2005. Heat stress effects on ribulose-1,5-bisphosphate carboxylase/oxygenase, Rubisco binding protein and Rubisco activase in wheat leaves. *Biologia Plantarum,* **49(4)**:,521–525.

Desikan, R., Hancock, J.T. and Neill, S.J. 2004. Oxidative stress signaling. In, "*Plant responses to abiotic stress*," Eds. H. Hirt and K. Shinozaki, Berlin Heidelberg: Springer-Verlag; pp. 73–93.

Dickson, R., Weiss, C., Howard, R.J. Alldrick, S.P., Ellis, R.J., Lorimer, G., Azem, A. and Vitanen, P.V. 2000. Reconstitution of higher plant chloroplast chaperonin 60 tetradecamers active in protein folding. *Journal of Biological Chemistry*, **275(16):** 11829–11835.

Ding, D., Zhang, L., Wang, H., Liu, Z., Zhang, Z. and Zheng, Y. 2009. Differential expression of miRNAs in response to salt stress in maize roots. *Annals of Botany*, **103:** 29-38.

Dobson, C.M., Sali, A. and Karplus, M. 1998. Protein folding: a perspective from theory and experiment. *Angew Chem. Int. Ed.*, **37:** 868–893.

Dooki, A.D., Mayer-Posner, F.J., Askari, H., Zaiee, A.A. and Salekdeh, G.H. 2006. Proteomic responses of rice young panicles to salinity. *Proteomics*, **6:** 6498–6507.

Downs C.A., Heckathorn, S.A., Bryan, J.K. and Coleman, J.S. 1997. The methionine-rich low-molecular-weight chloroplast heat-shock protein: evolutionary conservation and accumulation in relation to thermotolerance. *American Journal of Botany*, **85:** 175-183.

Downs, C.A. and Heckathorn, S.A. 1998. The mitochondrial small heat shock protein protects NADH: ubiquinone oxidoreductase of the electron transport chain during heat stress in plants. *Febs Lett.*, **430:** 246–250.

Downs, C.A., Heckathorn, S.A., Bryan, J.K., Coleman, J.S. 1998. The methionine-rich low molecular weight chloroplast heat shock protein: Evolutionary conservation and accumulation in relation to thermotolerance. *American Journal of Botany*, **85(2):** 175-183.

Downs, C.A., Ryan, S.L. and Heckathorn, S.A. 1999. The chloroplast small heat-shock protein: evidence for a general role in protecting photosystem II against oxidative stress and photoinhibition. *Journal of Plant Physiology*, **155:** 488-496.

Drzymalla, C., Schroda, M. and Beck, C.F. 1996. Light-inducible gene HSP70B encodes a chloroplast- localized heat shock protein in Chlamydomonas reinhardtii. Plant Molecular Biology, **31***:* 1185-*1194.*

Duan, Y.H., Guo, J., Ding, K., Wang, S.J., Zhang, H., Dai, X.W., Chen, Y.Y., Govers, F., Huang, L.L. and Kang, Z.S. 2011. Characterization of a wheat HSP70 gene and its expression in response to stripe rust infection and abiotic stresses. *Molecular biology reports*, **38:** 301-307.

Duble, R.L. 2001. Turfgrasses: Their Management and Use in the Southern Zone. College Station, TX: Texas A*and*M University Press.

Ebrahim, A.M.H. and Quick, J.S. 2001. Genetic control of high temperature tolerance in wheat as measured by membrane thermal stability, *Crop. Sci.*, **41:** 1405-1407.

Efeoglu, B. 2009. Heat shock proteins and heat shock response in plants. *G.U. Journal of Science,* **22(2):** 67-75.

Eriksson, M.J. and Clarke, A.K. 1996. The heat shock protein ClpB mediates the development of thermotolerance in the cyanobacterium *Synechococcus* sp. strain PCC 7942. *J. Bacteriol.*, **178:** 4839–4846.

Essemine, J., Ammar, S. and Bouzid, S. 2010. Impact of heat stress on germination and growth in higher plants: Physiological, biochemical and molecular repercussions and mechanisms of defense. *J. Biol. Sci.,* **10:** 565-572.

Falk, S. and Sinning, I. 2010. cpSRP43 is a novel chaperone specific for light-harvesting chlorophyll a, b-binding proteins. *J. Biol. Chem.*, **285:** 21655–21661.

Fatehi, F., Hosseinzadeh, A., Alizadeh, H., Brimavandi, T. and Struik, P.C. 2012. The proteome response of salt-resistant and salt-sensitive barley genotypes to long-term salinity stress. *Mol. Biol. Rep.*, **39:** 6387-6397.

Feder, E.M. 1999. Organismal, ecological, and evolutionary aspects of heat shock proteins and the stress response: established conclusions and unresolved issues. *American Zoologist*, **39(6):** 857-864.

Feder, E.M. and Hofman G.E. 1999. Heat-shock proteins, molecular chaperons, and the stress response. *Annual* Review of Physiology, **61:** 243-282.

Feder, M.E., 2006. Integrative biology of stress: molecular actors, the ecological theater, and the evolutionary play. *International Symposium on Environmental Factors, Cellular Stress and Evolution, Varanasi, India, October 13–15, 2006*, p. 21.

Fender, S.E. and O'Connell, M.A. 1990. Expression of the heat shock response in a tomato interspecific hybrid is not intermediate between the two parental responses. *Plant Physiol.*, **93:**1140-1146.

Ferguson, D.L., Guikema, J.A. and Paulsen, G.M. 1990. Ubiquitin pool modulation and protein degradation in wheat roots during high temperature stress. *Plant Physiol.*, **92:** 740-746.

Fernandes, C., Fontainhas-Fernendes, A., Cabral, D. and Salgado, M.A. 2008. Heavy metals in water, Sediment and tissues of Liza saliens from Esmoriz-Paramos lagoon, Portugal. *Environ. Monit. Assess*, **136:** 267-275.

Ferreira, S., Hjernø, K., Larsen, M., Wingsle, G., Larsen, P., Fey, S., Roepstorff, P. and Salome Pais, M. 2006. Proteome profiling of *Populus euphratica* Oliv. upon heat stress. *Annals of Botany*, **98(2):** 361–377.

Fink, A.L. 1999. Chaperone-mediated protein folding. *Physiological Reviews*, **79(2):** 425–449.

Finka, A., Cuendet, A.F.H., Maathuis, F.J., Saidi, Y. and Goloubinoff, P. 2012. Plasma membrane cyclic nucleotide gated calcium channels control land plant thermal sensing and acquired thermotolerance. *The Plant Cell Online,*, **24:** 3333-3348.

Flaherty, K.M., DeLuca-Flaherty, C. and McKay, D.B. 1990. Three-dimensional structure of the ATPase fragment of a 70K heat-shock cognate protein. *Nature*, **346(6285):** 623–628.

Flom, G., Weekes, J., Williams, J.J. and Johnson, J.L. 2006. Effect of mutation of the tetratrico peptide repeat and asparatate proline 2 domains of Sti1 onHSP90 signaling and interaction in *Saccharomyces cerevisiae. Genetics*, **172(1):** 41–51.

Flowers, T.J. and Yeo, A.R. 1995. Breeding for salinity resistance in crop plants: Where next? *Aust. J. Plant Physiol.*, **22:** 875–884.

Franca, S., Vinagre, C., Cacador, I. and Cabral, H.N. 2005. Heavy metal concentrations in sediment, benthic invertebrates and fish in three salt marsh areas subjected to different pollution loads in the Tagus Estuary (Portugal). *Marine Pollut. Bull.*, **50:** 993-1018.

Frank, G., Pressman, E., Ophir, R., Althan, L., Shaked, R., Freedman, M., Shen, S. and Firon, N. 2009. Transcriptional profiling of maturing tomato (*Solanum lycopersicum* L.) microspores reveals the involvement of heat shock proteins, ROS scavengers, hormones, and sugars in the heat stress response. *J. Exp. Bot.*, **60:** 3891–3908.

Frova, C. 1997. Genetic dissection of thermotolerance in maize. In, "*Physical Stress in Plants,*" Eds. S. Grillo and A. Leone, Springer-Verlag, New York , pp. 31–38.

Fry, J. and Huang, B. 2004. Applied Turfgrass Science and Physiology, John Wiley *and* Sons, Hoboken, NJ, USA.

Fu, J.M. and Huang, B.R. 2003. Effects of foliar application of nutrients on heat tolerance of creeping bentgrass. *J Plant Nutrition*, **26:** 81–96.

García-Cardeña, G., Fan, R., Shah, V., Sorrentino, R., Cirino, G., Papapetropoulos, A. and Sessa, W.C. 1998. Dynamic activation of endothelial nitric oxide synthase by HSP90. *Nature.*, **392:** 821-824.

Geissler, N., Hussin, S. and Koyro, H.W. 2010. Elevated atmospheric CO_2 concentration enhances salinity tolerance in *Aster tripolium* L. *Planta,* **231:** 583–594.

Gilliham, M., Dayod, M., Hocking, B.J., Xu, B., Conn, S.J., Kaiser, B.N., Leigh, R.A. and Tyerman, S.D. 2011. Calcium delivery and storage in plant leaves: exploring the link with water flow. *Journal of experimental botany*, **62(7):** 2233-2250.

Giorno, F., Wolters-Arts, M., Grillo, S., Scharf, K , Vriezen, W.H. and Mariani, C. 2010. Developmental and heat stress-regulated expression of HsfA2 and small heat shock proteins in tomato anthers. *J. Exp. Bot.*, **61:** 453–462.

Glaczinski, N. and Kloppstech, K. 1988. Temperature-dependent binding to the thylakoid membranes of nuclear-coded chloroplast heat shock proteins. *European J Biochem.*, **173:** 579-583.

Glatz, A., Horvath, I., Varvasovszki, V., Kovacs, E., Torok, Z. and Vigh, L. 1997. Chaperonin genes of the Synechocystis PCC 6803 are differentially regulated under light-dark transition during heat stress. *Biochem and Biophy Res Comm*, **239:** 291-*297.*

Gomathi, R. and Vasantha, 2006. Change in nucleic acid content and expression of salt shock proteins in relation to salt tolerance in sugarcane. *Sugar Tech.*, **8:** 124-127.

Gorski, S.M., Chittaranjan, S., Pleasance, E.D., Freeman, J.D., Anderson, C.L., Varhol, R.J., Coughlin, S.M., Zuvderduvn, S.D., Jones, S.J. and Marra, M.A. 2003. A SAGE approach to discovery of genes involved in autophagic cell death. *Current Biology*, **13:** 358-363.

Grigorova, B., Vaseva, I., Demirevska, K. and Feller, U. 2011. Combined drought and heat stress in wheat: changes in some heat shock proteins. *Biologia Plantarum*, **55:** 105-111.

Guan, J.C., Jinn, T.L., Yeh, C.H., Feng, S.P., Chen, Y.M. and Lin, C.Y. 2004. Characterization of the genomic structures and selective expression profiles of nine class I small heat shock protein genes clustered on two chromosomes in rice (Oryza sativa L.). *Plant Molecular Biology*, **56(5):** 795– 809.

Guo, F., Maurizi, M.R., Esser, L. and DI. Xia, D.I. 2002. Crystal structure of C1pA, an HSP100 chaperone and regulator of C1pAP protease. *Journal of Biological Chemistry*, **277(48):** 46743–46752.

Guo, S., Zhou, H., Zhang, X., Li, X. and Meng, Q. 2007. Overexpression of CaHSP26 in transgenic tobacco alleviates photoinhibition of PSII and PSI during chilling stress under low irradiance. *J. Plant Physiol.*, **164(2):** 126-136.

Gupta, S.C., Sharma, A., Mishra, M., Mishra, R.K. and Chowdhuri, D.K. 2010. Heat shock proteins in toxicology: How close and how far? *Life Sciences*, **86:** 377-384.

Gurley, W.B. 2000. HSP101: a key component for the acquisition of thermotolerance in Plants. *Plant Cell.*, **12(4):** 457–460.

Guyomarc'h, S., Vernoux, T., Traas, J., Zhou, D.X. and Delarue, M. 2004. MGOUN3, an *Arabidopsis* gene with tetratrico peptide-repeat-related motifs, regulates meristem cellular organization. *J. Exp. Bot.*, **55:** 673-684.

Guy, C.L. and Li, Q. 1998. The organization and evolution of the spinach stress 70 molecular chaperone gene family. *The Plant Cell Online,* **10:** 539-556.

Haddadi, F. 2009. Development of a plant regeneration system and analysis of 101 heat shock protein in strawberry Cv.Camarosa following gene bombardment. *M. Sc thesis*, **2009,** 30-35

Hahn, J.S., Hu, Z.., Thiele, D.J. and Iyer, V.R. 2004. Genome-wide analysis of the biology of stress responses through heat shock transcription factor. *Mol. Cell. Biol.*, **24:** 5249–5256.

Hahn, A., Bublak, D., Schleiff, E. and Scharf, K.-D. 2011. Crosstalk between HSP90 and HSP70 chaperones and heat stress transcription factors in tomato. *Plant Cell*, **23:** 741–755.

Hall, J.L. 2002. Cellular mechanisms for heavy metal detoxification and tolerance. *Journal of Experimental Botany*, **53:** 1-11.

Hamilton III, E.W. and Heckathorn, S.A. 2001. Mitochondrial adaptations to NaCl stress: Complex I is protected by anti-oxidants and small heat shock proteins, whereas Complex II is protected by proline and betaine. *Plant Physiology*, **126:** 1266–1274.

Hamilton II, E.W., McNaughton, S.J. and Coleman, J.S. 2001. Molecular, physiological, and growth responses to sodium stress in C4 grasses from a soil salinity gradient in the Serengeti ecosystem. *Am. J. Bot*., **88(7):** 1258-1265

Hashimoto, M., Kisseleva, L., Sawa, S., Furukawa, T., Komatsu, S. and Koshiba, T. 2004. A novel rice PR10 protein, RSOsPR10, specifically induced in roots by biotic and abiotic stress, possibly via the jasmonic acid signalling pathway. *Plant Cell Physiol.*, **45:** 550–559.

Haslbeck, M., Franzmann, T., Weinfurtner, D. and Buchner, J. 2005. Some like it hot: the structure and function of small heat shock proteins. *Nature Structural and Molecular Biology,* **12(10):** 842–846.

Hare, P.D. Cress, W.A. and Van Staden, J. 1998. Dissecting the roles of osmolyte accumulation during stress. *Plant, Cell and Environment*, **21:** 535-553.

Hartl, F.U. 1999. Molecular chaperones in cellular protein folding. *Nature*, **381(6583):** 571–580.

Hartl, F.U. and Hayer-Hartl, M. 2002. Molecular chaperones in the cytosol: From nascent chain to folded protein. *Science.*, **295:** 1852-1858.

Hartl, F.U. and Hayer-Hartl, M. 2009. Converging concepts of protein folding in vitro and in vivo. *Nat. Struct. Mol. Biol.*, **16:** 574-581.

He, Y., Liu, X. and Huang, B. 2005. Protein changes in response to heat stress in acclimated and non-acclimated creeping bentgrass. *Journal of the American Society for Horticultural Science*, **130(4):** 521–526.

He, Y. and Huang, B. 2007. Protein changes during heat stress in three Kentucky bluegrass cultivars differing in heat tolerance. *Crop Science*, **47(6):** 2513–2520.

Heckathorn, S.A. Downs, C.A. and Coleman, J.S. 1998. Nuclear-encoded chloroplast proteins accumulate in the cytosol during severe heat stress. *International Journal of Plant Science*, **159:** 39-45.

Hemmingsen, S.M., Woolford, C., van der Vies, S.M. Tily, K., Dennis, D.T., Georgopoulos, C.P., Hendrix, R.W. and Ellis, R.J. 1988. Homologous plant and bacterial proteins chaperone oligomeric protein assembly. *Nature*, **333(6171):** 330–334.

Hernandez, E. and Vierling, E. 1993. Heat shock protein expression in the field. *Plant Physiology*, **101:** 1209–1216.

Hendrick, J.P. and Hartl, F.U. 1993. Molecular chaperone functions of heat-shock proteins. *Annu. Rev. Biochem,* **62:** 349-384.

Hester, M.W., Mendelssohn, I.A. and Mckee, K.L. 1996. Intraspecific variation in salt tolerance and morphology in the coastal grass *Spartina patens* (Poaceae). *American Journal of Botany*, **83:** 1521-1527.

Hijmans, R.J. 2003. The effect of climate change on global potato production. *Am. J. Potato Res.*, **80:** 271-280

Hihara, Y., Kamei, A., Kanehisa, M., Kaplan, A. and Ikeuchi, M. 2001. DNA microarray analysis of cyanobacterial gene expression during acclimation to high light. *Plant Cell*, **13:** 793-806.

Holt, S.E., Aisner, D.L., Baur, J., Tesmer, V.M., Dy, M., Ouellette, M. and Shay, J.W. 1991. Functional requirement of p23 and HSP90 in telomerase complexes. *Genes Dev.*, **13:** 817-826.

Hong, S.W. and Vierling, E. 2000. Mutants of Arabidopsis thaliana defective in the acquisition of tolerance to high temperature stress. *Proceedings of the National Academy of Sciences of the United States of America*, **97(8):** 4392–4397.

Horwich, A.L., Low, K.B., Fenton, W.A., Hirshfield, I.N. and Furtak, K. 1993. Folding *in vivo* of bacterial cytoplasmic proteins: role of GroEL. *Cell*, **74(5):** 909–917.

Horwich, A.L. 1995. Molecular chaperones: resurrection or destruction? *Current Biology*, **5(5):** 455–458.

Howarth, C.J. 1989. Heat shock proteins in *Sorghum bicolor* and *Pennisetum americanum*.I. genotypic and developmental variation during seed germination. *Plant Cell and Environment*, **12(5):** 471–477.

Howarth, C.J. 1990. Heat shock proteins in sorghum and pearl millet – ethanol, sodium arsenite, sodium malonate and the development of thermotolerance. *Expt. Bot.*, **41:** 877-883.

Howarth, C.J. and Skot, K.P. 1994. Detailed characterization of heat shock protein synthesis and induced thermotolerance in seedlings of *Sorghum bicolor* L. *Expt Bot.*, **45:** 1353-1363.

Hsieh, M.H., Chen, J.T., Jinn, T.L., Chen, Y.M. and Lin, C.Y. 1992. A class of soybean low molecular weight heat shock proteins. Immunological study and quantitation. *Plant Physiol.,* **99:** 1279-1284.

Hu, W., Hu, G. and Han, B. 2009. Genome-wide survey and expression profiling of heat shock proteins and heat shock factors revealed overlapped and stress specific response under abiotic stresses in rice. *Plant Sci.* **176:** 583-590.

Huang, B., Liu, X. and Fry, J.D. 1998. Shoot physiological responses of two bentgrass cultivars to high temperature and poor soil aeration. *Crop Science*, **38(5):** 1219–1224.

Huang, S., Spielmeyer, W., Lagudah, E.S., James, R.A., Platten, J.D., Dennis, E.S. and Munns, R. 2006. A sodium transporter (HKT7) is a candidate for *Nax1*, a gene for salt tolerance in durum wheat. *Plant Physiology*, **142:** 1718–1727.

Huang, B. and Xu, C. 2008. Identification and characterization of proteins associated with plant tolerance to heat stress. *Journal of integrative plant biology*, **50:** 1230-1237.

Iannucci, A., Rascio, A., Russo, M., Di Fonzo, N. and Martiniello, P. 2000. Physiological responses towater stress following a conditioning period in berseem clover. Plant Soil, **223:** 217-227. IPCC, 2007. Contribution of Working Groups I, II and III to the Fourth Assessment Report of the Intergovernmental Panel on Climate Change Core Writing Team (Eds R.K. Pachauri, A. Reisinger) IPCC, Geneva, Switzerland.

Iba, K. 2002. Acclimative response to temperature stress in higher plants: Approaches of genetic engineering for temperature tolerance, *Annu. Rev. Plant. Biol.*, **53:** 225-245.

Ibok, U.J., Udosen, E. D. and Udoidiong, O.M. 1989. Heavy metals in fishes from some streams in Ikot Ekpene area of Nigeria. *Nig. J. Tech. Res*, **1:** 61-68.

Igarashi, Y., Yoshiba, Y., Sananda, Y., Yamaguchishinozaki, K., Wada, K. and Shinozaki, K. 1997. Characterization of the gene for delta (l)-pyrroline-5-carboxylate synthetase and correlation between the expression of the gene and salt tolerance in (*Oryza sativa* L.). *Plant Mol. Biol.*, **33:** 857-865.

Ireland, H.E., Harding, S.J., Bonwick, G.A., Jones, M., Smith, C.J. and J.H. Williams, H.W. 2004. Evaluation of heat shock protein 70 as a biomarker of environmental stress in *Fucus serratus* and *Lemna minor*. *Biomarkers*, **9:** 139-155.

Iqbal, N., Farooq, S., Arshad, R. and Hameed, A. 2010. Differential accumulation of high and low molecular weight heat shock proteins in Basmati rice (*Oryza sativa* L.) cultivars. *Genetic Resources and Crop Evolution*, **57:** 65-70.

Izhaki, A., Swain, S.M., Tseng, T.S., Borochov, A., Olszewski, N.E. and Weiss, D. 2001. The role of SPY and its TPR domain in the regulation of gibberellin action throughout the life cycle of *Petunia hybrid* plants. *Plant J.*, **28:** 181-190.

Jackson-Constan, D., Akita, M. and Keegstra, K. 2001. Molecular chaperones involved in chloroplast protein import. *Biochimica Et Biophysica Acta (BBA)-Mol. Cell Res.*, **1541:** 102-113.

Jain, S., Srivastava, S., Sarin, N.B. and Kav, N.N.V. 2006. Proteomics reveals elevated levels of PR10 proteins in saline-tolerant peanut (*Arachis hypogaea*) calli. *Plant Physiol. Biochem.*, **44:** 253–259.

James, R.A., Davenport, R.J. and Munns, R. 2006. Physiological characterisation of two genes for Na^+ exclusion in durum wheat, *Nax1* and *Nax2* . *Plant Physiology*, **142:** 1537–1547.

Jaru-Ampornpan, P., Shen, K., Lam, V.Q., Ali, M., Doniach, S., Jia, T.Z. and Shan, S.O. 2010. ATP-independent reversal of a membrane protein aggregate by a chloroplast SRP subunit, *Nat. Struct. Mol. Biol.,* **17:** 696–702.

Jaya, N., Garcia, V. and Vierling, E. 2009. Substrate binding site flexibility of the small heat shock protein molecular chaperones. *Proc. Natl. Acad. Sci.*, **106:** 15604-15609.

Jiang, Y. and Huang, B. 2001. Drought and heat stress injury to two cool-season turf-grasses in relation to antioxidant metabolism and lipid peroxidation. *Crop Science*, **41(2):** 436–442.

Jiang, J., Prasad, K., Lafer, E.M. and Sousa, R. 2005. Structural basis of interdomain communication in the Hsc70 chaperone. *Molecular Cell*, **20(4):** 513–524.

Jiang, Y., Yang, B., Harris, N.S. and Deyholos, M.K. 2007. Comparative proteomic analysis of NaCl stress-responsive proteins in *Arabidopsis* roots. *J. Exp. Bot.*, **58:** 3591–3607.

Jellouli, N., Ben Jouira, H., Skouri, H., Ghorbel, A., Gourgouri, A. and Mliki, A. 2008. Proteomic analysis of Tunisian grapevine cultivar Razegui under salt stress. *J. Plant Physiol.*, **165:** 471–481.

Jorgensen, J.A., J. Weng, J., Ho, T.H.D. and Nguyen, H.T. 1992. Genotype-specific heat shock proteins in two maize inbreds. *Plant Cell Reports,* **11(11):** 576–580.

Joseph, B. and Jini, D. 2010. Proteomic Analysis of Salinity Stress-responsive Proteins in Plants. *Asian Journal of Plant Sciences,* **9:** 307-313.

Jovanoviæ, S.M., Tuciæ, B. and Matiæ, G. 2011. Differential expression of heat-shock proteins HSP70 and HSP90 in vegetative and reproductive tissues of *Iris pumila. Acta physiologiae plantarum*, **33:** 233-240.

Kadýoðlu, A., "Bitki Fizyolojisi", Esen Ofset, Trabzon, 217p. (2004).

Kalmar, B. and Greensmith, L. 2009. Induction of heat shock proteins for protection against oxidative stress. *Advanced drug delivery reviews*, **61:** 310-318.

Kalemba, E.M. and Pukacka, S. 2008. Changes in late embryogenesis abundant proteins and a small heat shock protein during storage of beech (*Fagus sylvatica* L.) seeds. *Environ.Exp. Bot.*, **63:** 274–280

Katiyar-Agarwal, S., Agarwal, M. and Grover, A. 2003. Heat-tolerant basmati rice engineered by over-expression of HSP101. *Plant Mol. Biol.*, **51:** 677–686.

Kavi Kishor, P.B., Sangam, S. and Amrutha, R.N. 2005. Regulation of proline biosynthesis, degradation, uptake and transport in higher plants: Its implications in plant growth and a biotic stress tolerance. *Curr. Sci.*, **88:** 424-438.

Keeler, S.J., Boettger, C.M., Haynes, J.G., Kuches, K.A., Johnson, M.M., Thureen, D.L. and Kitto, S.L. 2000. Acquired thermotolerance and expression of the HSP100/ClpB genes of lima bean. *Plant Physiol.*, **123:** 1121-1132.

Key, J.L., Lin, C.Y. and Chen, Y.M. 1981. Heat shock proteins of higher plants. *P Natl. Acad. Sci. USA*, **78:** 3526–3530.

Kilian, J., Whitehead, D., Horak, J., Wanke, D., Weinl, S., Batistic, O., Bornberg-Bauer. E., D'Angelo, Kudla, J. and K. Harter, K. 2007. The AtGenExpress global stress expression data set: protocols, evaluation and model data analysis of UV-B light, drought and cold stress responses. *Plant J.*, **50:** 347–363.

Kim, B. and Schöffl, F. 2002. Interaction between arabidopsis heat shock transcription factor 1 and 70 kDa heat shock proteins. *J. Exp. Bot.*, **53:** 371-375.

Kim, D.W., Rakwal, R., Agrawal, G.K., Jung, Y.H., Shibato, J., Jwa, N.S., Iwahashi, Y., Iwahashi, H., Kim, D.H., Shim, I.S. and Usii, K. 2005. A hydroponic rice seedling culture model system for investigating proteome of salt stress in rice leaf. *Electrophoresis*. **26:** 4521–4539.

Kloppstech, K., Mayer, G., Schuster, G. and Ohard, I. 1985. Synthesis, transport and localization of a nuclear coded 22-kd heat-shock protein in the chloroplast membranes of peas and *Chlamydomonas reinhardi. EMBO J.* **4(8):** 1902-1909.

Knack, G. and Kloppstech, K. 1992. The heat-shock response in a photoautotrophic cell-culture of Chenopodium rubrum: the effects of temperature and light. *Journal of Plant Physiology*, **140:** 489-493.

Kotak, S., Larkindale, J., Lee, U., Koskull-Döring, P., Vierling, E. and Scharf, K.-D. 2007. Complexity of the heat stress response in plants. *Curr. Opin. Plant Biol.*, **10:** 310–316.

Krishnan, M., Nguyen, H.T. and Burke, J.J. 1989. Heat shock protein synthesis and thermal tolerance in wheat. *Plant Physiology*, **90(1):** 140–145.

Krishna, P. 2003. Plant responses to heat stress. *Topics in Current Genetics*. **4:** 73-101.

Krishna, P. 2004. In, "*Plant responses to abiotic stress*", Springer, Berlin, pp. 73-93.

Krishnan, M., Nguyen, H.T. and Burke, JJ. 1989. Heat shock protein synthesis and thermal tolerance in wheat. *Plant Physiol.*, **90:** 140-145.

Krivokapich, S.J., Molina, V., Bergagna, H.F.J. and Guarnera, E.A. 2006. Epidemiological survey of *Trichinella* infection in domestic, synanthropic and sylvatic animals from Argentina. *Journal of helminthology*, **80:** 267-269.

Kropat, J., Oster, U., Rudiger, W. and Beck, C.F. 1997. Chlorophyll precursors are signals of chloroplast origin involved in light induction of nuclear heat-shock genes. *Proceedings of the National Academy of Sciences, USA*, **94:** 14168-*14172.*

Kumar, R.R., Goswami, S., Sharma, S.K., Pathak, H., Rai, G.K. and Rai, R.D. 2012. Genome wide identification of target heat shock protein90 genes and their differential expression against heat stress in wheat. *Int. J. Biochem. Res. Rev*, **2(1):** 12-30.

Kurek, I., Dulberger, R., Azem, A., Tzvi, B.B., Sudhakar, D., Christou, P. and Beriman, A. 2002. Deletion of the C-terminal 138 amino acids of the wheat FKBP73 abrogates calmodulin binding, dimerization and male fertility in transgenic rice. *Plant Mol. Biol.*, **48:** 369-381.

Larkindale, J. and Vierling, E. 2008. Core genome responses involved in acclimation to high temperature. *Plant Physiol.*, **146:** 61.

Lee, Y.R., Nagao, R.T. and Key, J.L. 1994. A soybean 101-kD heat shock protein complements a yeast HSP104 deletion mutant in acquiring thermotolerance. *Plant Cell*, **6:** 1889–1897

Lee, J.H., Hubel, A., and Schöffl, F. 1995. Derepression of the activity of genetically engineered heat shock factor causes constitutive synthesis of heat shock proteins and increased thermotolerance in transgenic *Arabidopsis*. *Plant J.*, **8:** 603–612.

Lee, G.J., Roseman, A.M., Saibil, H.R. and Vierling, E. 1997. A small heat shock protein stably binds heat-denatured model substrates and can maintain a substrate in a folding-competent state. *EMBO Journal*, **16:** 659–671.

Lee G.L. and Vierling, E. 2000. A small heat shock protein cooperates with heat shock protein 70 systems to reactivate a heat-denatured protein. *Plant Physiology*, **122(1):** 189–197.

Lee, B.H., Won, S.H., Lee, H.S., Miyao, M., Chung, W.I., Kim, I.J. and Jo, J. 2000. Expression of the chloroplast-localized small heat shock protein by oxidative stress in rice. *Gene.*, **245:** 283–290.

Lee, D.G., Ahsan, N., Lee, S.H., Kang, K.Y., Bahk, J.D., Lee, I.J. and Lee, B.H. 2007. A proteomic approach in analyzing heat-responsive proteins in rice leaves, *Proteomics*, **7(18):** 3369–3383.

Lerner, H.R. 1985. Adaptation to salinity at the plant cell level. *Plant and Soil*, **89:** 3-14.

Levitt, J. 1972. Responses of plants to environmental stresses, Physiological Ecology, Academic Press, New York, NY, USA.

Levitt, J. 1980. In, "*Responses of plants to environmental stresses: Chilling, freezing and high temperature stresses*", 2nd Ed., Ed. T.T. Kozlowski, Academic Press Inc., New York, pp. 388- 497.

Levitt, M., Gerstein, M., Huang, E., Subbiah, S. and Tsai, J. 1997. Protein folding: The endgame. *Annu. Rev. Biochem.*, **66:** 549-579.

Lewis, S., Donkin, M.E. and Depledge, M.H. 2001. HSP70 expression in *Enteromorpha intestinalis* (Chlorophyta) exposed to environmental stressors. *Aquatic Toxicology*, **51:** 277–291.

Li, Q.-B., and Guy, C.L. 2001. Evidence for non-circadian light/dark-regulated expression of HSP70s in spinach leaves. *Plant Physiology*, **125:** 1633-*1642.*

Li, G.C. and Mak, J.Y. 2009. Re-induction of HSP70 synthesis: an assay for thermotolerance. *Int. J. Hyperthermia*, **25:** 249-257.

Li, Z., Zhang, L., Wang, A., Xu, X. and Li, J. 2013. Ectopic overexpression of SlHsfA3, a heat stress transcription factor from tomato, confers increased thermotolerance and salt hypersensitivity in germination in transgenic *Arabidopsis. PLoS ONE,* **8(1):** e54880-10.1371/journal.pone.0054880.

Liming, Y., Qian, Y., Pigang, L. and Sen, L. 2008. Expression of the HSP24 gene from *Trichoderma harzianum* in *Saccharomyces cerevisiae. J. Therm. Biol.*, **33:** 1–6.

Lin, C.Y., Roberts, J.K. and Key, J.L. 1984. Acquisition of thermotolerance in soybean seedlings. *Plant Physiol.*, **74:** 152-160.

Lin, C.Y., Chen, Y.M. and Key, J.M. 1985. Solute leakage in soybean seedling under various heat shock regimes. *Plant Cell Physiol.*, **26:** 1493-1498.

Lindquist, S. 1986. Heat shock response. *Annual Review of Biochemistry*, **55:**, 1151-1191.

Lindquist, S. and Craig, E. 1988. The heat-shock proteins. *Annu. Rev. Genet.*, **22:** 631-677.

Lipschitz, N. and Waisel, Y. 1974. Existence of salt glands in various genera of the gramineae. *New Phytologist*, **73:** 507-513.

Liu, X. and Huang, B. 2000. Heat stress injury in relation to membrane lipid peroxidation in creeping bentgrass. *Crop Science*, **40(2):** 503–510.

Liu, X. and Huang, B. 2002. Cytokinin effects on creeping bentgrass response to heat stress: II. Leaf senescence and antioxidant metabolism, *Crop Science*, **42(2):** 466–472.

Liu, X., Huang, B. and Banowetz, G. 2002. Cytokinin effects on creeping bentgrass in responses to heat stress: I. Shoot and root growth, *Crop Science*, **42(2):** 457–465.

Liu, D.L., Zhang, X.X., Cheng, Y.X., Takano, T. and Liu, S.K. 2006a. Cloning and characterization of the rHSP90 gene in rice (*Oryza sativa.* L) under environmental stress. *Mol. Plant Breed.*, **4:**317–322.

Liu, D., Zhang, X., Cheng, Y., Takano, T. and Liu, S. 2006b. RHSP90 gene expression in response to several environmental stresses in rice (*Oryza sativa* L.). *Plant Physiol. Biochem.*, **44:** 380–386

Liu, H., Baldwin, C.M., Luo, H. and Pessarakli, M. 2007. Enhancing turfgrass nitrogen use under stresses. In, "*Handbook of Turfgrass Management and Physiology*", Ed. M. Pessarakli, Taylor and Francis, New York:. pp. 555–599.

Liu, H.C., Liao, H.T., and Charng, Y.Y. 2011. The role of class A1 heat shock factors (HSFA1s) in response to heat and other stresses in *Arabidopsis. Plant Cell Environ.,* **34:** 738–751.

Lohmann, C., Schumacher, G.E., Wunderlich, M., Schoffl, F. 2004. Two different heat shock transcription factors regulate immediate early expression of stress genes in *Arabidopsis. Mol Gen Genomics*, **271:** 11-21.

Long, S., Humphries, P.S. and Falkowski, P.G. 1994. Photoinhibition of photosynthesis in nature. *Annual Review of Plant Physiology and Plant Molecular Biology*, **45:** 633-662.

Löw, D., Brändle, K., Nover, L., and Forreiter, C. 2000. Cytosolic heat-stress proteins HSP17. 7 class I and HSP17. 3 class II of tomato act as molecular chaperones *in vivo. Planta.* **211:** 575-582.

Lüders, J., Demand, J. and Höhfeld, J. 2000. The ubiquitin-related BAG-1 provides a link between the molecular chaperones HSC70/HSP70 and the proteasome. *J. Biol. Chem.*, **275:** 4613-4617.

Ludwig-Müller, J., Krishna, P. and Forreiter, C.A 2000. Gucosinolate mutant of *Arabidopsis* is thermo-sensitive and defective in cytosolic HSP90 expression after heat stress, *Plant Physiol.*, **123:** 949-958.

Luo, G.Z., Wang, H.W., Huang, J., Tian, A.G., Wang, Y.J., Zhang, J.S. and Chen, S.Y. 2005. A putative plasma membrane cation/proton antiporter from soybean confers salt tolerance in *Arabidopsis. Plant Mol. Biol.*, **59:** 809–820.

Lyons, E.M., Pote, J., DaCosta, M. and B. Huang, B. 2007. Whole plant carbon relations and root respiration associated with root tolerance to high soil temperature for *Agrostis* grasses, *Environmental and Experimental Botany*, **59(3):** 307–313.

Lyons, J. 2012. In, "*Low temperature stress in crop plants: the role of the membrane*". Edited *Elsevier*, Ma, C., Haslbeck, M., Babujee, L., Jahn, O. and Reumann, S. 2006. Identification and characterization of a stress-inducible and a constitutive small heat-shock protein targeted to the matrix of plant peroxisomes. *Plant physiology*, **141:** 47-60.

Ma, H., Wang, C., Yang, B., Cheng, H., Wang, Z., Mijiti, A., Ren, C., Qu, G., Zhang, H. and Ma, L. 2016. CarHSFB2, a class B heat shock transcription factor, is involved in different developmental processes and various stress responses in chickpea (*Cicer arietinum* L.). *Plant Mol. Biol. Report.,* **34:** 1–14.

Macnair, M.R., Tilstone, G.H. and Smith, S.E. 2000. The genetics of metal tolerance and accumulation in higher plants. In, "*Phytoremediation of contaminated soil and water*",Eds. N. Terry, G. Banuelos, CRC Press LLC, pp. 235–250.

Maestri, E., Klueva, N., Perrotta, C., Gulli, M., Nguyen, H.T. and N. Marmiroli, N. 2002. Molecular genetics of heat tolerance and heat shock proteins in cereals. *Plant Molecular Biology*, **48(5-6):** 667–681.

Majoul, T., Bancel, E., Triboý, E., Hamida, J.B. and G. Branlard, G. 2003. Proteomic analysis of the effect of heat stress on hexaploid wheat grain: characterization of heat-responsive proteins from total endosperm. *Proteomics*, **3(2):** 175–183.

Majoul, T., Bancel, E., Triboý, E., Hamida, J.B. and Branlard, G. 2004. Proteomic analysis of the effect of heat stress on hexaploid wheat grain: characterization of heat-responsive proteins from non-prolamins fraction. *Proteomics*, **4(2):** 505–513.

Malik, M.K., Slovin, J.P., Hwang, C.H. and Zimmerman, J.L. 1999. Modified expression of a carrot small heat shock protein gene, HSP17.7, results in increased or decreased thermotolerance. *Plant J.*, **20:** 89-99.

Mamedov, T.G. and Shono, M. 2008. Molecular chaperone activity of tomato (*Lycopersicon esculentum*) endoplasmic reticulum-located small heat shock protein. *J. Plant Res.* **121:** 235-243.

Manaa, A., Ahmed, H.B., Valot, B., Bouchet, J.P., Aschi-Smiti, S., Causse, M. and Faurobet, M. 2011. Salt and genotype impact on plant physiology and root proteome variations in tomato. *J. Exp. Bot.* **62:** 2797–2813.

Marcum, K.B. and C. L. Murdoch 1992. Salt tolerance of the coastal salt marsh grass, *Sporobulus virginicus* (L.) kunth. *New Phytologist*, **120:** 281-288.

Marcum, K.B., Anderson, S.J. and Engelke, M.C. 1998. Salt gland ion secretion: a salinity tolerance mechanism among five *Zoysiagrass* species. *Crop Science*, **38:** 806-810.

Martin, J., Horwich, A.L. and Hartl, F.U. 1992. Prevention of protein denaturation under heat stress by the chaperonin HSP60. *Science*, **258(5084):** 995–998.

Mayer, M. and Bukau, B. 2005. HSP70 chaperones: Cellular functions and molecular mechanism. *Cell. Mol. Life Sci.*, **62:** 670-684.

Mayer, M.P. 2010. Gymnastics of molecular chaperones. *Mol. Cell,* **39:** 321–331.

McElwain, E.F. and Spiker, S. 1992. Molecular and physiological analysis of a heat shock response in wheat. *Plant Physiool.*, **99:** 1455-1460.

McLaughlin, S.H., Sobott, F., Yao, Z.P., Zhang, W., Nielsen, P.R., Grossmann, J.G., Laue, E.D., Robinson, C.V. and Jackson, S.E. 2006. The cochaperone p23 arrests the HSP90 ATPase cycle to trap client proteins. *Journal of Molecular Biology*, **356(3):** 746–758.

Medina, C. and Cardemil, L. 1993. *Prosopis chilensis* is a plant highly tolerant to heat shock. *Plant, Cell and Environment*, **16:** 305–310.

Merquiol, E., Pnueli, L., Cohen, M., Simovitch, M., Rachmilevitch, S., Goloubinoff, P., Kaplan, A. and Mittler, R. 2002. Seasonal and diurnal variations in gene expression in the desert legume *Retama raetam. Plant Cell Environ.*, **25:** 1627–1638

Miernyk, J.A. 1997. The 70 kDa stress-related proteins as molecular chaperones. *Trends Plant Sci.* **2:** 180-187.

Miller, L. and Qureshi, M. 1992. Induction of heat-shock proteins and phagocytic function of chicken macrophage following in vitro heat exposure. *Veterinary Immunology and Immunopathology.* **30:** 179-191.

Miller, G. and Mittler, R. 2006. Could heat shock transcription factors function as hydrogen peroxide sensors in plants? *Ann. Bot.*, **98:** 279–288.

Minhas, J.S., Singh, B., Kumar, D., Joseph, T.A. and Prasad, K.S.K. 2001. Selection of heat tolerant potato genotypes and their performance under heat stress. *J. Indian Potato Assoc.,* **28:** 132-134.

Mirus, O. and Schleiff, E. 2009. The evolution of tetratricopeptide repeat domain containing receptors involved in protein translocation. *Endocytobiosis Cell Res.*, **19:** 31-50.

Mishra, S.K., Tripp, J., Winkelhaus, S., Tschiersch, B., Theres, K., Nover, L. and Scharf, K. 2002. In the complex family of heat stress transcription factors, HsfA1 has a unique role as master regulator of thermotolerance in tomato. *Genes Dev.*, **16:** 1555-1567.

Mitra, R. and Bhatia, C.R. 2004. Bioenergetic cost of heat tolerance in wheat crop. *Curr. Sci.*, **94:** 1049–1053.

Mittler. R. 2006. Abiotic stress, the field environment and stress combination. *Trends Plant Sci.*, **11:** 15–19.

Miroshnichenko, S., Tripp, J., Nieden, U.Z., Neumann, D., Conrad, U. and Manteuffel, R. 2005. Immuno modulation of function of small heat shock proteins prevents their assembly into heat stress granules and results in cell death at sub-lethal temperatures. *Plant J.*, **41:** 269–281.

Mogensen, J.E., Wimmer, R., Larsen, J.N. and Spangfort, M.D. 2002. The major birch allergen, Bet v 1, shows affinity for a broad spectrum of physiological ligands. *J. Biol. Chem.*, **277:** 684–692.

Mogk, A., Schlieker, C., Friedrich, K.L., Schönfeld, H., Vierling, E. and Bukau, B. 2003. Refolding of substrates bound to small HSPs relies on a disaggregation reaction mediated most efficiently by ClpB/DnaK. *J. Biol. Chem.*, **278:** 31033-31042.

Morimoto, R.I. 1993. Cells in stress: The transcriptional activation of heat shock genes. *Science,* **259:**1409–1410.

Monjardino, P., Smith, A.G. and Jones, R.J. 2005. Heat stress effects on protein accumulation of maize endosperm. *Crop Science*, **45(4):** 1203–1210.

van Montfort, R.L.M., Basha, E., Friedrich, K.L., Slingsby, C. and E. Vierling, E. 2001. Crystal structure and assembly of a eukaryotic small heat shock protein. *Nature Structural Biology*, **8(12):** 1025–1030.

Monica, R., Consortium, L. and Consortium, L. 2007. Breeding temperate legumes: Advances and challenges. *Lotus Newsl.*, **37(3):** 105.

Morimoto. R.I. 1993. Cells in stress: THE transcriptional activation of heat shock genes. *Science*, **259:** 1409–1410.

Morimoto, R.I., Tissieres, A. and Georgopoulos, C. 1994. Heat Shock Proteins: Structure, Function and Regulation. Cold Spring Harbor Lab. Press, Cold Spring Harbor, NY.

Morimoto, R.I. and Santoro, M.G. 1998. Stress–inducible responses and heat shock proteins: New pharmacologic targets for cytoprotection. Nature Biotech. 1998, 16, 833-838.

Morimoto, R.I. 2008. Proteotoxic stress and inducible chaperone networks in neurodegenerative disease and aging. *Genes and Development*, **22:** 1427-1438.

Mu, C., Wang, S., Zhang, S., Pan, J., Chen, N., Li, X., Wang, Z. and Liu, H. 2011. Small heat shock protein LimHSP16.45 protects pollen mother cells and tapetal cells against extreme temperatures during late zygotene to pachytene stages of meiotic prophase I in David Lily. *Plant Cell Reports*, **30:** 1981–1989.

Mu, C., Zhang, S., Yu, G., Chen, N., Li, X. and Liu, H. 2013. Overexpression of small heat shock protein LimHSP16.45 in *Arabidopsis* enhances tolerance to abiotic stresses. *PLoS ONE*, **8(12):** e82264. doi:10.1371/journal.pone.0082264

Mullineaux, P.M., Karpinski, S. and Baker, N.R. 2006. Spatial dependence for hydrogen peroxide-directed signaling in light-stressed plants. *Plant Physiol.*, **141:** 346–350.

Munns, R., Hore, R.A.,,James, R.A. and Rebetzke, G.J. 2000. Genetic variation for improving the salt tolerance of durum wheat. *Australian Journal of Agricultural Research*, **51:** 69–74.

Munns, R. 2002. Comparative physiology of salt and water stress. *Plant, Cell and Environment*, **25:** 239–250.

Munns, R. 2005. Genes and salt tolerance: bringing them together. *New Phytol.*, **167:** 645–663.

Munns, R. and James, R.A. 2003. Screening methods for salinity tolerance: a case study of with tetraploid wheat. *Plant and Soil*, **253:** 201–218.

Murakami, T., Matsuba, S., Funatsuki, H., Kawaguchi, K., Saruyama, H., Tanida, M. and Sato, Y. 2004. Over-expression of a small heat shock protein, sHSP17.7, confers both heat tolerance and UV-B resistance to rice plants. *Mol. Breeding*, **13:** 165–175.

Musatenko, L.I., Vedenicheva, N.P., Vasyuk, V.A., Generalova, V.N., Martyn, G.I. and Sytnik, K.M. 2003. Phytohormones in seedlings of maize hybrids differing in their tolerance to high temperature. *Russian Journal of Plant Physiology*, **50(4):** 444–448.

Naidoo, G. and Naidoo, Y. 1998. Salt tolerance in *Sporobolus virginicus*: the importance of ion relations and salt excretion. *Flora*, **193:** 337-344.

Nakamoto, H., Suzuki, N. and Roy, S.K. 2000. Constitutive expression of a small heat-shock protein confers cellular thermotolerance and thermal protection to the photosynthetic apparatus in cyanobacteria. *FEBS Letters*, **483:** 169-174.

Nakamoto, H. and Vígh, L. 2007. The small heat shock proteins and their clients. *Cell. Mol. Life Sci.*, **64:** 294-306.

Nathan, D.F., Vos, M.H. and Lindquist, S. 1997. *In vivo* functions of the *Saccharomyces cerevisiae* HSP90 chaperone. , *USA,* **94:** 12949-12956.

Ndimba, B.K., Chivasa, S., Simon, W.J. and Slabas, A.R. 2005. Identification of *Arabidopsis* salt and osmotic stress responsive proteins using two-dimensional difference gel electrophoresis and mass spectrometry. *Proteomics,* **5:** 4185–4196.

Necchi, A., Pogna, N.E. and Mapelli, S. 1987. Early and late heat shock proteins in wheat and other cereal species. *Plant Physiology*, **84(4):** 1378–1384.

Neilson, K.A., Gammulla, G., Mirzaei, M., Imin, N. and Haynes, P.A. 2010. Proteomic analysis of temperature stress in plants. *Proteomics*, **10:** 828–845.

Neta-Sharir, I., Isaacson, T., Lurie, S. and Weiss, D. 2005. Dual role for tomato heat shock protein 21: Protecting photosystem II from oxidative stress and promoting color changes during fruit maturation. *The Plant Cell Online*, **17:** 1829-1838.

Neumann, D.M., Emmermann, M., Thierfelder, J.M., ZurNieden, U., Clericus, M., Braun, H.P., Nover, L., and Schmitz, U.K. 1993. HSP68-a DNAK-like heat-stress protein of plant mitochondria. *Planta*, **190:** 32-43.

Neumann, D., Lichtenberger, O., Günther, D., Tschiersch, K. and Nover, L.1994. Heat shock proteins induce heavy metal tolerance in higher plants. *Planta*, 194: 360–*367*.

Neumann, D., Nieden, U.Z., Lichtenberger, O. and Leopold, I.1995. How doesArmeria maritima tolerate high heavy metal concentrations? *Journal of Plant Physiology*, **146:** 704–717.

Neznanov, N., Komarov, A.P., Neznanova, L., Stanhope-Baker, P. and Gudkov, A.V. 2011. Proteotoxic stress targeted therapy (PSTT): induction of protein misfolding enhances the antitumor effect of the proteasome inhibitor bortezomib. *Oncotarget*, **2:** 209.

Nguyen, H.T., Joshi, C.P., Klueva, N., Weng, J., Hendershot, K.L. and Blum, A. 1994. The Heat-shock response and expression of heat-shock proteins in wheat under diurnal heat stress and field conditions *Australian Journal Plant Physiology*, **21:** 857-867.

Nishizawa, A., Yabuta, Y., Yoshida, E., Maruta, T., Yashimura, K., Shigeoka, S. 2006. *Arabidopsis* heat shock transcription factor A2 as a key regulator in response to several types of environmental stress. *The Plant Journal*, **48:** 535-547.

Nishizawa-Yokoi, A., Tainaka, H., Yoshida, E., Tamoi, M., Yabuta, Y. and Shigeoka, S. 2010. The 26S proteasome function and HSP90 activity involved in the regulation of HsfA2 expression in response to oxidative stress. *Plant Cell Physiol.*, **51:** 486–496.

Nishizawa-Yokoi, A., Nosaka, R., Hayashi, H., Tainaka, H., Maruta, T., Tamoi, M., Ikeda, M., Ohme-Takagi, M., Yoshimura, K., Yabuta, Y. and Shigeoka, S. 2011. HSFA1d and HSFA1e involved in the transcriptional regulation of HSFA2 function as key regulators for the Hsf signaling network in response to environmental stress. *Plant Cell. Physiol*, **52:** 933–945.

Nover, L., Neuman, D. and Scharf, K.D. 1989. In. "*Heat shock and other stress response system of plants*", Springer-Verlag, Berlin, pp. 155.

Nover, L. 1991. In, "*Heat shock response*", CRC Press Inc., Florida, pp. 509.

Nover, L., Bharti, K., Döring, P., Mishra, S.K., Ganguli, A. and Scharf, K. 2001. *Arabidopsis* and the heat stress transcription factor world: How many heat stress transcription factors do we need? *Cell Stress Chaperon*. **6(3):** 177-189.

Nover, L. and Baniwal, S.K. 2006. Multiplicity of heat stress transcription factors controlling the complex heat stress response of plants. In: *International Symposium on Environmental Factors, Cellular Stress and Evolution, Varanasi, India,* October 13–15, 2006, p. 15.

Nowak, J., and Colborne, D. 1989. *In vitro* tuberization and tuber proteins as indicators of heat-stress tolerance in potato. *Am. Potato J.*, **66:** 35-45.

Ogawa, D., Yamaguchi, K. and Nishiuchi, T. 2007. High-level overexpression of the *Arabidopsis* HsfA2 gene confers not only increased themotolerance but also salt/osmotic stress tolerance and enhanced callus growth. *J. Exp. Bot.* **58:** 3373–3383.

Ohad, I., Adir, N., Koike, H., Kyle, D. and Inoue, Y. 1990. Mechanism of photoinhibition *in vivo. A* reversible light-induced conformational change of reaction center II is related to an irreversible *modification of the D1 protein. Journal of Biological Chemistry*, **265***:* 1972-1979.

Ortiz, C., Bravo, L., Pinto, M. and Cardemil, L.1995. Physiological and molecular responses of *Prosopis chilensis* under field and simulation conditions. *Phytochemistry*, **40:** 1375-1382.

Ortiz, C. and Cardemil, L. 2001. Heat shock responses in two leguminous plants: a comparative study. *J. Experimental Botany,* **52(361):** 1711-1719.

Queitsch, C., Hong, S.-W., Vierling, E. and Lindquist, S. 2000. Heat shock protein 101 plays a crucial role in thermotolerance in *Arabidopsis*. *Plant Cell*, **12(4):** 479–492.

Panaretou, B. and Zhai, C. 2008. The heat shock proteins: Their roles as multi-component machines for protein folding. *Fungal Biol. Rev.* **22:**110-119.

Pang, Q., Chen, S., Dai, S., Chen, Y., Wang, Y. and Yan X. 2010. Comparative proteomics of salt tolerance in*Arabidopsis thaliana* and *Thellungiella halophila*. *J. Proteome Res.*, **9:** 2584–2599.

Parida, A.K., Das, A.B., Mittra, B. and Mohanty, P. 2004. Salt-stress induced alterations in protein profile and protease activity in the mangrove (*Bruguiera parviflora*). *Z. Naturforsch*, **59:** 408-414.

Park, H.-S., Jeong, W.-J., Kim, E.-C., Jung, Y., Lim, J.M., Hwang, M.S., Park, E.-J., Ha, D.-S. and Choi, D.-W. 2012. Heat shock protein gene family of the *Porphyra seriata* and enhancement of heat stress tolerance by PsHSP70 in *Chlamydomonas*. *Mar. Biotechnol.*, **14:** 332 342.

Parsell, D.A. and Lindquist, S. 1994. Heat shock proteins and stress tolerance. In, "*The biology of heat shock proteins and molecular chaperones*", Eds. R. Morimoto, A. Tissieres, and C. Georgopoulos, Cold Spring Harbor Press, New York, USA, 457-494.

Parsell, D.A., Kowal, A.S., Singer, M.A., and Lindquist, S. 1994. Protein disaggregation mediated by heat-shock protein HSP104. *Nature*, **372:** 475–478.

Pareek, A., Singla, S.L. and Grover, A. 1995. Immunological evidence for accumulation of two high-molecular-weight (104 and 90 kDa) HSPs in response to different stresses in rice and in response to high temperature stress in diverse plant genera. *Plant Molecular Biology*, **29**(2): 293–301.

Pareek, A., Singla, S.L. and Grover, A. 1998a. Plant HSP90 family with special reference to rice. *J. Biosci.* **23:** 361–367.

Pareek, A., Singla, S.L., and Grover, A. 1998b. Proteins alterations associated with salinity, desiccation, high and low temperature stresses and abscisic acid application in seedlings of Pusa 169, a high-yielding rice (*Oryza sativa* L.) cultivar. *Curr. Sci.*, **75:** 1023-1035.

Park, S.Y., Shivaji, R., V.Krans, J.V. and Luthe, D.S. 1996. Heat-shock response in heat-tolerant and non-tolerant variants of *Agrostis palustris* Huds. *Plant Physiology,* **111(2):** 515–524.

Park, C.J., Kim, K.J., Shin, R., Park, J.M., Shin, Y.C. and Paek K.H. 2004. Pathogenesis-related protein 10 isolated from hot pepper functions as a ribonuclease in an antiviral pathway. *Plant J.*, **37:** 186–198.

Patsikka, E., Kairavuo, M., Sersen, F., Aro, E.M. and Tyystjarvi, E. 2002. Excess copper predisposes photosystem II to photoinhibition in vivo by outcompeting iron and causing decrease in leaf chlorophyll. *Plant Physiology*, **129:** 1359-1367.

Pearl, L.H. and Prodromou, C. 2006. "Structure and mechanism of the HSP90 molecular chaperone machinery. *Annual Review of Biochemistry*, **75:** 271–294.

Peng, S., Zhu, Z., Zhao, K., Shi, J., Yang, Y., He, M. and Wang, Y. 2013. A novel heat shock transcription factor, VpHsf1, from Chinese wild *Vitis pseudoreticulata* is involved in biotic and abiotic stresses. *Plant Mol. Biol. Report.* **31:** 240–247.

Perez, D.E., Hoyer, J.S., Johnson, A.I., Moody, Z.R., Lopez, J. and Kaplinsky, N.J. 2009. BOBBER1 is a noncanonical *Arabidopsis* small heat shock protein required for both development and thermotolerance. *Plant Physiol*, **151:** 241–252.

Personat, J. M., Tejedor-Cano, J., Prieto-Dapena, P., Almoguera, C. and Jordano, J. 2014. Co-overexpression of two Heat Shock Factors results in enhanced seed longevity and in synergistic effects on seedling tolerance to severe dehydration and oxidative stress. *BMC Plant Biol.,* **14:** 56

Polenta, G.A., Calvete, J.J. and González, C.B. 2007. Isolation and characterization of the main small heat shock proteins induced in tomato pericarp by thermal treatment. *FEBS J.*, **274:** 6447-6455.

Porter, J.R. 2005. Rising temperatures are likely to reduce crop yields, *Nature*, **436(7048):**,174.

Prändl, R., Hinderhofer, K., Eggers-Schumacher, G., and Schöffl, F. 1998. HSF3, a new heat shock factor from *Arabidopsis thaliana*, derepresses the heat shock response and confers thermotolerance when overexpressed in transgenic plants. *Mol. Gen. Genet.*, **258:** 269–278.

Prasinos, C., Krampis, K., Samakovli, D. and Hatzopoulos, P. 2005. Tight regulation of expression of two *Arabidopsis* cytosolic HSP90 genes during embryo development. *J. Exp. Bot.*, **56:** 633-644.

Pratt, W.B., Krishna, P. and Olsen, L.J. 2001. HSP90-binding immunophilins in plants: The protein movers. *Trends Plant Sci.*, **6:** 54-58.

Prieto-Dapena, P., Castano, R., Almoguera, C. and Jordano, J. 2006. Improved resistance to controlled deterioration in transgenic seeds. *Plant Physiol*, **142:** 1102–1112.

Pritchard, J., Wyn Jones, R.G. and Tomos, A.D. 1991. Turgor, growth and rheological gradients of wheat roots following osmotic stress. *Journal of Experimental Botany*, **42:** 1043–1049.

Queitsch, C., Hong, S., Vierling, E. and Lindquist, S. 2000. Heat shock protein 101 plays a crucial role in thermotolerance in *Arabidopsis. Sci. Signal.*, **12:** 479-492

Rachmilevitch, S., Lambers, H. and Huang, B. 2006a. Root respiratory characteristics associated with plant adaptation to high soil temperature for geothermal and turf-type *Agrostis* species. *Journal of Experimental Botany*, **57(3):** 623–631.

Rachmilevitch, S. Huang, B. and Lambers, H. 2006b. Assimilation and allocation of carbon and nitrogen of thermal and nonthermal *Agrostis* species in response to high soil temperature. *New Phytologist,* **170(3):** 479–490.

Rachmilevitch, S., Xu, Y., Gonzalez-Meler, M.A. Huang, B. and Lambers, H. 2007. Cytochrome and alternative pathway activity in roots of thermal and non-thermal *Agrostis* species in response to high soil temperature. *Physiologia Plantarum*, **129(1):** 163–174.

Rajoriya, S., Rajoriya, R., Vandre, R., Gupta, V., Shukla, S. and Meshram, S. 2014. Heat shock proteins: A review. *International Journal of Biomedical and Life Sciences,* **5(3):** 398-411.

Rampino, P., Giovannoni, M., Stefano, P., Mariarosaria, D.P., Natale, D.F. and Carla, P. 2009. Acquisition of thermotolerance and HSP gene expression in durum wheat (*Triticum durum* Desf.) cultivars. *Environ. Exp. Bot.* **66:** 257–264.

Renaut, J., Hausman, J. F. and Wisniewski, M.E. 2006. Proteomics and low-temperature studies: bridging the gap between gene expression and metabolism. *Physiol. Plant*, **126:** 97-109.

Richter, K., Haslbeck, M. and Buchner, J. 2010. The heat shock response: Life on the verge of death. *Mol. Cell*, **40:** 253-266.

Ristic, Z., Gifford, D.J. and Cass, D.D. 1991. Heat shock proteins in two lines of *Zea mayse* L. that differ in drought and heat resistance. *Plant Physiol.*, **97:** 1430-1434.

Ristic, Z., Bukovnik, U. and Prasad, P.V.V. 2007. Correlation between heat stability of thylakoid membranes and loss of chlorophyll in winter wheat under heat stress. *Crop Science,* **47(5):** 2067–2073.

Rizhsky, L., Liang, H., Shuman, J., Shulaev, V., Davletova, S. and Mittler, R. 2004. When defense pathways collide: the response of Arabidopsis to a combination of drought and heat stress. *Plant Physiol.*, **134:** 1683–1696.

Rodríguez, J.G. and Cardemil, L.1994. Cell wall proteins in seedling cotyledons of *Prosopis chilensis. Phytochemistry*, **35:** 281–286.

Roshandel, P. and Flowers, T. 2009. The ionic effects of NaCl on physiology and gene expression in rice genotypes differing in salt tolerance. *Plant and Soil*, **315:** 35–147.

Rossel, J.B., Wilson, I.W. and Pogson, B.J. 2002. *Global changes in gene expression in response to high light in Arabidopsis.* Plant Physiology, **130***:* 1109-*1120.*

Saidi, Y., Finka, A., Muriset, M., Bromberg, Z., Weiss, Y.G., Maathuis, F.J. and Goloubinoff, P. 2009. The heat shock response in moss plants is regulated by specific calcium-permeable channels in the plasma membrane. *The Plant Cell Online,* **21:** 2829-2843.

Sairam, R.K., Srivastava, G.C. and Saxena, D.C. 2000. Increased antioxidant activity under elevated temperatures: a mechanism of heat stress tolerance in wheat genotypes. *Biologia Plantarum,* **43(2):** 245–251.

Sanchez, Y., Taulien, J., Borkovich, K.A. and Lindquist, S. 1992. HSP104 is required for tolerance to many forms of stress. *EMBO J.* **11:** 2357–2364

Sanchez, D.H., Lippold, F., Redestig, H., Hannah, M.A., Erban, A., Krämer, U., Kopka, J. and Udvardi, M.K. 2008. Integrative functional genomics of salt acclimatization in the model legume *Lotus japonicus* . *The Plant Journal*, **53:** 973–987.

Sanmiya, K., Suzuki, K., Egawa, Y. and Shono, M. 2004. Mitochondrial small heat-shock protein enhances thermotolerance in tobacco plants. *FEBS Lett.*, **557:** 65–268.

Sato, Y. and Yokoya, S. 2008. Enhanced tolerance to drought stress in transgenic rice plants over expressing a small heat-shock protein, sHSP17.7. *Plant Cell Reports*, **27:** 329-334.

Scarpeci, T.E., Zanor, M.I., Carrillo, N., Mueller-Roeber, B. and Valle, E.M. 2008. Generation of superoxide anion in chloroplasts of *Arabidopsis thaliana* during active photosynthesis: a focus on rapidly induced genes. *Plant Mol Biol.*, **66:** 361–378.

Scharf, K., Rose, S., Zott, W., Schöffl, F., Nover, L. and Schöff, F. 1990. Three tomato genes code for heat stress transcription factors with a region of remarkable homology to the DNA-binding domain of the yeast HSF. *EMBO J.*, **9:** 4495.

Scharf, K., Höhfeld, I. and Nover, L. 1998. Heat stress response and heat stress transcription factors. *J. Biosciences*, **23:** 313-329.

Scharf, K.D., Siddique, M. and Vierling, E. 2001. The expanding family of *Arabidopsis thaliana* small Heat stress proteins and a new family of proteins containing α-crystallin domains (Acd proteins). *Cell Stress and Chaperones*, **6(3):** 225–237.

Scharf, K.D., Berberich, T., Ebersberger, I., and Nover, L. 2012. The plant heat stress transcription factor (Hsf) family: structure, function and evolution. *Biochimica et Biophysica Acta.*, **1819:** 104–119.

Schat, H., Llugany, M. and Bernhard, R. 2000. Metal specific patterns of tolerance, uptake and transport of heavy metals in hyperaccumulating and nonhyperaccumulating metallophytes. In, "*Phytoremediation of contaminated soil and water*", Eds. N. Terry, G. Banuelos, CRC Press LLC, pp. 171–188.

Schirmer, E.C.; Glover, J.R.; Singer, M.A. and Lindquist, S. 1996. HSP100/Clp proteins: A common mechanism explains diverse functions. *Trends Biochem. Sci.*, **21:** 289-296.

Schmid, D., Baici, A., Gehring, H. and Christen, P. 1994. Kinetics of molecular chaperone action. *Science*, **263(5149):** 971–973.

Schramm, F., Ganguli, A., Kiehlmann, E., Englich, G., Walch, D. and von Koskull-Doring, P. 2006. The heat stress transcription factor HSFA2 serves as a regulatory amplifier of a subset of genes in the heat stress response in *Arabidopsis*. *Plant Mol. Biol.*, **60:** 72.

Schroda, M., Vallon, O. Wollman, F.-A. and Beck, C.F. 1999. *A chloroplast-targeted Heat shock protein* 70 (HSP70) contributes to the photoprotection and repair of photosystem II during and after *photoinhibition. Plant Cell*, **11:** 1165-*1178.*

Schulze-Lefert, P. 2004. Plant immunity: The origami of receptor activation. *Curr. Biol.*, **14:** R22-R24. Schuster, G., Even, D. Kloppstech, K. and Ohard, I. 1988. Evidence of protection by heat shock proteins against photoinhibition during heat shock. *EMBO J.*, **7:** 1-8.

Scott, A., Heckathorn, Kathleen Mueller, J., Stephanie laguidice, Bin zhu,Tara barrett, Brian blair, Yan dong. 2004. Chloroplast small heat-shock proteins protect photosynthesis during heavy metal stress. *American Journal of Botany*, **91:** 1312-1318.

Senthil-Kumar, M., Kumar, G., Srikanthbabu, V. and Udayakumar, M. 2007. Assessment of variability in acquired thermotolerance: potential option to study genotypic response and the relevance of stress genes. *Journal of Plant Physiology*, **2007,** 164, 111-125.

Shah, F., Huang, J., Cui, K., Nie, L., Shah, T., Chen, C. and Wang, K. 2011. Impact of high-temperature stress on rice plant and its traits related to tolerance. *The Journal of Agricultural Science*, **149:** 545-556.

Shao, H.B., Guo, Q.J., Chu, L.Y., Zhao, X.N., Su, Z.L., Hu, Y.C. and Cheng. J.F. 2007a. Understanding molecular mechanism of higher plant plasticity under abiotic stress *Colloids Surf. B: Biointerfaces*, **54:** 37–45.

Shao, H.B., Jiang, S.Y., Li, F.M., Chu, L.Y., Zhao, C.X., Shao, M.A., Zhao, X.N. and Li. F. 2007b. Some advances in plant stress physiology and their implications in the systems biology era. *Colloids Surf B: Biointerfaces*, **54:** 33–36.

Shavrukov, Y., Bowner, J., Langridge, P. and Tester, M. 2006. Screening for sodium exclusion in wheat and barley.. *In:* Proceedings of the 13th Australian Agronomy Conference Perth, The Regional Institute.

Shavrukov, Y., Langridge, P, and Tester, M. 2009. Salinity tolerance and sodium exclusion in genus *Triticum . Breeding Science*, **59:** 671–678.

Sharukov, Y. 2012. Salt stress or salt shock: Which gene are we studying? *J. Expt. Bot.,* **64:** 119-127.

Siddique, M., Port, M., Tripp, J., Weber, C., Zielinski, D., Calligaris, R., Winkelhaus, S. and Scharf, K.D. 2003. Tomato heat stress protein HSP16.1-CIII represents a member of a new class of nucleocytoplasmic small heat stress proteins in plants. *Cell Stress Chaperon*, **8:** 381–394.

Siddique, M., Gernhard, S., Von Koskull-Doring, P., Vierling, E. and K. D. Scharf, K.D. 2008. The plant sHSP superfamily: five new members in Arabidopsis thaliana with unexpected properties. *Cell Stress and Chaperones*, **13(2):** 183– 197.

Singh, R.P., Prasad, P.V.; Sunita, K., Giri, S.N. and Raja Reddy, K. 2007. Influence of high temperature and breeding for heat tolerance in cotton: a review. *Adv. Agron.,* **93:** 313-385.

Singla, S.L., Preek, A. and Grover, A. 1997. High temperature. In, "*Plant Ecophysiology*", Ed. M.N.V. Prasad, John Wiley, New York. pp. 101–127

Singla, S.L., Pareek, A., Kush, A.K. and Grover, A. 1998. Distribution patterns of 104 kDa stress-associated protein in rice. Plant Molecular Biology, **37(6):** 911–919.

Skylas, D.J., Cordwell, S.J., Hains, P.G., Larsen, M.R., Basseal, D.J., Walsh, B.J., Blumenthal, C., Rathmell, Copeland, L. and Wrigley, C.W. 2002. Heat shock of wheat during grain filling: Proteins associated with heat tolerance. *Journal of Cereal Science*, **35:** 175-188.

Snider, J.L., Oosterhuis, D.M., Skulman, B.W. and Kawakami, E.M. 2009. Heat stress induced limitations to reproductive success in *Gossypium hirsutum. Physiologia plantarum*, **137:** 125-138.

Snyman, M. and Cronje, M. 2008. Modulation of heat shock factors accompanies salicylic acid-mediated potentiation of HSP70 in tomato seedlings. *J. Exp. Bot.*, **59:** 2125-2132.

Solomon, J.M., Rossi, J.M., Golic, K., McGarry, T. and Lindquist, S. 1991. Changes in *HSP70*alter thermotolerance and heat-shock regulation in *Drosophila. New Biol.*, **3:** 1106–1120.

Soransen, J.G., Kristensen, T.N. and Loeschcke, V. 2003. The evolutionary and ecological role of heat shock proteins. *Ecology Letters*, **6:** 1025-1037.

Stapel, D., Kruse, E. and Kloppstech, K. 1993. The protective effect of heat shock proteins against photoinhibition under heat shock in barley (Hordeum vulgare). *Journal of Photochemistry and Photobiology B*, **21:** 211-*218.*

Struik, P.C. 2007. Responses of the potato plant to temperature. In, "*Plant Biology and Biotechnology: Advances and Perspectives*", Ed. D. Vreugdenhil, Elsevier, Amsterdam, pp. 367-393.

Stout R.G. and Al-Niemi, T.S. 2002. Heat-tolerant flowering plants of active geothermal areas in Yellowstone National Park. *Annals of Botany*, **90(2):** 259–267.

Sugimoto, M. and Takeda, K. 2009. Proteomic analysis of specific proteins in the root of salt-tolerant barley. *Biosci. Biotech. Biochem.*, **73:** 2762–2765.

Sule, A., Vanrobaeys, F., Hajos, G.Y., van Beeumen, J. and Devreese, B. 2004. Proteomic analysis of small heat shock protein isoforms in barley shoots. *Phytochemistry,* **65(12):** 1853–1863.

Sun, W., Bernard, C., Van De Cotte, B., Van Montagu, M. and Verbruggen, N. 2001. At HSP17. 6A, encoding a small heat shock protein in *Arabidopsis*, can enhance osmotolerance upon overexpression. *Plant J.*, **27:** 407-415.

Sun, W., Van Montagu, M. and Verbruggen, N. 2002. Small heat shock proteins and stress tolerance in plants. *Biochimica Biophysica Acta.*, **1577(1):** 1-9.

Sun, Y. and MacRae, T. H. 2005. Small heat shock proteins: molecular structure and chaperone function. *Cellular and Molecular Life Sciences CMLS*, **62:** 2460-2476.

Sung, D.Y., Vierling, E. and Guy, C.L. 2001. Comprehensive expression profile analysis of the *Arabidopsis* HSP70 gene family. *Plant Physiology*, **126(2):** 789–800.

Sung, D.Y., Kaplan, F., Lee, K.J. and Guy, C.L. 2003. Acquired tolerance to temperature extremes,. *Trends in Plant Science*, **8(4):** 179–187.

Sung, D.Y. and Guy, C.L. 2003. Physiological and molecular assessment of altered expression of Hsc70-1 in*Arabidopsis*. Evidence for pleiotropic consequences. *Plant Physiol.*, **132:** 979–987.

Swindell, W.R., Huebner, M. and Weber, A.P. 2007. Transcriptional profiling of *Arabidopsis* heat shock proteins and transcription factors reveals extensive overlap between heat and non-heat stress response pathways. *BMC Genomics*, **8:** 1–15.

Szabo, A., Langer, T., Schroder, H., Flanagan, J. Bukau, B. and Hartl. F.U. 1994. The ATP hydrolysis-dependent reaction cycle of the Escherichia coli HSP70 system - DnaK, DnaJ, and GrpE. *Proceedings of the National Academy of Sciences* , USA, **91(22):**, 10345–10349.

Tada, Y. and Kashimura, T. 2009. Proteomic analysis of salt-responsive proteins in the mangrove plant *Bruguiera gymnorhiza*. *Plant Cell Physiol.*, **50:** 439–446.

Taipale, M., Jarosz, D.F. and Lindquist, D.S. 2010. HSP90 at the hub of protein homeostasis: Emerging mechanistic insights. *Nat. Rev. Mol. Cell Biol.*, **11:** 515-528.

Taiz, L. and Zaiger, E. 1988. In, "*Plant Physiology*", Sinauer Associates Inc. Publishers, Sunderland, Massachusetts, pp. 792.

Tang, D., Khaleque, M.A., Jones, E. L.; Theriault, J.R.; Li, C.; Wong, W.H., Stevenson, M.A. and Caldwood, S.K. 2005. Expression of heat shock proteins and heat shock protein messenger ribonucleic acid in human prostate carcinoma in vitro and in tumors in vivo. *Cell stress and chaperones*, **10(1):** 46-58.

Tawfik, A.A., Kleinhenz, M.D. and Palta, J.P. 1996. Application of calcium and nitrogen for mitigating heat stress effects on potatoes. *Am Potato J*, **73:** 261–273.

Tercek, M.T., Hauber, D.P. and Darwin, S.P. 2003. Genetic and historical relationships among geothermally adapted *Agrostis* (bentgrass) of North America and Kamchatka:evidence for a previously unrecognized, thermally adapted taxon. *American Journal of Botany*, **90(9):** 1306–1312.

Tian, J., Belanger, F.C. and Huang, B. 2009. Identification of heat stress-responsive genes in heat-adapted thermal *Agrostis scabra* by suppression subtractive hybridization. *Journal of Plant Physiology*, **166(6):** 588–601.

Timperio, A.M., Egidi, M.G. and Zolla, L. 2008. Proteomics applied on plant abiotic stresses: Role of heat shock proteins (HSP). *J. Proteomics.*, **71:** 391-411.

Tompa, P. and Kovacs, D. 2010. Intrinsically disordered chaperones in plants and animals Canadian Society of Biochemistry, Molecular *and* Cellular Biology 52nd Annual Meeting-Protein Folding: *Biochemistry and Cell Biology*, **88:** 167-174.

Treshow, M. 1970. In, "*Environment and plant response*", Mcgraw-Hill Company, pp. 421.

Tripp, J., Mishra, S.K. and Scharf, K. 2009. Functional dissection of the cytosolic chaperone network in tomato mesophyll protoplasts. *Plant Cell Environ.*, **32:** 123-133.

Tsan, M.F. and Gao, B. 2004. Heat shock protein and innate immunity. *Cell Mol Immunol*, **1:** 274-279.

Tseng, T.S., Tzeng, S.S., Yeh, C.H., Chang, F.C., Chen, Y.M. and Lin, C.Y.1993. The heat shock response in rice seedlings—isolation and expression of cDNAs that encode class I low molecular weight heat shock proteins. *Plant and Cell Physiology* **34:** 165–168.

Veilleux, R.E., Paz, M.M. and Levy, D. 1997. Potato germplasm development for warm climates: genotypic enhancement of tolerance to heat stress. *Euphytica*, **98:** 83-92.

Velanquez, J.M. and Lindquist, S. 1984. HSP 70: nuclear concentration during environmental stress and cytoplasmic storage during recovery. *Cell,* **36:** 655-662.

Veerasamy, M., He, Y. and Huang, B. 2007. Leaf senescence and protein metabolism in creeping bentgrass exposed to heat stress and treated with cytokinins," *Journal of the American Society for Horticultural Science*, **132(4):** 467–472.

Veselov, A.P., Lobov, V.P. and Olyunina, L.N. 1998. Phytohormones during heat shock and Recovery. *Russian Journal of Plant Physiology*, **45(5):** 611–616.

Vierling, E. 1991. The roles of heat shock proteins in plants. *Annual Review of Plant Physiology and Plant Molecular Biology,* **42(1):** 579–620.

Vierling, R.A. and Nguyen, H.T. 1992. Heat shock protein gene expression in diploid gene genotypes differing in thermal toerance. *Crop Sci.*, **32:** 370-377.

Vierling, E. 1997. The small heat shock proteins in plants are members of an ancient family of heat induced proteins. *Acta Physiologiae Plantarum*, **19(4):** 539-547.

Vigh, L., Horváth, I., Maresca, B. and Harwood, J.L. 2007. Can the stress protein response be controlled by membrane-lipid therapy'? *Trends in Biochemical Sciences*, **32:** 357-363.

Vincent, D., Ergül, A., Bohlman, M.C., Tattersall, E.A.R., Tillett, R.L., Wheatley, M.D., Woolsey, R., Quilici, D.R., Joets, J., Schlauch, K., Schooley,, D.A., Cushman,, J.C. and Cramer, G.R. 2007. Proteomic analysis reveals differences between *Vitis. vinifera* L. cv. Chardonnay and cv. *Cabernet auvignon* and their responses to water deficit and salinity. *J. Exp. Bot.*, **58:** 1873–1892.

Vinit-dunand, F., Epron, D., Alaoui-sosse, B. and Badot, P.M. 2002. Effects of copper on growth and on photosynthesis of mature and expanding leaves in cucumber plants. *Plant Science,* **163:** 53–58.

Vinocur, B. and Altman, A. 2005. Recent advances in engineering plant tolerance to abiotic stress: Achievements and limitations. *Curr. Opin. Biotechnol,* **16:** 123–132.

Viitanen, P.V., Schmidt, M., Buchner, J., Suzuki, T., Vierling, E., Dickson, R., Lorimer, G.H., Gatenby, A. and Soll, J. 1995. Functional characterization of the higher plant chloroplast Chaperonins. *Journal of Biological Chemistry*, **270(30):** 18158– 18164.

Viswanathan, C. and Khanna-Chpra, R. 1996. Heat shock proteins – Role in thermotolerance of crop plants. *Current Science*, **71(4):** 275-284.

Volkov, R.A., Panchuk, .I.I and Schoffl, F. 2005. Small heat shock proteins are differentially regulated during pollen development and following heat stress in tobacco. *Plant Mol Biol,* **57(4):** 487–502.

Volkov, R.A., Panchuk, I.I., Mullineaux, P.M. and Schöffl, F. 2006. Heat stress-induced H_2O_2 is required for effective expression of heat shock genes in *Arabidopsis. Plant Mol. Biol.*, **61:** 733-746.

von Gromoff, E. Treier, D.U. and Beck, C.F. 1989. *Three light-inducible heat shock genes of Chlamydomonas reinhardtii. Molecular and Cellular Biology*, **9:** 3911-*3918.*

Wahid, A. Gelani, S., Ashraf, M. and Foolad, M.R. 2007. Heat tolerance in plants: an Overview. *Environmental and Experimental Botany*, **61(3):** 199–223.

Wahid, A., and Close, T.J. 2007. Expression of dehydrins under heat stress and their relationship with water relations of sugarcane leaves. *Biologia Plantarum,* **51(1):** 104-109.

Waters, E.R., Lee, G.J. and Vierling, E. 1996. Evolution, structure and function of the small heat shock proteins in plants. *Journal of Experimental Botany*, **47(296):** 325-338.

Weng, J. and Nguyen, H.T. 1992. Differences in the heat shock response between thermotolerant and thermosusceptible cultivars of hexaploid wheat . *Theor. Appl. Genet.,* **84:** 941-946.

Wang, H., Goffreda, H. and Leustek, T. 1993. Characteristics of an HSP70 homolog localized in higher plant chloroplasts that is similar to DnaK, the HSP70 of eukaryotes. *Plant Physiology*, **102:** 843- 850.

Wang, W., Vinocur, B. and Altman, A. 2003. Plant responses to drought, salinity and extreme temperatures: towards genetic engineering for stress tolerance. *Planta*, **218:** 1–14.

Wang, W., Vinocur, B., Shoseyov, O., Altman, A. 2004. Role of plant heat shock proteins and molecular chaperons in the abiotic stress response. *Trends in Plant Science*, **9(5):** 1360-1385.

Wang, Y., Ying, J., Kuzma, M., Chalifoux, M., Sample, A., McArthur, C., Uchacz, T., Sarvas, C., Wan, J., Dennis, D.T., McCourt, P. and Huang, Y. 2005. Molecular tailoring of farnesylation for plant drought tolerance and yield protection. *Plant J.*, **43:** 413–424.

Wang, D., Heckathorn, S.A., Mainali, K. and Hamilton, E.W. 2008. Effects of N on plant response to heat-wave: a field study with prairie vegetation. *J Integr Plant Biol*, **50:** 1416–1425.

Wang, X., Yang, P., Gao, Q., Liu, X., Kuang, T., Shen, S. and He, Y. 2008. Proteomic analysis of the response to high-salinity stress in *Physcomitrella patens*. *Planta*. **228:** 167–177.

Wang, K., Zhang, X., Goatley, M. and Ervin, E. 2014. Heat shock proteins in relation to heat stress tolerance of creeping bentgrass at different N levels. *PLOS One*, **9(7):** 1-10.

Waters, E.R., Lee, G.J. and Vierling, E. 1996. Evolution, structure and function of the small heat shock proteins in plants. *Journal of Experimental Botany*, **47(296):** 325-338.

Weber, C., Nover, L. and Fauth, M. 2008. Plant stress granules and mRNA processing bodies are distinct from heat stress granules. *The Plant Journal*, **56:** 517-530.

Wells, D.R., Tanguay, R.L., Le, H. and Gallie, D.R. 1998. HSP101 functions as a specific translational regulatory protein whose activity is regulated by nutrient status. *Genes Dev.* **12:** 3236–3251.

Welte, M.A., Tetrault, J.M., Dellavalle, R.P. and Lindquist, S. 1993. A new method for manipulating transgenes: Engineering heat tolerance in a complex, multicellular organism. *Curr. Biol.*, **3:** 842–853.

Weng, J. and Nguyen, H.T. 1992. Differences in the heatshock response between thermotolerant and thermosusceptible cultivars of hexaploid wheat, *Theor Appl. Genet.*, **84:** 941-946.

Westerheide, S D., Anckar, J., Stevens, S M., Sistonen, L. and Morimoto, R I. 2009. Stress-inducible regulation of heat shock factor 1 by the deacetylase SIRT1. *Science*, **323:** 1063-1066.

Whaibi, M.H. 2011. Plant heat-shock proteins: A mini review. Journal of King Saud University-*Science*, **23:** 139-150.

Wollgiehn, R. and Neumann, D.1999. Metal stress response and tolerance of cultured cells from *Silene vulgaris* and *Lycopersicon peruvianum*: role of heat stress proteins. *Journal of Plant Physiology,* **154:** 547–553.

Wu, M. T. and Wallner, S.1983. Heat stress responses in cultured plant cells. *Plant Physiology*, **72:** 817–820.

Wu, H.C., Hsu, S.F., Luo, D.L., Chen, S.J., Huang, W.D., Lur, H.S. and Jinn, T.L. 2010. Recovery of heat shock-triggered released apoplastic Ca^{2+} accompanied by pectin methylesterase activity is required for thermotolerance in soybean seedlings. *Journal of Experimental Botany*, **61:** 2843-2852.

Xing, J., Xu, Y., Tian, J., Gianfagna, T. and Huang, B. 2009. Suppression of shade- or heat-induced leaf senescence in creeping bentgrass through transformation with the ipt gene for cytokinin synthesis. *J. Am. Soc. Hortic. Sci.*, **134:** 602–609.

Xu, Z., Horwich, A.L. and Sigler, P.B. 1997. The crystal structure of the asymmetric GroEL-GroES-(ADP)7 Chaperon in complex. *Nature*, **388(6644):** 741–750.

Xu, Q. and Huang, B. 2006. Seasonal changes in root metabolic activity and nitrogen uptake for two cultivars of creeping bentgrass. *Hort Science*, **41(3):** 822–826.

Xu, Y. and Huang, B. 2007. Heat-induced leaf senescence and hormonal changes for thermal bentgrass and turf-type bentgrass species differing in heat tolerance. *Journal of the American Society for Horticultural Science*, **132(2):** 185–192.

Xu, C. and Huang, B. 2008. Root proteomic responses to heat stress in two *Agrostis* grass species contrasting in heat tolerance. *Journal of Experimental Botany*, **59(15):** 4183–4194.

Xu, Y. and Huang, B. 2009. Effects of foliar-applied ethylene inhibitor and synthetic cytokinin on creeping bentgrass to enhance heat tolerance, *Crop Science,* **49(5):** 1876–1884.

Xu, C. and Huang, B. 2010. Differential proteomic response to heat stress in thermal *Agrostis scabra and* heat-sensitive *Agrostis stolonifera. Physiologia Plantarum*, **139(2):**192–204.

Xu, Y., Zhan, C. and Huang, B. 2011. Heat shock proteins in association with heat tolerance in grasses. *Int. J. Proteomics*, **2011:** Article ID 529648, 11 pages.

Yamada, K., Fukao, Y., Hayashi, M., Fukazawa, M., Suzuki, I. and Nishimura, M. 2007. Cytosolic HSP90 regulates the heat shock response that is responsible for heat acclimation in *Arabidopsis thaliana. J. Biol. Chem.,* **282:** 37794–37804.

Ye, S., Yu, S., Shu, L., Wu, J., Wu, A. and Luo, L. 2012. Expression profile analysis of 9 heat shock protein genes throughout the life cycle and under abiotic stress in rice. *Chinese Science Bulletin*, **57:** 336-343.

Yildiz, M. and Terzi, H. 2008. Small heat shock protein responses in leaf tissues of wheat cultivars with different heat susceptibility. *Biologia*, **63:** 521–525.

Yokotani, N., Ichikawa, T., Kondou, Y., Matsui, M., Hirochika, H., Iwabuchi, M., and Oda, K. 2008. Expression of rice heat stress transcription factor OsHsfA2e enhances tolerance to environmental stresses in transgenic *Arabidopsis*. *Planta,* **227(5):** 957–967.

Yoshiba, Y. Kiyosue, T., Nakashima, K., Yamaguchi-Shinozaki, K. and Shinozaki, K. 1997. Regulation of levels of proline as an osmolyte in plants under water stress. *Plant Cell Physiology*, **38:** 1095- 1102.

Young, R.A. 1990. Stress proteins and immunology. *Annu. Rev. Immunol.*, **8:** 401-420.

Young, R.A. and Elliott, T.J. 2002. Stress proteins, infection and immune surveillance. *Cell*, **59:** 5-8.

Young, J.C., Agashe, V.R., Siegers, K. and Hartl, F.U. 2004. Pathways of chaperone-mediated protein folding in the cytosol. *Nature Reviews Molecular Cell Biology*, **5:** 781-791.

Young, J. C. 2010. Mechanisms of the HSP70 chaperone system. *Biochemistry and Cell Biology*, **88:** 291-300.

Yu, J.H., Kim, K.P., Park, S.M. and Hong, C. 2005. Biochemical analysis of a cytosolic small heat shock protein, NtHSP18. 3, from *Nicotiana tabacum. Mol. Cells*, **19:** 328-333.

Zabaleta, E., Oropeza, A., Nacyra, A., Mandel, A., Salemo, G. and Herrera-Estrella, L. 1994. Artisense expression of chaperonin 60â in transgenic tobacco plants leads to abnormal phenotypes and altered distribution of photoassimilates. *The Plant Journal*, **6(3):** 425-432.

Zahur, M., Maqbool, A., Irfan, M., Barozai, M.Y.K., Qaiser, U., Rashid, B. and Riazuddin, S. 2009. Functional analysis of cotton small heat shock protein promoter region in response to abiotic. stresses in tobacco using agrobacterium-mediated transient assay. *Mol. Biol. Reports*, **36:** 1915-1921.

Zhang, W., Peumans, W.J., Barre, A., Astoul, C.H., Rovira, P., Rougé, P., Proost, P., Truffa-Bachi, P., Jalali, A.A.H. and van Damme, E.J.M. 2000. Isolation and characterization of a jacalin-related mannose-binding lectin from salt-stressed rice (*Oryza sativa*) plants. *Planta.* **210:** 970–978.

Zhang, X. and Glaser, E. 2002. Interaction of plant mitochondrial and chloroplast signal peptides with the HSP70 molecular chaperone. *Trends Plant Sci.*, **7:** 14-21.

Zhang, Y., Mian, M.A.R. Chekhovskiy, K., So, S., Kupfer, D., Lai, H. and Roe, B.A. 2005. Differential gene expression in *Festuca* under heat stress conditions. *Journal of Experimental Botany*, **56(413):** 897–907.

Zhang, X. and Ervin, E.H. 2008. Impact of seaweed extract-based cytokinins and zeatin riboside on creeping bentgrass heat tolerance. *Crop Science*, **48(1):** 364–370.

Zhang, Q. and Denlinger, D. L. 2010. Molecular characterization of heat shock protein 90, 70 and 70 cognate cDNAs and their expression patterns during thermal stress and pupal diapause in the corn earworm. *Journal of Insect Physiology,* **56:** 138-150.

Zhang, J., Li, J., Liu, B., Chen, J. and Lu, M. 2013. Genome-wide analysis of the Populus HSP90 gene family reveals differential expression patterns, localization, and heat stress responses. *BMC Genomics,* **14(1):** 1-14.

Zhao, R. and Houry, W.A. 2005. HSP90: a chaperone for protein folding and gene Regulation. *Biochem and Cell Biology*, **83(6):** 703–710, 2005.

Zhao, W.Y., Xu, S., Li, J.L., Cui L.J., Chen Y.N. and Wang, J.Z. 2008. Effects of foliar application of nitrogen on the photosynthetic performance and growth of two fescue cultivars under heat stress. *Biol Plantarum*, **52:** 113–116.

Zlatev, Ż. and Lidon, F.C. 2012. An overview on drought induced changes in plant growth, water relations and photosynthesis. *Emir. J. Food Agric.*, **24:** 57-72.

Zou, J., Liu, A.L., Chen, X.B., Zhou, X.Y., Gao, G.F., Wang, W. and Zhang, X. 2009. Expression analysis of nine rice heat shock protein genes under abiotic stresses and ABA treatment. *J Plant Physiol*, **166(8):** 851–861.

Zubo, Y.O., Lysenko, E.A., Aleinikova, A.Y., Kusnetsov, V.V. and Pshibytko, N.L. 2008. Changes in the transcriptional activity of barley plastome genes under heat shock. *Russian Journal of Plant Physiology*, **55:** 293-300.